BIOMECHANICS OF THE FEMALE PELVIC FLOOR

BIOMECHANICS OF THE FEMALE PELVIC FLOOR

Edited by

LENNOX HOYTE

Director, Urogynecology and Female Pelvic Reconstructive Surgery, USF College of Medicine, Tampa, FL, United States of America;
Medical Director, Urogynecology and Robotic Surgery, Tampa General Hospital, Tampa, FL, United States of America

MARGOT DAMASER

Professor of Molecular Medicine, Cleveland Clinic Lerner College of Medicine, Case Western Reserve University, Cleveland, OH, United States of America;
Lerner Research Institute, Cleveland Clinic, Cleveland, OH, United States of America;
Advanced Platform Technology Center, Cleveland VA Medical Center, Cleveland, OH, United States of America

ELSEVIER

AMSTERDAM • BOSTON • HEIDELBERG • LONDON
NEW YORK • OXFORD • PARIS • SAN DIEGO
SAN FRANCISCO • SINGAPORE • SYDNEY • TOKYO
Academic Press is an imprint of Elsevier

Academic Press is an imprint of Elsevier
125 London Wall, London EC2Y 5AS, UK
525 B Street, Suite 1800, San Diego, CA 92101-4495, USA
50 Hampshire Street, 5th Floor, Cambridge, MA 02139, USA
The Boulevard, Langford Lane, Kidlington, Oxford OX5 1GB, UK

British Library Cataloguing in Publication Data
A catalogue record for this book is available from the British Library

Library of Congress Cataloging-in-Publication Data
A catalog record for this book is available from the Library of Congress

ISBN: 978-0-12-803228-2

For information on all Academic Press publications
visit our website at www.elsevier.com

Publisher: Joe Hayton
Acquisitions Editor: Fiona Geraghty
Editorial Project Manager: Maria Convey
Production Project Manager: Lisa Jones
Cover designer: Maria Ines Cruz

CONTENTS

CONTRIBUTORS

S. Abramowitch
University of Pittsburgh, Pittsburgh, PA, United States of America

A. Akhondi-Asl
Department of Radiology, Boston, MA, United States of America

J.B. Alford
University of South Florida, Tampa, FL, United States of America

M. Alperin
University of California, San Diego, CA, United States of America

J.A. Ashton-Miller
University of Michigan, Ann Arbor, MI, United States of America

A. Borazjani
Lerner Research Institute, Cleveland Clinic, Cleveland, OH, United States of America

M.A.T. Bortolini
Federal University of São Paulo, São Paulo, Brazil

S. Brandão
University of Porto, Porto, Portugal

B. Brazile
Mississippi State University, Starkville, MS, United States of America

L. Chen
University of Michigan, Ann Arbor, MI, United States of America

D. Christiansen
Mississippi State University, Starkville, MS, United States of America

C.E. Constantinou
Stanford University, Stanford, CA, United States of America

T. Da Roza
University of Porto, Porto, Portugal

M. Damaser
Professor of Molecular Medicine, Cleveland Clinic Lerner College of Medicine, Case Western Reserve University; Lerner Research Institute, Cleveland Clinic; Advanced Platform Technology Center, Cleveland VA Medical Center, Cleveland, OH, United States of America

J.O.L. DeLancey
University of Michigan, Ann Arbor, MI, United States of America

K. Downing
Albert Einstein College of Medicine/Montefiore Medical Center, Bronx, NY, United States of America

D. Easley
University of Pittsburgh, Pittsburgh, PA, United States of America

V. Egorov
Artann Laboratories, Trenton, NJ, United States of America

J. Fielding
UTSouthwestern, Dallas, TX, United States of America

L. Hoyte
Director, Urogynecology and Female Pelvic Reconstructive Surgery, USF College of Medicine; Medical Director, Urogynecology and Robotic Surgery, Tampa General Hospital, Tampa, FL, United States of America

K. Knight
University of Pittsburgh, Pittsburgh, PA, United States of America

R.K. Kotarinos
University of South Florida, Morsani College of Medicine, Tampa, FL, United States of America

J.A. Kruger
University of Auckland, Auckland, New Zealand

X. Li
University of Sheffield, Sheffield, United Kingdom

J. Liao
Mississippi State University, Starkville, MS, United States of America

C. Lisle
KnowledgeVis, LLC, Maitland; University of Central Florida, Orlando, FL, United States of America

V. Lucente
The Institute for Female Pelvic Medicine & Reconstructive Surgery, Allentown, PA, United States of America

J. Luo
University of Michigan, Ann Arbor, MI, United States of America

S. Madill
School of Physical Therapy, University of Saskatchewan, Saskatchewan, Canada

T. Mascarenhas
University of Porto, Porto, Portugal

H.C. Ming
Singapore General Hospital; Duke-National University of Singapore, Singapore

P.A. Moalli
University of Pittsburgh, Pittsburgh, PA, United States of America

M.P. Nash
University of Auckland, Auckland, New Zealand

R. Natal Jorge
University of Porto, Porto, Portugal

P.M.F. Nielsen
University of Auckland, Auckland, New Zealand

M. Parente
University of Porto, Porto, Portugal

S.S. Patnaik
Mississippi State University, Starkville, MS, United States of America

P.L. Ryan
Mississippi State University, Starkville, MS, United States of America

A. Sarvazyan
Artann Laboratories, Trenton, NJ, United States of America

D.C. Simkins Jr.
University of South Florida; Formerics, LLC, Tampa, FL, United States of America

C.H. van der Vaart
University Medical Center Utrecht, Utrecht, Netherlands

H. van Raalte
Princeton Urogynecology, Princeton, NJ, United States of America

S.K. Warfield
Department of Radiology, Boston, MA, United States of America

B. Weed
Mississippi State University, Starkville, MS, United States of America

X. Yan
University of Auckland, Auckland, New Zealand

EDITORS' BIOGRAPHY

Lennox Hoyte, MD, MSEECS is Professor of Obstetrics and Gynecology at the University of South Florida (USF). He is the founding Division and Fellowship director of Female Pelvic Medicine and Reconstructive Surgery (FPMRS) at USF. Board certified in OB/Gyn, and fellowship-trained in FPMRS, he received a BSEE from WPI, an MSEECS from MIT, and his MD from Stanford University. He completed OB/Gyn residency training at Brigham and Womens Hospital/Mass General Hospital, Fellowship training in FPMRS at Loyola University, and Surgical Planning at Harvard. He performs advanced robotic and vaginal surgery to correct prolapse, and treats women with all types of pelvic floor dysfunction. He invents and commercializes medical devices for simplifying surgery and enhancing patient safety. He has received 4 patents, with several more pending. He spent a decade designing and building advanced computer circuits, and tools for automating their design and production. This prepared him for a research career in MRI based 3D pelvic reconstruction. Dr. Hoyte has published a large number of peer-reviewed manuscripts, including the first to demonstrate differences in 3D pelvic floor structure between normal women, and those with prolapse and urinary incontinence. His current research is geared to developing imaging markers to predict the success of different surgical approaches to correct pelvic organ prolapse.

Margot Damaser, PhD is Professor of Molecular Medicine at the Cleveland Clinic Lerner College of Medicine with appointment as Staff in the Department of Biomedical Engineering and the Glickman Urological and Kidney Institute of the Cleveland Clinic. She also has an appointment as Senior Research Career Scientist in the Advanced Platform Technology Center of the Louis Stokes Cleveland Department of Veterans Affairs Medical Center. She received an A.B. in Engineering Science from Harvard and Radcliffe Colleges and a PhD in Bioengineering from the University of California at Berkeley and San Francisco. She completed two postdoctoral fellowships, at Lund University, Lund, Sweden, and in the Department of Urology at the Hospital of the University of Pennsylvania, Philadelphia, PA. The focus of her 20 year research career has been on improving knowledge of pelvic floor function and biomechanics and translating this knowledge to improve diagnosis and treatment of pelvic floor disorders. She has had research funding from NIH, VA, private foundations, and corporations. She has five patents pending and over 120 peer-reviewed publications. She is considered an internationally recognized expert in the fields of laboratory-based and preclinical investigations in pelvic floor disorders, as well as development of novel therapies and methods of diagnosing dysfunction.

FOREWORD

This cutting-edge textbook of female pelvic floor biomechanics brings an integrated, multidisciplinary focus to the study of female pelvic floor disorders. With contributions from recognized experts in the fields of biomedical, structural, materials, and tissue engineering, as well as anatomy, imaging, visualization, biochemistry, tissue ultrastructure, pelvic surgery, and radiology, the book is designed to garner new insights into the female pelvic floor. While this innovative textbook encourages novel solutions to female pelvic floor dysfunction, it also serves as a primer on pelvic floor biomechanics.

Thus, unlike previous books on the pelvic floor, the text covers a wide range of topics vital to understanding the biomechanics of the pelvic floor, including physiology, anatomy, ultrastructure, medicine, engineering analysis, imaging, and visualization. The impact of pregnancy and childbirth on pelvic floor biomechanics is covered in detail. Central to such a biomechanical analysis is the need to generate realistic biomechanical models and this, too, is extensively covered. Each chapter builds on the previous ones, such that a reader can read the book in its entirety from front to back, as one might in a course on biomechanics.

I am confident that readers of this book will see opportunities to evaluate theories about childbirth injury in silico thus, bypassing the ethical and logistic complexities arising from in vivo attempts to study the effects of labor, and maternal and fetal variables on pelvic floor outcomes in large cohorts of women. In addition, readers will be able to use the concepts presented herein to test patient-specific scenarios for surgical repair of prolapse.

In short, this book fills an important gap in our knowledge base of female pelvic floor disorders. It is required reading for those who wish to fully grasp relevant biomechanics and make an important contribution to the field of female pelvic floor dysfunction.

Charles J. Lockwood, MD
Professor of Obstetrics, Gynecology, and Public Health
Senior Vice President, USF Health
Dean, Morsani College of Medicine
University of South Florida

PREFACE

There are in fact two things, science and opinion; The first begets knowledge, the second ignorance.

Hippocrates, Laws, Book IV

This book on the biomechanics of the pelvic floor disorders that you are reading marks a milestone in the development of this exciting new field. It signals the transition from opinion to science. The path between these two points of view is not always an easy one to take and depends on developing novel tests that can provide relevant measurements so that competing hypotheses can be tested. The important new work presented here on understanding the mechanical nature of the pelvic floor marks a distinct shift from a historically observation-based approach to these problems that have dominated the field during the last century.

Early in the 20th century, outstanding observational research was conducted that resulted in two landmark books: Joseph Halban's and Julius Tandler's Anatomy and Etiology of Genital Prolapse in Women [1] and R.H. Paramore's the Statics of the Female Pelvic Viscera [2]. These two works contain, in the first case, the results of meticulous dissections and documentation of the anatomy seen in scores of cadavers with prolapse. In the second case, Paramore provides a thoughtful and detailed theoretical analysis of the biomechanical factors involved in normal support and a consideration of how they are altered in women with prolapse. Unfortunately, the only tools available for these early investigators were acute observation, clear thinking, and simple devices for measuring distances and pressure. The observations made by these pioneers are important, but scientifically testing the ideas that they and subsequent investigators proposed was not possible at that time.

Great strides occur when a relevant basic science discipline is linked to a clinical problem. For example, the revolution in infertility treatments that occurred with development of in vitro fertilization occurred by applying insights made in cell biology, endocrinology, and genetics; sciences that are at the essence of fertilization, implantation, and embryogenesis. For the pelvic floor and its disorders, the primary basic science is biomechanics; the understanding of complex structural interaction between biological tissues. Certainly cell and molecular biology and genetics have a role to play, but they depend on identifying the relevant disease processes and questions that are involved in structural failure.

Three major developments have allowed pelvic floor science to advance. First is the availability of advanced imaging in the form of MRI and ultrasound, second is the ability to construct measureable anatomically accurate geometric models from real women, and third the advent of computational methods including finite element modeling so that simulations of pelvic floor mechanics and defects can be made. MRI and ultrasound, along with the 3D models that can be made from symptomatic and asymptomatic women, allow direct measurements to be made of morphological features to compare women with and without disease. The power of this scientific strategy cannot be over emphasized. As William Thomson (Lord Kelvin of temperature fame), astutely pointed out,

> *When you can measure what you are speaking about, and express it in numbers, you know something about it; but when you cannot measure it, when you cannot express it in numbers, your knowledge is of a meager and unsatisfactory kind: It may be the beginning of knowledge, but you have scarcely, in your thoughts, advanced to the stage of science.*
>
> ***Ref. [3]***

Computational models allow structural hypotheses to be examined in ways that were not previously possible. The elegantly complex structure of the pelvic floor in humans is unique. It involves the complex interaction between muscle, connective tissue, and visceral walls under a sophisticated neural control mechanism. Isolating the effect of changes in a single element is difficult because no animal model has the same geometry and construction as is found in women. Certainly, animal models can be used productively to answer specific questions about tissue response to injury or interactions between implantable meshes and the vagina for example, but the basic understanding of pelvic floor structural mechanics must be done in women. Computational biomechanical modeling allows us to make specific structural perturbations and see the consequences in ways not ethically possible in living women. Once identified in silico, specific experiments can be developed in women to test their importance in vivo.

The fact that it is now possible to fill a book with new observations concerning pelvic floor biomechanics marks a coming of age for this new science. Certainly this field is in its infancy and full maturity is some years in the future but it is certainly an exciting phase where new discoveries are made and patterns are established on which subsequent work must be based.

John O. L. DeLancey, MD
Norman F. Miller Professor of Gynecology
Professor of Urology
University of Michigan Medical School

REFERENCES

[1] J. Halban, J. Tandler, Anatomie und Äetiologie der Genitalprolapse beim Weibe, Wilhelm Braumuller, Vienna and Leipzig, 1907.
[2] R.H. Paramore, The Statics of the Female Pelvic Viscera, vol. I, H.K. Lewis & Co. Ltd, London, 1918.
[3] W. Thomson (Lord Kelvin), Popular Lectures and Addresses, Cambridge University Press, London, 1889.

Introduction to the First Edition of *Biomechanics of the Female Pelvic Floor*: How to Use This Book

Pelvic floor disorders (PFDs) may occur due to weakness or injury to the muscles and connective tissues in the pelvic cavity. Some of these muscles and tissues span the distal opening of the pelvis, and others course through the upper part of the pelvis. Together, they provide support for the uterus and vagina, which in turn holds the bladder, rectum, and bowels in place.

Urinary incontinence, fecal incontinence, and pelvic organ prolapse are the three main PFDs. Myofascial pain syndromes may also be considered as PFDs. Urinary and fecal incontinence refers to the involuntary loss of urine or stool, respectively. Pelvic organ prolapse refers to the condition in which the pelvic organs such as the uterus, bladder, and bowel slide into the vagina. The resulting pressure may cause a bulge or protrusion in the vaginal canal. Often, this bulge is uncomfortable for afflicted women, and can make physical activity difficult, in addition to interfering with sexual function.

Nearly 24% of U.S. women are afflicted with one or more PFDs, and this cluster of health problems can cause physical discomfort as well as limiting activity, often diminishing quality of life. The frequency of PFDs increases with age, affecting more than 40% of women from 60 to 79 years of age, and about 50% of women 80 and older.

Risk factors for PFD include childbirth, and conditions that involve chronic straining such as heavy lifting and chronic constipation. Evidence continues to emerge in support of a substantial genetic component to the risk factors for PFD.

When conservative therapy for prolapse and incontinence fails, surgical therapy is the next step in the treatment of these conditions. Many surgical options exist for the treatment of prolapse; these include vaginal and abdominal endoscopic (and open) approaches. Surgical interventions are often offered to patients based on surgeon preference and skill rather than being optimized for individual patient pathology and biomechanical status. Surgical success rates are often initially reported as "high," but under deeper scrutiny, when strict anatomic and patient symptom parameters are considered, success rates are often found to be lower than initially thought. Furthermore, early and late surgical complications and unintended side effects can alter the risk/benefit profile of surgical interventions for POP and stress urinary incontinence (SUI).

Engineering analysis methods are commonly applied to study the behavior of buildings, bridge, and aircraft designs, to name a few, enabling the refinement of each design well prior to actual construction. Currently, there is no such systematic methodology for preoperatively studying the utility of surgical procedures contemplated for the repair of POP or SUI. The present textbook is a step toward bringing engineering analysis methods to bear in the risk stratification, prevention, diagnosis, evaluation, and therapy for female PFDs.

This book represents a significant departure from previous texts and heralds a new era in the investigation of biomechanics of the pelvic floor with application to PFDs. Previous books on the female pelvic floor have focused on only one or a few aspects of the pelvic floor, such as anatomy, pelvic pain, childbirth trauma, physical therapy, colorectal surgery, or electrical stimulation, and therapy for these conditions. However, the increasing pace of clinical innovations to address pelvic organ prolapse and SUI has highlighted the need for a more integrated understanding of the biomechanical underpinnings of female pelvic floor function under normal and diseased conditions. This integrated biomechanical understanding can also shed light on the biomechanical aspects of vaginal childbirth, providing insight into ways to reduce childbirth related injury and future PFDs.

Since most research in pelvic floor dysfunction to date has been epidemiological with outcomes focused on anatomy and quality of life, commercial products addressing pelvic floor dysfunction have been released for use in the absence of thorough preclinical engineering evaluation and testing. The resulting complications attributed to the design, commercialization, and surgical implantation of prostheses to treat pelvic organ prolapse, stress urinary, and fecal incontinence have pointed to the need for an interdisciplinary approach to understanding the biomechanics of the female pelvic floor. Prostheses designed for permanent implantation into the pelvic floor need to have material characteristics, textile, and structural characteristics optimally matched to those of the tissues that they are designed to reinforce or support since they participate in the structural configuration of the pelvic floor tissues, and often become load bearing. This requires that materials and structural engineers collaborate with anatomists, surgeons, and musculoskeletal specialists to ensure that prosthetic designs are appropriate for insertion into the planned areas, and will not cause untoward disruptions of otherwise normal functions.

Preoperatively understanding the impact of a surgical intervention on the specific structural anatomy and function of the area of interest will require appropriate engineering analysis, driven by realistic geometric information obtained from pelvic imaging, as well as appropriate engineering models of the tissues. This requires contributions from radiologists, biomedical engineers, imaging and visualization specialists, and experts in engineering analysis. Basic scientists with intimate knowledge of tissue ultra-, micro-, and macro-structure, as well as the impact of biochemistry and genetics on structural biomechanics should also contribute to the development of realistic tissue models suitable for use in engineering analysis. All of these experts need to "speak the same language" and share vital information with each other in order to create appropriate and useful interventions for women with PFDs. Those who seek to become experts in female pelvic floor dysfunction will need to understand these diverse aspects of pelvic floor biomechanics.

This text is meant to provide a "go-to" reference for those interested in studying pelvic floor biomechanics, but who may have expertise in one area, such as anatomy or medicine, but not in another relevant area, such as methods of biomechanical modeling. One goal of this book is to develop a common language for the different disciplines to discuss, research, and drive solutions to pressing problems related to female pelvic floor dysfunction. This text is also designed to get the different stakeholders on the same page regarding anatomy, physiology, musculoskeletal, and tissue characteristics of the pelvic floor tissues, and to establish the basic principles of biomechanical analysis as it applies to PFDs. We expect this text will foster a common language for continued conversations and collaboration to improve design, conduct, and reporting of research study outcomes.

Another goal of this first-ever text is that it could (and should) be used to develop a course on biomechanics with pelvic floor examples used to illustrate the concepts presented. In current texts, biomechanical theories are illustrated, for instance, with examples from orthopedic and cardiovascular systems. Therefore students of biomechanics can gain knowledge and a head start in those fields from their classroom texts. Since no such text for pelvic floor biomechanics has been published previously, students wishing to model the pelvic floor must learn the defining aspects of this field outside of their classes. We hope this book will herald a new era that includes course instruction in biomechanics of the pelvic floor, inspiring a new generation to create careers investigating pelvic floor biomechanics in health and disease, improving healthcare for millions of women worldwide.

With this in mind, in contrast to previous books on the pelvic floor, we present many, if not all, aspects of physiology, anatomy, and medicine that impact pelvic floor biomechanics, including chapters on the influence of biochemistry (Chapter 8) and genetics (Chapter 9) of pelvic floor tissues, development of animal models for PFDs (Chapter 6), and biomechanics of pelvic floor prostheses (Chapter 7), in addition to chapters on pelvic floor anatomy (Chapters 2 and 3), and biomechanical characterization of pelvic floor organs and tissues (Chapter 5). We also provide a comprehensive history of biomechanics (Chapter 1) and an introduction to the terminology and structure of classical biomechanics (Chapter 4) for those without a background in that field. Also important to completing knowledge about pelvic floor biomechanics are chapters on the impact of pregnancy and childbirth on pelvic floor biomechanics (Chapter 10) and the biomechanical environment of the pelvic floor (Chapter 11), which explores the interactions between different muscles and connective tissue structures, particularly when one or more are dysfunctional.

Anatomically realistic models of pelvic floor structures must be based on high quality clinical images with ultrasound and magnetic resonance proving to be the most useful imaging methods in this regard. Chapters 12 and 13 review these imaging methods, respectively, and define the current state of the art of methods of clinical imaging for biomechanical outcomes and modeling. In addition, tactile imaging is a new method not yet in clinical practice that holds great potential for investigation of biomechanics of the pelvic floor and is presented in Chapter 16.

The latter portion of the book is focused on methods of generating biomechanical models, starting with methods of geometric representation of pelvic floor tissues (Chapter 14) and image segmentation (Chapter 15). Once the images are segmented, pelvic floor models and simulations of specific scenarios can be built, as in Chapter 19. Computational tools that can be used to model the pelvic floor are presented and described (Chapter 17), and utilized to model the pelvic floor (Chapter 18). Lastly, the state of the field of pelvic floor biomechanical models is reviewed from the perspective of the application of these models to simulate the biomechanics of human childbirth (Chapter 21) and address the biomechanics of pelvic floor dysfunction (Chapter 20).

Thus, each reader can potentially read the book in its entirety from front to back, as one might in a course on biomechanics. However, those with some knowledge in one or a few areas, but not all, could alternatively pick and choose chapters to read in order to gain from the book only what they are missing from their own background and expertise. Thus armed with new knowledge, they could then begin designing experiments, models, and/or simulations to investigate different novel aspects of pelvic floor biomechanics.

We would like to thank the incredible expert authors of each chapter who devoted great amounts of time to this project and very patiently addressed the continual stream of comments and feedback from the editors. We would also like to thank the fine editorial and publishing staff at Elsevier for their hard work and great attention to detail. We hope this book brings as much knowledge and joy to the readers as it did to the editors in assembling it.

LENNOX HOYTE and MARGOT DAMASER

SECTION 1

Principles of Pelvic Floor Anatomy and Biomechanics

S. Abramowitch
University of Pittsburgh, Pittsburgh, PA, United States of America

CHAPTER ONE

What Biomechanics Has to do With the Female Pelvic Floor — A Historical Perspective

M. Alperin
University of California, San Diego, CA, United States of America

> *They had different backgrounds and temperaments and perspectives, and if you gave them something to think about...you were guaranteed a fresh set of eyes.*
>
> ***Malcolm Gladwell (2008)***

The pelvic floor consists of bones, connective tissue, smooth and skeletal muscles, and their innervation. The complex mechanical function of the female pelvic floor is to provide support to the pelvic organs and contribute to continence by counteracting the forces generated by intra-abdominal pressure, gravity, and inertia, without interfering with micturition, defecation, sexual functions, and parturition. Biomechanics is a branch of bioengineering, which applies engineering principles and the methods of mechanics to the studies of biological systems. In other words, by means of biomechanics we can investigate how physical forces interact with living systems. As its name implies, one of the central characteristics of biomechanics is that it is highly interdisciplinary. This is highlighted by the American Society of Biomechanics mission statement: "To foster the exchange of information and ideas among biomechanists working in different disciplines, biological sciences, exercise and sports science, health sciences, ergonomics and human factors, and engineering, and to facilitate the development of biomechanics as a basic and applied science" [1]. Biomechanics includes the study of motion, material deformation and load bearing, tissue remodeling, flow of bodily fluids, transport of chemical constituents across biological and synthetic membranes, and tissue engineering. Thus, the applications of biomechanics are diverse, extending from expanding our understanding of biological systems through engineering sciences, to the development of novel clinically relevant technologies. Following the advancement of knowledge pertaining to the biological systems and human physiology through history presents a compelling story of long-standing symbiotic relationships between medical and biomechanical sciences.

EVOLUTION OF BIOMECHANICS THROUGH THE CENTURIES

Human interest in biomechanics is long-standing and can be traced back to antiquity (650 BC–AD 200). Aristotle was fascinated by anatomy and the structure of living things and is considered to be the first biomechanician. In his book titled "De Motu Animalium" — "On the Movement of Animals," he described animal bodies as mechanical systems [2]. After the fall of Greece, the new Roman Empire became the world's scientific center. One of the dominant Roman figures in medicine in the 2nd century was Galen, an anatomist and a personal physician of the Roman Emperor, Marcus Aurelius. Galen promoted the application of Hippocrates' bodily humors theory to the understanding of human diseases. As Roman law had prohibited the dissection of human cadavers, Galen performed vivisections and anatomical dissections on dead animals to determine biomechanical function of internal organs based on their structure. He summarized his observations in his monumental work, "On the Function of the Parts," which served as the world's medical text for close to 1500 years [3].

The advancement of all sciences, including biomechanics, was frozen until the Renaissance, which started in Italy in the late medieval period (1300–1500) and subsequently spread through the rest of Europe. One of the most prominent Renaissance figures, Leonardo da Vinci (1452–1519), who was an accomplished artist and an engineer, studied anatomy in the context of mechanics, effectively learning biomechanics. Through military and civil engineering projects and his inventions, ranging from water skis to hang gliders, Da Vinci developed an understanding of components of force vectors, friction coefficients, and the acceleration of falling objects [4]. He applied these principles to analyze force vectors of skeletal muscles and joint function. Soon afterward, the groundwork for subsequent mechanical advancements was also laid on other areas. In 1543, Copernicus published his manuscript, titled "Dē revolutionibus orbium coelestium" — "On the Revolutions of the Heavenly Spheres" [5]. In opposition to Aristotelian common-sense physics, his derivations promoted mathematical reasoning to explain orbital motion of "heavenly spheres."

Two decades after the death of Copernicus, Galileo Galilee (1564–1642) was born in Italy. Galileo made significant contributions to observational astronomy and played an important role in the scientific revolution of the Renaissance period. He is considered to be the father of modern science. Indeed, Galileo's investigative approach was identical to what we refer to today as the scientific method: he examined facts critically and reproduced known phenomena experimentally to determine cause and effect [6]. Galileo is also referred to as the father of biomechanics. Based on his observation that animals' masses increased disproportionately to their sizes, Galileo concluded that their bones must also disproportionately increase in girth, adapting to the load-bearing conditions rather than merely to size. This was one of the first documented examples demonstrating the principles of biological optimization and allometry, which examines size of an organism and its consequences.

Another significant contributor to the development of biomechanics in the 17th century was Galileo's contemporary, a French mathematician Rene Descartes (1596–1650). He suggested a philosophical system whereby all living systems, including the human body (but not the soul), are simply machines ruled by the same mechanical laws [7]. Descartes was instrumental in establishing the iatrophysical approach to medicine, which attempted to explain physiological phenomena in mechanical terms. The idea that mechanics was the key to understanding the function of the human body further promoted the evolution of biomechanics and was later embraced by the Italian physiologist and physicist Giovanni Borelli (1608–79). Borelli studied walking, running, jumping, the flight of birds, the swimming of fish, and determined the position of the human center of gravity [8,9]. In addition, he calculated and measured tidal volumes and showed that inspiration is muscle-driven and expiration is due to the elastic recoil of tissue. Similarly, he studied the piston action of the heart within a mechanical framework. Borelli also made significant contributions to astronomy, predicting that planets follow parabolic orbits due to the forces exerted on them by the sun. This work preceded Isaac Newton's (1642–1727) law of universal gravitation, which states that a force of interaction between any two bodies in the universe is directly proportional to the product of their masses and inversely proportional to the square of the distance between them. Newton's law of universal gravitation, published in 1687, is based on the classical mechanics principle that the motion of macroscopic objects is determined by the forces exerted upon them [10].

The second half of the Enlightenment era (1620–1780) was marked by wars in Europe, the Indian subcontinent, and the Americas, culminating in the French and American revolutions. Following the advancements in biomechanics during the 1600s, the wars and the turmoil halted major developments in the field until the early 19th century. In 1807, Thomas Young (1773–1829), an English physician and scientist whose discoveries span many areas, defined the modulus of elasticity that bares his name today [11]. Young studied fluid flow in pipes and the propagation of impulses in elastic vessels. When he applied his observations to the analysis of blood flow in the arteries, he deduced that it was the heart and not the peristaltic motion of arterial walls that mainly contributed to blood circulation. Young also engineered a device for determining the size of a red blood cell, which he remarkably accurately measured to be 7.2 μm [12].

The late 18th and the 19th centuries were the times of the Industrial Revolution, marked by the transition from hand production to new manufacturing processes using machines. The demands placed by the industrialization fostered studies of mechanics of various materials, leading to one of the most striking historical examples of significant advancement in biology made possible by application of engineering science to the human body. Toward the end of the 19th century, extensive studies of human movement were conducted in Europe. One of leaders in the field of locomotion was Christian Wilhelm Braune (1831–92), a German anatomist who determined the center of gravity of the human body through cadaveric dissections. Braune did extensive work on analysis of human gait and calculation of resistive forces placed on skeletal muscles during movement. The end of the 19th century was also marked by the start of bone biomechanics after a German engineer, Karl Culmann (1821–81), saw a presentation by the anatomist Hermann von Meyer (1801–69), in which von Meyer

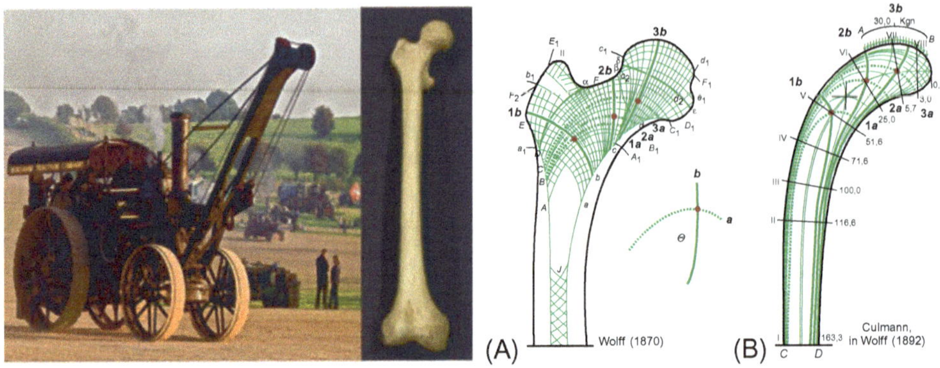

Figure 1 Bone biomechanics: comparison of stress patterns in (A) human femur and (B) a similarly shaped crane. Fig. 1A composed of the image from http://www.flickriver.com/photos/anguskirk/2817575022/ and Fig. 1B composed of the image from http://www.informance-design.com/?p=593.

described the internal architecture of the bone in the femoral head. Culmann was struck by the similarities between the pattern of the cancellous bone of the femur and the stress trajectories in a crane that he was designing with the same shape (Fig. 1). Their discussions led to von Meyer's publication, describing the above similarities [13]. Fortuitously, a German surgeon, Julius Wolff (1836–1902), read von Meyer's manuscript, which served as the basis for Wolff's theory that bone in a healthy person adapts to the loads under which it is placed. In 1892 he postulated the "law of transformation of bone" [14], later known as the Wolff's law of bone remodeling. The biomechanical phenomenon of bone remodeling in response to new loading conditions was responsible for the emergence of orthopedics as a distinct surgical subspecialty and shaped orthopedic surgery in the 20th century.

BIOMECHANICS IN THE 20TH CENTURY AND BEYOND AND APPLICATIONS TO MEDICINE

In the 20th century the field of biomechanics saw a tremendous expansion. For example, the study of bone biomechanics is based on the notion that remodeling occurs in response to changes in loading conditions. Thus, we can deduce that decrease in mechanical load exerted on the bone will lead to bone loss. In fact, loss of bone mass continues to be an inevitable outcome of spaceflight to date and is one of the greatest challenges in space medicine [15]. The widespread use of resistance training, designed to deliver mechanical loads to skeletal sites suffering decline of bone mass due to spaceflight, osteoporosis, or other causes, is an example of the direct application of Wolff's law [16]. This law is also very relevant to the field of female pelvic medicine. Liang et al. demonstrated that vaginal tissue remodels in response to loading conditions [17] such that implantation of high stiffness meshes results in decreased collagen and elastin content in the fibromuscular layer of the vagina and significantly decreased contribution of tissue to the mechanical properties of the graft-vagina complex [18]. This maladaptive response, consistent with tissue degeneration, is caused by stress-shielding of the vagina by the graft from the physiological loads normally experienced by this tissue [17]. Application of the biomechanical concept of tissue remodeling in response to the altered mechanical load provided us with a glimpse into the pathophysiology of the most common mesh-related complication: vaginal mesh exposure.

In the late 19th century, a German physician and physiologist, Otto Frank (1865–1944), found that the strength of left ventricular contraction increased when the ventricle was stretched prior to contraction in frog hearts [19]. In the early 20th century, this observation was further extended by an English physiologist, Ernest Starling (1866–1927). Starling found that increasing venous return, and therefore the filling pressure of the ventricle, led to increased stroke volume in the mammalian heart. The Frank-Starling law of the heart and *heart mechanics* emerged from Starling's 1918 hypothesis that the "mechanical energy set free in the passage from the resting to the active state is a function of the length of the fiber" [20]. In order to appreciate the Frank-Starling mechanism, one must understand the biomechanics of cardiac muscle, which possesses features of both striated and smooth muscles. The ability of muscle fibers to generate force is dictated by the sliding filament model, which states that active muscle force is proportional to the amount of overlap between thin actin and thick myosin filaments within the sarcomere, the functional unit of striated muscle [21]. One of the most fundamental properties of striated muscle is that the amount of force it generates depends on its length [22]; thus, the structure-function that relates sarcomere length to active force production is the length-tension relationship, characterized by three regions on the sarcomere length-tension curve (Fig. 2). The plateau region is associated with optimal overlap with the greatest number of actin-myosin binding sites available and maximal force production. As sarcomere length increases (the descending limb), filament overlap decreases, and force is reduced. Similarly, as sarcomere length decreases (the ascending limb), actin filaments interfere with each other, resulting in rapidly diminishing force [23]. The whole muscle force generation results from the aggregate force produced by all sarcomeres. The Frank-Starling law of the heart refers to the heart's intrinsic ability to increase systolic force and cardiac output in response to a rise in ventricular filling [24]. This occurs because cardiac muscle normally operates across the ascending limb of the length-tension curve. In other words, as muscle fibers and consequently the sarcomeres are stretched, the heart produces more force [24].

As in the case of cardiac muscle, the sarcomere length at which various muscles in the body operate suits their specific functional needs and can be quite different from the optimal sarcomere length. Fecal incontinence is a pelvic floor disorder with significant detrimental impact on quality of life [25]. Unfortunately, currently available treatments for this condition are associated with high failure rates [26]. Understanding the biomechanics of the striated external anal sphincter (EAS), responsible for the continence mechanism, is important

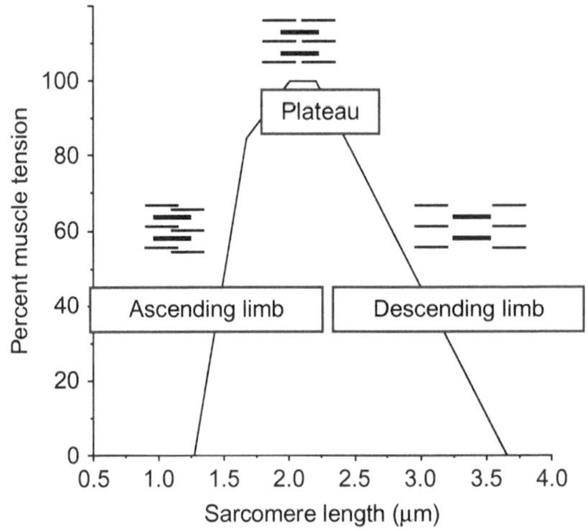

Figure 2 The sarcomere length-tension curve with schematic drawings of overlap of thin and thick myofilaments in the three regions of the curve. *Adopted from R.L. Lieber, Skeletal Muscle Structure, Function, and Plasticity: The Physiologic Basis of Rehabilitation, third ed. The figure in the chapter is a reproduction of Figure 2-4, Ch. 2, p. 47.*

for improving treatment outcomes. In the experiments conducted in a rabbit model, it was determined the optimal sarcomere length of the EAS is 2.59 μm and that the EAS generates more force as the muscle lengthens [27,28]. Thus, analogous to cardiac muscle, the EAS operational sarcomere length is on the ascending limb of the length-tension curve. In a follow-up study, it was demonstrated that during anal sphincteroplasty, an overlap of only 20% resulted in optimal sarcomere length, corresponding to the highest force production [29]. Given that the EAS operational sarcomere length is less than the optimal length, these findings suggest that an increase in force generation can be achieved by minimal overlap and even by the plication of intact EAS. The above has important potential implications for anal sphincteroplasty, as currently the degree of overlap is arbitrarily chosen by the surgeon.

Another contemporary area of biomechanics that has been especially relevant to medicine is *computational biomechanics*, which employs finite-element modeling to determine the effects of mechanical stresses and strains as well as interaction of different structural components such as bone, muscle, and fascia on each other, and on clinical outcomes. Direct anatomic assessments are often precluded due to ethical constraints associated with procurement of human tissue. Finite-element modeling uses mesh discretization to transfer continuous shapes with complex geometry into simple elements, allowing computer simulation of their interactions. This method has become an invaluable and ubiquitous alternative to in vivo assessment. Applications of computational biomechanics in female pelvic medicine have significantly improved our understanding of loads and associated deformations exerted on the individual components of the pelvic floor supportive complex during vaginal childbirth [30,31].

A branch of biomechanics, referred to as *fluid mechanics*, enabled significant advancements in the understanding of human hemodynamics. Due to its incompressibility, the total volume of fluid moving past any point along a pipe per unit time is the same, otherwise the law of conservation of mass would be violated. Bernoulli's principle, that the sum of all forms of mechanical energy is the same at all points along the stream of flow, is similarly derived from the law of conservation of energy [32]. Thus, a reduction in pipe diameter, accompanied by an increase in fluid velocity, is associated with a decrease in pressure. In the late 1920s, Robin Fåhraeus (1888–1968), a Swedish professor of pathology, and Torsten Lindqvist (1906–2007), a physician working in his laboratory, used the above principles to determine that viscosity of blood decreases with the decreasing diameter of the blood vessel, increasing the velocity, and resulting in decreased shear stress [33]. This phenomenon, known today as the Fåhraeus-Lindqvist effect, reduces microvascular resistance to maintain tissue perfusion at a relatively low blood pressure [34]. The concept unites the relationships between hematocrit that measures the concentration of red blood cells, vessel diameter, red blood cell deformability, and resistance to blood flow.

The application of fluid mechanics to the field of female pelvic medicine allowed modeling of the relationship between urinary flow rate, intravesicular pressure, and anatomy of the bladder and the urethra. Bush, Petros, and the coauthors determined the geometry of the urethra based on a video cystourethrogram, a procedure for visualization of urinary bladder and urethra. They then used Bernoulli's equation, modified to account for energy lost to friction as a result of turbulent urethral flow, to predict the effect of varying bladder pressure on urethral flow rates [35]. This biomechanical model enables physicians to use clinical measurements of urinary flow parameters to detect and treat voiding dysfunction [36].

Another area of biomechanics with a wide application to biological systems is *continuum biomechanics*, which focuses on the mechanical behavior of materials modeled as a continuous mass, rather than discrete particles [37]. Structural properties are determined from load-elongation curves and reflect properties of tissue complexes. In contrast, mechanical properties are described by the stress-strain curve, which shows the relation between stress (average restorative internal force per unit area) and strain (deformation relative to non-stressed length). Mechanical properties are intrinsic to the material and determine its behavior independent of size or geometry, allowing comparisons of different materials. Application of structural and mechanical analyses from continuum biomechanics to biological materials provides insight into functional tissue capacity and helps elucidate mechanisms of tissue injury [38,39]. The field of continuum biomechanics is extremely relevant to female pelvic medicine. The mechanical behavior of human pelvic organs and their supportive structures, including tensile strength, elasticity, deformation, and load-bearing capacity, are critical for maintaining pelvic floor integrity and protecting against pelvic floor disorders [40,41].

Mechanical material properties are also important in determining biomechanical biocompatibility, which refers to the need for artificial devices, implants, and prostheses to have similar biomechanical properties as the tissues nearest their clinical site of application. A more recently developed branch of biomechanics called *tissue engineering* refers to the growth of new tissues from cells and collagen scaffolds to produce functional materials for implantation into a living host. Tissue engineering relies extensively on the knowledge of mechanical and material properties of the native tissue that is in need of replacement or augmentation [42,43]. From a clinical point of view, the biomechanical approach to tissue engineering is essential [44]. In order to produce functional site-appropriate tissue, biomechanical similarity between native tissue and the implant must be optimized.

The above necessity for biomechanical similarity of implants to native tissues is extremely relevant in female reconstructive surgery. The mechanical incompatibility of host tissue and synthetic meshes for the treatment of pelvic organ prolapse, is one of the main culprits behind mesh-related complications that prompted the recently issued Federal Drug Administration warning regarding synthetic graft use for vaginal prolapse reconstruction [45–47]. Multiple companies produce synthetic meshes for prolapse repair, all of which are made out of macroporous polypropylene [48]. Biomechanical testing of different grafts reveals that, despite being made from the same material, commercially available synthetic meshes are not equivalent [48,49]. On the contrary, meshes demonstrate marked variation in their textile properties, pore geometry, anisotropic behavior, and stiffness, all of which have important clinical implications [50]. The above exemplifies the importance of the interdisciplinary approach to the evolving science of graft and tissue engineering for prolapse reconstruction between female pelvic surgeons and biomechanical engineers. The biomechanical properties of newly developed graft materials must be fully characterized before their clinical utilization. The above will, in turn, enable clinicians to make scientifically rationale choices for their patients and decrease complication rates.

As evident throughout the history, bridging the fields of biomechanics and biological sciences is vital for significant progress to be realized in prevention and treatment of human disease. To facilitate a shared approach to current clinical and research questions in female pelvic medicine, a common language between biomechanical engineers and health care providers must be created. I would like to conclude this chapter with another quote from Malcolm Gladwell that highlights the importance of interdisciplinary interactions: "Surgeons had all kinds of problems that they didn't realize had solutions, and physicists… (*OR BIOMECHANICAL ENGINEERS*)… had all kinds of solutions to things that they didn't realize were problems" [51].

REFERENCES

[1] American Society of Biomechanics. http://www.asbweb.org.
[2] M.C. Nussbaum, Aristotle's De Motu Animalium, Princeton University Press, Princeton, NJ, 1986.
[3] A. Galenus, Galen on the Usefulness of the Parts of the Body, Cornell University Press, Ithaca, NY, 1969. 802 pp.
[4] L.d. Vinci, Leonardo da Vinci: The Complete Works, David & Charles, London, 2006.

[5] N. Copernicus, H. Stephen, On the Revolutions of Heavenly Spheres, Running Press, Philadelphia, PA, 2004.
[6] G. Galileo, Two New Sciences. The Essential Galileo (M.A. Finocchiaro, Trans.), Hackett Publishing Company, Inc., Indianapolis, IN, 2008. p. 320.
[7] R. Descartes, Descartes: Discourse on Method, Tuttle Publishing, North Clarendon, VT, 2002.
[8] G. Borelli, Borelli's on the Movement of Animals — On the Natural Motions Resulting from Gravity, 2015 ed., Springer, Cham, 2014.
[9] G. Borelli, The Flight of Birds, Hardpress Publishing, Lenox, MA, 2013.
[10] I. Newton, Philosophiae Naturalis Principia Mathematica, Watchmaker Publishing, Seaside, OR, 2010.
[11] T. Young, A course of Lectures on Natural Philosophy and The Mechanical Arts, Nabu Press, Charleston, SC, 2011.
[12] F. Tice, Diseases of the Circulatory System, Bones and Joints, Muscles, Blood, Kidneys, Practice of Medicine, vol. VI, W.F. Prior Company, Inc., New York, 1927.
[13] H.v. Meyer, Die architektur der spongiosa, Reichert Du Bois-Reymond's Archiv 8 (1867).
[14] J. Wolff, Das Gesetz der Transformation der Knochen, Verlag von August Hirschwald, Berlin, 1892.
[15] S.M. Smith, et al., Fifty years of human space travel: implications for bone and calcium research, Annu. Rev. Nutr. 34 (2014) 377–400.
[16] S.M. Smith, et al., Benefits for bone from resistance exercise and nutrition in long-duration spaceflight: evidence from biochemistry and densitometry, J. Bone Miner. Res. 27 (9) (2012) 1896–1906.
[17] R. Liang, et al., Vaginal degeneration following implantation of synthetic mesh with increased stiffness, BJOG 120 (2) (2013) 233–243.
[18] A. Feola, et al., Deterioration in biomechanical properties of the vagina following implantation of a high-stiffness prolapse mesh, BJOG 120 (2) (2013) 224–232.
[19] B. Lohff, 1899: the first mathematical description of the pressure-volume diagram by Otto Frank (1865–1944), Sudhoffs Arch. 83 (2) (1999) 131–151.
[20] C. Michel, One hundred years of Starling's hypothesis, Physiology 11 (5) (1996) 229–237.
[21] H.E. Huxley, Fifty years of muscle and the sliding filament hypothesis, Eur. J. Biochem. 271 (8) (2004) 1403–1415.
[22] R.L. Lieber, J. Friden, Functional and clinical significance of skeletal muscle architecture, Muscle Nerve 23 (11) (2000) 1647–1666.
[23] M. Takahashi, S.R. Ward, R.L. Lieber, Intraoperative single-site sarcomere length measurement accurately reflects whole-muscle sarcomere length in the rabbit, J. Hand. Surg. [Am.] 32 (5) (2007) 612–617.
[24] E.G. Lakatta, Length modulation of muscle performance: Frank-Starling law of the heart, in: H.A. Fozzard (Ed.), The Heart and Cardiovascular System, Raven Press Publishers, New York, NY, 1992, pp. 1325–1351.
[25] I. Meyer, H.E. Richter, Impact of fecal incontinence and its treatment on quality of life in women, Women's Health (Lond. Engl.) 11 (2) (2015) 225–238.
[26] E.R. Trowbridge, et al., Sexual function, quality of life, and severity of anal incontinence after anal sphincteroplasty, Am. J. Obstet. Gynecol. 195 (6) (2006) 1753–1757.
[27] M.R. Rajasekaran, et al., Length-tension relationship of the external anal sphincter muscle: implications for the anal canal function, Am. J. Physiol. Gastrointest. Liver Physiol. 295 (2) (2008) G367–G373.
[28] R.K. Mittal, et al., The external anal sphincter operates at short sarcomere length in humans, Neurogastroenterol. Motil. 23 (7) (2011) 643–647. e258.
[29] M.R. Rajasekaran, et al., Novel applications of external anal sphincter muscle sarcomere length to enhance the anal canal function, Neurogastroenterol. Motil. 23 (1) (2011) 70–75. e7.
[30] K.C. Lien, et al., Levator ani muscle stretch induced by simulated vaginal birth, Obstet. Gynecol. 103 (1) (2004) 31–40.
[31] D. Jing, J.A. Ashton-Miller, J.O. DeLancey, A subject-specific anisotropic visco-hyperelastic finite element model of female pelvic floor stress and strain during the second stage of labor, J. Biomech. 45 (3) (2012) 455–460.
[32] O. Darrigol, Worlds of Flow: A History of Hydrodynamics from the Bernoullis to Prandtl, Oxford University Press, Oxford, 2009.

[33] R.L. Fåhraeus, T. Lindqvist, The viscosity of the blood in narrow capillary tubes, Am. J. Physiol. 96 (1931) 562–568.
[34] L.N. Toksvang, R.M. Berg, Using a classic paper by Robin Fahraeus and Torsten Lindqvist to teach basic hemorheology, Adv. Physiol. Educ. 37 (2) (2013) 129–133.
[35] M.B. Bush, P.E. Petros, B.R. Barrett-Lennard, On the flow through the human female urethra, J. Biomech. 30 (9) (1997) 967–969.
[36] P.E. Petros, U.I. Ulmsten, An integral theory and its method for the diagnosis and management of female urinary incontinence, Scand. J. Urol. Nephrol. Suppl. 153 (1993) 1–93.
[37] M. Epstein, The Elements of Continuum Biomechanics, Wiley, Hoboken, NJ, 2012. 392 pp.
[38] J.L. Lowder, et al., Biomechanical adaptations of the rat vagina and supportive tissues in pregnancy to accommodate delivery, Obstet. Gynecol. 109 (1) (2007) 136–143.
[39] M. Alperin, et al., Collagen scaffold: a treatment for simulated maternal birth injury in the rat model, Am. J. Obstet. Gynecol. 202 (6) (2010) 589.e1–589.e8.
[40] C. Rubod, et al., Biomechanical properties of human pelvic organs, Urology 79 (4) (2012) 968.e17–968.e22.
[41] G. Rivaux, et al., Comparative analysis of pelvic ligaments: a biomechanics study, Int. Urogynecol. J. 24 (1) (2013) 135–139.
[42] N.I. Osman, et al., The effect of ascorbic acid and fluid flow stimulation on the mechanical properties of a tissue engineered pelvic floor repair material, Proc. Inst. Mech. Eng. H 228 (9) (2014) 867–875.
[43] C.M. Kolb, L.M. Pierce, S.B. Roofe, Biocompatibility comparison of novel soft tissue implants vs. commonly used biomaterials in a pig model, Otolaryngol. Head Neck Surg. 147 (3) (2012) 456–461.
[44] S. Gupta, et al., Stiffness- and wettability-dependent myoblast cell compatibility of transparent poly (vinyl alcohol) hydrogels, J. Biomed. Mater. Res. B Appl. Biomater. 101 (2) (2013) 346–354.
[45] J.L. Clemons, et al., Impact of the 2011 FDA transvaginal mesh safety update on AUGS members' use of synthetic mesh and biologic grafts in pelvic reconstructive surgery, Female Pelvic Med. Reconstr. Surg. 19 (4) (2013) 191–198.
[46] FDA public health notification: serious complications associated with transvaginal placement of surgical mesh in repair of pelvic organ prolapse and stress urinary incontinence. http://www.fda.gov/medicaldevices/safety/alertsandnotices/publichealthnotifications/ucm061976.htm, 2008.
[47] FDA safety communication: UPDATE on serious complications associated with transvaginal placement of surgical mesh for pelvic organ prolapse. http://www.fda.gov/medicaldevices/safety/alertsandnotices/ucm262435.htm, 2011.
[48] A. Feola, et al., Characterizing the ex vivo textile and structural properties of synthetic prolapse mesh products, Int. Urogynecol. J. 24 (4) (2013) 559–564.
[49] J.P. Shepherd, et al., Uniaxial biomechanical properties of seven different vaginally implanted meshes for pelvic organ prolapse, Int. Urogynecol. J. 23 (5) (2012) 613–620.
[50] A. Feola, et al., Varying degrees of nonlinear mechanical behavior arising from geometric differences of urogynecological meshes, J. Biomech. 47 (11) (2014) 2584–2589.
[51] M. Gladwell, In the air, in: The New Yorker — Archive. http://www.newyorker.com/magazine/2008/05/12/in-the-air, 2008.

CHAPTER TWO

Pelvic Floor Anatomy and Pathology

J.O.L. DeLancey
University of Michigan, Ann Arbor, MI, United States of America

OVERVIEW

In the pelvis, there are two compartments: anterior and posterior. They are separated by the vaginal walls that are attached to the pelvic sidewalls (Fig. 1). The anterior compartment is occupied by the bladder and urethra. The pubic bones and pelvic sidewalls form inflexible anterior and lateral boundaries to the anterior compartment so that

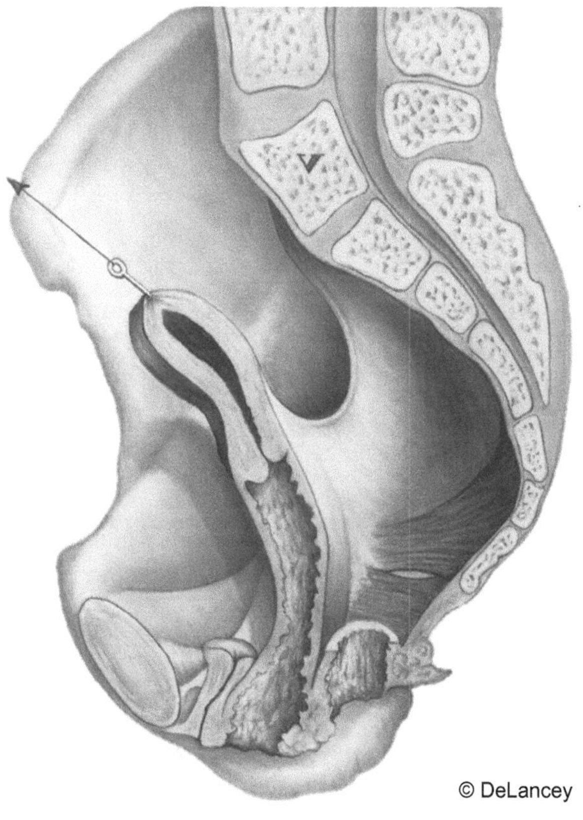

Figure 1 Sagittal view of the pelvic organs after removal of the bladder and rectum showing the anterior and posterior compartments separated by the uterus and vagina. © *DeLancey.*

only the anterior vaginal wall is a flexible component. The posterior compartment contains the rectum and anus. In contrast to the anterior compartment, the only inflexible portion of the posterior compartment is the sacrum. The levator ani muscles, the perineal body, and the posterior vaginal wall are flexible. The external openings to the bladder and rectum are closed by muscles; the urethral sphincter and anal sphincter, respectively. The vagina and the entire pelvic cavity are closed by the actions of the levator ani muscles. At rest, the levator ani muscles squeeze to hold the pelvic organs closed and support the weight placed from above by the abdominal and pelvic organs. They also resist the forces created by differences between abdominal and atmospheric pressure, both at rest and during pressure increases that occur during activities such as coughing, straining, and jumping.

It is possible to propose a broad "unified theory" of pelvic floor mechanics. The muscles and connective tissue supports of the pelvic floor are interdependent and consist of the pelvic organs, their connections to the pelvis, and a neuromuscular system that adjusts to changing forces placed on the pelvic floor by the weight of the pelvic organs and by pressure differentials that can exist between abdominal and atmospheric pressure (Fig. 2). The century-old argument about whether it is failure of the muscles or failure of the

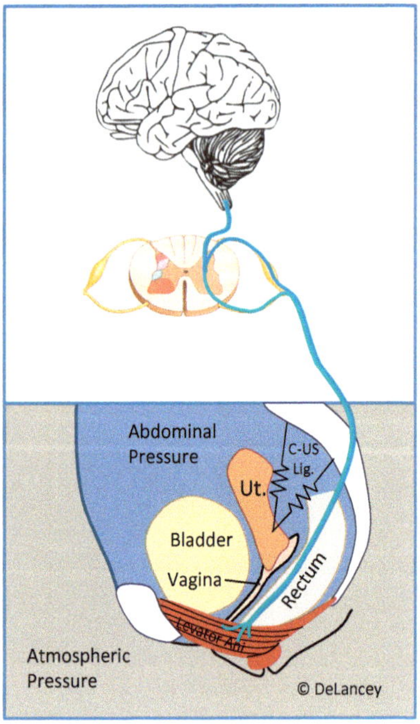

Figure 2 Schematic of the pelvic floor support structures including neural control comprised of cortical, brainstem, spinal cord, and peripheral nerve factors in control of the levator ani muscles. The blue shading represents the domain of abdominal pressure while the light gray denotes atmospheric pressure. N.B. paravaginal attachments of the vagina to the lateral walls not shown. C-US Lig., cardinal and uterosacral ligaments; Ut, uterus. © *DeLancey.*

connective tissue that causes prolapse has fortunately been replaced by discussion of how these two types of tissues interact to provide pelvic organ support and where individual failure sites occur [1]. The several structural failures associated with prolapse can occur singly or in combination with cause prolapse.

Let us begin by considering the normal situation and how it is altered in prolapse. Under normal conditions, the actions of the levator ani muscles hold the pelvic floor closed and provide a lifting force to prevent pelvic floor descent (Fig. 3) [2]. In this situation, the pressures in the anterior and posterior compartments are equal and balanced; canceling each other out; making the pelvis, in essence, an isobaric chamber. When the muscles are damaged or weakened, the hiatus in the levator can be pushed open and the pelvic organs descend. In this situation, one or both of the vaginal walls will descend. The resultant misalignment creates a pressure differential between abdominal pressure and atmospheric pressure. This pressure, acting on the vaginal walls, places abnormal stresses on the tissues that attach the uterus and vagina to the pelvic walls. Therefore, it is this combination of muscle activity and connective tissue attachments that determines the state of pelvic organ support. The following sections will provide additional detail to these general observations to provide more detailed information.

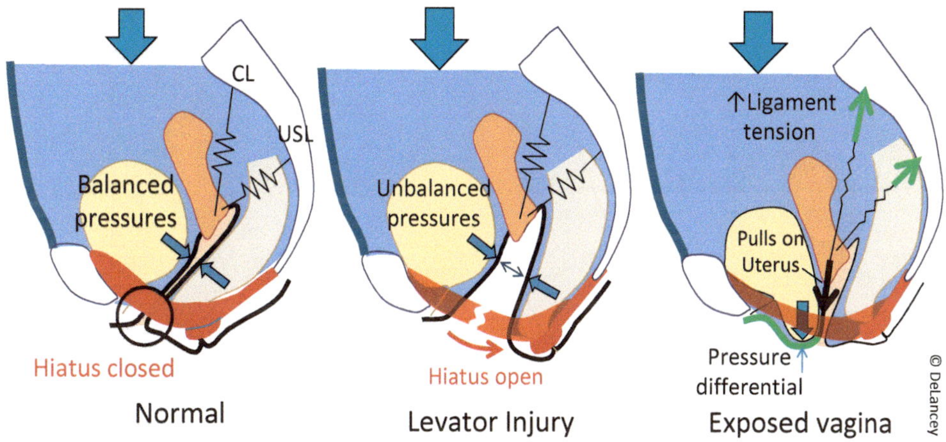

Figure 3 Diagrammatic representation of interactions between levator ani muscle, anterior vaginal wall prolapse, and cardinal/uterosacral ligament suspension. With normal levator function, (A) the vaginal walls are in apposition, and anterior and posterior pressures are balanced. Levator damage (B) results in hiatal opening, and the vagina becomes exposed to a pressure differential between abdominal and atmospheric pressures. This pressure differential (C) creates a traction force on the cardinal ligament (CL) and uterosacral ligament (USL). *© DeLancey. Modified from J.O. Delancey, Surgery for cystocele III: do all cystoceles involve apical descent? Observations on cause and effect, Int. Urogynecol. J. 23 (6) (2012) 665–667.*

LEVATOR ANI AND ANAL SPHINCTERS

Levator Ani Muscle and Its Three Subdivisions

Overall Anatomy of the Levator Ani Muscle

The levator ani muscle consists of three primary subdivisions: pubococcygeal (or pubovisceral), iliococcygeal, and puborectal (Fig. 4). Both pubococcygeal and its synonym have been used in illustrations from our published works and so will both occur in this chapter. The pubovisceral muscle, in turn, has three components: pubovaginal, puboperineal, and puboanal. The origin and insertion as well as muscle action for these elements are listed in Table 1, based on a review of anatomical descriptions available in the literature [3]. Although these origin-insertion pairs are simple and described consistently by all authors that have personally studied the muscle, a profusion of conflicting terms has been applied to this region, making it somewhat difficult to interpret the literature. It is our hope that the present chapter, grounded in the nomenclature of Terminilogica Anatomica [4], will help establish the terminological consistency that will

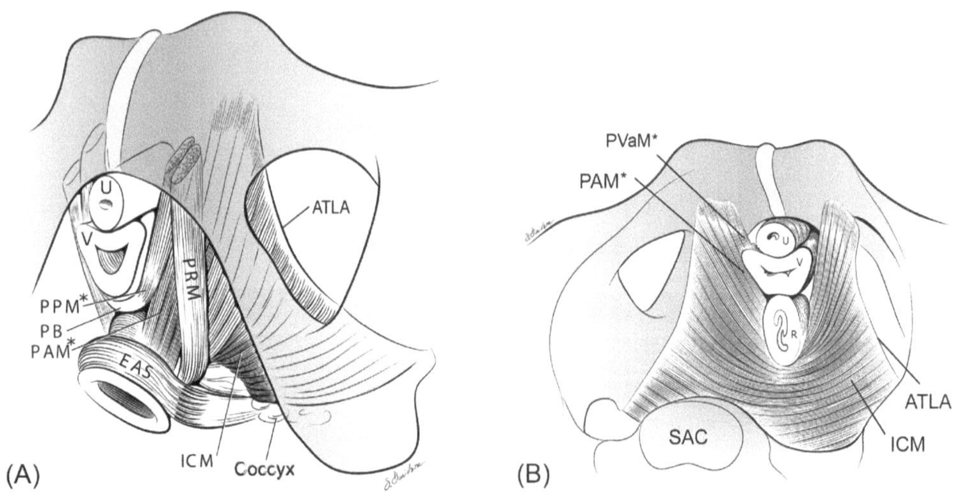

Figure 4 Panel A shows a schematic view of the levator ani muscles from below after the vulvar structures and perineal membrane have been removed showing the arcus tendineus levator ani (ATLA); external anal sphincter (EAS); puboanal muscle (PAM); perineal body (PB) uniting the two ends of the puboperineal muscle (PPM); iliococcygeal muscle (ICM); puborectal muscle (PRM). Note that the urethra and vagina have been transected just above the hymenal ring. Panel B shows the levator ani muscle seen from above looking over the sacral promontory (SAC) showing the pubovaginal muscle (PVM). The urethra, vagina, and rectum have been transected just above the pelvic floor. PAM, puboanal muscle; ATLA, arcus tendineus levator ani; ICM, iliococcygeal muscle. (The internal obturator muscles have been removed to clarify levator muscle origins.) © *DeLancey. R. Kearney, R. Sawhney, J.O. DeLancey, Levator ani muscle anatomy evaluated by origin-insertion pairs, Obstet. Gynecol. 104 (1) (2004) 168–173.*

Table 1 Overview of the nomenclature and functional anatomy of the levator ani

Terminologia anatomica	Origin	Insertion	Function
Pubococcygeal (=pubovisceral)			
Puboperineal	Pubis	Perineal body	Tonic activity pulls perineal body ventrally toward pubis
Pubovaginal	Pubis	Vaginal wall at the level of the mid-urethra	Elevates vagina in region of mid-urethra
Puboanal	Pubis	Intersphincteric groove between internal and external anal sphincter to end in the anal skin	Inserts into the intersphineteric groove to elevate the anus and its attached anoderm
Puborectal	Pubis	Sling behind rectum	Forms sling behind the rectum forming the anorectal angle and closing the pelvic floor
Iliococcygeal	Levator arch	Two sides fuse in the iliococcygeal raphe	The two sides form a supportive diaphragm that spans the pelvic canal

allow the pelvic anatomy to become the central focus as we expand our understanding of the structures in this region, rather than the source of controversy of what labels to use.

The pubovisceral muscle (PVM) arises from the pubic bone and passes beside the pelvic organs. We favor the pubovisceral term coined by Lawson [5] because it more accurately describes the origin and insertion, whereas the older term, pubococcygeal, is based on evolutionary considerations rather than human anatomy. Within the pubovisceral muscle are three areas of insertion to (1) attach to the perineal body (puboperineal), (2) insert into the anal canal and skin (puboanal), and (3) attach the vaginal wall to the medial portion of the pubovisceral muscle (pubovaginal). These are not distinct muscles, but rather are three aspects of one muscle.

The iliococcygeal muscle (ICM) begins at the lateral margin of the pubovisceral muscle with the transition usually occurring at approximately the level of the obturator canal. It is a thin sheet of muscle that spans the pelvic canal from the arcus tendineus levator ani or "levator arch" to the midline iliococcygeal raphe where it interdigitates with the muscle of the other side and connects with the superior surface of the sacrum and coccyx. Details of this anatomy will be provided in the next sections.

The puborectal muscle (PRM) arises near the perineal membrane and courses lateral to the other parts of the levator ani muscles. It forms a sling behind the rectum and is distinct from the pubovisceral muscle. The puborectal muscle creates an angulation in the rectum and the pubovisceral muscle elevates the anus, perineal body, and vagina. In the literature, the terms pubovisceral and puborectal are often confused or used interchangeably; an incorrect practice because they are two distinct muscles with different origins, insertions, and lines of action.

Levator Ani and Anal Sphincter Lines of Action

The action of pelvic floor muscles is determined by the direction of its muscle fiber bundles (fascicles), the muscle shape, and points of attachment. Within the levator ani, each muscle element has an origin and insertion that determines its unique mechanical action. Therefore, injury to one component may have different mechanical effects than damage to another. The angle of each subdivision has been established in living women (Fig. 5) [6] and forms the basis for understanding the mechanical effect of muscle contraction and tone. As shown in Fig. 5A, the pubovisceral muscle fibers course 41 degrees above the horizontal in the standing posture in a direction close the iliococcygeal muscles (33 degrees). By contrast, the puborectal muscles actually act below the horizontal (−19 degrees). The anal sphincter is also below the horizontal in its line of action. The subcutaneous portion of the external anal sphincter constricts the anal verge while the superficial portion pulls dorsally, opposing the puborectal muscle.

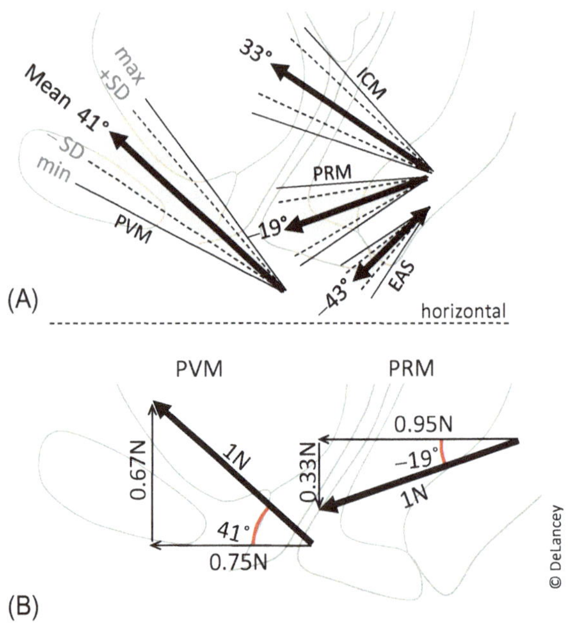

Figure 5 (Panel A) The thick arrow displays the mean direction to the horizontal line in a two-dimensional graphic. The dashed line is the horizontal line from which the angles are measured. Angles above the horizontal line have a "+" sign and those below the horizontal line a "−" sign. On MRI, the PVM was found medial to the PRM; for graphical reasons their lines of actions are depicted in the same plane. (Panel B) Horizontal and vertical components of the PVM and PRM in the standing position. The thick arrows show the average direction of the lines of action of the PVM and PRM muscles relative to the horizontal with a theoretical 1 N force. Thin lines indicate the portion of each force related to a closing and lifting function. (Note: vectors are shown larger than the background anatomy to avoid an overlap in the display.) *From C. Betschart, et al., Comparison of muscle fiber directions between different levator ani muscle subdivisions: in vivo MRI measurements in women, Int. Urogynecol. J. 25 (9) (2014) 1263–1268.*

The large difference (60 degrees) in mean fiber angle between the pubovisceral and the puborectal shows that they have two different mechanical actions. These can be understood within the context of the standing position (Fig. 5B). The diagonal vectors of the muscles can be resolved into two physiological vectors. The first, a vertically oriented "lifting" vector, acts against gravity, lifting the perineal structures and can be attributed to the pubovisceral muscle. Clinically, descent of the perineal structures is found in women with prolapse. This descent is associated with injury to the pubovisceral portion of the levator either with or without prolapse [7,8]. The second vector is oriented horizontally toward the pubic bone. Both the puborectal and pubovisceral muscle have a horizontal component. This vector helps to develop a "closing" force, acting in a horizontal direction so as to close the levator hiatus. This horizontal action creates a vaginal high pressure zone [9,10]. As shown in Fig. 5B, the small vertical puborectal component actually acts in a downward (caudal) direction so that it has no "lifting" action.

External and Internal Anal Sphincters

The external anal sphincter (EAS) is closely associated with the levator ani muscles and has three parts [11,12]. It is not a doughnut-shaped muscle encircling the anal canal as is often erroneously described. The subcutaneous portion of the sphincter is circular and does surround the anal orifice, creating the radially oriented creases in the skin. However, this small muscle is the only part that is circular. The superficial portion lies just cranial to the subcutaneous part (Fig. 6). It crosses the midline ventrally in the perineal body. Dorsally,

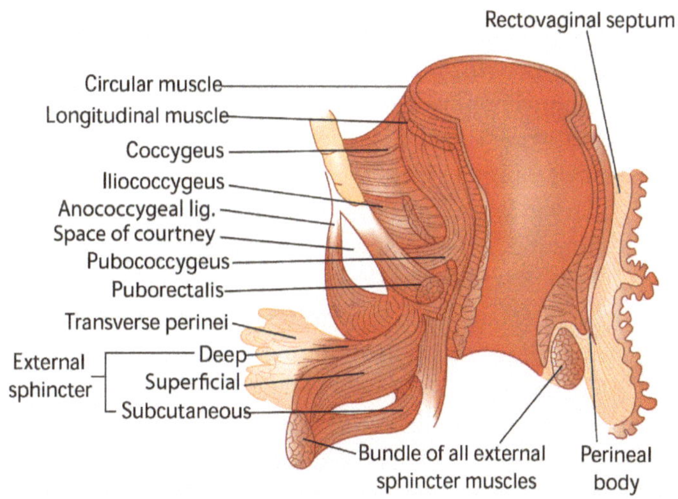

Figure 6 Semidiagrammatic dissection of the anorectal region in the female with the external sphincter cut in the anterior mid-sagittal plane and reflected posteriorly (mucosa removed). The origin of the anterior muscle bundle is clarified and the remaining anterolateral portion of the external sphincters is interdigitated into the transverse perinei. *From C. Oh, A.E. Kark, Anatomy of the external anal sphincter, Br. J. Surg. 59 (9) (1972) 717–723.*

however, it does not cross the midline. Its fibers are directed dorso-cranially, being attached by the anococcygeal ligament to the coccyx. A deep portion of the sphincter has extensions that diverge from the anus ventrally pointing in the same direction as the fibers of the puborectal muscle. The internal anal sphincter is a thickening of the circular smooth muscle of the bowel wall and extends between the anal verge and the dentate line.

Levator Ani Muscle Injury and Pelvic Organ Prolapse

Levator Ani Muscle Injury Associated with Prolapse

The concept that injury to the levator ani muscles is important to pelvic organ prolapse has been discussed for over a century. The meticulous dissections of cadavers with prolapse presented in Halban and Tandler's landmark work [13] on the etiology of genital prolapse at the turn of the last century clearly make this point. Since that time, debate has been polarized between those who believe that pelvic organ support was determined by connective tissue ligaments and fascia and those who believe in the importance of pelvic muscle support [14]. Not surprisingly, both factors are important and they interact with one another. It is this interaction that makes the pelvic floor such an interesting biomechanical structure. Our understanding of muscle-ligament interactions is improving with the ability to independently assess the relative contributions of both components [1].

Injury to the pubovisceral muscle is found in 55% of women with prolapse but only 16% of women with normal pelvic organ support, thus proving the association between levator ani muscle injury and pelvic organ prolapse [15]. A few individuals also sustain some visible change in the iliococcygeal muscle [16], but the clinical significance of this is uncertain. This injury specifically involves the pubovisceral portion of the muscle. It does not involve the puborectal portion [17], a point made somewhat confusing because some authors have used the term puborectal and pubovisceral muscle interchangeably. However, levator ani muscle injury does not appear to be associated with stress incontinence in middle-aged women in the absence of pelvic organ prolapse [18].

Because birth-related injury involves the pubovisceral muscle origin from the pubic bone [19], a clear understanding of the detailed anatomy of this region is needed in order to help elucidate its function. An overview of the relationship between the muscle near the pubis is given in Fig. 7. Specifically, the pubovisceral muscle's attachment to the pubic bone is a thin and transparent aponeurosis that attaches tangentially to the pubic periosteum, shown in cadaveric dissection (Fig. 8). Its morphology changes from the medial to lateral aspect [20]. Medially, fibers of the thick muscle belly coalesce toward multiple narrow points of bony attachment for individual fascicles. In the central portion, there is an aponeurosis between the muscle and bone and the distance between muscle and periosteum is wider (3 mm) than in the medial region. Laterally, the LA fibers attach to the levator arch where the transition from pubovisceral muscle to the iliococcygeal muscle occurs. The thin nature of this aponeurosis makes it subject to injury when the forces in the fiber direction exceed the strength of the muscle origin.

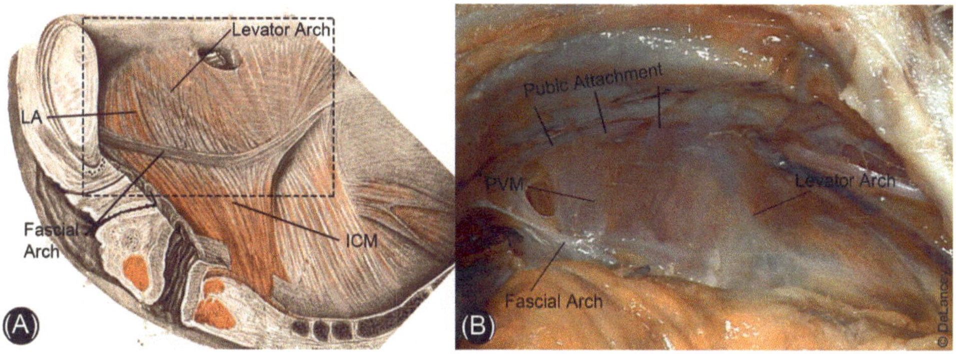

Figure 7 (Panel A) Left mediolateral view of the anterior pelvic sidewall, with pelvic organs removed, showing the characteristic features of the origin of the pubovisceral muscle. The dotted quadrangle shows the anatomic area of study. (Panel B) A similar view of the right anterior pelvic sidewall of a 50-year-old nulliparous fresh cadaver showing the pubovisceral attachments to the pubic bone. The fascial arch lies across the pubovisceral muscle, and the levator arch forms the lateral margin of the pubovisceral muscle as it attaches onto the pelvic sidewall. *(A) Reproduced from J. Halban, J. Tandler, Anatomie und Atiologie der Genital prolapse biem Weibe, Wilhelm Braunmuller, Vienna and Leipzig, 1907. Figure 2, p. 34. (B) From J. Kim, et al., Anatomy of the pubovisceral muscle origin: macroscopic and microscopic findings within the injury zone, Neurourol. Urodyn. 33 (7) (2014) 1073.*

There are two distinct types of pubovisceral injury: (a) loss of part of the pubovisceral muscle fibers, but where the overall attachment of the muscle remains intact (this is defined as a Type I injury) and (b) detachment of the muscle from its origin along with the levator arch from the pubic rami (defined as a Type II injury) (Fig. 9). This latter injury causes "architectural distortion" of the vagina in the axial plane, wherein the normal spatial relationships of the vaginal and structures of the pelvic wall are altered [21,22].

As one might expect with injury to a muscle, the amount of muscle involved matters. It would be easy for the body to compensate for loss of a few muscle fascicles, and this would not be expected to cause problems. However, with the loss of the majority of the muscle; especially if the muscle origin is detached, loss of muscle function can be expected to occur. When levator injury results in loss of more than 50% of the muscle substance (herein called "major loss"), prolapse is found more often than when it is not. As previously mentioned, major loss is found in 55% of women with prolapse compared to 16% of age-matched women who have normal support [15]. Subsequent analysis of a large dataset involving 503 subjects evenly divided between those with and without prolapse has supported the idea that prolapse is only associated with damage of this extent [23]. This finding has been independently confirmed using ultrasound [24], establishing the certainty that this injury is clinically associated with prolapse. Levator injury involving major loss is associated with a 40% reduction in the force that the levator ani muscles can generate during a pelvic muscle contraction [15].

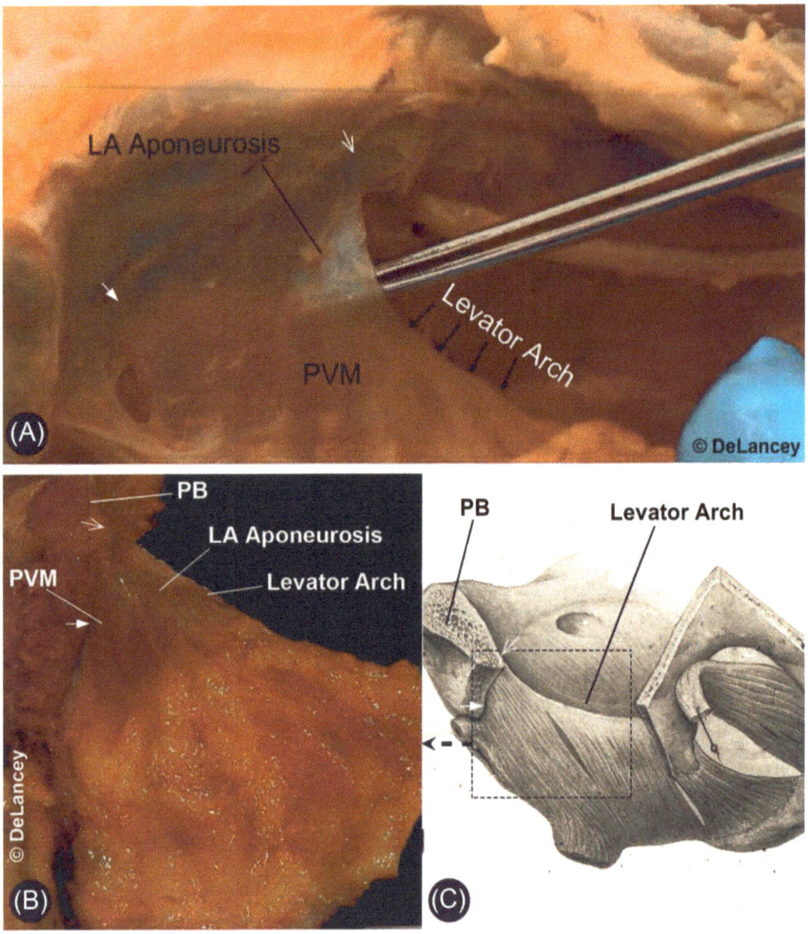

Figure 8 Anatomy of the pubovisceral muscle origin (Fig. 3). (A) A view of the inner surface of the right pubic origin of the PVM similar in orientation with Fig. 2B. Note how thin the semitransparent levator ani (LA) aponeurosis is (tips of a clamp being visible through the aponeurosis) as it attaches to the superior pubic ramus. This aponeurosis poses a smooth transition from the muscular area of the PVM to the periosteum of the pubic bone. Medially the muscle can be seen to extend closer to the bone than it is laterally where the aponeurosis widens. (B, C) Left lateromedial view of the left PVM origin area seen from outside the pelvis after the lateral pelvic bone was removed to reveal the outside of the aponeurotic levator ani muscle. Note that the medial portion of the left PVM extends to the bone while the more lateral portion has a wide aponeurosis between the muscle and bone. The rectangle shows the view shown in Fig. 3B. In this and succeeding figures, closed arrows designate medial origin, and open arrows designate lateral LArch attachment that form the origin of the PVM. *(C) From Kim, et al., Anatomy of the pubovisceral muscle origin: macroscopic and microscopic findings within the injury zone, Neurourol. Urodyn. 33 (7) (2014) 1073.*

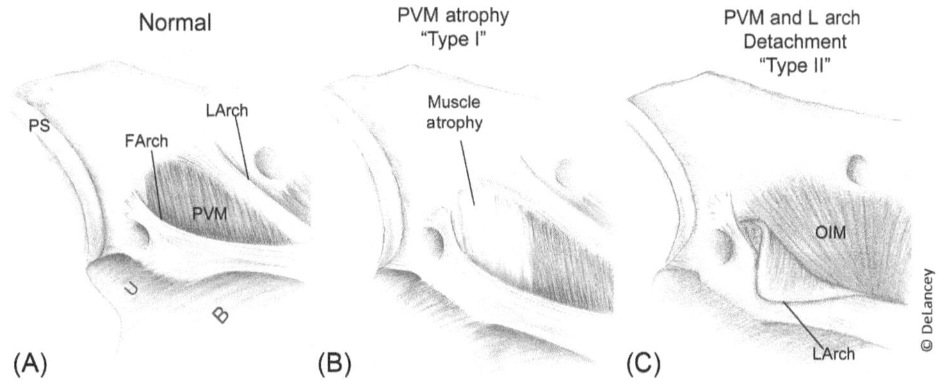

Figure 9 Illustrations of the right inner pelvic sidewall showing the fascial arch (Farch) and levator arch (Larch) originating from the pubic bone (A). The pubovisceral muscle (PVM), pubic symphysis (PS), urethra (U), and bladder can also be seen. In panel B, atrophy in a portion of the PVM is shown; a Type I injury. In Panel C, the levator arch and PVM have become detached from the pubic bone exposing the obturator internus muscle (OIM) in a Type II injury. Since there is only a single point of pubic origin at the LArch for the lateral margin of the PVM, detachment of that point will result in complete offloading of that region of the PVM. *Figure modified from Figure 2A of J.O. DeLancey, Fascial and muscular abnormalities in women with urethral hypermobility and anterior vaginal wall prolapse, Am. J. Obstet. Gynecol. 187 (1) (2002) 93–98.*

How Does Childbirth Cause Injury?

There have been many competing theories proposed to explain the cause of birth-related pelvic floor injury. Muscle damage from compression or tearing and neuropathy from compression or stretch have all been proposed. There is clear neurophysiological evidence of alterations in neural function (for example, [25]) and also evidence from blood samples suggesting ischemia/reperfusion [26] from compression. Computational simulations of birth revealed remarkable elongation of the levator ani muscles where portions of the muscle more than tripled their lengths [27]. These observations, however, establish the mechanisms responsible for the levator defect that is associated with prolapse. Since efforts at prevention must be based on a proper understanding of why the injury occurs, it is necessary to decide between these hypotheses. For example, if compression is the proposed mechanism, lessening the duration of time during which the tissues are compressed is logical. If tearing of the muscle is the mechanism involved, then allowing a slow and gradual delivery, that may take a somewhat longer time, would be the right solution. Because the clinical actions suggested by these two theories are opposite, knowing the specific mechanism of injury matters.

By using fluid sensitive MR scan techniques that document muscle anatomy as well as tissue edema due to injury and evaluating women at early and later time points during recovery, it is possible to test these hypotheses. Injuries that would occur with muscle tearing would be evident immediately and would not resolve. On the other hand, muscle injury due to neurological injuries would develop over time. Compression would cause edema that would involve both the internal obturator muscle and the levator muscle since they are both immediately adjacent and would be equally compressed between the pubic bones and fetal head. Therefore, if muscle compression is responsible for injury, the edema would be expected to affect both internal obturator and pubovisceral muscle muscles where they lie between the fetal head and pelvic bones.

By imaging women with risk factors for pelvic floor injuries both early (1–2 months) and late (7 months) in the postpartum period, it is possible to study these injury mechanisms [28]. Among 17 women in this mechanistic study, 3 women had unilateral high-grade tears (more than 50% of the muscle involved), 3 had unilateral low-grade tears, and 1 had bilateral high- and low-grade tears of the levator ani muscles (Fig. 10). All tears were located at the pubic portion of the levator ani muscles. Regarding the compression hypothesis, levator edema was present in all women but the internal obturator muscle was never involved. Edema resolved at the later time point. None of the women showed a pattern where an initially normal appearing muscle became atrophic that would support atrophy due to neural injury. In a larger subsequent group of women judged to be at high

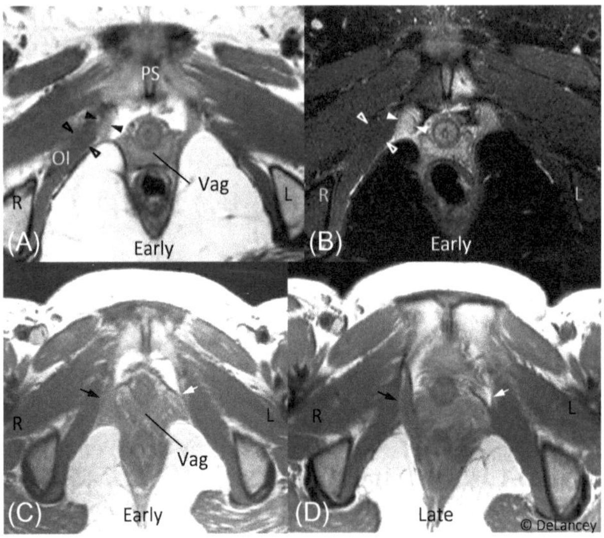

Figure 10 Mid-urethral axial MR images in the region were the pubococcygeal muscle is normally seen lateral to the vagina. Panel A shows a proton density scan where the solid arrow heads mark the pubococcygeal muscle and the open arrow heads show the obturator internus. Signal intensity is lower in the pubococcygeal (arrow head) than the adjacent internal obturator (*open arrow head*). Panel B, a fluid sensitive scan, this difference is more apparent. In panel C, a normal pubococcygeal muscle (*black arrow*) is seen between the vagina and internal obturator (*black arrow*) while it is absent on the left. This pattern persists in the late scan (D). Panels C and D also show the "architectural distortion" that involves the vaginal wall protruding laterally on the side (subject's left) of the injury. © *DeLancey.*

risk for the levator injuries, injuries to the pubic bones were seen as well [29]. There were 29% (17/59) who had subcortical pubic bone fractures and 66% (39/59) were found to have pubic bone marrow edema attesting to the remarkable forces placed on the bony pelvis during birth. Similar to the original study, 90% (53/59) had LA muscle edema and 41% (28/68) had low-grade or greater LA tear 7 weeks postpartum. The magnitude of LA muscle tear did not substantially change by 8 months postpartum, but LA muscle edema and bone injuries showed total or near total resolution at 8 months postpartum ($p < 0.05$). The magnitude of unresolved musculoskeletal injuries correlated with magnitude of reduced LA muscle force and posterior vaginal wall descent ($p < 0.05$).

Alteration in Pelvic Sidewall Anatomy Due to Injury

It is important to consider the different structural implications of Type I muscle injury where there is a loss of a portion of the muscle tissue, but the levator arch remains intact and Type II injuries where the levator arch and pubovisceral muscle becomes detached from its origin. Both the fascial and levator arches lie in this region and since the fascial arch lies on the medial portion of the pubovisceral muscle (Fig. 9), its location could be affected by pubovisceral detachment from the pubic bone.

Using MR-based 3D reconstruction, it is possible to study the effects of this architectural distortion in women with asymmetric levator anatomy, defined as normal levator architecture on one side and levator arch injury on the contralateral side. This methodology permitted comparison of structural alterations on the abnormal side to normal architecture in the same women (Fig. 11) [22].

The structural levator failure involves displacement in both of the fascial and levator tendinous arches. The most pronounced differences in location were seen in the ventral

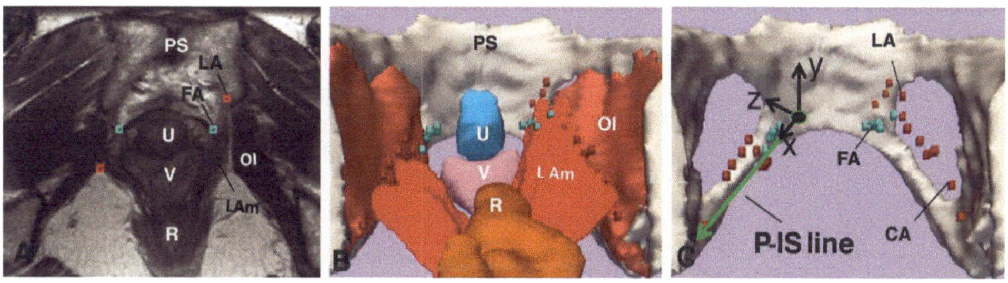

Figure 11 Building 3D model: (A) Axial image showing fascial arch (FA) and levator arch (LA). 3D models made from MRI showing arches with (B) and without (C) adjacent structures. P-IS reference line as X-axis in local pelvic sidewall coordinate system visible in C on defect side. LAm, levator ani; OI, obturator internus; PS, pubic symphysis; R, rectum; U, urethra; V, vagina. © *DeLancey. From K.A. Larson, et al., Measurement of the 3D geometry of the fascial arches in women with a unilateral levator defect and "architectural distortion", Int. Urogynecol. J. 23 (1) (2012) 57–63; K.A. Larson, et al., 3D analysis of cystoceles using magnetic resonance imaging assessing midline, paravaginal, and apical defects, Int. Urogynecol. J. 23 (3) (2012) 285–293.*

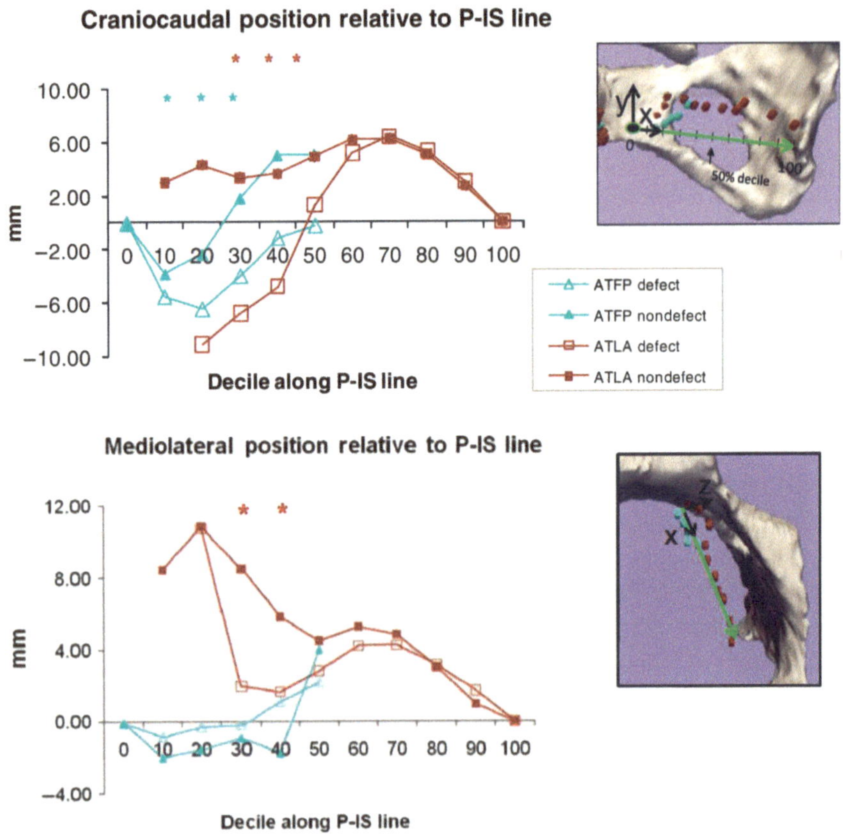

Fig. 12 Craniocaudal position (*top*) relative to P-IS line seen from inside the pelvis looking at the right pelvic sidewall. "0" is pubic end and "100" is at ischial spine and positive numbers on the *y*-axis indicating a cephalad direction. Mediolateral position (*bottom*) seen from above with positive numbers indicating a more lateral direction. Asterisks reveal statistically significant differences between normal and abnormal sides. © *DeLancey. From K.A. Larson, et al., Measurement of the 3D geometry of the fascial arches in women with a unilateral levator defect and "architectural distortion", Int. Urogynecol. J. 23 (1) (2012) 57–63; K.A. Larson, et al., 3D analysis of cystoceles using magnetic resonance imaging assessing midline, paravaginal, and apical defects, Int. Urogynecol. J. 23 (3) (2012) 285–293.*

region near the pubic bone. They occurred in the cranio-caudal direction in the location of the pubovisceral muscle (Fig. 12). These changes in location have two effects: (1) they affect the ability of the muscle to close the urogenital hiatus and levator ani muscles due to loss of muscle origin and (2) they redirect the vectors of the fascial connections to the tendineus arch of the pelvic fascia. This is demonstrated in Fig. 12. The levator arch (arcus tendineus levator ani or ATLA) is lower on the defect side than the nondefect side in the region nearest to the pubic bone. A similar, though smaller, change is seen for the fascial arch. In looking at the medial to lateral displacements, the levator arch is more lateral than the normal side on the side with architectural distortion.

CONNECTIVE TISSUE SUPPORTS OF THE PELVIC ORGANS

Apical Supports: Level I

Downward prolapse of the cervix and vagina are important aspects of pelvic organ prolapse. Understanding how the tissues that connect the uterus and upper vagina to the pelvic sidewalls is important; knowing about uterine/upper vaginal supports and defining how they fail is relevant to understanding the pathophysiology of prolapse. There are important interactions between these supports and the supports of the anterior and posterior vaginal walls that are involved in cystocele and rectocele.

Overview

The connective tissues of the pelvis connect the vagina and uterus to the pelvic walls as outlined in Fig. 1. These can be divided into three levels (I, II, III, numbered respectively from cephalad to caudad) reflecting the changing nature of support (Fig. 13) [30]. In Level I, the upper third of the vagina and the cervix are attached to the pelvic walls by mesenteric structures that suspend these organs. Next is Level II wherein the middle third of the vagina it is attached laterally to fascial structures (arcus tendineus fascia pelvis, ATFP, or fascial arch). The most distal, or caudad, is Level III, wherein the distal third of the vagina is fused with the surrounding structures, namely, the levator ani muscles and the perineal body. The following paragraphs discuss the connective tissue supports at each level in detail.

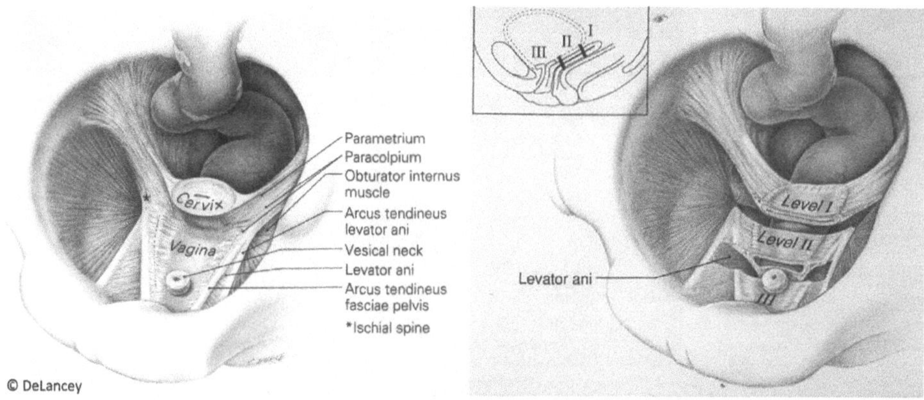

Figure 13 The left panel shows a view of the pelvic organs from above looking over the pubic symphysis showing structures of the pelvic sidewall in relationship to the vagina (outlined by dotted line) after removal of the bladder and uterine corpus. Right panel shows different levels of support in post-hysterectomy cadaver. The parametrium and paracolpium are two portions of the cardinal ligament. Arcus tendineus fasciae pelvis and arcus tendineus levator ani refer to the fascial and levator arches, respectively. *From J.O. DeLancey, Anatomic aspects of vaginal eversion after hysterectomy, Am. J. Obstet. Gynecol. 166 (6 Pt. 1) 1992 1717–1724; discussion 1724–1728.*

Level I: Cardinal and Uterosacral Ligaments

In Level I, there are two aspects to the suspending tissues: (a) the cardinal ligaments (CL) and (b) the uterosacral ligaments (USL). These are unlike knee ligaments for example, which are uniform dense regular connective tissue that connect two bony structures. Instead, the cardinal/uterosacral ligaments are bilateral mesenteries of the genital tract that consist of blood vessels, lymphatics, and nerves embedded in adipose (fatty) tissue with intervening septa of connective tissue (Fig. 14).

These mesenteries have mechanical properties that resist downward displacement of the vagina and uterus while allowing a great range of normal movement. But they have quite different mechanical characteristics when compared to classical "ligaments" in the body. Thus, considering them to be analogous to, for example, knee ligaments, is fundamentally incorrect, and decidedly unhelpful in aiding an understanding of pelvic floor

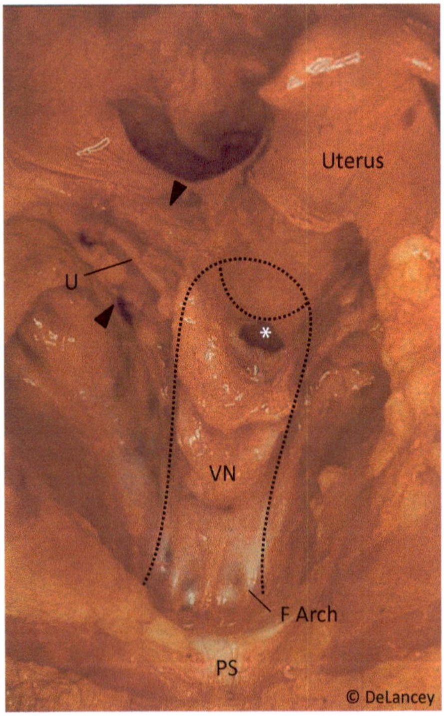

Figure 14 View of the pelvic organs from the front looking over the pubic symphysis (PS) in a normal 17 y.o. nulliparous cadaver after removal of the bladder above the vesical neck (VN). The cardinal ligament (between the arrow heads) can be seen as well as the ureter (U). The round ligament and upper peritoneal portion of the broad ligament have been excised to the level of the uterine artery. An incision has been made into the vagina at the cervico-vaginal junction (*). The lumen of the vagina and the location of the cervix are shown in outline. Note that the cardinal ligament goes to both the cervix and upper vagina. The distal attachment of the arcus tendineus fascia pelvis (ATFP) can be seen attaching to the pubic bone. © *DeLancey*.

function. By way of comparison, the stiffness of the cardinal-uterosacral complex tested in vivo is 0.49 N/mm [31], which is three orders of magnitude less than that of the anterior cruciate ligament of the knee (200 N/mm) [32]. It is also important to recognize that the cardinal/uterosacral tissues have a heterogeneous, nonuniform composition. This contrasts with a piece of a rubber band (for example) which has a uniform, homogeneous composition. This fundamental difference has important implications for biomechanical modeling of the pelvic tissues. The high stiffness implied by the term "ligament" may lead to development of surgical strategies that are not normally compliant. Mechanical testing results from a small piece of rubber would be representative of a larger piece of the same material. However, results obtained from testing a single small excised piece of cardinal/uterosacral tissue would not be representative of the whole structure in the same way.

The structural effect of the cardinal and uterosacral ligaments is most evident when the uterine cervix is pulled downwards with a tenaculum applying traction on the cervix. After several centimeters of descent within the elastic range of the "ligaments" (usually about 3–5 cm), the cardinal-uterosacral complex becomes tight and arrests the further downward cervical movement. Similarly, these same tissues resist downward descent of the vaginal apex after hysterectomy. The fact that these ligaments do not determine the position of the uterus in normal healthy women is attested to by the observation that the cervix may be easily pulled down to the level of the hymen [33] or pushed up to the level of the sacrum.

Tissue Composition of Cardinal and Uterosacral Ligaments

A recent review of the literature indicates the following consensus concerning the structure of the cardinal-uterosacral complex [34]. The cardinal ligament consists mainly of vascular, neural, and areolar tissue containing fat and a delicate meshwork of connective tissue between different elements. The cardinal ligament tends to be more vascular in the cranial portion with more neural components caudally. The vascular section is an extension of the perivascular sheath of visceral branches of the internal iliac vessel going to the genital tract, whereas the neural section contains parts of the inferior hypogastric plexus.

The cardinal ligament takes its origin at the upper border of the greater sciatic foramen where the vascular aspect of this structure is fixed to the pelvic sidewall where the superior gluteal artery exits from the pelvis and fixes the structure in place. This observation has been made on structural grounds previously [35], and we have confirmed it by cutting a pelvis in half and moving the cervix in all directions so that the center of its movement can be determined and confirmed in this location (unpublished data).

There is general consensus that the uterosacral ligament originates from tissues in the region of S2–S4 sacral vertebrae, with no direct insertion to the bone. Insertion into the genital tract occurs at the dorsal margin of the uterine cervix and to the upper third of the posterior vaginal wall. The uterosacral ligament is positioned lateral to the rectum and medial to the ureter. It is a multifaceted, mesentery-like structure containing loose connective tissue, smooth muscle, vessels, and autonomic nerve fibers from the inferior hypogastric plexus, with contributions from sacral nerves. The superficial portion typically seen during abdominal surgery is composed primarily of smooth muscle and attaches at the cervico-vaginal junction. This anatomy will be illustrated below in considering the posterior compartment. The deep uterosacral ligament attaches to the cervix and upper vagina and has a substantial neural content, explaining its origin from S2 to S4.

Ligament Geometry

It is important to view these ligaments in the standing posture because it is in this orientation that they have their mechanical effect during normal activities of daily living. Most clinical and anatomical drawings show the cardinal and uterosacral ligaments with the body supine as they are usually encountered in the outpatient clinic, operating room, and dissecting lab. In addition, these drawings are often made while the uterus is being elevated rather than drawn downward. Both of these conditions give rise to a false impression of their structural mechanics. To illustrate, consider a picture of a hammock shown perpendicular to its usual orientation. This would be out of context and misrepresent the hammock's function.

Modern imaging allows the appearance of the cardinal and uterosacral ligaments [20,36] and their structural orientation [37] to be directly evaluated and displayed as if the individual is standing. Although it would be ideal to make the images of individuals in the upright posture, the low magnet strengths of open MRI units and the difficulties of keeping the subject motionless while standing preclude routine use of this posture.

Fortunately there are no significant shifts in position between upright and supine postures [38], presumably because the levator ani adjusts its activity to the loads placed on it so as to maintain a constant position. Therefore, acquiring images supine but evaluating them in the upright posture is reasonable as a first-order approximation.

The cardinal ligaments are in a relatively vertical orientation in the upright posture, a logical direction for them to resist downward forces (Fig. 15) [39]. The uterosacral is more dorsally directed in an orientation where it can prevent the uterus and upper vagina from sliding down the inclined plane of the levator plate toward the opening in the levator muscles through which prolapse occurs.

The lines of action for the cardinal and uterosacral vary among different women as measured in MR images (Fig. 16) [37]. In connecting the origin and insertion points, cardinal ligaments are 18.1 ± 6.8 degrees (SD) from the cephalic-caudal body axis with a slight dorsal inclination from their origin to insertion and their curved-line length from origin to the middle of their insertion in the genital tract is $5.7 + 1$ cm. The deep uterosacral ligaments are dorsally directed and 92.5 ± 13.5 degrees from the body axis and $2.7 + 9.5$ cm in length, about half the length of the cardinal ligament. Using the lines of action for the cardinal and uterosacral ligament and the inclination of the levator plate (the portion of the levator ani muscles in the midline behind the rectum), it is possible to

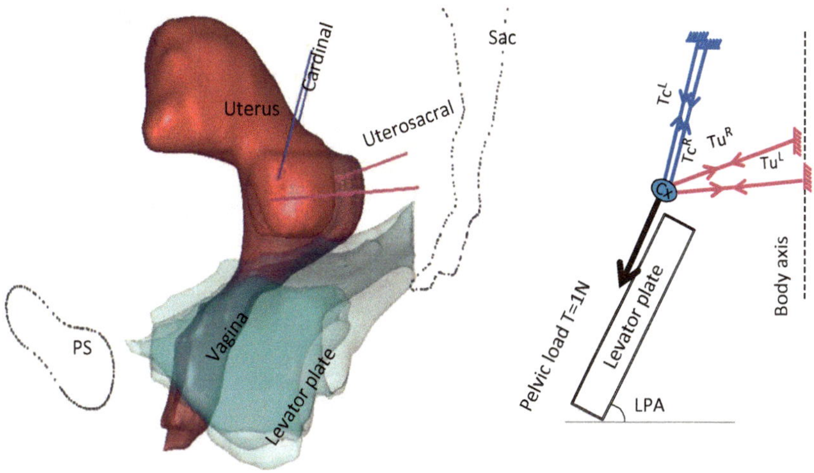

Figure 15 Left panel shows the MR-based 3D model of the uterus and vagina and transparent levator ani muscle (blue). (Right) Four-cable suspension biomechanical model showing lines of action for the cardinal and uterosacral ligaments. PS, pubic sympyhsis; Sac, sacrum. *From L. Chen, et al., Cardinal and deep uterosacral ligament lines of action: MRI based 3D technique development and preliminary findings in normal women, Int. Urogynecol. J. 24 (1) (2013) 37–45.*

make a theoretical calculation of the tensions on the two ligaments for a given 1 N unit load (Fig. 15). Analysis shows a cardinal ligament load that is 52% larger than the load on the uterosacral ligament.

Ligament Changes with Prolapse

It has been possible to examine, in vivo, the resting length and force-displacement behavior of the cervix and supporting ligaments in women undergoing surgery and compare it with the clinical assessment of apical support that is measured as the location of the cervix (Point C on the clinical pelvic organ prolapse quantitation — POP-Q examination) during a maximal Valsalva maneuver. This was done in a study of 17 women with varying degrees of uterine support from normal to prolapse, whose intraoperative measurements were made using a tripod-mounted computer-controlled linear servoactuator [40]. The device produced a caudally directed force on a tenaculum that was attached to the cervix at 4 mm/s velocity until the traction force reached 17.8 N (4 lbs). Cervix location at rest was assessed to serve as a proxy for ligament length. POP-Q point C correlated most strongly with cervix location at maximum force ($r=+0.68$; $p=0.003$) and at rest ($r=+0.62$; $p=0.009$). There was a trend between cervix location and ligament stiffness ($r=-0.44$; $p=0.079$), but this was not statistically significant at this sample size. Cervix location at rest and at maximum traction force indicates that ligament length predicts

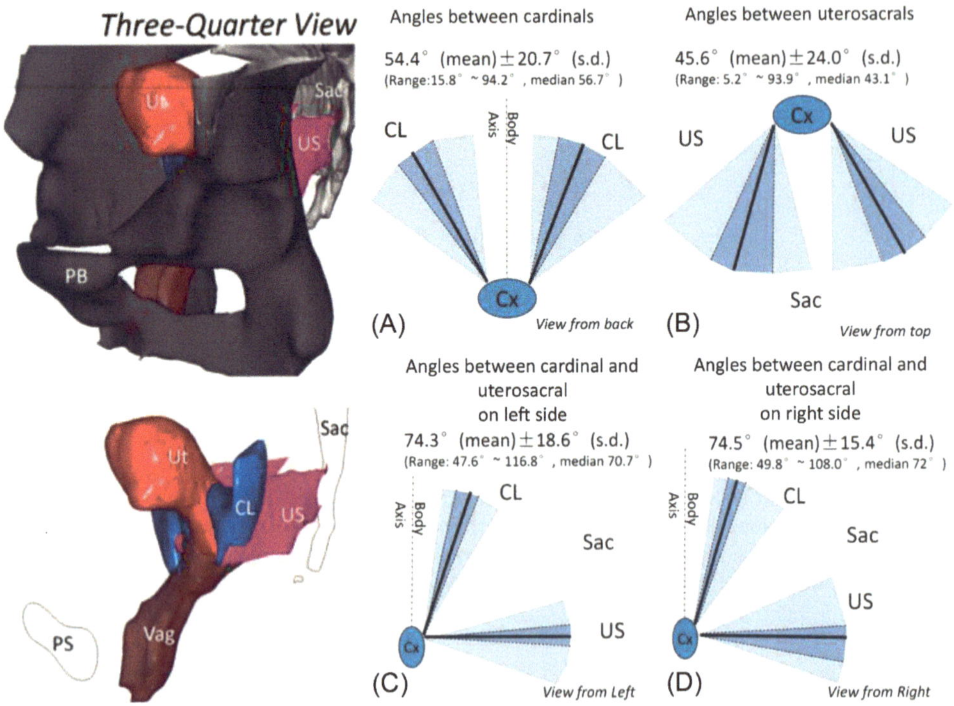

Figure 16 On the top left are shown 3D models of pubic bone (PB), sacrum (Sac), uterus (Ut), vaginal (Vag), and cardinal ligaments (CL) and deep uterosacral ligament (US) in three-quarter view. Bottom left shows the same structures without the bones. The four panels at right show the mean angle for the lines of action for the cardinal ligament seen from the top (A) and uterosacral ligament seen from the bottom (B). Lateral views and data for the left (C) and right (D) ligaments are also shown. © *DeLancey. From L. Chen, et al., Cardinal and deep uterosacral ligament lines of action: MRI based 3D technique development and preliminary findings in normal women. Int. Urogynecol. J. 24 (1) (2013) 37–45.*

about 38–46% of cervix location during maximal Valsalva. By contrast, only 19% of the variation in POP-Q point C location was explained by ligament stiffness.

Using data in these women from three "ramp and hold" trials, the tissue stiffness and energy absorbed during the ramp phase and normalized final force during the hold phase were assessed and a simplified four-cable model was used to analyze the material behavior of each ligament. The mean stiffness of the USL increased with successive trials from 0.49 ± 0.13 (s.d.) to 0.61 ± 0.22, and 0.59 ± 0.2 N/mm from trials 1 to 3, with the latter two values differing significantly from the first. The energy absorbed decreased from trials 1 (0.27 ± 0.07) to 2 (0.23 ± 0.08) and 3 (0.22 ± 0.08 J) but not from trials 2 to 3. The normalized final relaxation force increased significantly with trial 1. Modeling results for trials 1 showed that the stiffness of CL and USL was 0.20 ± 0.06 and 0.12 ± 0.04 N/mm, respectively. Under the maximum load applied in this study, the strain in the CL and USL approached about 100%. In the relaxation phase, the peak force decreased by 44 ± 4% after 60 s. This serves to demonstrate that the CL and USL exhibit visco-hyperelastic behavior. These findings have implications for the repetitive loading that would occur during daily activities and may help to explain changes over time.

The cardinal and uterosacral ligaments change their length and angle in response to the forces applied to the pelvic organs during increases in abdominal pressure. Using 3D reconstructions from MR images of women with and without pelvic organ prolapse at rest and during maximal straining, ligament length and orientation have been measured (Fig. 17) [41]. The cardinal ligament measured at rest is longer in women with prolapse compared to women with normal support (71 ± 16 vs. 59 ± 9 mm, $p=0.051$). By contrast, the deep uterosacral ligament at rest has a similar length in women with prolapse and normal support (38 ± 16 vs. 36 ± 11 mm, $p=0.8$), respectively. During a maximal Valsalva maneuver, the cardinal ligament lengthens by 30 ± 16 mm in normal women versus 15 ± 9 mm in women with prolapse ($p=0.03$). During Valsalva, deep uterosacral ligament elongation was 15 ± 12 mm versus 7 ± 4 mm ($p=0.09$). By contrast, the cardinal ligaments remain the same from rest to strain while the deep uterosacral ligament

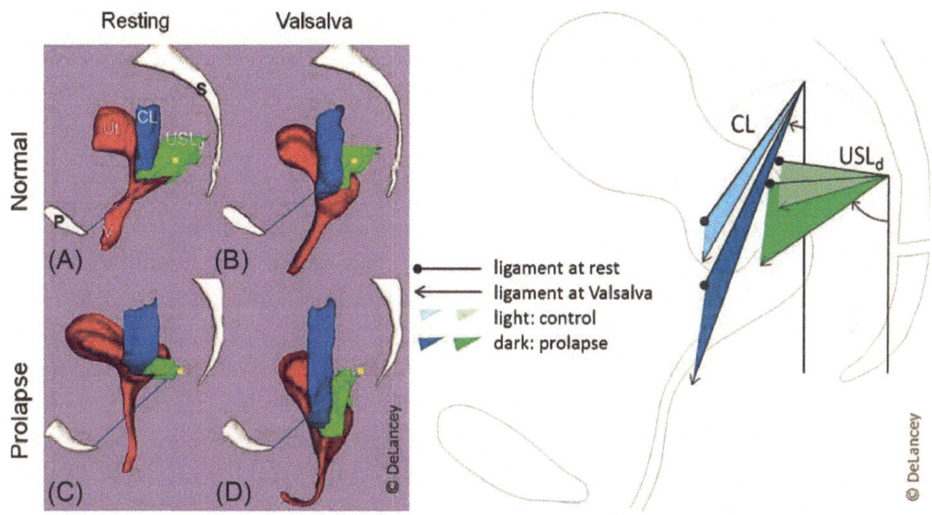

Figure 17 (Left panel) Normal and prolapse ligament comparison under resting and maximal Valsalva. Left side view of 3D models of one healthy individual at rest (A) and at maximal Valsalva (B), with their relationship to the normalized arcus tendineus fascia pelvis (ATFP) [turquoise lines extending from the public symphysis (P) to the ischial spines (yellow squares)], or the pubic symphysis to the ischial spine line (P-IS) for spatial reference. Left-side view of models of one individual with prolapse at rest (C) and at maximal Valsalva (D). PS and sacrum (S) are shown in the midsagittal plane. Ut, uterus; V, vagina; CL, cardinal ligament; USLd, deep uterosacral ligament. (Right panel) Straight length and angle of ligaments at rest and maximal Valsalva. Average straight length and angle of cardinal ligament (CL) and deep uterosacral ligament (USLd) are shown based on one average, midsagittal image from a healthy woman, with outline for uterus, vagina, pubic symphysis, and sacrum. © *DeLancey. J. Luo, et al., Using stress MRI to analyze the 3D changes in apical ligament geometry from rest to maximal Valsalva: a pilot study. Int. Urogynecol. J. 25 (2) (2014):197–203.*

angle was significantly different from the cardinal ligament angle change from rest to strain ($p < 0.001$). To summarize, cardinal ligament elongation was greater with Valsalva in women with prolapse than those with normal support, whereas, under the same conditions, the deep uterosacral ligament exhibited greater changes in ligament inclination angle.

Anterior Compartment: Level II

In Level II, in the mid-vagina, the endopelvic fascia attaches the vagina laterally to the pelvic sidewalls at the anterior and posterior fascial arches (Fig. 13). This attachment stretches the vagina transversely between the bladder and rectum and has functional significance. The structural layer that supports the bladder is composed of (a) the anterior vaginal wall and (b) its attachment to the fascial arch by the endopelvic fascia to the pelvic wall. What surgeons dissect and call the "fascia" is the fibromuscular layer of the vagina, which consists of the smooth muscle layer of the vaginal wall that contains a great deal of interstitial connective tissue. Similarly, the posterior vaginal wall and endopelvic fascia (rectovaginal fascia) form the restraining layer that prevents the rectum from protruding forwards, blocking formation of a rectocele. This will be discussed in more detail later in the chapter. The concept that there is a separate layer between the vagina and bladder or rectum that is responsible for holding the organs in place is not consistent with detailed analysis of the histological structure of this area.

Anterior Fascial and Levator Arches

The fascial arch and the levator arch can be seen in the anterior compartment lying on the medial surface of the pelvic sidewall (Fig. 18). The fascial arch has a fine tendinous attachment to the inner surface of the pubic bone 1 cm from the midline and 1 cm above the bottom of the pubic symphysis. There is a short (1 cm) area of tendon, and then, this structure broadens into a node that consists of the vaginal wall, attachment of the pubovaginal portion of the pubovisceral muscle, and endopelvic fascia. Extending dorsally from this "node," the fascial arch continues to its dorsal attachment 1–2 cm above the ischial spine. The arch is actually just the visible margin of this connective tissue and not actually a cord-like band as often described. The connective tissue fibers in this area run parallel to the edge of the fascial arch and not transversely from the vagina to the arch.

The levator arch starts at the inside of the superior pubic ramus just anterior to the obturator canal. Its other end inserts at the ischial spine near the insertion of the fascial arch. In the dorsal half of the pelvic sidewall, the courses of the two arches overlie one another. Although the levator arch is subtle in appearance, it can readily be palpated as a band from the superior pubic ramus to the ischial spine.

Why do the fascia and levator arches exist? These structures serve as bands that are stretched over the internal obturator muscle. They form critical structural locations of attachment to the bony pelvis in an area where the internal obturator muscle lies between

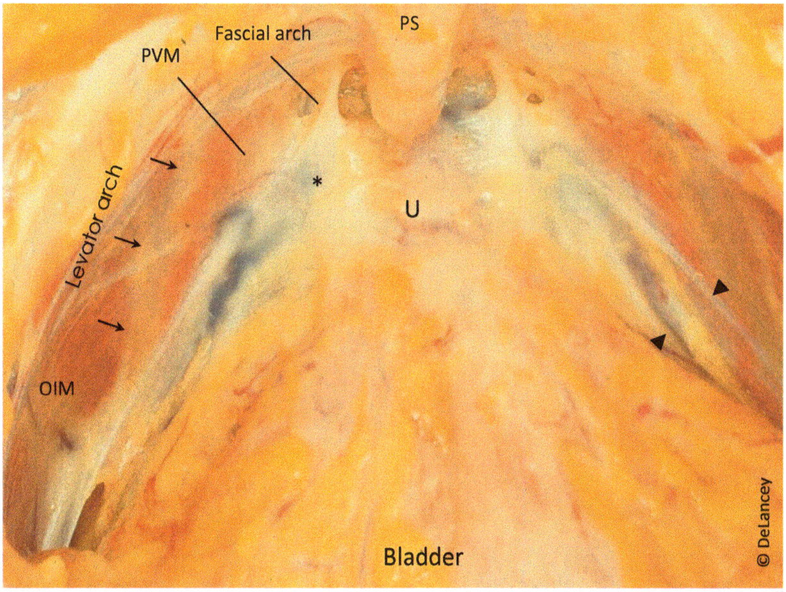

Figure 18 View into space of Retzius looking into the pelvis from above with the pubic symphysis (PS) in the midline showing, on the left side showing fascial arch as well as the inner surface of the of the pubovisceral muscle (PVM). The bladder and urethra (U) have been pulled back away from the inside of the pubic bones for exposure. The edge of the levator arch is marked with arrows as it lies on the inner surface of the obturator internus muscle. Approximately 1 cm from its origin, the fascial arch enters a "node" marked with an asterisk (*) where there is fusion with the vaginal wall periurethrally and also an attachment to the medial portion of the levator ani (pubovaginal muscle). On the right side, a small paravaginal defect can be seen whose margins are marked with arrowheads. © *DeLancey.*

the vagina and the pelvic bones. Because of the obturator's position, it is not possible for the lateral vagina to have a direct connection to the bones in this region. These two fibrous bands, therefore, stretch over the internal obturator muscle to span the distance between the pubic bones and the ischial spines to form a structure to which the vagina and levator are attached.

Anterior Compartment: Level III

In Level III, the vagina is attached to its surrounding tissues. It is connected to the levator ani muscles on either side just above the hymenal ring. This attachment is referred to as the pubovaginal portion of the pubovisceral component of the levator ani muscles. The attachment of the fascial arch, as described earlier in this chapter, forms the ventral connection at this level as a ventral extension of the perineal membrane. Dorsally, the vagina is fused with the connective tissue of the perineal body. These connections are all covered in the sections concerning these structures.

3D Reconstructions of the Anterior Compartment and What Happens During Prolapse

The anterior vaginal wall and its suspension in Level I to the cardinal and uterosacral ligaments and the lateral attachment in Level II play a critical role in anterior compartment support. Understanding the impact of changes in these structures and relationships is a key to understanding anterior vaginal wall prolapse (Fig. 19). 3D stress MRI, where the structures are imaged directly or indirectly in living women during a maximal Valsalva maneuver, has provided the first opportunity to study and compare the spatial changes in anterior vaginal wall prolapse to women with normal support.

The separation between the vagina and the fascial arch seen during surgery always starts at the end near the ischial spine and proceeds toward the pubic end much as a zipper separates its two elements from one end to the other [42]. This can be seen in Fig. 18, where a small right-sided paravaginal defect is visible and compared with the left side, that is intact. Descent of the anterior vaginal wall involves this paravaginal detachment, which is usually (but not always) associated with descent of cervix [43,44].

3D stress MRI in women shows that the normal configuration and deformation of the anterior vaginal wall at rest and under load is different than previously thought [45] (Fig. 20). In women with normal support, the upper vagina is usually above the fascial

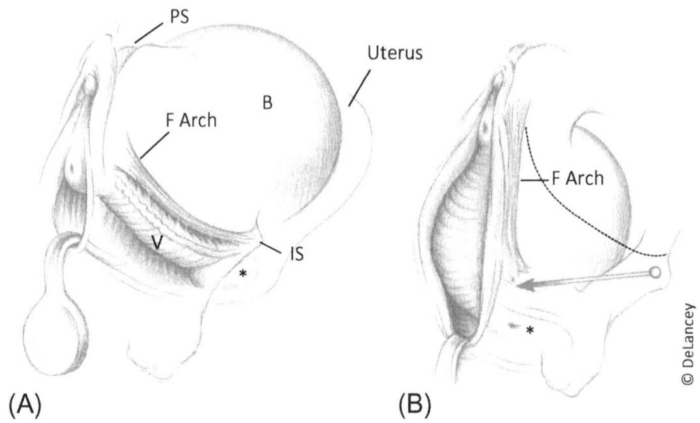

Figure 19 Conceptual diagram showing (A) the normal location of the anterior vaginal wall (V) and descent and appearance through the vaginal opening, (B) in anterior vaginal prolapse or cystocele. Structures seen from the side with a woman in the lithotomy position and a weighted speculum in place including the bladder (B), uterus, cervix (*), ischial spine (IS), and pubic symphysis (PS). The mechanical effect of apical descent (*arrow*) and concomitant downward movement of the fascial arch (F Arch) whose normal position is indicated with a dashed line can be appreciated. Note that it is not changes in the vaginal wall that allow the vaginal wall to protrude but rather loss of its normal attachments. © *DeLancey. From J.O. DeLancey, Fascial and muscular abnormalities in women with urethral hypermobility and anterior vaginal wall prolapse, Am. J. Obstet. Gynecol. 187 (1) (2002) 93–98.*

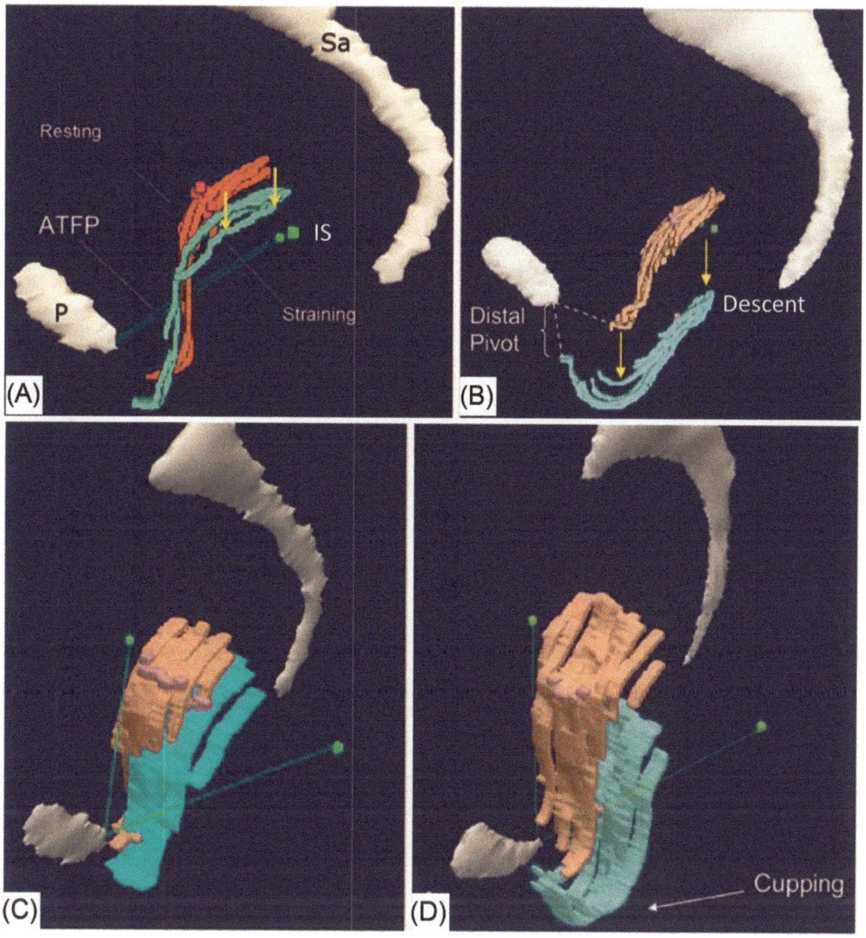

Figure 20 Lateral view of 3-D model of the vaginal wall of control (A) versus case (B). Resting (pink) and straining (turquoise) cross-sections of the vagina and ATFP (green). This shows descent along the length of the vagina with Valsalva as well as the distal pivot seen in cases (B) in comparison with a representative control (A). P, pubic symphysis; Sa, sacrum. Oblique view of the same models. Note cupping of the posterior vaginal wall in (d) Vaginal "cupping." The cupping occurs in both the sagittal and frontal planes. © *DeLancey. From K.A. Larson, et al., Magnetic resonance imaging-based three-dimensional model of anterior vaginal wall position at rest and maximal strain in women with and without prolapse, Int. Urogynecol. J. 21 (9) (2010) 1103–1109.*

arch and the anterior wall is not parallel to the fascial arcus [46]. Although they are attached as described in the earlier sections, elevation of the vaginal wall by the levator ani muscles places the vaginal wall in a position oblique to the line of the fascial arch. In women with cystocele, there is an overall downward movement of the anterior vaginal wall and there is "cupping" of the lower portion where the vaginal wall protrudes through the vaginal introitus. The distal vaginal wall pivots around the point of attachment of the fascial arch to the pubic bone in much the same way as a child's swing pivots around its attachment. As part of this change, there is an overall downward translation of the anterior vaginal wall.

Quantitative assessment of vaginal wall support, comparing women with and without anterior vaginal wall prolapse, has allowed structural hypotheses to be tested concerning the changes associated with prolapse [47]. In this study we measured the width and length of the vagina as well as the degree of apical descent and the size of the paravaginal defect that was measured as the distance from the mean location of the normal lateral vaginal wall and the vaginal wall in an individual. The widths and the paravaginal distance were assessed at five locations equally spaced along the length of the vagina. The parameter that showed the greatest difference between women with cystocele and those with normal support is the lower position of the lateral anterior vaginal wall's lateral margin (paravaginal gap) with effect sizes in the mid-vagina of 2.2–2.8 at the five different locations (Fig. 21). The apex was lower and the effect size for that parameter was 1.7. Strong correlations between apical and paravaginal support were evident in the mid- and upper-vagina (r=0.77–0.93 for the five locations of paravaginal gap). The vagina is also somewhat longer in women with prolapse (effect size 1.2), but vaginal widths only differed at one location in the mid-vagina with an effect size of 1.0. Therefore, these key findings show that, in women with prolapse, changes in the attachments of the vagina to the pelvis (apical and paravaginal) are much greater than changes in the vaginal wall itself (width or length), in both number of sites affected and magnitude of the difference between cases and controls.

Traditionally, it had been assumed that cystocele was associated with changes to the fascia of the anterior vaginal wall because a bulging anterior vaginal wall was what clinicians could see. Descriptions of the bladder "herniating" through defects in the vaginal wall are common in the literature. This traditional assumption is not supported by these current data. It is evident from current data that the bulging occurs because the vagina descends due to weakened attachments and failure of the levator ani muscles to hold the vaginal introitus closed.

Cystocele and Exposed Vaginal Length

A key factor in the formation of a cystocele is the degree to which the anterior vaginal wall is exposed to the pressure differential between abdominal and atmospheric pressure [48]. As outlined in Fig. 3, the combination of impaired muscles and weakened vaginal attachments act together to allow downward anterior vaginal wall descent so that it becomes exposed to the difference between abdominal and atmospheric pressure. The greater the area of the vaginal wall that is exposed to this pressure difference, the

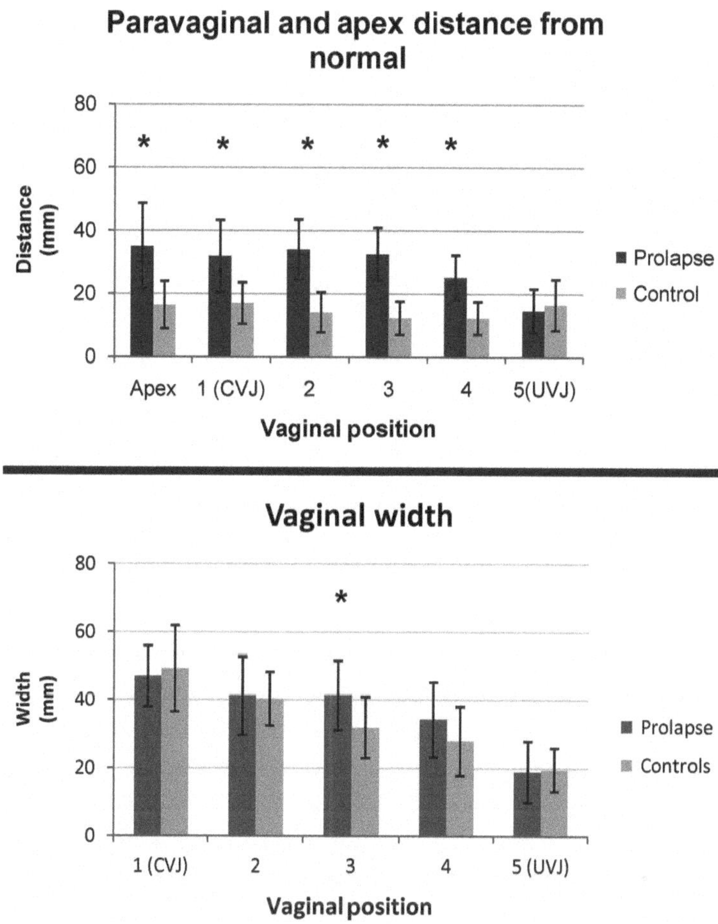

Figure 21 Top panel compares apical and paravaginal distances from normal and bottom panel; vaginal width. Note that "1" represents the apex at the cervico-vaginal junction (CVJ) and position "5," the urethrovaginal junction (UVJ) at the distal end. Asterisks mark statistically significant differences ($p < 0.05$). Note that right and left paravaginal distance means have been combined to one overall mean. Standard deviation shown. © *DeLancey. From K.A. Larson, et al., Measurement of the 3D geometry of the fascial arches in women with a unilateral levator defect and "architectural distortion", Int. Urogynecol. J. 23 (1) (2012) 57–63; K.A. Larson, et al., 3D analysis of cystoceles using magnetic resonance imaging assessing midline, paravaginal, and apical defects, Int. Urogynecol. J. 23 (3) (2012) 285–293.*

greater the force placed on the structures that attach the vagina to the pelvis. Supine, dynamic mid-sagittal MR imaging shows important relationships between the area of the vaginal wall that is exposed to this pressure differential and to descent of the bladder and apex, and to the size of the urogenital hiatus through which this prolapse occurs (Fig. 22).

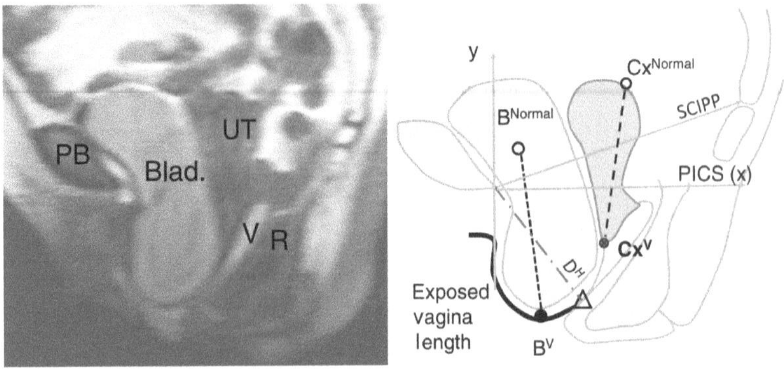

Figure 22 Mid-sagittal MRI at maximal straining and measurement scheme based with sacrococcygeal inferior pubic point (SCIPP) line-oriented axes; cervix location at maximal straining (CxV); most independent bladder point at maximal straining (BV); normal cervix location (CxNormal), and normal bladder rest location (BNormal). Δ transition point where the anterior vaginal wall (AVW) contacts the perineal body. DH, R, rectum; UT, uterus; V, vagina; PB, pubic bone; PICS, pelvic inclination correction system. © *DeLancey. From A. Yousuf, et al., The length of anterior vaginal wall exposed to external pressure on maximal straining MRI: relationship to urogenital hiatus diameter, and apical and bladder location, Int. Urogynecol. J. 25 (10) (2014) 1349–1356.*

A strong bilinear relationship exists between the length of the exposed vaginal wall and most dependent bladder locations ($r^2 = 0.91$, $p \leq 0.001$) (Fig. 23) [48]. When the bladder descends up to the inflection point (about 4 cm away from its normal position), there is little change in the exposed anterior vaginal wall length. The vagina descends but stays in contact with the posterior vaginal wall. With further descent, the exposed vaginal wall length increases significantly, with a 2 cm increase in exposed anterior vaginal wall (AVW) length for every additional 1 cm of descent in bladder location. A similar bilinear relationship exists between exposed AVW and apical location. Exposed vaginal wall length is also highly correlated with hiatus diameter ($r^2 = 0.85$, $p < 0.001$).

There are clinical implications of these observations, when bladder descent was less than 4 cm, very little of the vagina is exposed. In the descent that occurs up to this inflection point, the anterior vaginal wall is still in contact with the posterior vaginal wall, so only a small portion is exposed to the pressure differential. With further bladder downward movement, the exposed vaginal wall length starts to increase linearly. Similarly, a significant but weaker bilinear relationship was found between exposed vaginal wall length and apical location, the inflection point is at about 5 cm below the normal apex location. Both of these observations suggest a threshold effect wherein some degree of descent is not associated with much exposure of the vaginal wall to a pressure differential [48].

The symptoms of bulging and pressure experienced by women with anterior vaginal wall prolapse occur when the anterior vaginal wall descends to the level of the hymenal

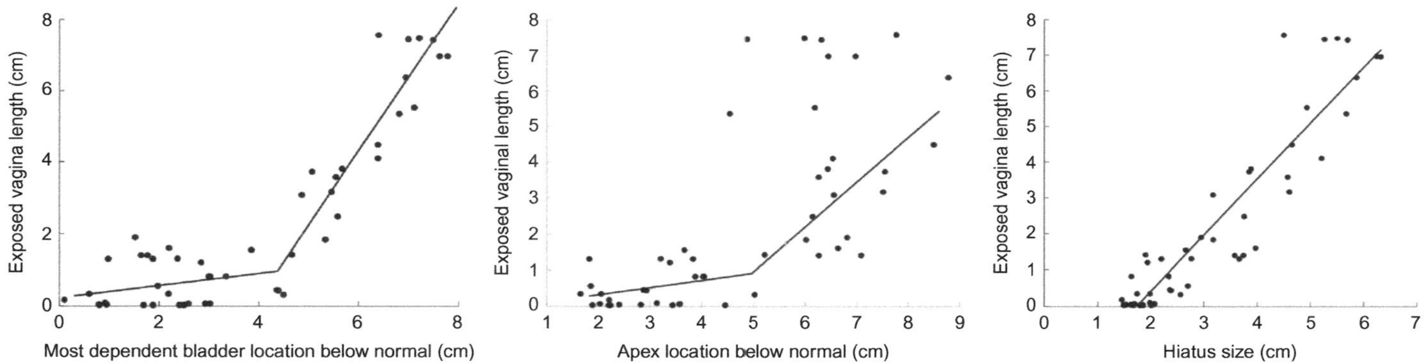

Figure 23 The relationships between the exposed vaginal length and most dependent bladder location, apex location, and hiatus size are shown. A bilinear relationship was used to describe the relationship between the position of the most dependent bladder point below normal and the length of exposed vagina. The correlation coefficient $r^2=0.91$. The estimated transition point is when the bladder location is 4.4 cm below the normal position with a 95% confidence interval (3.9 cm, 4.9 cm) A bilinear relationship between exposed vaginal wall length and apex location was also seen. The correlation coefficient $r^2=0.78$. The estimated transition point is at the apical descent at 5.0 cm below the normal position with a 95% confidence interval 3.4–6.5 cm. Hiatus size has a strong linear relationship between hiatus diameter and exposed vaginal wall length with a correlation coefficient $r^2=0.85$ but it is not bilinear. © *DeLancey. From A. Yousuf, et al., The length of anterior vaginal wall exposed to external pressure on maximal straining MRI: relationship to urogenital hiatus diameter, and apical and bladder location, Int. Urogynecol. J. 25 (10) (2014) 1349–1356.*

remnants [49]. The hymen lies at the lower border of the levator ani and so marks the level of the levator hiatus through which prolapse occurs. It is the level when the vagina becomes exposed to the pressure differential. A descent of 4 cm would bring the vaginal wall just below the hymen. The fact that the vagina becomes increasingly exposed to the pressure differential at the level of the hymen helps explain why women experience a pressure sensation when the vaginal wall reaches this point. It supports the clinical position that the hymen be considered an appropriate level both for defining symptomatic prolapse and for assessing surgical cure [50]. It is also consistent with the observation that women's dissatisfaction with surgical results increases once the vaginal wall is below the hymen. They are not, however, dissatisfied with results when the anterior vaginal wall is at or above the hymen [51].

Posterior Compartment

Anatomy of the Posterior Vaginal Wall Support as It Applies to Rrectocele

Level I: The support of the upper one-third of the vagina has previously been described as being suspended by the cardinal-uterosacral complex and it is generally similar for the posterior compartment as for the anterior. There is one important difference, however. The uterosacral complex, dorsal aspect of this support, is of particular significance to the posterior compartment. The vaginal portion of the uterosacral ligaments attaches to the upper posterior vaginal wall and holds it dorsally. This part of the uterosacral ligament lies deep below the peritoneum and cannot be seen without dissection. It can, however, be palpated when the apex is displaced ventrally (toward the abdominal wall during vaginal surgery). This portion of the suspensory system holds the upper posterior vaginal wall dorsally over the levator plate; a vector that cannot be operational for the more vertically oriented cardinal ligament. This is a critical element of the mechanism that prevents development of enterocele where the space between the vagina and the rectum becomes distended (Fig. 24) [52].

The mid-portion of the posterior vaginal wall (Level II) is connected laterally to the posterior fascial arch by the endopelvic fascia [53] in a way that is analogous to the anterior vaginal wall. The posterior fascial arch is stretched between an attachment near the ischial spine and the perineal body. The distal attachment to the moveable perineal body differs from the anterior fascial arch whose distal end has an immovable attachment to the pubic bone. These lateral connections of the vaginal wall to the posterior fascial arch prevent ventral movement of the vagina during increases in abdominal pressure.

A fascial layer lies between the rectum and vagina called the rectovaginal fascia or Denonvillier's fascia. This structure is frequently described in articles concerning posterior vaginal wall support. Its existence as a separate layer has been documented as the presence of a visible layer in women with Müllerian agenesis [21]. The visible

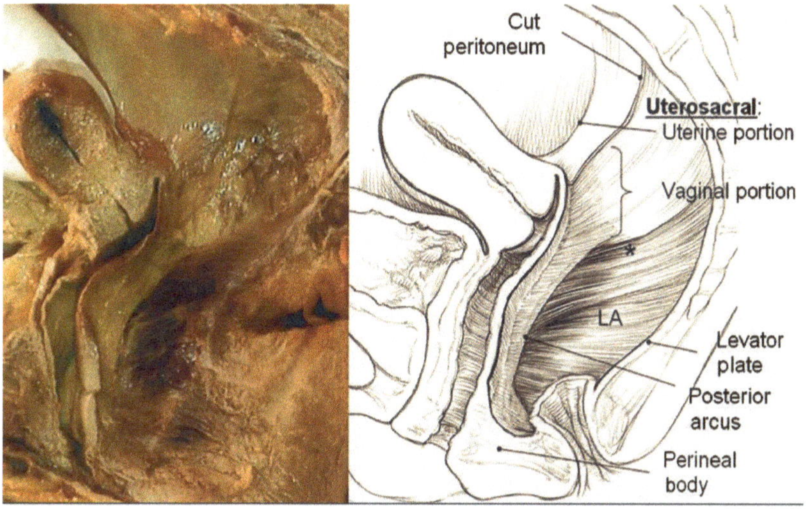

Figure 24 Cadaver dissection (*left*) and illustration (*right*) of posterior compartment of a 56-year-old multiparous female showing structural relationships after the rectum has been removed. Note the apical connections of the upper posterior vagina to the inside of the pelvic wall in a retroperitoneal position. These lie below the peritoneum and are dorsal and caudal to what is traditionally referred to as the uterosacral ligament. These structures are continuous with the posterior arcus tendineus fascia pelvis. At the distal end of the vagina, the wall merges into the top of the perineal body. The lateral and dorsal margins of the compartment are formed by the levator ani muscles (*LA*) and the levator plate. The asterisk denotes the region of the sacrospinous ligament overlain by the coccygeus muscle. *© DeLancey. From Y. Hsu, C. Lewicky-Gaupp, J.O. DeLancey, Posterior compartment anatomy as seen in magnetic resonance imaging and 3-dimensional reconstruction from asymptomatic nulliparas, Am. J. Obstet. Gynecol. 198 (6) (2008) 651e1-7.*

presence of this layer, however, does not mean that it has sufficient mechanical strength to be used during surgery.

Level III: The lower one-third of the vagina is fused laterally with the perineal membrane and medial aspect of the levator ani muscles and distally to the perineal body (Fig. 25A).

The perineal membrane (formerly known as the urogenital diaphragm) is a dense triangular membrane with a central opening through which the vagina and urethra pass. It is composed of two regions, one dorsal and one ventral [54]. The dorsal portion consists of bilateral transverse fibrous sheets that attach the lateral wall of the vagina and perineal body to the ischiopubic ramus. This portion is devoid of striated muscle. The ventral portion is part of a solid three-dimensional tissue mass in which several structures are embedded. It is intimately associated with the compressor urethrae and the urethrovaginal

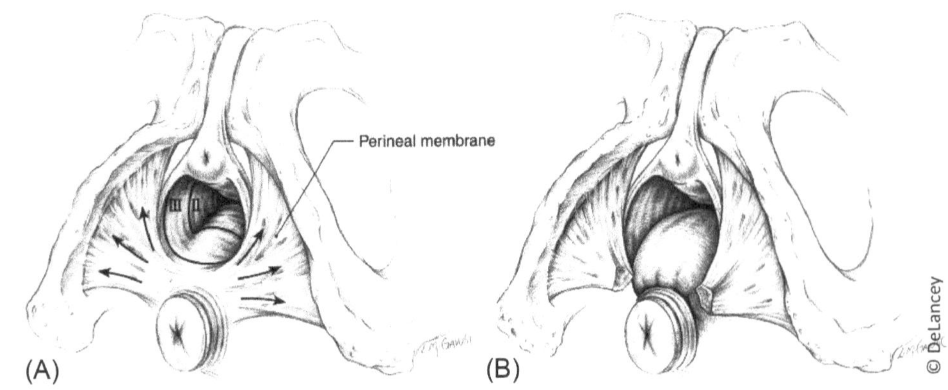

Figure 25 Peripheral attachments of perineal membrane to ischiopubic rami (A) and direction of tension on fibers uniting through perineal body (*arrows*). The U-shaped contour of posterior vaginal wall in Level III and W-shaped contour in Level II demonstrated. Loss of connection (B) shown as would occur due to obstetrical trauma that could lead to distal rectocele. © *DeLancey. From J.O. DeLancey, Structural anatomy of the posterior pelvic compartment as it relates to rectocele, Am. J. Obstet. Gynecol. 180 (4) (1999) 815–823.*

sphincter muscle of the distal urethra. In this region, the perineal membrane is continuous with the insertion of the anterior fascial arch. The levator ani muscles are connected with the cranial surface of the perineal membrane. The vestibular bulb and clitoral crus are fused with the membrane's caudal surface.

How Does Posterior Vaginal Wall Support Fail?

There has been a century of conjecture concerning the actual structural defects that are involved in rectocele formation, and many competing hypotheses have been put forward. The rectal shape and distortion has long been studied using defecography. More recently MRI and 3D stress MRI has allowed the objective documentation of posterior vaginal wall support itself to be described. At present the scientific foundation for structural changes that are associated with posterior vaginal wall prolapse are not as well developed as they are for anterior compartment support, but this is rapidly changing with new techniques that allow changes in the anatomical structures associated with posterior prolapse to be objectively documented and quantified.

The most common form of posterior vaginal prolapse is manifest as rectocele. 3D stress MRI reconstructions demonstrate that rectocele involves a forward protrusion of the posterior wall vaginal wall and anterior rectal wall through the urogenital hiatus in the levator ani muscle (Fig. 26) [55]. In addition, there is overall descent of the posterior vaginal wall and perineal body. It is not evident that there is any localized defect in a fascial layer through which the rectum herniates as is often suggested. It can be presumed that this is associated with damage to the levator ani muscles but whether

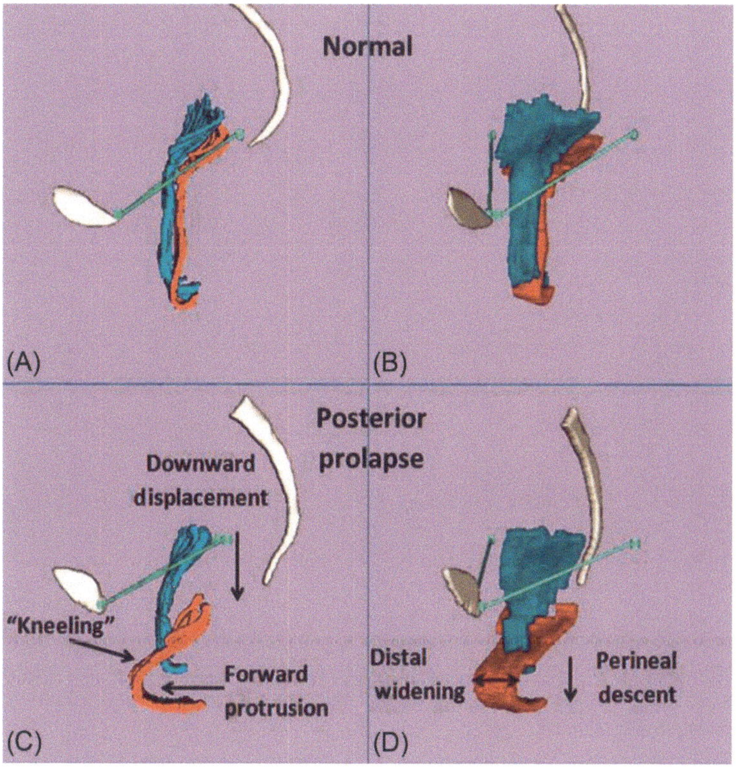

Figure 26 Posterior vaginal prolapse shape and position changes (Fig. 3). Characteristics of posterior prolapse. Comparison of control (A, B) and case (C, D) in lateral view (A, C) and oblique view (B, D) showing five characteristic features (C, D) during rest (blue) and Valsalva (pink): (1) Increased folding (kneeling), (2) downward displacement in the upper two-thirds part of the vagina, (3) forward protrusion, (4) perineal descent, and (5) distal widening in the lower third part of the vagina. Pubis and sacrum are shown in white. The P-IS line is shown in turquoise. © *DeLancey 2011; From J. Luo, et al., Posterior vaginal prolapse shape and position changes at maximal Valsalva seen in 3-D MRI-based models, Int. Urogynecol. J. 23 (9) (2012) 1301–1306.*

or not this association is as strong as it is with anterior compartment prolapse has not yet been assessed.

The perineal membranes on either side are connected to each other through the perineal body. This connection helps limit the degree to which the perineal body can descend. If the fibers that connect one side with the other rupture, then the bowel may protrude downward resulting in a posterior vaginal wall prolapse. In addition, because the levator ani muscles are fused with the other tissues in this area, distention of the vaginal opening is also associated with enlargement of the levator hiatus through which prolapse occurs.

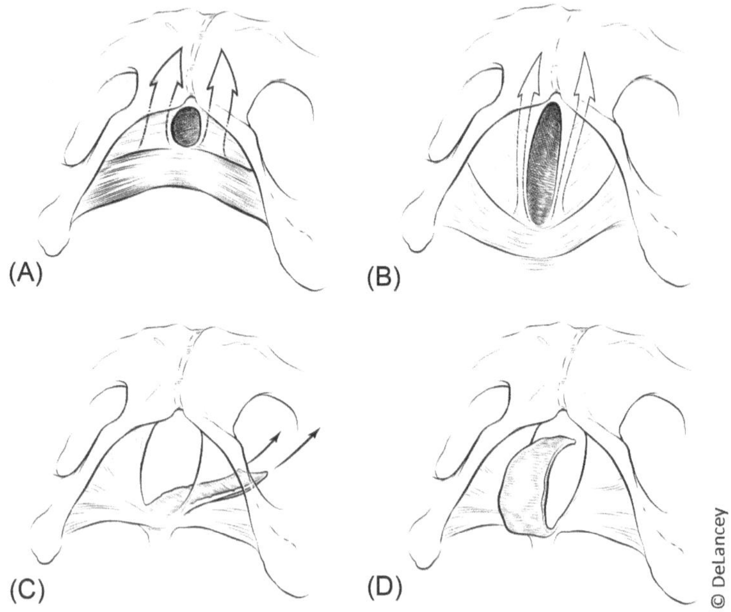

Figure 27 Factors involved in rectocele and enterocele. In (A, B), the effect of levator ani on the hiatus and perineal structures is shown. Loss of levator tone results in an enlarged hiatus. The effect of loss of apical support is demonstrated in C, D where loss of apical support allows the apex to descend as occurs in enterocele. © *DeLancey.*

Failure of the posterior vaginal wall's upper suspension is associated with enterocele and prolapse of the vaginal vault (Fig. 27). This loss of attachment to the sacrum allows forward protrusion of the posterior vaginal wall. This often occurs with downward tipping of the levator plate that is associated with pelvic organ prolapse in general. This is an area currently under active investigation to so that the mechanics can be as well understood as the anterior compartment.

INTERACTIONS AMONG DIFFERENT COMPARTMENTS

Thus far we have discussed anterior and posterior pelvic organ support separately. Important interactions occur between the compartments and must also be considered. It is impossible for one vaginal wall to move without a compensating movement of the other wall because the vaginal wall is a potential space, not normally filled with air or fluid. Understanding the direction and magnitude of vaginal movement is important to our overall quantitative understanding of prolapse.

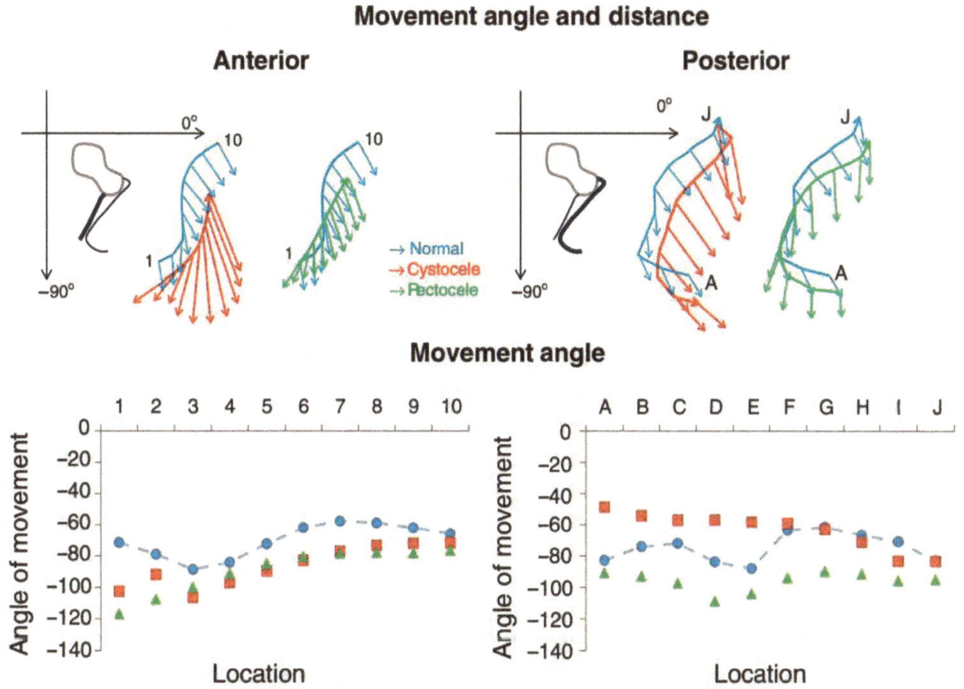

Figure 28 (*Top row*) Direction and magnitude (mm) of vaginal wall displacement of the anterior and posterior support system. (*Bottom row*) Angle of displacement (degrees from the horizontal) in normal women, women with cystocele, and women with rectocele. *From D.M. Spahlinger, et al., Relationship between intra-abdominal pressure and vaginal wall movements during Valsalva in women with and without pelvic organ prolapse: technique development and early observations, Int. Urogynecol. J. 25 (7) (2014) 873–881.*

The direction and magnitude of vaginal wall movement can be studied by analyzing 2D mid-sagittal dynamic MR images [56]. The direction and degree of movement as well as the compliance of both anterior and posterior vaginal wall support systems vary along different regions of vaginal wall for all three groups, with the highest compliance found near the vaginal apex and the lowest near the introitus (Fig. 28). Women with cystocele have more compliant anterior and posterior vaginal wall support systems than women with rectocele. The movement direction differs between cystocele and rectocele (Fig. 29). In cystocele, the anterior vaginal wall moves mostly toward the vaginal orifice in the upper vagina, but in a ventral direction in the lower vagina. In rectocele, the direction of the posterior vaginal wall movement is generally toward the vaginal orifice.

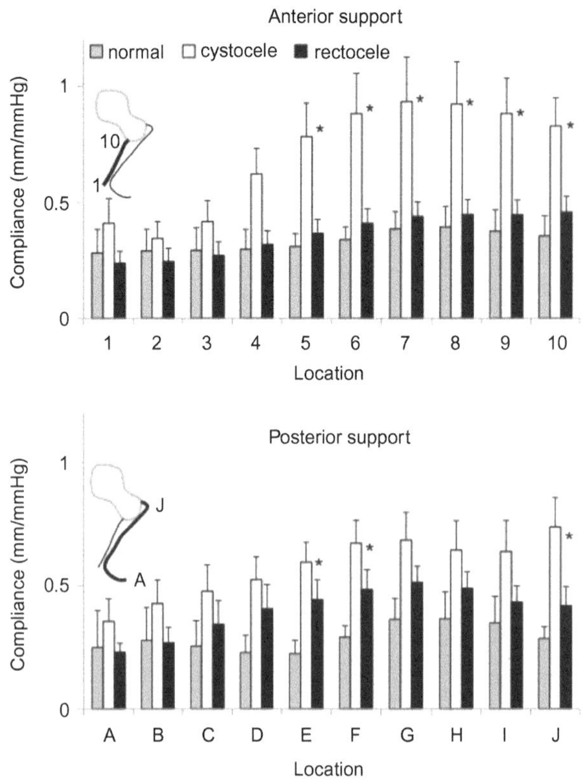

Figure 29 Mean regional compliance (mm/mm Hg) measured at the 10 locations along the anterior (*top*) and posterior wall (*bottom*). Error bars represent standard error. Light gray denotes women with normal support, white denotes women with anterior wall predominant prolapse, and black denotes women with posterior wall-predominant prolapse. Asterisks show the locations at which the compliance among groups is significantly different ($p < 0.05$). *From D.M. Spahlinger, et al., Relationship between intra-abdominal pressure and vaginal wall movements during Valsalva in women with and without pelvic organ prolapse: technique development and early observations, Int. Urogynecol. J. 25 (7) (2014) 873–881.*

CONCLUSION

Our understanding of pelvic organ prolapse is evolving rapidly. Techniques that can evaluate the relative merits and contributions of competing hypotheses are emerging so that opinion is slowly, but surely giving way to science. More importantly, tests that can make measurements of biomechanically important factors in individual women will soon be used to base treatment of each woman's specific situation. Treatment based on specific mechanical abnormalities is the norm in fields such as cardiology where functional imaging techniques such as echocardiography have evolved since its introduction in the 1960s. During that time, over 20,000 articles have been written based on findings derived with this technique. Many of these articles describe new ways of assessing cardiac functional derangements and how they alter treatment outcomes. It can be expected that similar progress will be possible as our understanding of how to measure pelvic floor functional anatomy evolves.

ACKNOWLEDGMENT

We gratefully acknowledge support from NIH Office for Research on Women's Health SCOR P50 HD 44406 and NICHD R01 HD 38665 for supporting this work.

REFERENCES

[1] L. Chen, et al., Interaction among apical support, levator ani impairment, and anterior vaginal wall prolapse, Obstet. Gynecol. 108 (2) (2006) 324–332.
[2] J.O. DeLancey, Surgery for cystocele III: do all cystoceles involve apical descent? Observations on cause and effect, Int. Urogynecol. J. 23 (6) (2012) 665–667.
[3] R. Kearney, R. Sawhney, J.O. DeLancey, Levator ani muscle anatomy evaluated by origin-insertion pairs, Obstet. Gynecol. 104 (1) (2004) 168–173.
[4] F.C.O.A Terminology, Terminologia Anatomica: International Anatomical Terminology, first ed., Thieme Stuttgart, Stuttgart, 1998. 292.
[5] J.O. Lawson, Pelvic anatomy. I. Pelvic floor muscles, Ann. R. Coll. Surg. Engl. 54 (5) (1974) 244–252.
[6] C. Betschart, et al., Comparison of muscle fiber directions between different levator ani muscle subdivisions: in vivo MRI measurements in women, Int. Urogynecol. J. 25 (9) (2014) 1263–1268.
[7] N.A. Clark, et al., Levator defects affect perineal position independently of prolapse status, Am. J. Obstet. Gynecol. 203 (6) (2010) 595 e17-22.
[8] C. Lewicky-Gaupp, et al., Structural position of the posterior vagina and pelvic floor in women with and without posterior vaginal prolapse, Am. J. Obstet. Gynecol. 202 (5) (2010). 497 e1-6.
[9] S.A. Jung, et al., Vaginal high-pressure zone assessed by dynamic 3-dimensional ultrasound images of the pelvic floor, Am. J. Obstet. Gynecol. 197 (1) (2007). 52 e1-7.
[10] N.M. Guaderrama, et al., The vaginal pressure profile, Neurourol. Urodyn. 24 (3) (2005) 243–247.
[11] C. Oh, A.E. Kark, Anatomy of the external anal sphincter, Br. J. Surg. 59 (9) (1972) 717–723.
[12] Y. Hsu, et al., Magnetic resonance imaging and 3-dimensional analysis of external anal sphincter anatomy, Obstet. Gynecol. 106 (6) (2005) 1259–1265.
[13] J. Halban, J. Tandler, Anatomie und Atiologie der Genital prolapse biem Weibe, Wilhelm Braunmuller, Vienna and Leipzig, 1907.
[14] R.H. Paramore, The supports-in-chief of the female pelvic viscera, Proc. R. Soc. Med. 1 (Obstet. Gynaecol. Sect.) (1908) 195–214.
[15] J.O. DeLancey, et al., Comparison of levator ani muscle defects and function in women with and without pelvic organ prolapse, Obstet. Gynecol. 109 (2 Pt. 1) (2007) 295–302.
[16] J.O. DeLancey, et al., The appearance of levator ani muscle abnormalities in magnetic resonance images after vaginal delivery, Obstet. Gynecol. 101 (1) (2003) 46–53.
[17] J.O. DeLancey, et al., Comparison of the puborectal muscle on MRI in women with POP and levator ani defects with those with normal support and no defect, Int. Urogynecol. J. 23 (1) (2012) 73–77.
[18] J.O. DeLancey, et al., Stress urinary incontinence: relative importance of urethral support and urethral closure pressure, J. Urol. 179 (6) (2008) 2286–2290. discussion 2290.
[19] R.U. Margulies, et al., Appearance of the levator ani muscle subdivisions in magnetic resonance images, Obstet. Gynecol. 107 (5) (2006) 1064–1069.
[20] J. Kim, et al., Anatomy of the pubovisceral muscle origin: macroscopic and microscopic findings within the injury zone, Neurourol. Urodyn. 34 (8) (2015) 774–780.

[21] M. Huebner, R.U. Margulies, J.O. DeLancey, Pelvic architectural distortion is associated with pelvic organ prolapse, Int. Urogynecol. J. Pelvic Floor Dysfunct. 19 (6) (2008) 863–867.
[22] K.A. Larson, et al., Measurement of the 3D geometry of the fascial arches in women with a unilateral levator defect and "architectural distortion," Int. Urogynecol. J. 23 (1) (2012) 57–63.
[23] M.B. Berger, D.M. Morgan, J.O. DeLancey, Levator ani defect scores and pelvic organ prolapse: is there a threshold effect? Int. Urogynecol. J. 25 (10) (2014) 1375–1379.
[24] H.P. Dietz, J.M. Simpson, Levator trauma is associated with pelvic organ prolapse, BJOG 115 (8) (2008) 979–984.
[25] A.C. Weidner, et al., Neuropathic injury to the levator ani occurs in 1 in 4 primiparous women, Am. J. Obstet. Gynecol. 195 (6) (2006) 1851–1856.
[26] E. Conner, et al., Vaginal delivery and serum markers of ischemia/reperfusion injury, Int. J. Gynaecol. Obstet. 94 (2) (2006) 96–102.
[27] K.C. Lien, et al., Levator ani muscle stretch induced by simulated vaginal birth, Obstet. Gynecol. 103 (1) (2004) 31–40.
[28] J.M. Miller, et al., Evaluating maternal recovery from labor and delivery: bone and levator ani injuries, Am. J. Obstet. Gynecol. 213 (2) (2015). 188 e1–188 e11.
[29] J.M. Miller, et al., MRI findings in patients considered high risk for pelvic floor injury studied serially after vaginal childbirth, AJR Am. J. Roentgenol. 195 (3) (2010) 786–791.
[30] J.O. DeLancey, Anatomic aspects of vaginal eversion after hysterectomy, Am. J. Obstet. Gynecol. 166 (6 Pt. 1) (1992) 1717–1724. discussion 1724–8.
[31] J. Luo, et al., In vivo properties of uterine suspensory tissue in pelvic organ prolapse, J. Biomech. Eng. 136 (2) (2014) 021016.
[32] S.L. Woo, et al., Tensile properties of the human femur-anterior cruciate ligament-tibia complex. The effects of specimen age and orientation, Am. J. Sports Med. 19 (3) (1991) 217–225.
[33] K.D. Bartscht, J.O. DeLancey, A technique to study the passive supports of the uterus, Obstet. Gynecol. 72 (6) (1988) 940–943.
[34] R. Ramanah, et al., Anatomy and histology of apical support: a literature review concerning cardinal and uterosacral ligaments, Int. Urogynecol. J. 23 (11) (2012) 1483–1494.
[35] R.L. Range, R.T. Woodburne, The gross and microscopic anatomy of the transverse cervical ligament, Am. J. Obstet. Gynecol. 90 (1964) 460–467.
[36] W.H. Umek, et al., Quantitative analysis of uterosacral ligament origin and insertion points by magnetic resonance imaging, Obstet. Gynecol. 103 (3) (2004) 447–451.
[37] L. Chen, et al., Cardinal and deep uterosacral ligament lines of action: MRI based 3D technique development and preliminary findings in normal women, Int. Urogynecol. J. 24 (1) (2013) 37–45.
[38] J.R. Fielding, et al., MR imaging of the female pelvic floor in the supine and upright positions, J. Magn. Reson. Imaging 6 (6) (1996) 961–963.
[39] J.O. DeLancey, Structural anatomy of the posterior pelvic compartment as it relates to rectocele, Am. J. Obstet. Gynecol. 180 (4) (1999) 815–823.
[40] T.M. Smith, et al., A novel technique to measure in vivo uterine suspensory ligament stiffness, Am. J. Obstet. Gynecol. 209 (5) (2013). 484 e1-7.
[41] J. Luo, et al., Using stress MRI to analyze the 3D changes in apical ligament geometry from rest to maximal Valsalva: a pilot study, Int. Urogynecol. J. 25 (2) (2014) 197–203.
[42] J.O. DeLancey, Fascial and muscular abnormalities in women with urethral hypermobility and anterior vaginal wall prolapse, Am. J. Obstet. Gynecol. 187 (1) (2002) 93–98.
[43] K. Rooney, et al., Advanced anterior vaginal wall prolapse is highly correlated with apical prolapse, Am. J. Obstet. Gynecol. 195 (6) (2006) 1837–1840.
[44] A. Summers, et al., The relationship between anterior and apical compartment support, Am. J. Obstet. Gynecol. 194 (5) (2006) 1438–1443.
[45] K.A. Larson, et al., Magnetic resonance imaging-based three-dimensional model of anterior vaginal wall position at rest and maximal strain in women with and without prolapse, Int. Urogynecol. J. 21 (9) (2010) 1103–1109.
[46] K.A. Larson, Y. Hsu, J.O. DeLancey, The relationship between superior attachment points for anterior wall mesh operations and the upper vagina using a 3-dimensional magnetic resonance model in women with normal support, Am. J. Obstet. Gynecol. 200 (5) (2009). 554 e1-6.

[47] K.A. Larson, et al., 3D analysis of cystoceles using magnetic resonance imaging assessing midline, paravaginal, and apical defects, Int. Urogynecol. J. 23 (3) (2012) 285–293.
[48] A. Yousuf, et al., The length of anterior vaginal wall exposed to external pressure on maximal straining MRI: relationship to urogenital hiatus diameter, and apical and bladder location, Int. Urogynecol. J. 25 (10) (2014) 1349–1356.
[49] S.E. Swift, S.B. Tate, J. Nicholas, Correlation of symptoms with degree of pelvic organ support in a general population of women: what is pelvic organ prolapse? Am. J. Obstet. Gynecol. 189 (2) (2003) 372–377. discussion 377–9.
[50] M.D. Barber, et al., Defining success after surgery for pelvic organ prolapse, Obstet. Gynecol. 114 (3) (2009) 600–609.
[51] K.A. Larson, et al., Long-term patient satisfaction with michigan four-wall sacrospinous ligament suspension for prolapse, Obstet. Gynecol. 122 (5) (2013) 967–975.
[52] Y. Hsu, C. Lewicky-Gaupp, J.O. DeLancey, Posterior compartment anatomy as seen in magnetic resonance imaging and 3-dimensional reconstruction from asymptomatic nulliparas, Am. J. Obstet. Gynecol. 198 (6) (2008) 651 e1-7.
[53] K.S. Leffler, et al., Attachment of the rectovaginal septum to the pelvic sidewall, Am. J. Obstet. Gynecol. 185 (1) (2001) 41–43.
[54] T.A. Stein, J.O. DeLancey, Structure of the perineal membrane in females: gross and microscopic anatomy, Obstet. Gynecol. 111 (3) (2008) 686–693.
[55] J. Luo, et al., Posterior vaginal prolapse shape and position changes at maximal Valsalva seen in 3-D MRI-based models, Int. Urogynecol. J. 23 (9) (2012) 1301–1306.
[56] D.M. Spahlinger, et al., Relationship between intra-abdominal pressure and vaginal wall movements during Valsalva in women with and without pelvic organ prolapse: technique development and early observations, Int. Urogynecol. J. 25 (7) (2014) 873–881.

CHAPTER THREE

Musculoskeletal Pelvic Anatomy

R.K. Kotarinos
University of South Florida, Morsani College of Medicine, Tampa, FL, United States of America

THE BONY PELVIS AND ITS ARTICULATIONS

The primary components of the bony female pelvis are the two mirror image innominate (hip) bones and the sacrum. Each innominate (hip bone) is formed by the fusion of the ilium, ischium, and the pubis. Fusion takes place during the teen years so that the individual bones are indistinguishable in the adult. The pelvic ring is formed by the sacrum articulating with the left and right hip bones at the sacroiliac joints posteriorly and the two hip bones articulating with each other at the symphysis pubis anteriorly. The coccyx forms the inferior tip of the sacrum and articulates with the sacrum at the sacrococcygeal joint. This is illustrated in Fig. 1.

Sacrum

Sometimes referred to as the "keystone of the pelvic ring," the sacrum is positioned at the end of the vertebral column as a wedge between the two innominate bones. This arrangement facilitates, locomotion, and the transfer of load or pressure between the ground, legs, and vertebral column [1] (see Fig. 2). The base of the sacrum is superior and articulates with the fifth lumbar vertebrae, while the inferior end is referred to as the apex articulating with the coccyx. Projection of the anterior surface of the first sacral vertebrae creates the sacral promontory. Five sacral vertebrae fuse together to form the sacrum. The pelvic surface is concave and smooth, while the dorsal surface is convex, extremely irregular, and rough. Four transverse ridges on the pelvic surface represent the fusion sites of the original sacral vertebrae. Sacral foramina are located at the end of the transverse ridges bilaterally, allowing the passage of the first four sacral nerves, sacral arteries, and veins. The median sacral crest, located in the midline of the dorsal surface, is formed by the fusion of the spinous processes. Lateral to the median sacral crest, fusion of the lamina creates the intermediate sacral crest. The sacral hiatus is formed when the fifth and sometimes the fourth laminae segments fail to fuse. Lateral to the intermediate sacral crest are

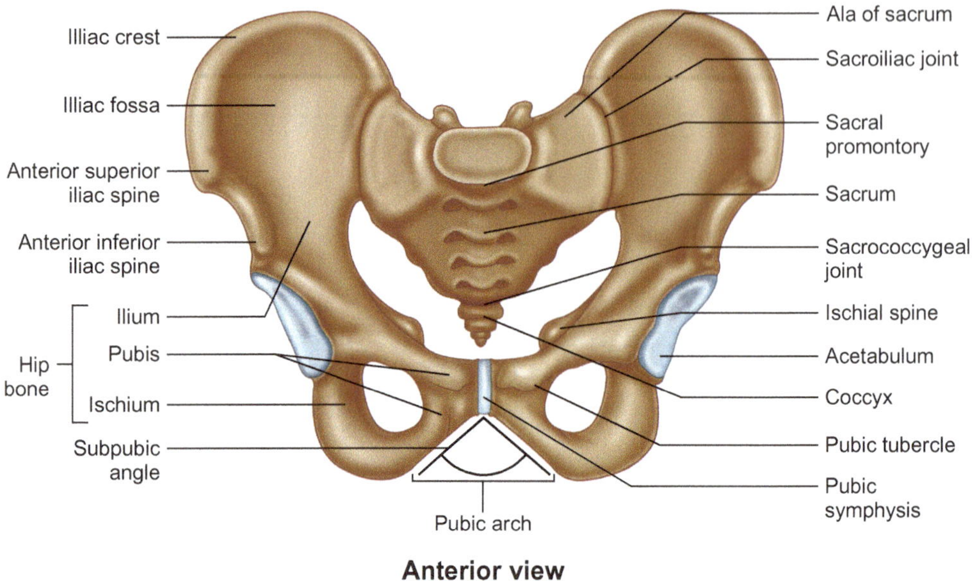

Figure 1 Anteroposterior bony pelvis: pubis, ischium, and ilium. *From A.M.R. Agur, A.F. Dalley, Grant's Atlas of Anatomy, Lippincott Williams & Wilkins, 2005.*

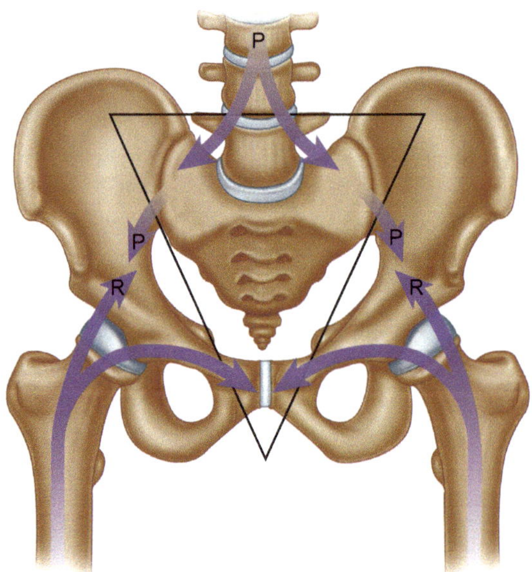

Figure 2 Transmission of trunk forces (P) and ground forces (R) into the self-locking system of the sacroiliac region. *From I.A. Kapandji, The Physiology of Joints, Vol. 3: The Vertebral Column, Pelvic Girdle and Head, Churchill Livingston, 2008.*

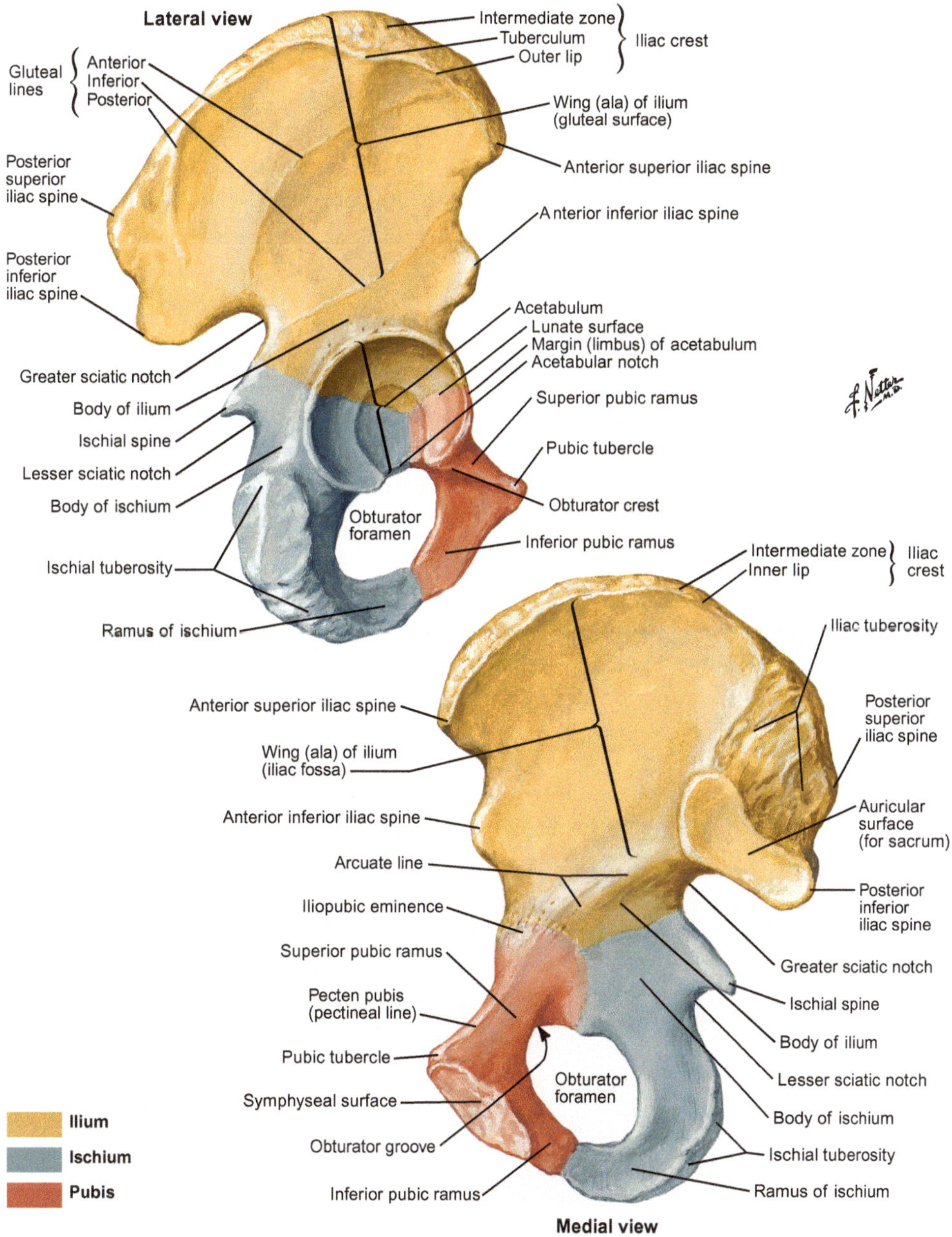

Figure 3 Medial and lateral view of the coxal/hip bone. *From F.H. Netter, Atlas of Human Anatomy, Ciba-Geigy Corporation, 1989.*

the four dorsal sacral foramina. Fusion of the transverse processes and costal features creates the lateral surface of the sacrum. The superior aspect of the lateral surface forms the ear-shaped area referred to as the auricular surface, articulating with the ilium to create the SI joint. This is illustrated in Fig. 3.

Coccyx

The coccyx may be present as the fusion of four rudimentary vertebrae or the first segment may be separated from the others. Occasionally there may be five or more, but rarely three, vertebral segments forming the coccyx. Similar to the sacrum, the pelvic surface of the coccyx is concave and fairly smooth. Orientation of the coccygeal joint to the sacral apex is inferior and anterior. Although the coccygeal vertebrae may be separate at birth, they fuse with aging and may fuse with the sacrum in advanced age.

Innominate Bones

The innominate (hip) bones are components of the pelvis as well as the lower extremity. This is shown in the coronal view where the femoral head articulates with the hip bone at the acetabulum. This is illustrated in Fig. 4. The hip bones form the anterior and lateral walls of the pelvis with the sacrum forming the posterior wall of the pelvis.

Ilium

The ilium is divided into a large upper portion referred to as a wing or ala and an inferior body. This is illustrated in Fig. 3. A significant landmark associated with the ilium is the iliac crest. It forms the upper border of the ilium and functionally is the lower edge of the waist providing attachment for the lateral musculature of the abdominal wall as well

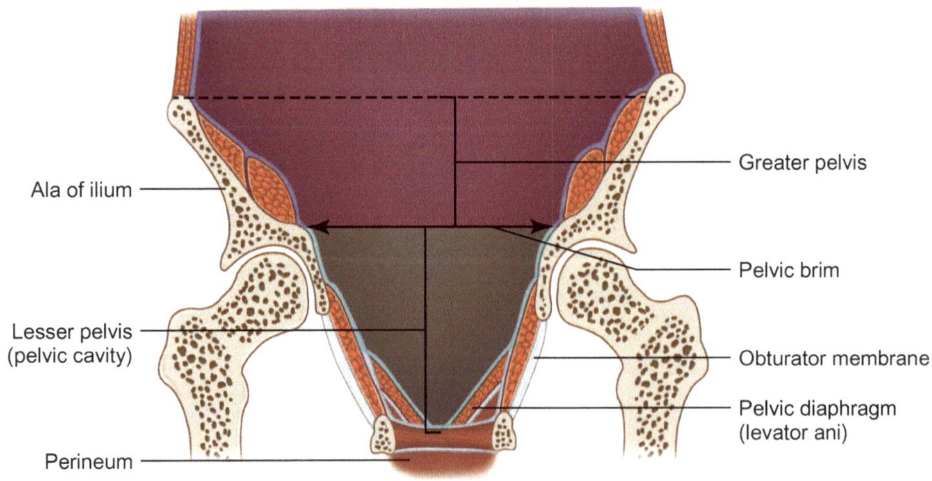

Figure 4 Divisions of the pelvis: greater (false) and lesser (true) pelvic separated by the pelvic brim. *From A.M.R. Agur, A.F. Dalley, Grant's Atlas of Anatomy, Lippincott Williams & Wilkins, 2005.*

as the fascia and muscles of the lower extremity and back. Projections of the anterior and posterior extremities of the crest create the anterior and posterior superior iliac spines. Both the sartorius muscle and inguinal ligament attach at the anterior superior iliac spine (ASIS), while the sacrotuberous and posterior SI ligaments and the multifidi attach at the posterior superior iliac spine (PSIS). The posterior inferior iliac spine (PIIS) is a wide, low protrusion where the posterior border makes a sharp bend. Continuation of this line to blend with the ischium creates the greater sciatic notch. Inferior to the ASIS is the anterior inferior iliac spine (AIIS) to which the rectus femoris muscle attaches.

The external or dorsal surface of the ilium is concave posteriorly, convex anteriorly, and has three gluteal lines associated with the gluteal muscle attachments. The internal surface of the alar portion of the ilium, which is concave anteriorly, is referred to as the iliac fossa and ends at the iliopectineal eminence, which is the junction of the ilium and pubis. The iliac fossa is a smooth internal concave aspect of the upper two-thirds of the ilium and is the origin of the iliacus muscle. The auricular surface of the ilium that articulates with the sacrum is posterior to the iliac fossa. The medial border on the internal surface of the ilium separates the iliac fossa from the sacropelvic surface. The posterior inferior extent of the iliac fossa is the arcuate line of the ilium that delineates the pelvic brim.

Ischium

The inferior posterior aspect of the innominate bone is the V-shaped ischium. This is illustrated in Fig. 3. Its component parts are the body and ramus. The body of the ischium is heavy as a result of its weight-bearing function. Three borders are associated with the body of the ischium: external, internal, and posterior. The external border of the ischium forms the prominent edge of the acetabulum. The internal border is sharp and forms the lateral margin of the obturator foramen. The posterior border is continuous with the posterior border of the ilium completing the edge of the greater sciatic notch. The sacrospinous ligament, resting slightly anterior to the sacrotuberous ligament, converts the sciatic notch into the greater and lesser sciatic foramen. Structures passing through the greater sciatic foramen include the piriformis muscle and the superior gluteal vessels and nerves, going through above the piriformis. Passing inferior to the piriformis muscle are the inferior gluteal vessels and nerves, sciatic and posterior femoral cutaneous nerves, the internal pudendal vessels and pudendal nerves, and the nerves to the obturator internus and quadratus femoris muscles. Four paths for the sciatic nerve have been documented: the usual as described above; the peroneal portion passes through the piriformis muscle with the tibial portion passing inferior to the piriformis, the peroneal passing above and the tibial portion passing below, and when the sciatic nerve does not divide then it may pierce the piriformis muscle [2]. The ischial spine, an important bony landmark of the ischium, is a blunt prominence that is directed posteromedially below the greater sciatic notch to which the sacrospinous ligament is attached, connecting the sacrum to the ischial spine. Muscular attachments to the ischial spine include the levator ani and coccygeus muscles on its internal surface and the superior gemellus attachment on its external surface. The sacrospinous ligament and coccygeus muscle have the same origin and insertion. The levator ani muscles attach to the ischial spine via the arcus tendineus levator ani (ATLA). Endopelvic fascial attachments to the ischial spine include the pubocervical fascia and the rectovaginal septum [3].

The ramus of the ischium is divided into a superior and inferior aspect [4]. The adductor magnus and obturator externus originate from the external surface of the ischial ramus [3]. As a component of the border of the inferior pelvic aperture, the internal surface provides support for the clitoris and provides for the origin of the ischiocavernosus muscle and the obturator internus muscle [3]. Other attachments to the inferior ischial ramus include the superficial perineal fascia (Colles fascia) and the superficial transverse perineal muscles [3]. The upper border of the ischial ramus completes the obturator foramen. The lower border with the medial border of the inferior pubic ramus composes the lateral border of the subpubic angle and a portion of the pubic arch (see Fig. 3).

Where the ischial tuberosity can be found depends upon the anatomical text that is being reviewed. Schaeffer describes the ischial tuberosity as a thickened projection of the posterior surface of the superior ramus of the ischium that is contiguous with the inferior ramus of the ischium [3]. More recent anatomical texts ascribe the location of the ischial tuberosity to the body of the ischium [5]. Because the medial aspect of the lower end of the ischial tuberosity is the weight-bearing surface in the sitting position they are commonly referred to as "sit bones." In standing the ischial tuberosity is covered by the gluteus maximus muscle. Located superior and medial to the tuberosity on the posterior surface a wide shallow groove provides passage of the obturator internus tendon, which exits the pelvis via the lesser sciatic foramen, just inferior to the sacrospinous ligament and the ischial spine. Muscular attachments to the ischial tuberosity include the hamstrings and the adductor magnus. The internal border of the ischial tuberosity provides for the attachment of the sacrotuberous ligament. This is illustrated in Fig. 5.

Pubis

The pubis is considered the anterior aspect of the innominate bone. The left and right pubic bones meet medially to form the pubic symphysis. Each pubic bone is composed of a body and a superior and inferior ramus. The body of the pubis is anterior, with the superior ramus above and behind the acetabulum and the inferior ramus directed posteriorly, inferiorly, and laterally to join with the ramus of the ischium on the medial side of the obturator foramen.

The body of the pubis has three surfaces: anterior, posterior, and medial. The anterior surface serves as an attachment site for the medial thigh muscles. The posterior surface forms the anterior wall of the pelvis minor and provides for the origin of the levator ani and obturator internus muscles. The pubic crest is the superior border of the body

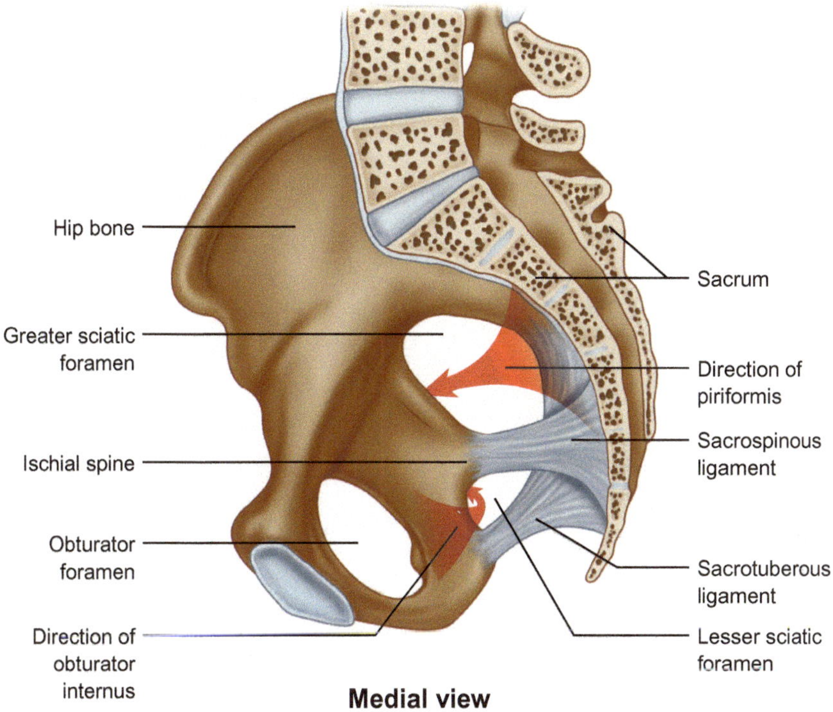

Figure 5 Hip bone and sacrum: sacrotuberous attachment. *From A.M.R. Agur, A.F. Dalley, Grant's Atlas of Anatomy, Lippincott Williams & Wilkins, 2005.*

of the pubis. The pubic tubercle, a bony prominence that is found at the lateral edge of the pubic crest, provides for the attachment of the inguinal ligament. The medial, also referred to as symphysial surface, is an elongated and oval-shaped surface connecting with the opposite side to establish the symphysis pubis (see Fig. 6).

The superior ramus is derived from the superior lateral aspect of the body of the pubis. It has three surfaces: pectineal, pelvic, and obturator. Significant to the pectineal surface is a bony ridge, the pectineal line, which is an important landmark for pelvic surgeons. The pectineal line is the location of the pectineal ligament, also referred to as Cooper's ligament. Associated with this ligament are the obturator vessels, which during retropubic surgery need to be located to avoid accidental injury. The pelvic surface has no features of significance other than a few fascicles of the obturator internus may attach in the area. On the underside of the superior ramus, the obturator surface has the obturator groove, providing passage of the obturator nerve and vessels. The obturator foramen is a large opening with its borders created by the bodies and rami of the ischium and the pubis. Thin margins of the foramen give attachment to the obturator membrane. The female obturator foramen is smaller and triangular in shape as compared to the larger oval shape in the male.

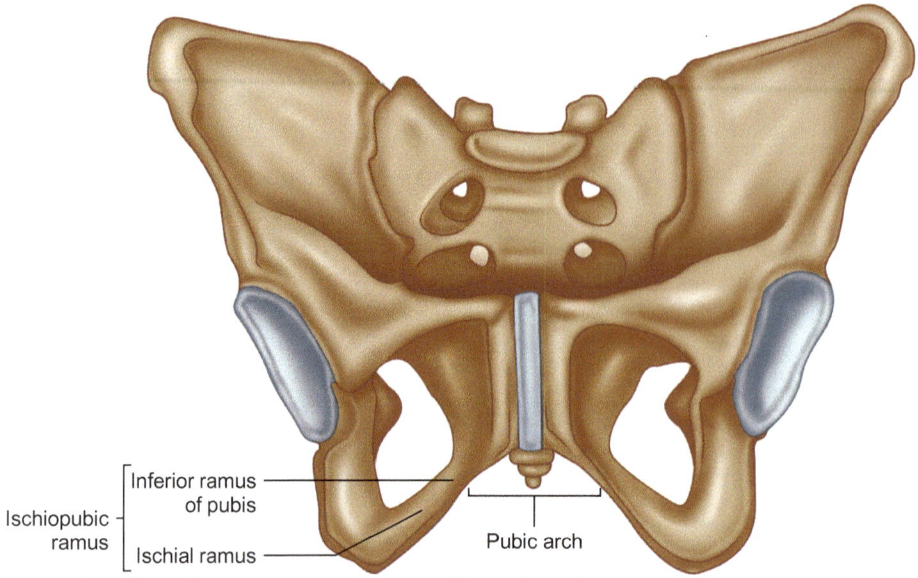

Figure 6 Symphysis pubis. *From A.M.R. Agur, A.F. Dalley, Grant's Atlas of Anatomy, Lippincott Williams & Wilkins, 2005.*

The inferior ramus is derived from the inferior lateral aspect of the body and joins the ischial ramus on the medial side of the obturator foramen. It has two surfaces: anterior or external and posterior or interior. Muscles attached to the anterior surface include the adductor brevis and magnus and obturator externus. The medial border of the inferior ramus has ridges that are continuous with the ischial ramus, giving attachment to the inferior fascia of the urogenital diaphragm and membranous layer of the subcutaneous connective tissue of the perineum. Attachments of the obturator internus and sphincter urethrae muscle are found on the posterior surface.

THE SKELETAL PELVIS AS A WHOLE

When considered as a whole, the pelvis has two divisions: the greater or false pelvis and the lesser or true pelvis. An oblique plane passing through the sacral promontory, the arcuate line of the ilium, the pectineal line of the pubis, and the superior border of the pubic symphysis marks this division. This in turn creates the terminal line ("linea terminalis" or pelvic brim). Primarily composed of the iliac fossa laterally and the sacrum posteriorly, the greater pelvis is more related to the abdominal cavity (see Fig. 7).

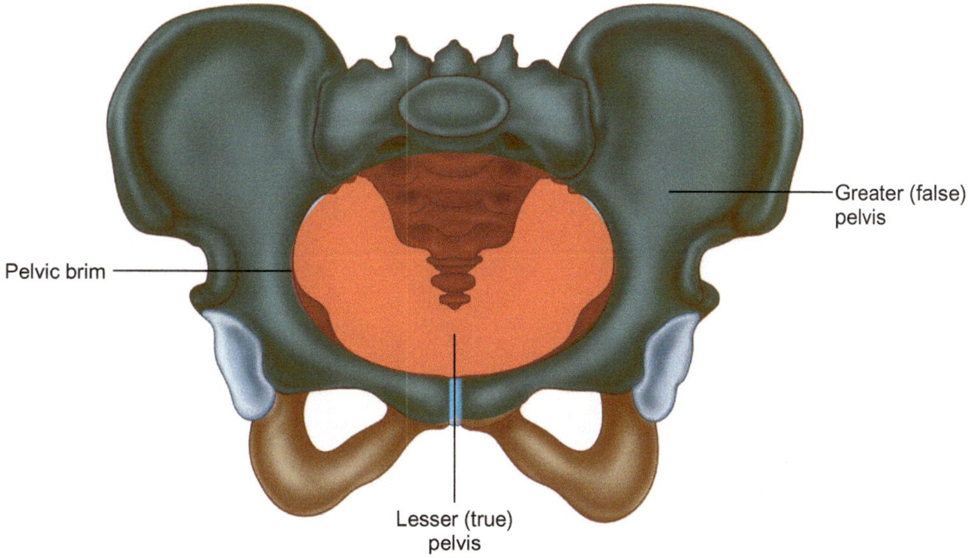

Figure 7 Pelvis divisions: greater (false) pelvis, which is superior to the pelvic brim, and the lesser (true) pelvis inferior to the pelvic brim. *From http://teachmeanatomy.com*

The lesser pelvis, housing the pelvic viscera, more closely meets the description of the pelvis as a basin because it has more complete bony walls. It is located inferior to the terminal line. Two openings are associated with the lesser pelvis: superior pelvic aperture or pelvic inlet and the inferior pelvic aperture or pelvic outlet. As stated above the boundary for the superior pelvic aperture is the pelvic brim. Three primary diameters, which have obstetric significance, are associated with the pelvic inlet: anteroposterior, transverse, and oblique (see Fig. 8). The inferior pelvic aperture is significantly more complicated than the superior pelvic aperture. Posteriorly, the coccyx, sacrum, and ischial tuberosities infringe upon its space while three notches interfere with the perimeter. The anterior notch is the pubic angle, formed by the pubic arch that is produced by the two ischiopubic rami laterally. The other two gaps are the greater and lesser sciatic notches, which are converted to foramina by the sacrotuberous and sacrospinous ligaments and bordered medially by the sacrum. Diamond shape is the usual description used when discussing the pelvic outlet. Boundaries to create this shape are the pubic symphysis anteriorly, anterolaterally the ischiopubic rami, and posteriorly the sacrotuberous ligaments with the coccyx at the midline. Diameters of the inferior pelvic aperture of the lesser pelvis are the anteroposterior, transverse, and oblique (see Fig. 9).

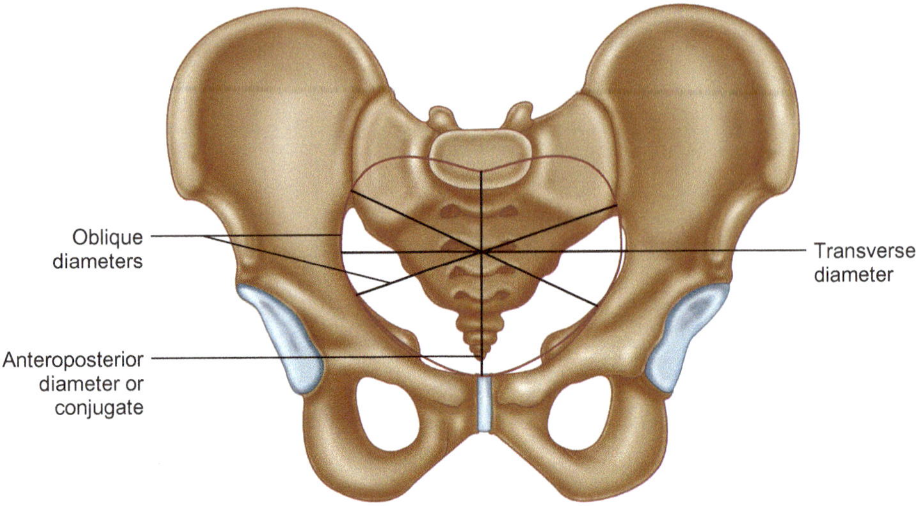

Figure 8 Diameters of the superior opening to the lesser (true) pelvis. *From J.P. Schaeffer, Morris' Human Anatomy, 11th ed., The Blakiston Division, 1953, p. 246.*

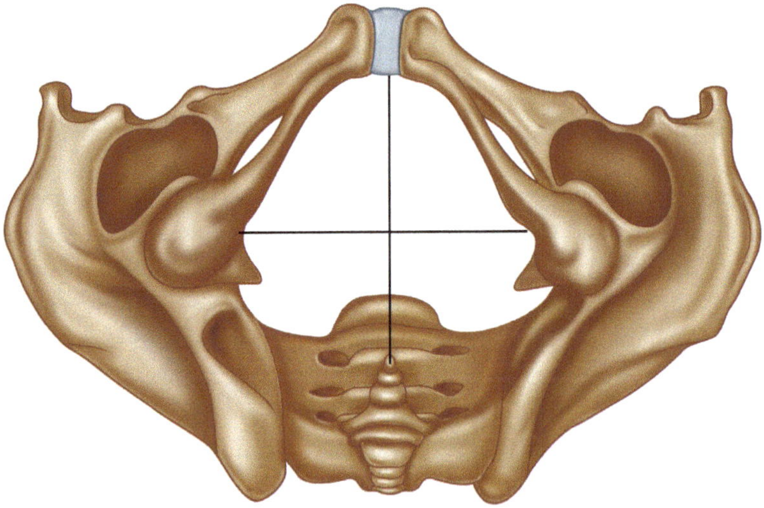

Figure 9 Diameters of the inferior opening of the lesser (true) pelvis. *From P.J. Williams, R. Warwick, Gray's Anatomy, 36th British ed., W.B. Saunders Company, Churchill Livingston, 1980, p. 386.*

The lesser pelvis cavity is short and curved. It is shallow in front and is longer on its posterior wall. The boundaries of the cavity anteriorly are the body, rami, and symphysis of the pubis, posteriorly by the coccyx and concave surface of the sacrum and laterally the pelvic portion created by the fusion of the ilium and ischium. This produces the pelvic cavity proper, which contains the terminations of the alimentary and urogenital system.

Articulations of the Pelvic Ring

The bony pelvic basin serves to support and protect the viscera that it contains. The individual bones articulate with each other as well as with the lower extremity. The surfaces of the bony pelvis provide for the attachment of muscles between the pelvis, trunk, and the lower extremity. There are 28 muscles that have attachments to the pelvis with most originating on the ilium [6]. The muscles are the rectus abdominis, internal oblique and external oblique and transversus abdominal muscles, tensor fascia latae, gluteus maximus, medius and minimus, hamstrings, piriformis, quadratus femoris, adductor magnus, longus and brevis muscles, gracilis, rectus femoris of the quadriceps group, sartorius, obturator externus and internus, superior and inferior gemellus, quadratus lumborum, and the iliacus. These muscles are primarily concerned with upright locomotion. The levator ani, coccygeus, ischiocavernosus, bulbocavernosus, and superficial transvers perinei complete the 28 muscles that attach to the pelvis. An important additional service of the pelvic ring is to transmit the weight of the upper body bidirectionally to the lower body by way of the femurs [1].

The main articulations of the pelvic bones are the SI joints and symphysis pubis. Minimal motion at these joints provides for the stability that is necessary to transmit the forces between the trunk and pelvic ring [7]. The other joints of the pelvis are the lumbosacral and the sacrococcygeal joints.

The Sacroiliac Joint

The paired SI joints, formed between the auricular surfaces of the sacrum and ilium, are the transition between the axial skeleton and the lower extremities. Normally the joint is formed within the sacral segments of S1–S3, except in females complete inclusion of S3 is not common [8]. The SI joint is divided into a short cranial and long caudal limb, with the lower aspect of both limbs synovial and the upper aspect of the cranial limb fibrous [9]. The joints are L-shaped with their surfaces divided into three parts: cranial, middle, and caudal segments, except in the erect posture where the sacrum is anteriorly tilted, the terms ventral, middle, and dorsal would be more correct [8]. The paired surfaces have interdigitating irregular elevations and depressions creating the highest coefficient of friction of any diarthrodial joint that contributes to the significant stability of the pelvic ring. The irregular surfaces of the SI joint develop in response to the stresses that it is subjected to and vary from individual to individual [10]. Even though the joint functions as a major stabilizer to the pelvic ring, studies have confirmed there is a screw-axis motion of simultaneous sagittal plane rotation and translation [11]. Mobility of the SI joint has been shown to be 40% greater in women than in men [8]. It is thought that the decreased curvature of the SI joint in women facilitates a greater degree of mobility [8]. This architectural difference combined with the hormonal changes of pregnancy produce a significant increase in SI joint mobility [5].

Very strong viscoelastic ligaments, fascia, and muscles provide additional stabilization of the SI joint. Muscles that stabilize and impact the movement at the SI joint include the latissimus dorsi through the thoracolumbar fascia, the gluteus maximus, and the piriformis [12]. Pel et al. found that activation of transversely oriented transversus abdominis and pelvic floor muscles increased the stability at the SI joint by decreasing the vertical shear: crossing the joint, the transversus abdominis holds the sacrum between the hip bones and the pelvic floor prevents lateral movement of the hip bones [11].

Ligaments associated with the SI joints are some of the largest and strongest in the human body [7]. Strong accessory SI ligaments include the iliolumbar, sacrotuberous, and the sacrospinous ligaments. The core ligaments of the SI joint include the ventral, dorsal, and interosseous ligaments. A thickening of the anterior and inferior aspect of the joint capsule makes up the ventral SI ligament. This ligament unites the base and lateral aspect of the sacrum to the ilium and is especially well developed at the level of the arcuate line and inferiorly at the level of the PIIS [3].

Forming the main bond between the sacrum and ilium, the interosseous ligament is massive and considered the strongest of the SI joint ligaments. It has the greatest bony attachment and the most overall volume of all the SI ligaments. The interosseous ligament is larger in females [13]. Multidirectional structural stability is its major function [8].

Filling the depression between the sacrum and the ischial tuberosity and overlying the interosseous ligament is the dorsal SI ligament. This ligament is divided into superficial and deep ligament layers, which are also referred to as short dorsal and long dorsal SI ligaments. The layer arrangement of the dorsal ligaments forms a multidirectional cross-hatched fiber direction that may reinforce the sacral attachment to the ilia [8].

The Pubic Symphysis

The pubic symphysis is a cartilaginous joint formed by the meeting of the superior rami of the pubic bones in the midline. As with the SI joints, surfaces of the pubic symphysis joint are reciprocally ridged and grooved to fit tightly together. Movement at this joint is minimal with small amounts of angulation, rotation, and displacement being possible. In the female the joint is shorter and broader. Most anatomy texts list the ligaments associated with the joint as the superior pubic ligament, arcuate pubic ligament, and the interpubic disc. Recent studies describe the symphysis pubis ligaments as superior, inferior (also know as the arcuate ligament), anterior, and posterior [14,15]. The superior pubic ligament attaches the rami together along their superior aspect, extends laterally to the pubic tubercle, and provides for the origin of the rectus abdominis. Inferiorly the arcuate ligament is a thick, arch band of closely packed fibers connecting the rami and filling in the pubic angle. Superiorly it blends with the interpubic disc, laterally it is attached to the rami, and the base is free. Connecting the pubic bones anteriorly, the anterior pubic ligament blends with the periosteum on its lateral edge, and is second to the interpubic disc for stabilization of the symphysis pubis [16]. Very little is known about the posterior pubic ligament, but there are a variety of descriptions as to its makeup and attachments [16]. This ligament has been reported to be larger in multiparous women [17].

The interpubic disc can vary in size, but is thicker in the female. A thin layer of hyaline cartilage covers each surface and is firmly adhered to the bone. The hyaline cartilaginous surfaces are attached via a lamina of fibrocartilage that frequently has an elongated narrow fissure or cavity. This cavity is better developed in the female. A recent tomographic study reported that the anterior and the middle parts of the symphysis pubis are wider in women, because the interpubic disc is thicker to enhance mobility [16]. Decussating rectus abdominis and external oblique muscle fibers pass anterior to the disc to strengthen the joint.

The Lumbosacral Joint

The lumbosacral joint is the connection between the fifth lumbar vertebrae and the first vertebrae of the sacrum. Components of this joint, as with other vertebral joints, include an intervertebral disc, anterior and posterior longitudinal ligaments, synovial facet joints at the articular processes, ligament flava, and interspinal and supraspinal ligaments. Unique to this vertebral joint is a disc that is very large and thicker anteriorly giving it a wedge shape. As a result of this structural variation of the disc, there is a great deal more motion possible at this articulation than elsewhere in the lumbar spine. Some of the stabilization of the SI joint is provided by the iliolumbar ligament, which has a main attachment to the transverse process of the fifth lumbar vertebrae and sometimes a weak attachment to the fourth. Insertion on the pelvis is by two bands: a lower band referred to as a lumbosacral ligament that blends with the ventral SI ligament and an upper band attaching to the iliac crest anterior to the SI joint to then blend with the thoracolumbar fascia.

The Sacrococcygeal Joint

The junction between the base of the coccyx and the apex of the sacrum creates the sacrococcygeal joint. Joining the two together are an intervertebral disc and the anterior, posterior, and lateral sacrococcygeal ligaments. The posterior ligament is divided into superficial and deep layers. Freely movable in early life, the joint is usually fused in the adult. Sacrococcygeal fusion occurs in 24% of women within the fourth decade and 44% in the age group 50–79 [8]. Fifty percent of the women with a sacrococcygeal fusion have a decreased posterior sagittal diameter; couple this with a narrow subpubic arch and one could have obstructive childbirth [18].

MUSCLES OF THE THORACOABDOMINAL PELVIC CAVITY

When discussing muscles of the pelvis the pelvic floor musculature is usually the primary focus. However it is important to realize that the pelvis is the lower portion of the thoracoabdominal cavity; therefore, muscles associated with the thoracoabdominal cavity must be considered as instrumental in the normal function of the pelvis as a whole. The ribs, sternum, vertebral column, and the pelvis provide skeletal support for the thoracoabdominal musculature [2]. Muscle groups associated with the thoracoabdominal cavity include the anterolateral and posterior abdominal wall muscles as well as the diaphragm. The diaphragm, the superior boundary of the abdominal cavity, separates the thoracic cavity from the abdominal cavity. The primary function of the diaphragm is as a muscle of respiration. The diaphragm does contract along with the abdominal muscles to increase intraabdominal pressure during expulsive acts such as sneezing, laughing, coughing, crying, vomiting, urination, defecation, and parturition [3].

Pelvic floor and diaphragmatic function are intimately related. Talasz et al. demonstrated with real-time dynamic MRI synchronous movement of the pelvic floor musculature with physiologic diaphragmatic alteration as with normal respiration or coughing [19]. It is therefore reasonable to hypothesize that a dysfunction in one system might cause or be associated with be a corresponding dysfunction in the other. For example, it is observed that pelvic pain patients resort to very shallow breathing to minimize pelvic floor motion [20]. Bordoni summarizes the interrelationship of the pelvic floor and diaphragm, as "the patient is never a symptom localized, but a system that adapts to a corporeal dysfunction" [21].

MUSCLES OF THE ANTEROLATERAL ABDOMINAL WALL

The abdomen is that area below the diaphragm, the upper portion referred to as the abdomen proper and the lower portion being the lesser pelvis [3]. The anterior border of the abdomen proper is the rectus abdominis muscles, pyramidalis, and the rectus sheath formed by the aponeuroses of the lateral abdominal musculature. The lateral border is made up of the muscle bellies of the external and internal oblique and transversus muscles, iliacus and iliac bones. The lumbar vertebrae, crura of the diaphragm, the psoas and quadratus lumborum muscles, and the posterior aspect of the iliac bones form the posterior boundary of the abdomen proper. Superiorly the diaphragm is the roof of the abdomen proper, while inferiorly the abdomen proper is continuous with the lesser pelvis. The extensive flat bellies of the abdominal wall muscles are arranged in reinforcing layers that connect the lower edge of the thoracic cage and a portion of the vertebral column to the bony pelvic ring. Varied fiber orientation of the abdominal muscles allows for the muscles to function in various combinations, providing strength to the wall and trunk stabilization [1,4,6] (see Fig. 10).

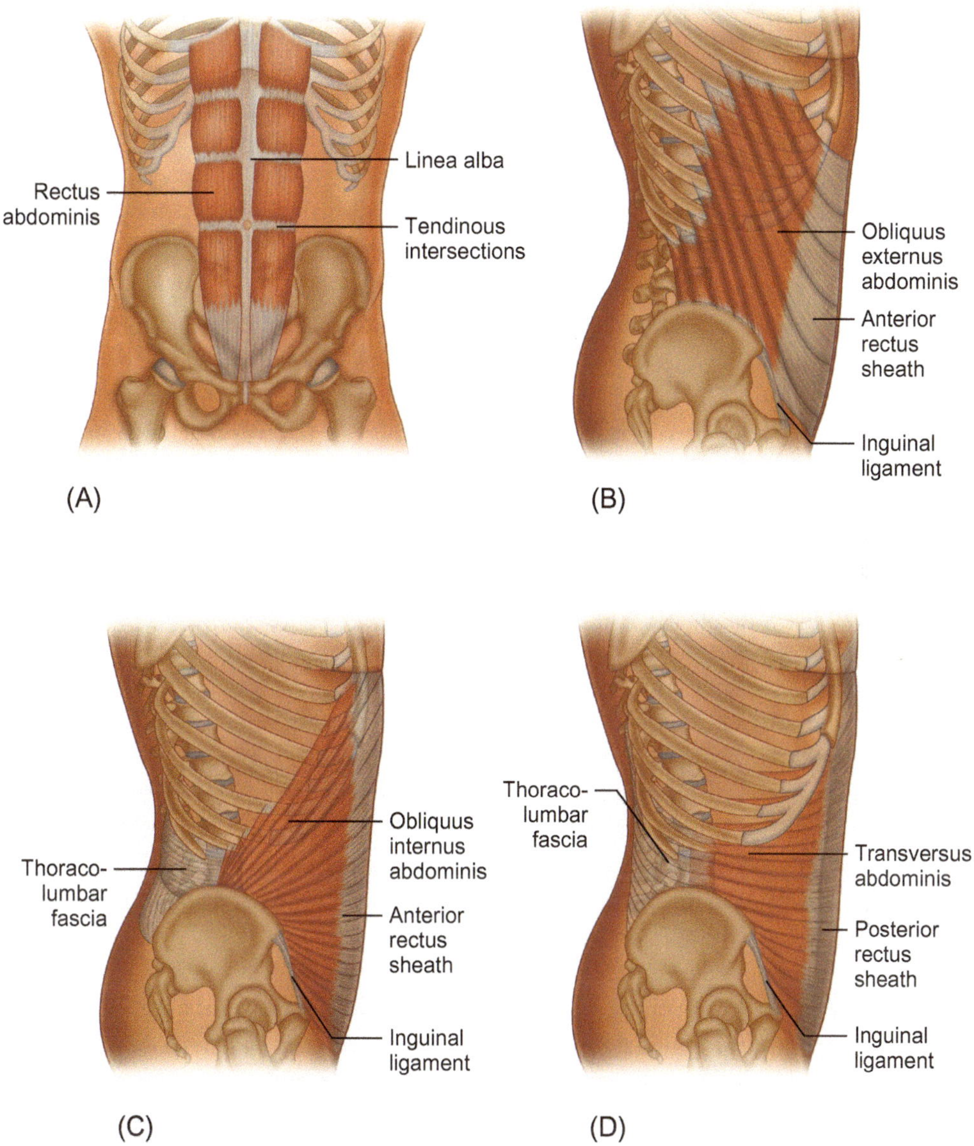

Figure 10 Abdominal muscles of the anterolateral trunk: (A) rectus abdominis, (B) external oblique, (C) internal oblique, and (D) transversus abdominis.

The abdomen is primarily muscular with the tone and integrity of these muscles instrumental in maintaining the abdominal and pelvic viscera in proper position [3]. Four paired muscle groups comprise the muscles of the abdominal wall: the rectus abdominis, external oblique, internal oblique, and transversus. The rectus abdominis originates at the pubic crest and symphysis and inserts into the costal cartilages of ribs 5–7 and the xiphoid process. The rectus sheath, formed by the aponeuroses of the external and internal oblique and the transversus muscles, surrounds the rectus bellies. Above the arcuate line the aponeurosis of the internal oblique divides. The anterior layer blends with the aponeurosis of the external oblique to form the ventral layer of the rectus sheath, while the posterior layer blends with the aponeurosis of the transversus to form the dorsal layer of the rectus sheath. Below the arcuate line all aponeuroses fuse to form the ventral layer of the rectus sheath (see Fig. 11). The aponeuroses join in the midline between the rectus bellies to form the linea alba. Tendinous intersections divide the rectus bellies transversely and are more developed on the ventral surface than on the dorsal surface. The action of the anterolateral muscles of the abdominal wall is to flex the vertebral column, moving the thorax toward the pelvis when the pelvis is fixed and moving the pelvis toward the thorax when the thorax is fixed (see Fig. 12).

The external oblique has two fiber orientations: anterior and lateral. The anterior fibers originate on the external surfaces of ribs 5–8 and insert into the linea alba by a flat aponeurosis. Action of the anterior fibers bilaterally is to flex the spine, support and compress the abdominal viscera, depress the thorax, and assist with respiration. When acting unilaterally they work with the anterior fibers of the internal oblique of the opposite side to rotate the spine. The lateral fibers originate on the external surface of ribs 9–12 and insert into the inguinal ligament, ASIS, pubic tubercle, and external lip of the anterior half of the iliac crest. Bilateral action of the lateral fibers of the external oblique impacts the lumbar spine producing a posterior pelvic tilt. Unilaterally they work with the lateral fibers of the internal oblique on the same side to flex the spine.

The internal oblique muscle has three fiber orientations: lower anterior, upper anterior, and lateral. Origin of the lower anterior fibers of the internal oblique is from the lateral two-thirds of the inguinal ligament and the iliac crest close to the ASIS with insertion into the pubic crest, medial aspect of the pectineal line, and by aponeurosis into the linea alba. The primary action of the lower anterior fibers working with the transversus muscle is to compress and support the lower abdominal organs. The upper anterior fibers originate on the anterior one-third of the intermediate line of the iliac crest and insert into the linea alba by aponeurosis. Bilateral action of the upper fibers is to flex the spine, support, and compress the abdominal viscera, depress the thorax, and assist with respiration. Unilateral action is with the anterior fibers of the external oblique of the opposite side to rotate the spine, thorax to pelvis or pelvis to thorax depending on which is fixed. Origin of the lateral fibers of the internal oblique is from the mid one-third of the intermediate line of the iliac crest and thoracolumbar fascia. Insertion is into the inferior borders of ribs 10–12 and by aponeurosis into the linea alba.

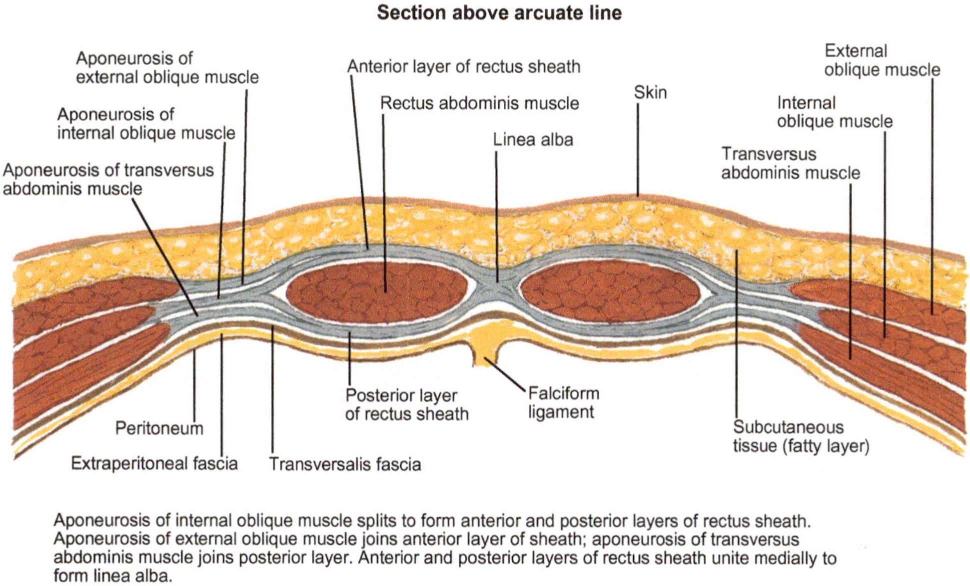

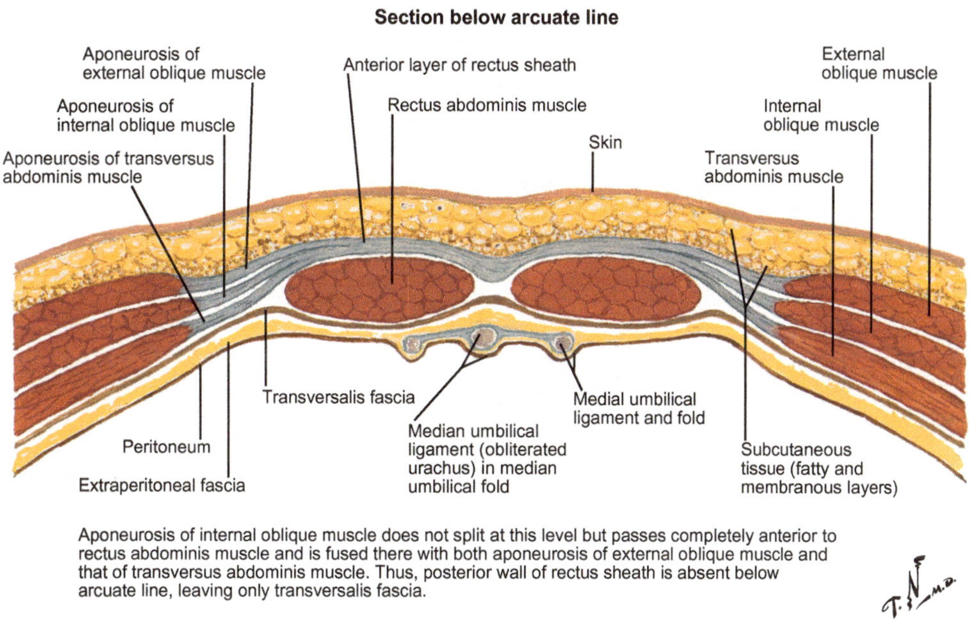

Figure 11 Rectus sheath cross section above the arcuate line: the aponeurosis of the external oblique muscle passes anterior to the rectus abdominis while the aponeurosis of the internal oblique muscle splits into an anterior and posterior layer to surround the rectus abdominis and the aponeurosis of the transversus muscle passes posterior to the rectus abdominis muscle. The anterior and posterior layers of the rectus sheath meet in the midline to form the linea alba. Rectus sheath cross section below the arcuate line: the aponeuroses of external oblique, internal oblique, and the transversus abdominis muscles all pass anterior to the rectus abdominis. Below the arcuate line there is no posterior wall of the rectus sheath. The rectus abdominis muscle is within the transversalis fascia.

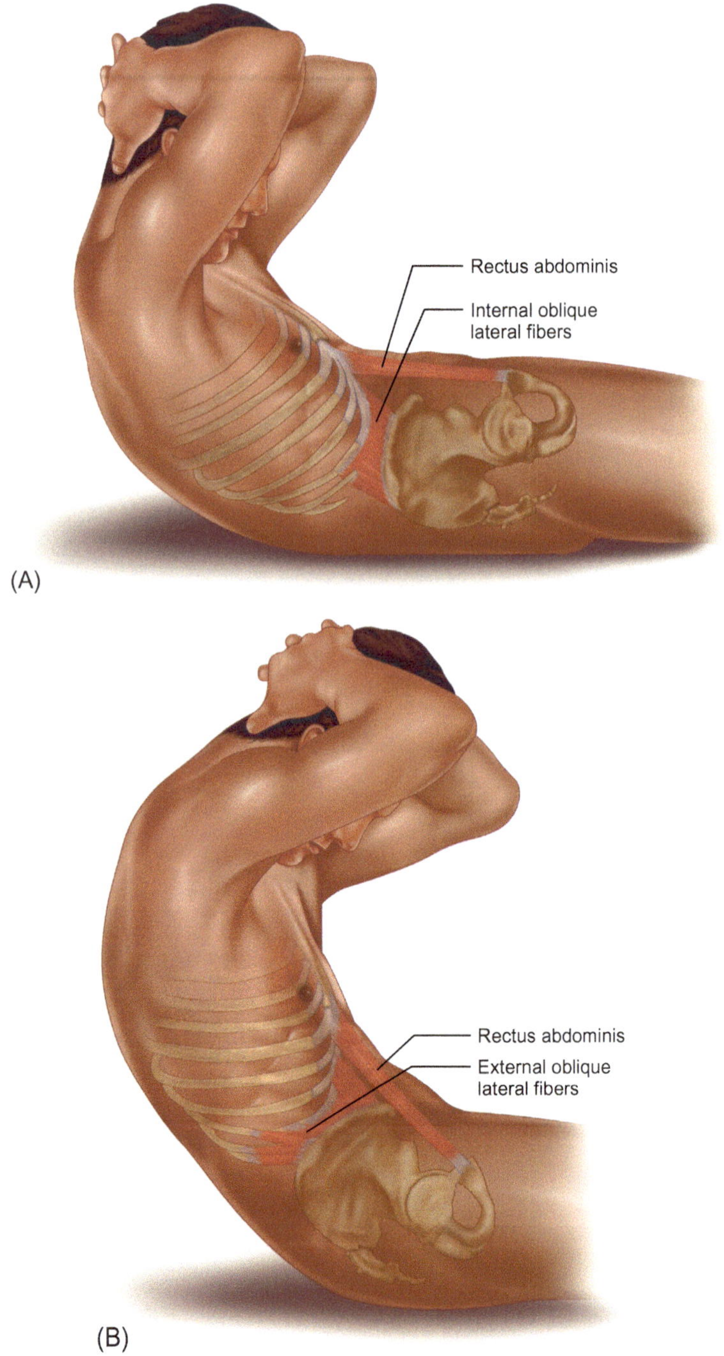

Figure 12 (A) During trunk raising the rectus abdominis, anterior fibers of the external oblique, and the superior anterolateral fibers of the internal oblique shorten, (B) while the posterolateral fibers of the external oblique lengthen.

Continued

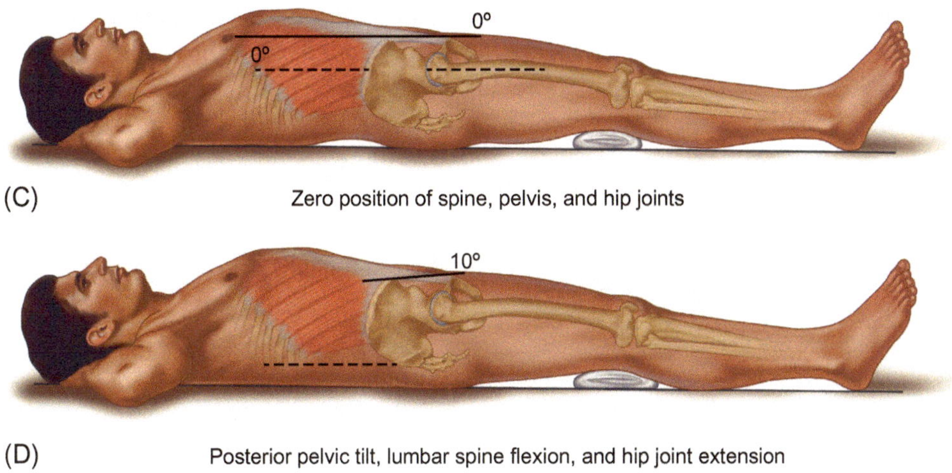

Figure 12—Cont'd (C) Neutral or zero position of the spine, pelvis, and hip joints in the supine position, (D) 10 degree posterior pelvic tilt prior to initiation of trunk raising where the hip flexors have lengthened to help stabilize the pelvis preventing further posterior pelvic tilt. *From F.P. Kendall, E.K. McCreary, P.G. Provance, Muscles Testing and Function, fourth ed., Williams and Wilkins, 1983, (A) p. 152, (B) p. 153, and (C & D) p. 172.*

The transversus abdominis originates from the inner aspect of the cartilages of the lower six ribs, thoracolumbar fascia, anterior three-fourths of the internal lip of the iliac crest, and the lateral one-third of the inguinal ligament with the insertion into the linea alba by aponeurosis, pubic crest, and pecten pubis. The primary function of the transversus is to act like a living girdle to support the abdominal viscera and to compress them to facilitate expelling their contents [2]. Decreasing the infrasternal angle of the ribs in expiration, especially during forced expiration, is a respiratory function of the upper portion of the transversus abdominis.

Posterior Muscles of the Abdominal Wall

The muscular components of the posterior abdominal wall include the psoas major and minor, iliacus, and quadratus lumborum. The psoas major and iliacus are muscles within the abdominal cavity that are commonly associated with lower extremity function. The psoas major originates from the last thoracic and all five lumbar vertebrae and inserts into the lesser trochanter of the femur. The psoas minor can be absent in as many as 40% of the subjects but when present originates from the last thoracic and first lumbar vertebrae and inserts into the pecten pubis, iliopectineal eminence, and iliac fascia [3]. The psoas minor does not cross the hip joint; therefore, it cannot be considered a muscle of the lower extremity. The iliacus arises from the iliac fossa, internal lip of the iliac crest, iliolumbar and ventral sacral ligaments, and the ala of the sacrum with its insertion at the lesser trochanter lateral to the psoas major attachment. Working together the iliacus and psoas major are referred to as the iliopsoas, functioning as flexors at the hip joint. If the origin is fixed, the iliopsoas flexes the femur on the trunk; if the insertion is fixed bilaterally, it will flex the trunk on the femurs. Without the iliacus, the psoas major acting bilaterally with the insertion fixed will increase the lumbar lordosis. An appropriate balance between the anterior and posterior muscles attached to the pelvis achieves proper alignment of the pelvis. The anterior abdominal muscles have an upward pull; while the iliopsoas, tensor fascia latae, and rectus femoris as hip flexors have a downward pull. Back extensors posteriorly have the upward pull while the hip extensors, gluteus maximus, and hamstrings create the downward pull. These muscles have a major impact on posture when weak or in a state of contracture.

The quadratus lumborum originates from the iliolumbar ligament, iliac crest and inserts into the last rib and the transverse processes of the top four lumbar vertebrae. The action of the quadratus lumborum is twofold: vertebral motion and respiratory. Bilateral contraction will provide for lumbar extension where as unilaterally it laterally flexes the lumbar vertebrae. Also, with bilateral function it works with the diaphragm to stabilize the last two ribs for respiration [1].

The piriformis, the one muscle that has a full attachment to the sacrum, and the obturator internus are muscles primarily concerned with lower extremity motion but do make up the lateral pelvic wall. Additionally, the superior aspect of the coccygeus also has an attachment to the sacrum. The piriformis arises from the pelvic surfaces of the second through fourth sacral vertebrae, posterior aspect of the greater sciatic notch, and the pelvic surface of the sacrotuberous ligament and inserts into the greater trochanter. Its primary action is lateral hip rotation but does abduct the hip when the hip is flexed to 90 degrees. The obturator internus has its origin from the bony margin of the obturator foramen, internal surface of the obturator membrane, and the pelvic surface of the posterior ischium and inserts on to the medial aspect of the greater trochanter. Like the piriformis it laterally rotates the femur and abducts the hip when flexed to 90 degrees (see Fig. 13). Dysfunction within the major muscles attaching to the pelvis, weakness or contracture, could be instrumental in creating the postural dysfunctions that have been associated with pelvic floor disorders [22–27].

MUSCLES OF THE PELVIS

The pelvic outlet is the inferior aperture associated with the thoracoabdominal pelvic cavity. The inferior boundary of the thoracoabdominal pelvic cavity is the pelvic floor musculature, which is divided into two diaphragms: pelvic and urogenital. The pelvic diaphragm consists of the levator ani and coccygeus muscles.

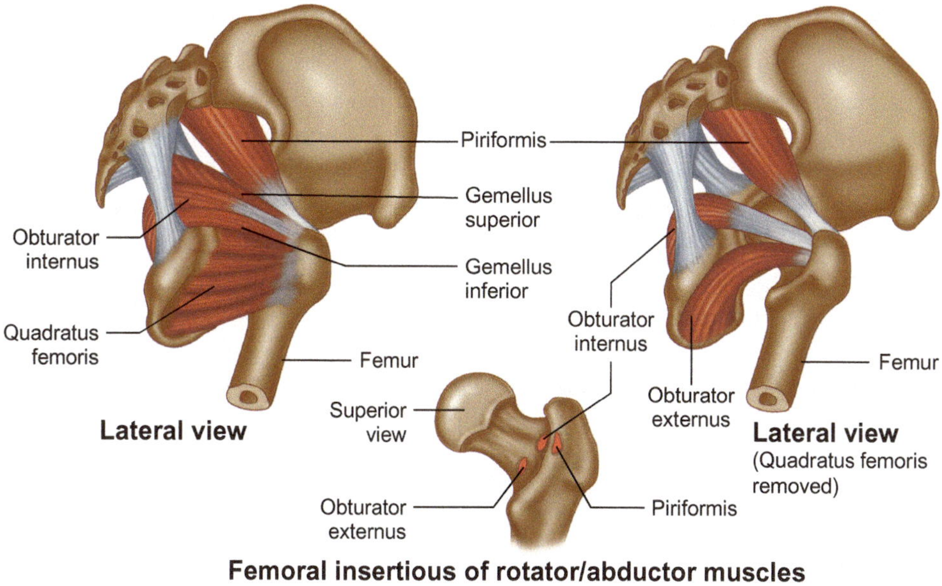

Figure 13 When the hip is flexed to 90 degrees the piriformis and obturator internus function as abductors of the femur.

The coccygeus can be found within the same tissue plane as the levator ani but is posterior and superiorly to the levator ani [3]. Triangular in shape, the coccygeus originates from the pelvic surface of the ischial spine and sacrospinous ligament moving medially to insert into the coccyx and the fifth sacral vertebrae. There is variability in its makeup of muscle to connective tissue. Essentially it lies over the pelvic surface of the sacrospinous ligament, which is often considered an aponeurosis of the coccygeus [4]. Movement of the coccyx by the coccygeus is flexion, causing elevation of the pelvic diaphragm and abduction. Bilateral action of the coccygeus muscles will help support abdominal and pelvic viscera, stabilize the coccyx, as well as return it to its resting position after defecation and parturition. The coccygeus, with the levator ani and piriformis muscles, close the posterior aspect of the pelvic outlet. An inconsistent muscle associated with the coccyx is the sacrococcygeus ventralis or anterior. It is present normally in the human only during fetal life. Located medial to the coccygeus it appears by 12 weeks and grows through the 18th week to finally blend with the dorsal aspect of the levator ani [28]. In animals it is well developed to control their tail, but can persist in the adult human on occasion [29].

The levator ani has a significant role in pelvic organ support, establishing the need to have a comprehensive understanding of its constitution and function. Kearney et al. examined published literature for consistency in the descriptions of the muscle origins and insertions [30]. The standardized anatomical terminology developed from this study describes three major subdivisions of the levator ani: pubovisceralis (pubococcygeus), puborectalis, and illiococcygeus [31]. These levator ani muscle divisions, their lines of actions, and other details of their normal and pathological function are described in detail in the previous chapter.

Perineum: Outlet of the Pelvis

The diamond shape of the perineum is created by its boundaries: anteriorly by the symphysis pubis, anterolaterally by the inferior pubic ramus and ischial tuberosity and ramus, posteriorly by the coccyx, and posterolaterally by the sacrotuberous ligaments (see Fig. 14). A line drawn between the anterior aspects of the ischial tuberosities further divides the perineum into two triangles: the anal triangle posteriorly and the urogenital triangle anteriorly. Muscles and fascia of the two triangles meet to form the perineal body, which separates the vagina from the anal canal. Soft tissue structures of the urogenital triangle formerly referred to as the urogenital diaphragm are now designated as the perineal membrane [32]. According to Barber, the urogenital triangle is divided into a superficial and deep perineal space by the perineal membrane [33].

Muscles within the superficial space include ischiocavernosus, bulbospongiosus, and superficial transverse perineal. Originating from the medial aspect of the ischial tuberosity, the ischiopubic ramus and from behind the crura of the clitoris, the ischiocavernosus muscle fibers insert by aponeurosis on the sides of and posterior surface of the crura of the clitoris. The function of the ischiocavernosus is to compress the crura, limiting venous return to maintain clitoral erection [5]. Depending on which anatomical text one refers to, the bulbospongiosus is also referred to as the bulbocavernosus [4]. Its muscle fibers arise from fibrous tissue posterior to the clitoris, from the tunica fibrosa of the corpus cavernosum, and from the surface of the tissue of the urogenital diaphragm in the angle formed by crura of the clitoris. The fibers surround the vaginal opening and travel posteriorly to form the central tendon of the perineum inserting into the perineal body to finally blend with the external anal sphincter fibers. It acts as a vaginal sphincter where the anterior fibers may facilitate clitoral erection by compression of the deep dorsal vein. The superficial transverse perineal muscles originate from the anterior medial aspect of the ischial tuberosity to move transversely medially to insert into the central tendon of the perineum at the perineal body. Bilateral contraction of the superficial transverse muscles act as stabilizers of the perineal body.

The fibromuscular perineal membrane is composed of an anterior and posterior compartment [34]. Triangular in shape, the perineal membrane has an anterior attachment at the pubic bone with a slant to the inferior posterior border of the ischiopubic ramus at the level of the perineal body [34]. In women the urethra and vagina perforate the perineal membrane, which is attached to the vaginal walls. In men it is a continuous sheet. Brandon et al. report that the muscular structures within the anterior compartment of the perineal membrane in the female are the compressor urethrae and the urethrovaginal sphincter [34]. This area was formerly referred to as the deep transverse perineal muscles. Barber describes the deep perineal space as lying deep to the perineal membrane with the muscles within being the external urethral sphincter, urethrovaginalis, and compressor urethrae [35]. Oelrich, because they function as a single unit, collectively calls these

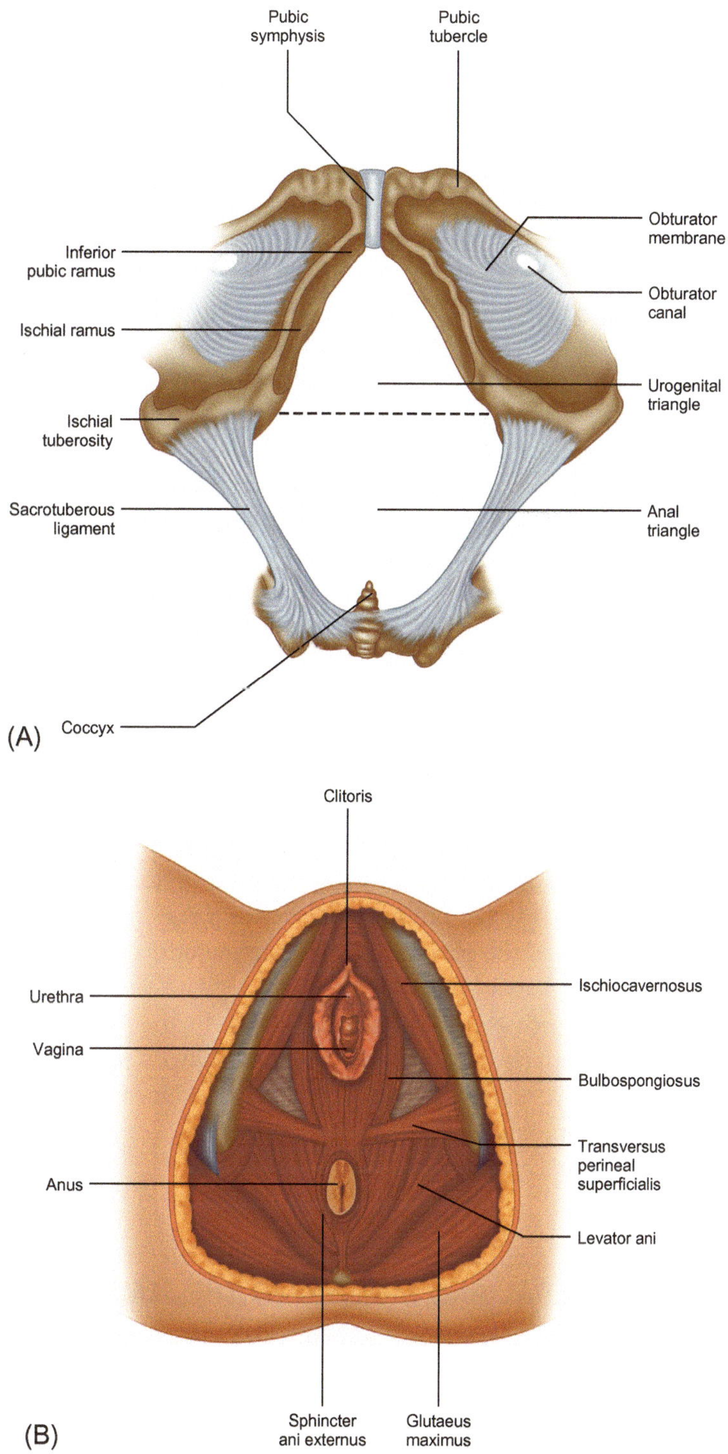

Figure 14 Boundaries, subdivisions, and muscles of the female perineum. *From (A) R. Woodburne, Essentials of Human Anatomy, fourth ed., Oxford University Press, 1969, p. 460; (B) P.J. Williams, R. Warwick, Gray's Anatomy, 36th British ed., W.B. Saunders Company, Churchill Livingston, 1980, p. 564.*

three muscles the striated urogenital sphincter [31]. The striated urogenital sphincter circumferentially covers the midurethra from below the bladder neck to the perineal membrane, providing it with one-third of its resting tone as well as being responsible for the reflex and voluntary increases in intraurethral pressure necessary for urinary continence [35]. Within the upper four-fifths of the urethra, there is a circular layer of smooth muscle that acts to constrict the urethra and may play a role in maintaining continence [36].

At the vaginal opening the posterior portion of the perineal membrane becomes a more fibrous structure with lateral extensions to the inferior pubic ramus [34]. Fat of the superior rectal fossa and the inferior perineal components are separated by the perineal membrane at the level of the perineal body. As a result of the muscular and connective tissue interconnections, the perineal membrane is thought to be a substantial support component of the pelvic floor [34].

There has been much controversy or at least confusion about the structure referred to as the perineal body. Even Gray's anatomy discusses the dilemma, stating that it is often called the central tendon, which it describes as "unsuitable; the body is not an insubstantial centre nor is it tendinous" [4]. A recent MRI study of the perineal body describes it as having three distinct regions: superficial, mid, and deep [37]. The superficial region has attachments of the bulbospongiosus, superficial transverse perineal muscle, and the external anal sphincter [38]. In the midregion, the puboperinealis muscle inserts at the lateral border of the perineal body, which also has the distal aspect of the internal anal sphincter. Also in the midregion, the puboanalis is present as it inserts into the intersphincteric groove of the internal and external anal sphincters [38]. The MRI study demonstrated at the level of the midurethra that the puboanalis and the internal anal sphincters were integral to the deep region of the perineal body. Fibers from the pubovaginalis muscle were noted on the MRI study to contribute to the deep region of the perineal body. The common description of the perineal body as a pyramidal fibromuscular structure was visualized on sagittal images [38].

The anal triangle is limited by the coccyx posteriorly, by the posterior border of the urogenital diaphragm anteriorly, and the sacrotuberous ligaments posterolaterally. Structures within the anal triangle include the anal opening, anal sphincter, and the ischiorectal fossa. The normal and pathologic functions of anal sphincter complex are described in detail in the previous chapter.

FASCIA: IN GENERAL AND SPECIFIC TO THE FEMALE PELVIS

There is much confusion and controversy that surrounds the general term of fascia. To date fascia has not been comprehensively studied as to its role and incorporation within the whole body and its organ systems [34]. Fascia is described as a continuous three-dimensional viscoelastic matrix that affords structural support throughout the entire body [35]. Fascial tissue is made up of collagenous connective tissue that encircles and permeates skeletal muscle, joints, organs, nerves, and vascular beds [39]. Sensory corpuscles, such as Pacini's and Ruffini's, permeate fascia providing the fascia with proprioceptive function that provides for significant peripheral information and possible nociceptive function [20]. Fascial tissue also contains contractile elements that could produce spasms with resultant pain and dysfunction [20].

The human fascial system encompasses the layers of tissue from the skin to the deepest layer that envelops the muscles: superficial and deep fascia [40]. Under the skin the superficial fascia is divided into two fibroadipose layers: superficial and deep. The deep fascia, encompassing the muscles, will vary according to its location within the body. The superficial fascia found throughout the body will vary in its arrangement and thickness depending on the area of the body, the surface, and gender. It is thicker in females, in the lower extremities, and on the posterior aspect of the body [37]. A membranous layer of connective tissue, the superficial fascia, is found within the subcutaneous tissue that divides it into superficial and deep adipose tissue [37]. The superficial fascia is composed of interwoven collagen fibers and many elastic fibers with muscle fibers found in the neck, face, anal region, scrotum, and areola. The function of the superficial fascia is to provide for the integrity of the skin and support subcutaneous structures [37].

As a fibrous membrane, the deep fascia encompasses and separates muscles, creates sheaths for nerves and vascular elements, reinforces ligaments, and generally speaking, adheres all structures together. This dense, deep fascia is composed of a single layer of undulated collagen fibers and elastin fibers independent of the muscle that it overlies [36]. Muscles freely slide under the deep fascia as a result of their superficial connective tissue layer the epimysium [36,37]. Coursing through the entire body, the deep fascia is maintained under basal tension because of many muscular expansions. According to Stecco, tension created by a muscular contraction is transmitted by the fascial expansions to surrounding areas thereby stimulating proprioceptors within the area [41].

In summary, fascial tissue have contractile elements that provide for a regulating function in force transmission and mechanosensory adjustments [36]. Developing a healthier respect of the fascial system will help with research and improved clinical management of the patient. Klinger et al. state, "... understanding the details of regional features of the fascia and their dynamic anatomy may help improve the results of manual and physical therapies which may affect the fascial system, and also of the various surgical operations which every day require to cut and suture the fascia to create fascio-cutaneous flaps or to operate in deeper structures." [36].

Fascia of the Female Pelvis

Fascia within the female pelvis can be divided into three groups: parietal pelvic fascia, visceral pelvic fascia, and extraserosal pelvic fascia [38]. Each of these can be further divided into various components characterized by their location, spatial orientation, consistency, and function [35].

Parietal pelvic fascia is the dense fascia that covers the structures of the boundaries of the pelvic cavity. It is further divided into the superior fascia of the pelvic diaphragm, obturator fascia, and the piriformis fascia.

The superior fascia of the pelvic diaphragm is that which covers the levator ani and coccygeus muscles. There are three prominent tissue thickenings seen bilaterally: the arcus tendinous fascia pelvis (ATFP, or fascial arch), the ATLA or levator arch, and the arcus tendinous rectovaginalis [42]. These connective tissue condensations serve to suspend the vagina and levator ani muscles across the midline. The ATFP is a taut line formed that runs from near the ischial spine to the posterior surface of the pubic bone. Albright et al. found that the length of the ATFP increased with the height of the individual [43]. The levator arch (ATLA) runs from the ischial spine to the posterior aspect of the pubic bone near the symphysis. At the pubic bone attachment, the levator arch provides for the attachment of the pubovisceral and puborectal fibers of the levator ani. The arcus tendinous rectovaginalis runs between the perineal body and the middle third of the vagina where it blends with the ATFP.

Visceral pelvic fascia is actually a mixed connective tissue layer that envelops the pelvic organs, attaching them to the pelvic sidewall. Despite the name, it is not considered to be fascia in the same way that, say, the rectus sheath is. Instead, this tissue consists of elements of collagen, muscle fibers, nerves, and blood vessels. Divisions of the visceral pelvic fascia include the pubocervical fascia, rectovaginal fascia, and uterosacral and cardinal ligaments. The normal and pathologic anatomy, structure, and function of these tissue condensations are described in detail in the previous chapter.

The pelvic organs are surrounded by a mass of dense connective tissue that lies between the parietal pelvic fascia and visceral pelvic fascia and is referred to as the extraserosal pelvic fascia. The extraserosal pelvic fascia has two functions: a mesenteric function providing support for the vascular and neural structures of the pelvis, and allowing for physiological volume variations of the pelvic viscera and freedom of movement of each viscera with regard to the other viscera [35]. The components of the extraserosal pelvic viscera include the parametrium, the paracervix, the superior vesical ligament, the lateral ligament of the rectum, and the presacral fascia. The presacral fascia is an extension of the visceral abdominal fascia. Laterally it covers the pelvic sidewall, dorsally the sacrum, ventrally it combines with the parametrium and paracervix, dorsolaterally it extends over the uterosacral and rectovaginal ligaments and rectal stalks, and caudally with the rectovaginal fascia. The presacral fascia provides support to the rectum, the adnexal vessels, the ureter, the hypogastric and sacral splanchnic nerves, and the inferior hypogastric plexus and efferent branches [35].

NEUROANATOMY OF THE FEMALE PELVIS

The human nervous system is divided into the central nervous system (CNS) and the peripheral nervous system (PNS). Components of the CNS are the brain and spinal cord, including the efferent (motor) and the afferent (sensory) nuclei and their neural paths. The PNS is made up of the nerve fibers that transmit information between the CNS and the somatic tissues of the body. Nerve fibers of the PNS are also divided into afferent and efferent divisions. The afferent fibers carry information from the periphery to the CNS, while the efferent fibers carry information from the CNS to the periphery. The efferent division is further divided into the somatic and autonomic nervous systems. Fibers of motor neurons from the somatic division supply the skeletal muscles of the body. Fibers from the autonomic nervous system (ANS) innervate the smooth muscle, cardiac muscle, and glandular tissue. A further division of the ANS is into the parasympathetic and sympathetic nervous systems, which innervate the organs of the autonomic system. Within the pelvis the PNS provides for the control of the muscles of the pelvic floor and perineum.

Autonomic Nervous System of the Female Pelvis

ANS function in the pelvis is to control the function of the pelvic organs. The parasympathetic system primarily controls the emptying functions of the pelvic organs, while the sympathetic system controls storage within the pelvic organs. As elimination is under voluntary control, the somatic nervous system works with the sympathetic nervous system to facilitate the storage functions of the bowel and bladder.

Upper motor neurons originating in the hypothalamus control the sympathetic and parasympathetic nervous systems [44]. The fibers from the hypothalamus create the hypothalamospinal tracts: anterior fibers form the parasympathetic system while the posterior fibers form the sympathetic system. The hypothalamospinal tract fibers associated with the sympathetic nervous system run in the white matter posterolateral to the dorsal horn of the spinal cord to terminate on the cell bodies inside the intermediolateral horns between the first thoracic and second lumbar vertebral levels. Nerve fibers from these cell bodies enter the sympathetic trunk by a white ramus communicans. After entering the sympathetic trunk the axon can synapse in a sympathetic ganglion above, below, or at the entrance level. After synapsing the postganglionic fiber may leave the sympathetic trunk on its way to its intended organ, usually moving along blood vessels within the body cavity. White ramus communicans transport visceral afferents from the body cavities.

The lowest section of the sympathetic trunk ganglia is the ganglion impar, which is found anterior to the coccyx connecting the two trunks. The lumbar and sacral segments of the sympathetic trunk ganglia each have four ganglia bilaterally. The inferior hypogastric plexus receives several sacral splanchnic nerve branches from the sacral segment of the trunk. The lumbar and sacral splanchnic nerves are preganglionic fibers of the sympathetic trunk that do not synapse. The intermesenteric plexus, inferior mesenteric plexus, and superior hypogastric plexus have nerve branches from the lumbar and sacral splanchnics.

Cell bodies for the parasympathetic nervous system are located in the brain stem and sacral spinal cord. The vagus nerve, a cranial nerve, sends parasympathetic fibers to the abdomen. Within the abdomen the vagal parasympathetic fibers and the sympathetic fibers comingle. The vagal fibers do not synapse within the sympathetic ganglia but will synapse with the enteric ganglia associated with its target organ.

The sacral portion of the parasympathetic nervous system is located in the sacral spinal cord segments two through four. Associated cell bodies are within the intermediolateral horn of the thoracic and lumbar spinal cord. The pelvic splanchnic nerves, which are postganglionic nerve fibers, serve the pelvis and abdomen. They exit the anterior sacral foramina at levels two through four with the third sacral spinal nerve providing the majority of the fibers.

Spinal nerves are created by dorsal and ventral rootlets that come from the dorsal and ventral horns. Dorsal rootlets contain somatic and visceral sensory fibers while the ventral rootlets contain somatic and autonomic motor fibers [38]. A mixed spinal nerve is formed by the fusion of dorsal and ventral roots. Spinal nerves are short and divide into dorsal and ventral rami. The dorsal rami serve only the deep muscles of the back, skin of the back, and the spinal facet joints. Ventral rami serve the rest of the body not innervated by cranial nerves.

Somatic Nervous System of the Female Pelvis

The lumbar and lumbosacral plexuses are somatic plexuses of the abdominopelvic region. The lumbar plexus is formed by the ventral rami of the spinal nerves from the first through the fourth lumbar levels. Nerves of this plexus include the iliohypogastric, ilioinguinal, lateral femoral cutaneous, genitofemoral, femoral, and obturator. Skin superior to the pubis is innervated by the iliohypogastric nerve, which will also innervate the pyramidalis muscle when present. The ilioinguinal nerve serves the skin of the labia majora, medial thigh skin, and areas of the transversus and internal oblique abdominal muscles. The second most common neuropathy associated with gynecological surgery is related to the ilioinguinal and iliohypogastric nerves [45]. The lateral femoral cutaneous nerve supplies the skin of the lateral thigh and is the only nerve of this plexus that does not have a somatic motor component. The genitofemoral nerve innervates the skin of the proximal anterior thigh by its femoral branch and the labia majora and skin of the medial thigh by its genital branch. It has been reported that the genitofemoral nerve is at risk of injury during a transobturator (TOT) midurethral sling producing perineal neuralgia

[46]. The femoral nerve innervates the anterior thigh muscles and the skin of the anterior thigh and medial aspect of the lower leg as well as the hip and knee joints. Self-retaining retractors utilized during abdominal gynecological surgical procedures could cause femoral neuropathy [38]. The obturator nerve provides for the motor innervation to the adductor muscles of the medial thigh and the hip and knee joints. The adductor muscles are responsible for moving the lower extremity toward the midline from an abducted position. Several reports have been made regarding injury to the obturator nerve during tension-free vaginal tape (TVT) and TOT antiincontinence surgical procedures [39,47]. The inside-outside technique is considered to have the greatest risk of nerve injury because it comes closest to the obturator canal [48].

A portion of the ventral ramus of the fourth lumbar spinal nerve and the whole ventral ramus of the fifth lumbar spinal nerve forms the lumbosacral trunk. The lumbosacral trunk contributes lumbar fibers to the sacral plexus creating the lumbosacral plexus. Of all the nerves entering the sacral plexus, the third sacral nerve is the only nerve that does not split into ventral and dorsal divisions [38]. Nerves of this plexus include the sciatic nerve, which divides into the tibial and common perineal nerves of the lower extremity. The fourth and fifth lumbar and first sacral ventral rami provide branches to the nerve of the quadratus femoris and the inferior gemellus muscles, both having a primary function of lateral hip rotation. The nerve to the obturator internus and superior gemellus is comprised of the ventral rami of the fifth lumbar, first and second sacral nerves, which also have a primary function of lateral hip rotation. The gluteal nerve, which divides into a superior and inferior branch, is made up of branches from the fourth and fifth lumbar and first sacral nerves. The superior branch terminates at the gluteus minimus, with its primary action as abduction and medial rotation of the hip joint. The inferior branch provides filaments to the gluteus medius, which functions primarily as a hip abductor with the anterior fibers acting as medial rotators and assisting with hip flexion. Both the gluteus medius and the gluteus minimus into the greater trochanter of the hip. The nerve to the piriformis is derived from either the second sacral nerve alone or from the first and second sacral nerves. The primary action of the piriformis is lateral hip rotation.

The coccygeus and levator ani are innervated by ventral sacral nerve rami. After years of debate, recent studies have determined that the levator ani nerve made up of ventral rami of the third through fifth sacral nerves is the primary nerve to supply the levator ani muscles [49–51]. Wallner et al. also reported that a communicating nerve branch connecting the levator ani nerve and the pudendal nerve was seen in 50–60% of the pelvic specimens that they studied [47].

Originating from the first through third sacral nerves, the posterior femoral cutaneous nerve serves the skin of the perineum, posterior surface of the thigh, and top portion of the lower leg and the posterolateral vulva. Branches of the posterior femoral nerve include the perineal and cluneal branches. The inferior cluneal nerve innervates the skin of the overlying the inferior aspect of the gluteus maximus. The perineal branch innervates the labia. A recent anatomical study found communicating branches between the perineal branch of the posterior femoral cutaneous nerve and the perineal branch of the pudendal nerve [52]. Knowledge of the posterior femoral cutaneous nerve is significant for the potential role it may have with some pelvic pain syndromes. Pain associated with this nerve and its branches is characterized by a burning sensation that is increased with sitting and can include the medial buttock, posterior thigh crease (sits bone pain), perianal tissue, and the labia [53]. Neuropathy of this nerve or its branches could occur as a result of an intragluteal injection [54] or from compression. Compression injuries can be the result of an acute hematoma [55] or could be the result of chronic ischemia associated with prolonged sitting [56]. Clinical experience has shown that the chronic ischemia from compression of the microvasculature from the tight elastic of underwear in this area can be a factor in this neuropathic condition. An additional source of chronic ischemic that could be associated with this neuropathy would be the reflex vasoconstriction associated with a viscero-somatic/somato-visceral reflex for central pelvic organ dysfunction [57,58]. A recent study has discovered a somato-inhibitory reflex associated with the posterior femoral cutaneous nerve [58]. Tai et al. found that electrical or mechanical stimulation of the dermatome of the posterior femoral cutaneous nerve inhibited bladder activity [58]. A pressing question might be, as there is a communication process between the posterior femoral cutaneous nerve and the bladder, is it possible that pathology such as chronic ischemia could be a source of some of the common irritative bladder symptoms? Soft tissue manipulation techniques to the viscero-somatic referral zone for the bladder and the dermatone for the posterior femoral cutaneous nerve were major treatment components of two randomized controlled studies investigating treatment of patients with urologic pelvic pain syndromes [59,60]. To date these two studies had the highest positive responder rate of any treatment study associated with this patient population. The physical therapy treatment studied consisted of soft tissue manual therapy to address myofascial trigger points, connective tissue restrictions, with the active exercise component focusing on pelvic floor lengthening.

The combination of the fifth sacral nerve, a small branch from the fourth sacral nerve, and the coccygeal nerve form the coccygeal plexus. From the coccygeal plexus the anococcygeal nerve passes through the sacrotuberous ligament to innervate the skin overlying the coccyx region [5].

The origin of the pudendal nerve is from ventral rami of the second, third, and fourth spinal nerves. The pudendal nerve leaves the pelvis passing through the greater sciatic foramen crossing the ischial spine. Along with the internal pudendal vessels the pudendal nerve reenters the pelvis through the lesser sciatic foramen between the sacrospinous and sacrotuberous ligaments passing into the pudendal (Alcock's) canal. This canal is formed by a split in the obturator internus fascia along the lateral wall of the ischiorectal fossa passing over the obturator internus muscle. It is located inferior to the levator arch, following the superior border of the ischiopubic ramus toward the posterolateral urogenital diaphragm where the exit of the canal is located. The inferior rectal nerve, the perineal nerve, and the dorsal nerve to the clitoris are the three branches of the pudendal nerve. Variations of the pathway of the branches of the pudendal nerve have been described [61–63]. The inferior rectal nerve arises from the third and fourth sacral nerves and may interconnect with the posterior femoral cutaneous and the posterior labial nerves. It provides the motor and sensory innervation to the striated external anal sphincter. The internal anal sphincter derives its visceral sensory innervation from the parasympathetic nerves. The perineal nerve has superficial and deep branches. The superficial branch, referred to as the posterior labial nerve, innervates the skin of the posterior labia. The deep branch of the perineal nerve is the motor supply to the muscles of the perineal membrane: superficial transverse perinei, ischiocavernosus, bulbospongiosus, external urethral sphincter, and the deep muscles of the perineal membrane. The terminal branch of the pudendal nerve, the dorsal nerve to the clitoris, is only sensory [47,48].

ABDOMINAL WALL, PELVIC FLOOR DYSFUNCTION, AND THE PROMISE OF COMPUTATIONAL MODELING

The interaction between the abdominal wall muscles and the pelvic floor was a cornerstone of the routine educational training of obstetrics and gynecology in the 19th and early 20th centuries [64]. The focus of this interaction was lost for decades, but has now returned to the fore. There has been a significant increase in the amount of research in recent years, investigating the interaction between the pelvic floor and abdominal wall muscles [64–68]. An historical perspective is helpful in showing how the relationship came to be understood and to argue for continued study of this interaction.

In the early 1900s the terms enteroptosis or visceroptosis were the used to describe what is currently referred to as pelvic organ prolapse. In 1907, Webster stated, "The nature of the normal relationships of the abdominal and pelvic contents and the methods by which they are altered is a somewhat complicated problem, involving a consideration of anatomic, physiologic, and physical factors which are not yet thoroughly understood. It is not my purpose to refer to these in detail, but rather to direct attention particularly to one factor, viz., the abdominal wall. This structure, composed of a variety of tissue, the most important of which are the fascial and muscular layers, normally exhibits a certain degree of tonicity helping to maintain the position of the trunk and to exert pressure on the abdominal contents. The latter function it shares with the diaphragm and the pelvic floor." [69].

Control of intraabdominal pressure is managed as a joint effort by the pelvic floor, abdominal wall, and the diaphragm [70]. Furthermore, conditions that chronically increase intraabdominal pressure (eg, COPD, straining at stool) are known to be risk factors for pelvic organ prolapse. As present day researchers seek to apply computational methods to study the biomechanical response of the pelvic organs to normal and extreme stimuli, they need to include models of the abdominal wall and the diaphragm to create simulations that reflect the true anatomic environment around the pelvic organs.

Such appropriately designed computational models can allow for hypothesis testing of conditions such as decreased abdominal wall strength, chronic straining, and diastasis recti (separation of the rectus abdominis muscles), to finally determine their impact on pelvic floor support and function.

REFERENCES

[1] R.T. Woodburne, Essentials of Human Anatomy, fourth ed., Oxford University Press, New York, 1971.

[2] J.G. Travel, D.G. Simons, Myofascial Pain and Dysfunction: The Trigger Point Manual, Vol. 2. The Lower Extremities, Williams & Wilkins, Baltimore, 1992.

[3] S.S. Retzky, R.M. Rogers, A.C. Richardson, Anatomy of female pelvic support, in: L.T. Brubaker, T.J. Saclarides (Eds.), The Female Pelvic Floor: Disorders of Function and Support, F.A. Davis Company, Philadelphia, 1996.

[4] J.P. Schaeffer (Ed.), Morris' Human Anatomy: A Complete Systematic Treatise, 11th ed., The McGraw-Hill Book Company, Inc., New York, 1953.

[5] P.L. Williams, R. Warwick (Eds.), Gray's Anatomy, 36th ed., W.B. Saunders Company, Philadelphia, 1980.

[6] F.L. Mitchell, P.S. Moran, N.A. Pruzzo, An Evaluation and Treatment Manual of Osteopathic Muscle Energy Procedures, first ed., Mitchell Moran and Pruzzo Associates, Valley Park, MO, 1979.

[7] J.A. Porterfield, C. DeRosa, Articulations of the Lumbo-Pelvic Region in Mechanical Low Back Pain: Perspectives in Functional Anatomy, second ed., W.B. Saunders Company, London, 1998.

[8] A. Vleeming, R. Stoeckart, The role of the pelvic girdle in coupling the spine and the legs a clinical-anatomical perspective on pelvic stability, in: A. Vleeming, V. Mooney, R. Stoeckart (Eds.), Movement, Stability and Lumbopelvic Pain: Integration and Research, Churchill Livingstone, Edinburgh, 2007, pp. 113–137.

[9] J.D. Cole, D.A. Blum, L.J. Ansel, Outcome after fixation of unstable posterior pelvic ring injuries, Clin. Orthop. Relat. Res. 16 (1996) 160–179.

[10] S.L. Forst, M.T. Wheeler, J.D. Fortin, J.A. Vilensky, The sacroiliac joint: anatomy, physiology and clinical significance, Pain Physician 9 (2006) 61–68.

[11] J.J.M. Pel, C.W. Spoor, A.L. Pool-Goudzwaard, A. Hoek van Dijke, C.J. Snijders, Biomechanical analysis of reducing sacroiliac joint shear load by optimization of pelvic muscle and ligament forces, Ann. Biomed. Eng. 36 (3) (2008) 415–424.

[12] A. Vleeming, A.L. Pool-Goudzwaard, E. Stoeckart, The posterior layer of the thoracolumbar fascia, Spine 20 (1995) 753–758.

[13] H. Steinke, N. Hammer, V. Slowik, et al., Novel insights into the sacroiliac joint ligaments, Spine 35 (2010) 257–263.

[14] M.J. Budak, T.B. Oliver, There's a hole in my symphysis — a review of disorders causing widening, erosion, and destruction of the symphysis pubis, Clin. Radiol. 68 (2013) 173–180.

[15] I. Becker, S.J. Woodley, M.D. Stringer, The adult human pubic symphysis: a systematic review, J. Anat. 217 (2010) 475–487.

[16] B. Alicioglu, O. Kartal, H. Gurbuz, N. Sut, Symphysis pubis distance in adults: a retrospective tomography study, Surg. Radiol. Anat. 30 (2008) 153–157.

[17] W.G.J. Putschar, The structure of the human symphysis pubis with special consideration of parturition and its sequelae, Am. J. Phys. Anthropol. 45 (1976) 589–594.

[18] R.G. Tague, Fusion of coccyx to sacrum in humans: prevalence, correlates, and effect on pelvic size, with obstetric and evolutionary implications, Am. J. Phys. Anthropol. 145 (2011) 426–437.

[19] H. Talasz, C. Kremser, M. Kofler, E. Kalchschmid, M. Lechleitner, A. Rudisch, Phase-locked parallel movement of diaphragm and pelvic floor during breathing and coughing — a dynamic MRI investigation in healthy females, Int. Urogynecol. J. 22 (2011) 61–68.

[20] P. Hodges, The abdominal canister and low back and pelvic pain: relationship between pelvic floor muscle dysfunction, breathing disorders and poor lumbopelvic control, in: B. Carriere, C.M. Feldt (Eds.), The Pelvic Floor, Georg Thieme Verlag, Stuttgart, 2006.
[21] B. Bordoni, E. Zanier, Anatomic connections of the diaphragm: influence of respiration on the body system, J. Multidiscip. Healthc. 6 (2013) 281–291.
[22] J.E. Goldthwaite, The relationship of posture to human efficiency and the influence of poise upon the support and function of the viscera, Boston Med. Surg. J. 161 (2) (1909) 839–849.
[23] T.F. Mattox, V. Lucente, P. McIntyre, J.R. Miklos, J. Tomezsko, Abnormal spinal curvature and its relationship to pelvic organ prolapse, Am. J. Obstet. Gynecol. 183 (2000) 1381–1384.
[24] J.K. Nguyen, L.R. Lind, J.Y. Choe, F. McKindsey, R. Sinow, N. Bhatia, Lumbosacral spine and pelvic inlet changes associated with pelvic organ prolapse, Obstet. Gynecol. 95 (2000) 332–336.
[25] M. Schimpf, P. Tulikangas, Evolution of the female pelvis and its relationships to pelvic organ prolapse, Int. Urogynecol. J. Pelvic Floor Dysfunct. 16 (4) (2005) 315–320.
[26] L.R. Lind, V. Lucente, N. Kohn, Thoracic kyphosis and the prevalence of advanced uterine prolapse, Obstet. Gynecol. 87 (4) (1996) 605–609.
[27] T. Sahinkanat, D.C. Arikan, E. Turgut, C. Ozkurkcugil, H. Yilmaz, H. Ekerbicer, Effects of lumbar lordosis and pelvic inlet orientation on the outcome of the transobturator tape sling operation in women, Arch. Gynecol. 284 (2011) 125–130.
[28] H. Niikura, Z.W. Jin, B.H. Cho, G. Murakami, N. Yaegashi, J.K. Lee, N.H. Lee, C.A. Li, Human fetal anatomy of the coccygeal attachments of the levator ani muscle, Clin. Anat. 23 (2010) 566–574.
[29] V. Nair, R.V. Nair, R.V. Mookambika, K.G. Mohandes Rao, S. Krishnaraja Somayaji, Persistent sacrococcygeus ventralis in an adult human pelvic wall: a variation for surgeons to note, J. Chin. Med. Assoc. 74 (12) (2011) 567–569.
[30] R. Kearney, R. Sawhney, J.O. Delancey, Levator ani muscle anatomy evaluated by origin-insertion pairs, Obstet. Gynecol. 104 (2004) 168–173.
[31] T.M. Oelrich, The striated urogenital sphincter muscle in the female, Anat. Rec. 205 (1983) 223–232.
[32] J.B. Brandon, C. Lewicky-Gaupp, K.A. Larson, J.P. DeLancey, Anatomy of the perineal membrane as seen in magnetic resonance images of nulliparous women, Am. J. Obstet. Gynecol. 200 (2009) 583.e1–583.e6.
[33] M.D. Barber, Contemporary views on female pelvic anatomy, Cleve. Clin. J. Med. 72 (Suppl. 4) (2005) S3–S11.
[34] E.H. Kwong, T.W. Findley, Fascia — current knowledge and future directions in physiatry: narrative review, J. Rehabil. Res. Dev. 51 (2014) 875–884.
[35] W. Klingler, M. Velders, K. Hoppe, M. Pedro, R. Schleip, Clinical relevance of fascial tissue and dysfunctions, Curr. Pain Headache Rep. 18 (2014) 439.
[36] J.A. Ashton-Miller, J.O. Delancey, Functional anatomy of the female pelvic floor, Ann. N. Y. Acad. Sci. 1101 (2007) 266–296.
[37] K.A. Larson, A. Yousuf, C. Lewicky-Gaupp, D.E. Fenner, J.O. DeLancey, Perineal body anatomy in living women: 3-dimensional analysis using thin-slice magnetic resonance imaging, Am. J. Obstet. Gynecol. 203 (2010) 494.e15–494.e21.
[38] A. Ercoli, V. Delmas, F. Fanfani, P. Gadonneix, M. Ceccaroni, A. Fagotti, S. Mancuso, G. Scambia, Terminologia Anatomica versus unofficial descriptions and nomenclature of the fascia and ligaments of the female pelvis: a dissection-based comparative study, Am. J. Obstet. Gynecol. 193 (2005) 1565–1573.
[39] R. Corona, C. De Cicco, R. Schonman, J. Verguts, A. Ussia, P.R. Koninckx, Tension-free vaginal tapes and pelvic nerve neuropathy, J. Minim. Invasive Gynecol. 15 (3) (2008) 262–267.
[40] C. Stecco, V. Macchi, A. Poorzionato, F. Duparc, R. De Caro, The fascia: the forgotten structure, Ital. J. Anat. Embryol. 116 (3) (2011) 127–138.
[41] C. Stecco, O. Gagey, A. Belloni, A. Pozzuoli, A. Porzionato, V. Macchi, R. Aldegheri, R. De Caro, V. Delmas, Anatomy of the deep fascia of the upper limb. Second part; study of innervation, Morphologie 91 (2007) 38–43.
[42] K.S. Leffler, J.R. Thompson, G.W. Cundiff, J.L. Buller, L.J. Burrows, M.A. Schon Ybarra, Attachment of the rectovaginal septum to the pelvic sidewall, Am. J. Obstet. Gynecol. 185 (2001) 41–43.
[43] T.S. Albright, A.P. Gehrich, G.D. Davis, F.L. Sabi, J.L. Buller, Arcus tendineus fascia pelvis: a further understanding, Am. J. Obstet. Gynecol. 193 (2005) 677–681.

[44] M. Shoja, A. Sharma, N. Mirzayan, C. Groat, K. Watanabe, M. Loukas, R.S. Tubb, Neuroanatomy of the female abdominopelvic region: a review with application to pelvic pain syndromes, Clin. Anat. 26 (2013) 66–76.

[45] J.L. Whiteside, M.D. Barber, M.D. Walters, T. Falcone, Anatomy of ilioinguinal and iliohypogastric nerves in relation to trocar placement and low transverse incisions, Am. J. Obstet. Gynecol. 189 (2003) 1574–1578.

[46] B.A. Parnell, E.A. Johnson, D.A. Zolnoun, Genitofemoral and perineal neuralgia after transobturator midurethral sling, Obstet. Gynecol. 119 (2012) 428–431.

[47] Z. Atassi, A. Reich, A. Rudge, R. Kreiienberg, F. Flock, Haemorrhage and nerve damage as complications of TVT-O procedure: case report and literature review, Arch. Gynecol. Obstet. 277 (2008) 161–164.

[48] C. Achtari, B.J. McKenzie, R. Hiscock, A. Tosamillia, L. Schierlitz, C.A. Briggs, P.L. Dwyer, Anatomical study of the obturator foramen and dorsal nerve of the clitoris and their relationship to minimally invasive slings, Int. Urogynecol. J. Pelvic Floor Dysfunct. 17 (4) (2006) 330–334.

[49] M.D. Barber, R.E. Bremer, K.B. Thor, P.C. Dolber, T.J. Kuehl, K.W. Coates, Innervation of the female levator ani muscles, Am. J. Obstet. Gynecol. 187 (2002) 64–71.

[50] C. Wallner, J. van Wissen, C.P. Maas, N.F. Dabhoiwala, M.C. DeRuiter, W.H. Lamers, The contribution of the levator ani nerve and the pudendal nerve to the innervation of the levator ani muscles; a study in human fetuses, Eur. Urol. 54 (2008) 1136–1144.

[51] S.A. Shobeiri, R.R. Chesson, R.F. Gasser, The internal innervation and morphology of the human female levator ani muscle, Am. J. Obstet. Gynecol. 199 (2008) 686.e1–686.e6.

[52] R.S. Tubbs, J. Miller, M. Loukas, M.M. Shoja, G. Shohouhi, A.A. Cohen-Gadol, Surgical and anatomical landmarks for the perineal branch of the posterior femoral cutaneous nerve: implications in perineal pain syndromes, J. Neurosurg. 111 (2009) 332–335.

[53] B. Darnis, R. Robert, J.J. Labat, T. Riant, C. Gaudin, A. Hamel, O. Hamel, Perineal and inferior cluneal nerves: anatomy and surgery, Surg. Radiol. Anat. 30 (2008) 177–183.

[54] V.G. Iyer, C.B. Shields, Isolation injection injury to the posterior femoral cutaneous nerve, Neurosurgery 25 (1989) 835–838.

[55] Y.B. Gomceli, A. Kapukiran, G. Kutlu, S. Kurt, A.I. Baysal, A case report of an uncommon neuropathy: posterior femoral cutaneous neuropathy, Acta Neurol. Belg. 105 (1) (2005) 43–45.

[56] W.J. Arnoldussen, J.J. Korten, Pressure neuropathy of the posterior femoral cutaneous nerve, Clin. Neurol. Neurosurg. 82 (1980) 57–60.

[57] H. Head, On disturbances of sensation with especial reference to the pain of visceral disease, Brain 16 (1893) 1–133.

[58] C. Tai, B. Shen, D.M. Abhijith, F. Zhang, S. Zhao, J. Wang, J.R. Roppolo, W.C. de Groat, Inhibition of the micturition reflex by activation of somatic afferents in posterior femoral cutaneous nerve, J. Physiol. 590 (19) (2012) 4945–4955.

[59] M.P. Fitzgerald, R.U. Anderson, J. Potts, et al., Randomized multicenter feasibility trial of myofascial physical therapy for the treatment of urological chronic pelvic pain syndromes, J. Urol. 182 (2) (2009) 570–580.

[60] M.P. Fitzgerald, C.K. Payne, E.S. Lukacz, et al., Randomized multicenter clinical trial of myofascial physical therapy in women with interstitial cystitis/painful bladder syndrome and pelvic floor tenderness, J. Urol. 187 (6) (2012) 2113–2118.

[61] G. Lazarou, B.A. Grigorescu, T.R. Olson, S.A. Downie, K. Powers, M.S. Mikhail, Anatomic variations of the pelvic floor nerves adjacent to the sacrospinous ligament: a female cadaver study, Int. Urogynecol. J. 19 (2008) 649–654.

[62] G.J. Furtmueller, C.A. McKenna, J. Ebmer, A.L. Dellon, Pudendal nerve 3-dimensional illustration gives insight into surgical approaches, Ann. Plast. Surg. 73 (2014) 670–678.

[63] T.I. Montoya, L. Calver, K.S. Carrick, J. Prats, M.M. Corton, Anatomic relationships of the pudendal nerve branches, Am. J. Obstet. Gynecol. 205 (2011) 504.e1–504.e5.

[64] H. Talasz, C. Kremser, M. Kofler, E. Kalchschmid, M. Lechleitner, R. Rudisch, Proof of concept: differential effects of Valsalva and straining maneuvers on the pelvic floor, Eur. J. Obstet. Gynecol. Reprod. Biol. 164 (2012) 227–233.

[65] P. Neumann, V.V. Gill, Pelvic floor and abdominal muscle interaction: EMG activity and intra-abdominal pressure, Int. Urogynecol. J. 13 (2002) 125–132.
[66] S.J. Madill, L. Mclean, Relationship between abdominal and pelvic floor muscle activation and intra-vaginal pressure during pelvic floor muscle contractions in healthy continent women, Neurourol. Urodyn. 25 (2006) 722–730.
[67] M.D. Smith, M.W. Coppieters, P.W. Hodges, Postural response of the pelvic floor and abdominal muscles in women with and without incontinence, Neurourol. Urodyn. 26 (2007) 377–385.
[68] B. Junginger, K. Baessler, R. Sapsford, P.W. Hodges, Effect of abdominal and pelvic floor tasks on muscle activity, abdominal pressure and bladder neck, Int. Urogynecol. J. 21 (2010) 69–77.
[69] J.C. Webster, A Textbook of Diseases of Women, W.B. Saunders Company, Philadelphia, 1907.
[70] G.D. Goldish, J.E. Quast, J.J. Blow, M.A. Kuskowski, Postural effects on intraabdominal pressure during Valsalva maneuver, Arch. Phys. Med. Rehabil. 75 (1994) 324–327.

CHAPTER FOUR

Introduction to Classical Mechanics

S. Abramowitch, D. Easley
University of Pittsburgh, Pittsburgh, PA, United States of America

INTRODUCTION

Mechanics of materials is the study of a material's response to a physical stressor. Generally, this is assumed to pertain to the study of how materials fail. However, this can also pertain to nonfailure experiments and analyses [1]. Mechanical characterization is the process by which scientists define an object's response to physical stressors. This response should be directly related to the intended function of the material being characterized. For example, traditional boat-making materials need to be workable, relatively low density, and robust. Historically, wood fit this role well because it was workable with period tools, it generally floated in water (low density), and it was strong enough to survive the loads encountered on the open seas. However, wood had the added, but critical, benefit of being porous and would swell when placed in water, creating a watertight seal between the interior and exterior of the boat. Just as wood proved to be a good material for boat making, each material should be specifically selected or designed, to optimally perform its intended function. With a comprehensive understanding of both the intended use of a material and the mechanical behavior of a large array of different materials, scientists and engineers can select materials that are most appropriate to meet a specific demand.

Traditionally, mechanical characterization is achieved by performing controlled mechanical tests on an object where either the applied force or deformation (change in shape and/or size) is controlled [2]. Force and deformation can then be related to each other through a mathematical equation or series of mathematical equations. The form of that mathematical relationship and ultimately its interpretation is dependent upon the object being tested, the conditions under which it is being tested, and the assumptions made about each.

In general, engineering materials are designed or selected to meet a specific set of predefined requirements. Likewise, biological materials have evolved in response to specific functional demands, and are thus "finely tuned" to their mechanical environment.

Disclosures: Deanna Easley — I have no potential conflicts of interests to disclose. Dr. Steven Abramowitch — I have no potential conflicts of interests to disclose.

For healthy load-bearing tissues and organs, their structure and composition is optimized to match their mechanical environment and their cells continually work to maintain an optimized performance even if the mechanical environment changes. For example, the femur is specifically designed to support the weight of the body, but also needs to be lightweight to allow for quick movement and agility. Muscles are a great example of how the body adapts to change. People who lift weights regularly are consistently engaging their muscles, causing them to grow through repetitive use. However, when muscles are not contracted regularly they will atrophy, such as in the case of a broken arm. This is a perfect example of how cells are in a constant state of balancing functional performance with the metabolic costs of maintaining a load-bearing tissue. It is the ability to actively remodel and heal in response to injury that makes tissue extremely unique from a material perspective and a very exciting field of study.

This chapter will focus on basic mechanics of materials, with some references to biological applications. It will describe the differences between structural and mechanical properties and how these properties are measured. The chapter will conclude with a focus on viscoelastic properties and related experimental methods.

STRUCTURAL VERSUS MECHANICAL PROPERTIES

Mechanical testing provides important information about the physical behavior of a specimen. In general, these tests are conducted via the application of force, which causes deformation (ie, a change in size or shape of the specimen), which is then measured. Alternatively, deformation can be applied and force can be measured. The parameters of the mathematical equations relating force and deformation are obtained and can be utilized for direct statistical comparisons between groups (eg, to identify the impact of a disease or disorder) and as inputs to computational models, giving researchers the capability to simulate situations that are difficult to recreate in the laboratory. Finally, as mentioned above, these parameters can be utilized to select the appropriate material for a specific application.

There are two major categories of information that can be obtained from mechanical tests. These categories are related, but their interpretation is very different.

(1) Structural properties describe how a material or collection of materials responds to an applied force or deformation. These properties are dependent on the intrinsic mechanical properties of each material, but also on the amount of each material, how they are shaped, how they are constrained and/or connected to other materials, and the direction(s) of the applied force(s). An analogy can be made to the crash test of an automobile. An automobile is made of many different components, each with its own intrinsic mechanical properties. They are arranged and organized in various ways to provide the shape and function of a car. By crashing that car head-on into a wall, we can measure the forces applied to the car and how much the car changed shape. We can directly affect its performance in this test by adding/subtracting material, changing how its components are arranged, etc. This is a structural test of the car. As the name implies, it provides an overall assessment of the structure's (ie, the car) physical response to this specific set of loading conditions. This alone is extremely valuable information. However, from just this test, it will be very difficult, if not impossible, to determine how the car will perform during a roll test, a side impact test, or a rear collision test. This is because structural properties are not solely intrinsic; instead they are dependent upon all of the characteristics mentioned above.

(2) Mechanical properties are used to describe the intrinsic mechanical behavior of a single material. The word "intrinsic" is used here because its mechanical behavior is a direct result of atomic and molecular interactions. This behavior is fundamental to the material and can only be changed if the material itself (ie, chemistry, bonding structure, etc.) is changed. For this reason, many people refer to these as "material properties." While not incorrect, that term includes the mechanical properties of the material and also properties that are not related to its mechanics (eg, electrical, magnetic, radiological, etc.). Thus, the term "mechanical properties" is more accurate. Unlike structural properties, these properties are independent of the size and shape of the material. Fig. 1 provides a summary table of structural and mechanical properties.

Structural properties and associated parameters	Mechanical properties and associated parameters
• Describe how a material or collection of materials respond to an applied force or deformation	• Intrinsic mechanical behavior of a material → Direct result of atomic and/or molecular interactions
• Load or force	• Stress
• Elongation or displacement	• Strain
• Ultimate load	• Ultimate stress or ultimate tensile strength
• Ultimate elongation	• Ultimate strain or ultimate tensile strain
• Stiffness	• Tangent modulus
• Energy absorbed	• Strain energy density

Figure 1 A conceptual comparison of structural vs. mechanical properties.

STRUCTURAL PROPERTIES

Structural and mechanical properties are often confused because the parameters that are used to describe them sound similar. Fig. 2A shows a representative load-elongation curve from a uniaxial test. The term "uniaxial" simply means that the pulling or pushing is being done along one axis of the sample. "Load" is interchangeable with "force" and is measured in units of force. Here it is the Newton, N (y-axis). "Elongation" is a nonnormalized measure of deformation and is measured in units of length. Here, it is millimeters, mm (x-axis). The curve itself represents the structural properties of the sample. Note its nonlinear shape with an initial toe region — typically associated with a large amount of elongation for small changes in force; the linear region, a region of the curve in

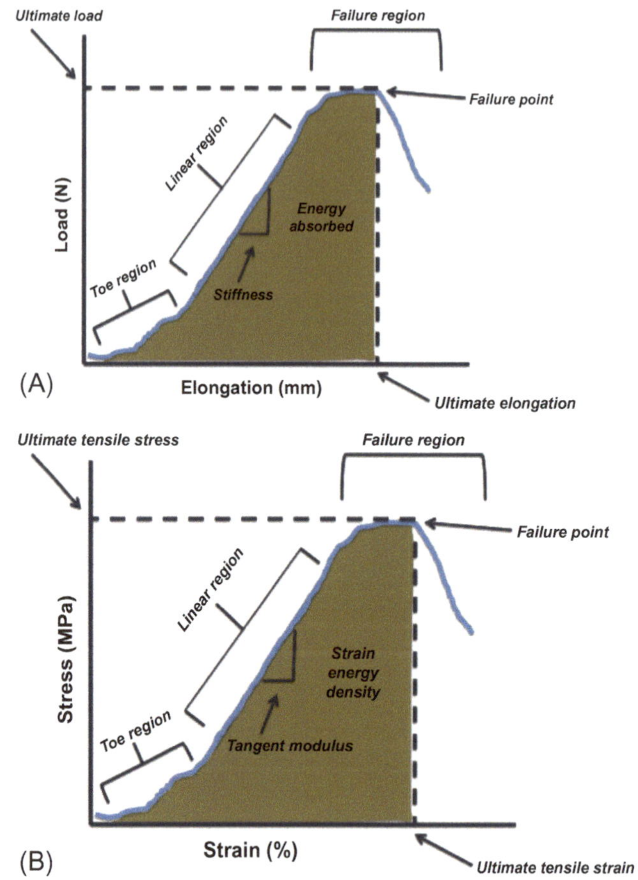

Figure 2 (A) Representative load-elongation curve from a uniaxial test. (B) Representative stress-strain curve from a uniaxial test.

which the relationship between load and elongation is more or less linear; and a failure region where the specimen is beginning to incur permanent structural damage. To describe these structural properties, it is common to utilize the following parameters. The *ultimate load* is the highest load the structure can bear before failure, where failure can be defined as the point at which the specimen begins to catastrophically come apart. *Ultimate elongation* is the maximum change in length of a structure before rupture/failure. *Stiffness* is simply defined as the slope of the linear region of the curve and is a measure of a structure's ability to resist elongation. The energy absorbed during elongation is represented by the area under the curve.

MECHANICAL PROPERTIES

The parameters describing the mechanical properties of a tissue are similar to those of the structural properties, but the interpretation is very different. First, it is important to recognize that these properties are independent of size and shape. Thus, if we have a very large piece of material and a very small piece of the same material, both should have the same mechanical properties even though their structural properties will be different. For example, the larger sample will be able to sustain a greater force and elongate much further than the smaller sample. To account for these differences, we introduce the concepts of stress and strain. These are normalized quantities that account for differences in the size and shape of the specimen. While there are, in fact, many different definitions for each of these quantities, the definitions that we are going to introduce here are that of engineering stress and strain. *Engineering stress* is the force or load applied to the system scaled by the original (before any load is applied) cross-sectional area of the specimen (F/A_0). Engineering stress is typically reserved to describe materials that experience small deformations. *Cauchy stress* is also known as true stress, or the load divided by the deformed cross-sectional area. Cauchy stress is used for calculating the stress of materials that undergo large deformations. The reason for this distinction is that the cross-sectional area tends to change significantly for materials that undergo large deformations. However, for small deformations, the change in the cross-sectional area is generally small, making engineering stress and cauchy stress roughly equal. *Engineering strain* is the change in length of the specimen normalized by its initial length ($\Delta L/L_0$). Strain is a unit-less parameter that is often reported as a %, $\Delta L/L_0 \times 100$. Most engineering materials undergo small deformations (ie, strains that are less than 5%); however, as mentioned above, many biological materials regularly undergo extremely large deformations (ie, >100%) [3–5]. An alternative measure of deformation is that of stretch. *Stretch* is simply the current length of the specimen, L, divided by its initial length, L_0. If there is no deformation, its value is 1, whereas the value for engineering strain would be 0. In fact, it can be easily shown that engineering strain is indeed equal to the stretch minus 1.

In Fig. 2B, a typical stress-strain curve is shown for a tissue tested uniaxially. Again, the curve itself is a representation of the mechanical properties of the tissue. The y-axis quantifies the engineering stress in MPa and the x-axis quantifies the deformation in terms of engineering strain (%). The same general toe, linear, and failure regions can be identified. For most collagenous tissues, the toe region reflects the contribution of elastin, realignment of collagen along the loading direction, and uncrimping of collagen from its natural wavy unstressed state [6,7]. As strain increases, there is a progressive recruitment of collagen as each individual fiber begins to uncrimp and resist elongation, as shown in Fig. 3. This is analogous to a string that has slack being pulled to the point where it becomes taught. Stress increases as more and more fibers are recruited. Once all are recruited, the fibers themselves begin to elongate, extending the curve into the linear region. If we continue to apply load, a single fiber will eventually break. When this happens, the load previously carried by the broken fiber will transfer to the surrounding fibers, which are already near their own breaking points. Generally, this triggers a domino effect causing all fibers to fail over a relatively small range of strain [8–10]. Depending on the orientation of the collagen in that particular specimen, its connections to other fibers, and its freedom to realign along the loading direction, the failure region may be more or less abrupt.

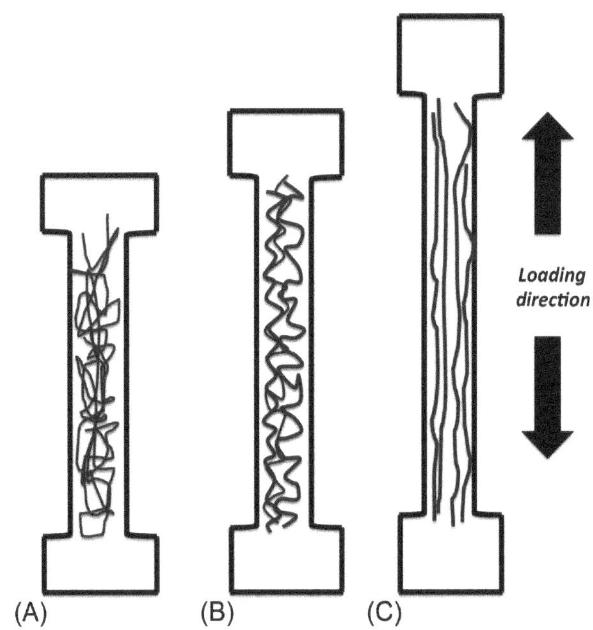

Figure 3 A depiction of collagen fibers within a dog-bone specimen. (A) The initial configuration of collagen fibers. (B) As the specimen is loaded, the collagen begins to uncrimp and align in the direction of loading. (C) Recruited collagen fibers.

To describe these mechanical properties, it is common to utilize the following parameters. The *ultimate tensile strength* is the highest stress the tissue can bear before failure. *Ultimate tensile strain* is the maximum strain achieved at failure. It is assumed that higher maximum strain indicates that a material is more "elastic." However, this is incorrect. The term "elasticity" refers to the ability of a material to recover its initial shape after it was deformed and the stress is removed. It is more appropriate to describe materials with a large maximum strain as "highly deformable." The *tangent modulus* is defined as the slope of the linear region of the curve. It describes the material's ability to resist deformation. Often, a tissue with a higher tangent modulus will be described as a "stiffer" tissue. This tends to cause confusion with the "stiffness" as defined for structural properties. In this sense, the term "stiffer" is referring to a "material stiffness," and not a "structural stiffness." Thus, it is often important to consider the context in which the term "stiffness" is being used.

Another point of confusion is that many authors will refer to the slope of the linear region of a stress-strain curve as the "Young's modulus" or "modulus of elasticity" [11]. These are terms that are generally used for a specific class of materials that are linear, elastic, isotropic, and homogenous or simply elastic. The use of these terms when referring to a tissue implies that the tissue has these qualities, which is rarely, if ever, a reasonable assumption. Thus, many in the tissue mechanics field prefer to use the term "tangent modulus" because it does not imply anything about the material. The parameter of *strain energy density* is represented by the area under the stress-strain curve and is a measure of the energy required to fail the material.

TYPES OF MECHANICAL TESTS

There are a number of mechanical tests that have been standardized for ascertaining the mechanical properties of materials. A few of the most common tests will be briefly described in this chapter, with special attention paid to tests that have been used to characterize the tissues within the female pelvis.

Uniaxial Tensile Test

The uniaxial tensile test is arguably the most common test used to begin to characterize the mechanical properties of materials. It is used not only for common materials such as metals, but also for biological tissues as well. The purpose of this test is to apply tension to a material along its primary axis, as shown in Fig. 4A, and measure its response [1]. The specimen is loaded at a constant rate, and is generally loaded until failure. Load and elongation data are collected. The load data are then normalized to obtain measures of stress, and strain is generally obtained using video measurements as will be described in more depth in the next chapter.

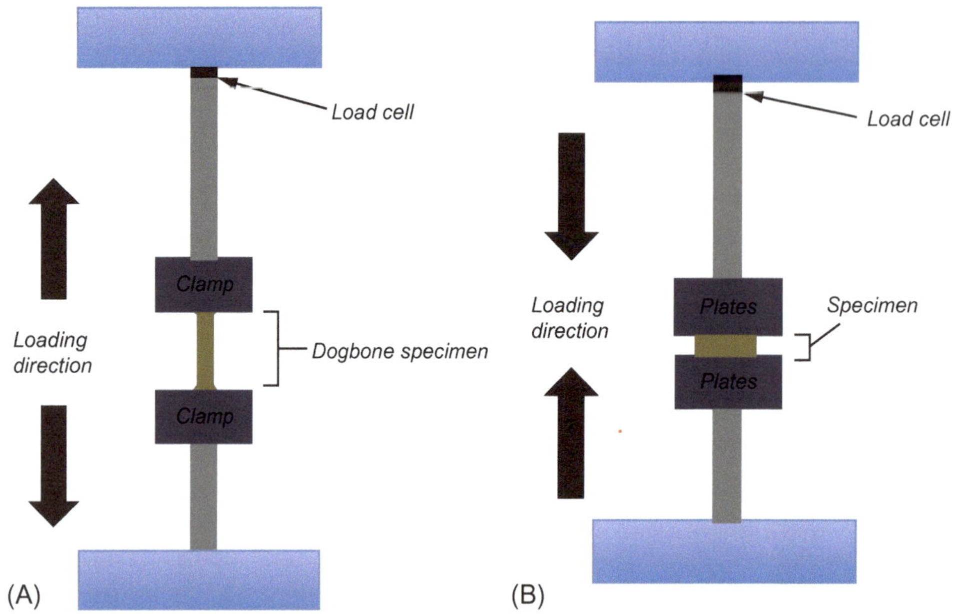

Figure 4 (A) Schematic of uniaxial tensile testing setup. Tension is being applied along the longitudinal axis. (B) Schematic of uniaxial compression testing setup. The specimen is being compressed between two plates. Note the primary differences between the tests are their loading directions.

One of the first challenges that scientists encounter when performing tensile testing on biologic materials is in securing the sample. The standard method is to clamp both ends of the sample; however, this can cause both damage to the sample and create boundary conditions that confound measurements [12,13]. An important strategy that investigators use to minimize the effects of clamping is to cut the sample into a dog-bone shape with the two wider ends used for clamping and the thinner middle region for testing. By ensuring that the length of the middle region is much longer than its width (the length to width ratio), scientists can ensure uniform tensile stresses in the sample at the location where strain is being measured, that is, in the middle region of the sample. In materials testing, it is not uncommon to see length to width ratios of 10 or more, which are ideal. However, for soft tissues, the overall size of the tissue sample often makes achieving large length to width ratios difficult without increasing the risk of accidentally damaging the specimen while placing it on the testing machine. Thus, it is not uncommon to see length to width ratios as low as 5 for soft tissue testing [14]. It should be noted that lower length to width ratios would significantly increase the errors in the mechanical properties that are calculated due to the fact that the region where strain is measured begins to approach the clamp. In addition to the concerns about damage in the sample impacting strain measurements, the artificial boundary causes stresses to be transferred from one corner of one clamp to the opposite corner of the other which results in the test no longer being one where only tensile stresses are applied, despite it being a uniaxial test.

One additional benefit of using the dog-bone geometry is that it allows failure of the specimen to occur in the middle of the sample where strain and cross-sectional area are measured. Otherwise, it is quite likely that the sample will fail prematurely at a location that is immediately adjacent to the clamp. The reason for this is also due to the effect of clamping the specimen. The artificial boundary imposed by the clamps causes stresses to rise more quickly near the clamps than further away from them. When this is combined with the high probability that the clamps cause local damage in the specimen, decreasing its tensile strength in that area, it becomes a very likely location of tissue failure that is not reflective of the actual tensile strength of the tissue. To avoid premature failures near the clamps, the width of the sample in this region is increased relative to the width in the middle of the sample. Since stress is equal to the force divided by cross-sectional area and the force is always equal at all points along the length of the specimen, the regional stress in the sample can be controlled by changing the cross-sectional area. Increasing the cross-sectional area near the clamps decreases the stress in this region, compensating for the regional increases in stress. The smaller cross-sectional area in the middle of the sample causes stresses to be higher in that region than near the clamp, increasing the likelihood the failure will occur there. This approach, therefore, allows for more accurate measurement of the tensile strength of the tissue sample.

Compression Tests

Compression tests are useful for characterizing materials that are meant to bear weight, such as support beams in a building, cartilage in the knee joint, or vertebrae in the spine [1,15,16]. The load is applied axially, but unlike with tension testing, the load is now directed in the opposite direction to compress the sample, as seen in Fig. 4B. Stresses and strains measured during compression testing are generally reported as negative because they are being applied in the opposite direction of tensile tests. In most cases, the test setup for compression tests mimics that of tensile tests, with the primary difference being the direction of the applied load and the use of plates instead of clamps. These tests can either be confined (all boundaries of the sample are enclosed in a confined space where the walls of one or more boundaries have pores to allow for fluid flow) or unconfined (only the top and bottom of the specimen are confined) depending on the specific goals of the test [1].

Shear Tests

Shear loads are defined as forces applied to a specimen tangent to the loading axis, but not through the center of the sample. For example, placing both hands together so that each finger is aligned and then sliding one hand up and one hand down is an example of shear

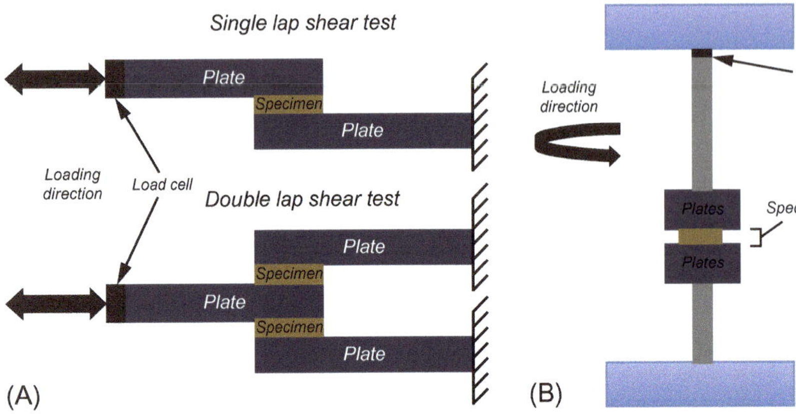

Figure 5 (A) Schematic of single- and double-lap shear tests. (B) Schematic of Torsion test.

motion. The initial motion of sliding hands past each other results in static friction, and that can be measured as a shear stress. Shear is typically tested using a single- or double-lap shear test, as shown in Fig. 5A [17]. The single-lap shear test is commonly used to determine the in situ strength of an adhesive by placing the adhesive between two plates and applying a load until failure. The double-lap shear test is very similar, but uses three plates and two adhesive regions to account for eccentric loading, as shown in Fig. 5A. These methods of testing are particularly common when dealing with adhesives, as a way of testing the strength of an adhesive, or to assess joint function within the human body [18]. Torsion is also an example of shear stress where a rotational motion is applied with respect to the long axis of a specimen, shown in Fig. 5B. While a common application of torsion tests are axles and drive shafts, the test can uniquely be adapted for biological tissues as well [19].

Bending Tests

Many building materials, such as beams, can be required to withstand constant heavy loads and not fail. However, if too much load is applied, the beams become unstable, bend, and eventually fail. Therefore, bending tests, such as the three-point flexural test shown in Fig. 6, are important for understanding how these failures occur. Bending tests can be adapted for biological applications as well, such as testing long bones (femur, humerus, etc.), cranial bones, or even the vertebral column to understand stability [20].

While this section highlights a few types of mechanical tests in a conceptually simple fashion, it should be understood that there is a significant amount of engineering theory upon which each of these tests are based. It is critical for individuals who want to perform these tests to have a thorough knowledge of the theory to perform the test correctly and to extract meaningful data.

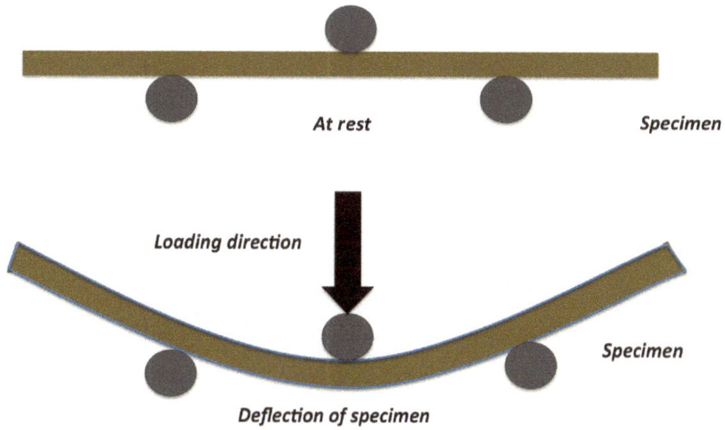

Figure 6 Three-point flexural test configuration.

CATEGORIZATION OF MATERIAL DEFORMATIONS

Materials generally undergo two types of deformations: elastic deformations and plastic deformations. At a fundamental level, elastic deformations are those from which the material can recover its initial size and shape, while plastic deformations are those deformations that result in permanent alterations in the size and shape of the specimen as a result of changes at the microstructure of a material.

Elastic Deformation

Elasticity is defined as the ability of a material to instantaneously return to its original shape after the application and removal of a load. For example, when a load is applied to a spring, the material deforms and the stress on the material increases. However, when the load is removed, the spring snaps back to its original shape, and all energy in the system is conserved. The elastic region for most classic materials is generally represented by the initial linear portion of a stress-strain curve, illustrated in Fig. 2B. This is known as a linear elastic response, elastic modulus, or Young's modulus for materials that can be assumed to be linearly elastic. This relationship can be represented by the following linear equation: $\sigma = E\varepsilon$, where σ represents engineering stress, E represents Young's modulus or the slope, and ε represents engineering strain.

Nonlinear elastic behavior is encountered when there is no longer a linearly proportional relationship between stress and strain. Most tissues display nonlinear behavior up to a point. While many assume that these tissues are elastic, they generally are not and therefore the terms describing the slope of the stress-strain curve are often inappropriately used. Fig. 7 is a representative graph, which exhibits the nonlinear response of a tissue.

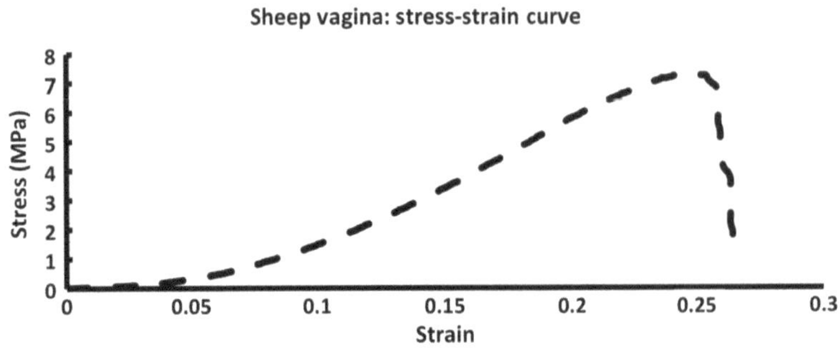

Figure 7 Representative plot that exhibits the nonlinear response of a tissue.

Plastic Deformation

Plastic deformation is permanent in that the material is unable to return to its original size and shape after the load is removed. Some materials plastically deform immediately and then fail without warning while others slowly deform before failure. Load to failure tests that are commonly associated with uniaxial tensile testing are performed to learn not only about how materials deform, but also to identify failure mechanisms. Plastic behavior can be viewed in Fig. 2B in the region beyond the elastic linear region of the curve. When a specimen undergoes plastic deformation, energy is consumed and expelled from the system as the microstructure is irreversibly changed [1]. Therefore, when the specimen is unloaded, its shape or volume remains permanently altered.

Materials that fail very rapidly with little to no warning are known as brittle materials. An example of this is shown in Fig. 8. We can observe that the edges at the point of rupture are jagged, which is a common characteristic of a sudden failure. On the opposite end of the spectrum are ductile materials, which are commonly identified by a phenomenon called necking. Necking occurs when a tensile load is being applied to a material, and the material elongates and stretches to achieve large deformations. The center portion of the specimen thins significantly prior to failure. Materials that experience necking are often described as being ductile (Fig. 9).

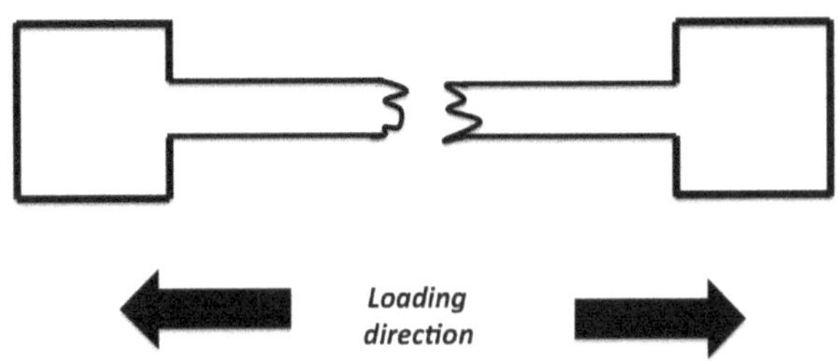

Figure 8 A depiction of a brittle fracture. Observe that the edges at the point of rupture are jagged, which is a common characteristic of sudden failure.

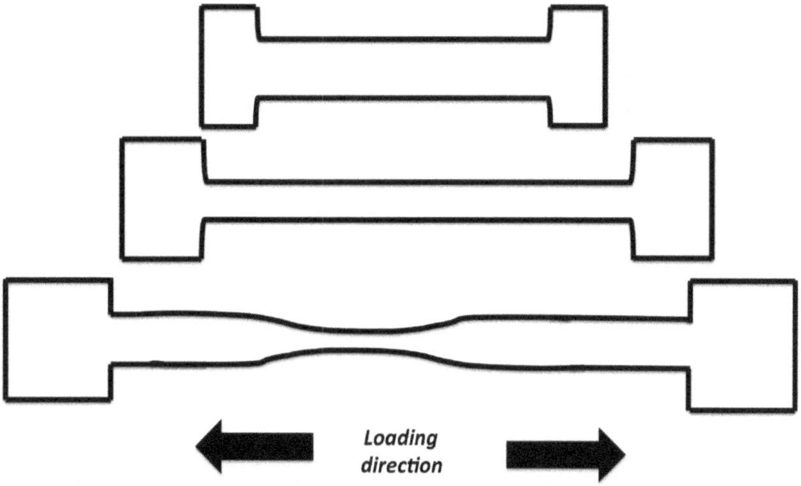

Figure 9 Representative dog-bone specimen undergoing plastic deformation. The loaded sample displays necking.

VISCOELASTICITY

An elastic material is strictly one whose stress is directly proportional to how much deformation that material has experienced. In fact, the relationship between stress and deformation is instantaneous. The instant a material is deformed is the exact instant the material experiences stress. Conversely, the instant a stress is removed is the exact instant that the material returns to its initial, undeformed shape. For example, this is what makes rubber more elastic than chewed gum. Both will deform if you pull upon them. However, only one (rubber) will return to its original shape when you stop pulling. This is why it is the ability to recoil that makes a material elastic.

While the term "viscosity" is commonly associated with everyday liquids, it is generally defined as a resistance to flow. If a material has low viscosity, it can flow very easily, like water. On the other hand, materials with a high viscosity, such as syrup, flow more slowly. Unlike elastic materials, whose stress is directly proportional to the amount of deformation, the stress in viscous materials is directly proportional to how quickly the material is deformed. The amount of deformation does not matter. For the example of pulling on rubber versus chewed gum, consider the following: if you pull on a rubber band with one end in each hand and hold your hands a set distance apart, the force required to maintain your hands at that distance never changes. There is always tension in the rubber band that is directly proportional to the distance between your hands. This is another characteristic of elastic materials. However, if you now do the same thing with gum, you should first notice that it requires more force to deform it quickly and less force to deform it slowly. It does not matter how far apart your hands are. Once your hands stop moving, you should feel that there is no tension in the gum. If anything is happening, it is likely that the middle of the gum is now flowing toward the floor due to the force of gravity.

Elasticity and viscosity highlight two different theoretical ends of a continuous spectrum of mechanical behaviors. In fact, most materials, including biological tissues, lie somewhere in the middle of these two definitions. That is, they have characteristics that are consistent with both elastic and viscous materials. In fact, both the rubber band and the gum are examples of materials that are called viscoelastic. The behavior of the rubber band is simply shifted more toward the elastic end of the spectrum and the behavior of the gum is shifted more toward the viscous end, as its viscoelastic properties depend upon the amount of saliva mixed within the polymer. However, each has a component of the other.

Viscoelastic behavior has important functional significance for biologic tissues, especially for the tissues in the female pelvis. For example, during the second stage of labor as the fetus enters the birth canal, the vagina and other pelvic supportive tissues will be deformed resulting in an increase in force [21]. At some point, the forces in the vagina and the surrounding tissues are greater than that being generated by the contractions of the uterus and the active pushing generated by the mother, preventing the baby from moving further into the vagina. Over time and with repeated contractions, the vagina and other supportive tissues soften like a viscous material allowing the fetal head to dilate the vagina further so labor can progress [22]. This is similar to the concept of plastic deformation, but here the tissue can recover its initial shape given enough time and in the absence of injury. If tissues lacked this ability, vaginal deliveries would be impossible.

VISCOELASTIC EXPERIMENTS

There are a number of important considerations when experimenting on viscoelastic materials.

Effect of Strain Rate

The term strain rate, often measured in units of %/second, refers to the speed at which a deformation is imposed upon a specimen. Strain rate is an important parameter to consider when dealing with viscoelastic materials because the speed at which a deformation is applied to a viscoelastic material will alter the way in which that material behaves. For example, chewed gum, which is a viscoelastic material, responds differently depending on the rate at which it is pulled apart because of its greater viscous component. When pulled apart quickly, the gum will break after only a small amount of deformation. However, when pulled apart slowly, the gum will break only after experiencing large deformations. Generally, tissues behave in a similar fashion, in that they exhibit a stiffer response when subjected to a higher strain rate [15]. However, this effect can vary significantly between different tissues. Engineers can utilize this property to isolate a tissue's elastic properties by using high strain rates, which minimize the impact of the tissue's time-based viscous properties.

Stress Relaxation

Tissues and other viscoelastic materials have the unique ability to gradually relieve stress when subjected to a constant deformation [23]. This phenomenon is known as stress relaxation. The specific ratios of constituents present in a tissue are key in determining the amount of stress relaxation experienced. Organs with higher amounts of proteoglycans and/or smooth muscle, such as the bladder, are able to relax more than a tendon, which is made up primarily of collagen [6,9,15,24,25]. The ability to reduce stress in response to prolonged deformation allows tissues and other viscoelastic materials to withstand large deformations without injury.

Stress relaxation testing is utilized as a method of characterizing the viscoelastic behavior of a material. Theoretically, the material is to be loaded instantaneously to a predefined strain and held at that strain for a period of time. The change in load required to hold the material at a constant strain is recorded over time. This information is especially useful for materials used in gaskets to observe how the tightness of the gasket decreases over time. Stress relaxation tests are dependent upon the initial strain rate and the length of time at which the material is held. However, given that loading can never truly be instantaneous, these tests are highly dependent upon the speed and precision of the test equipment. These tests can be performed in tension and compression, bending and torsion. Fig. 10A illustrates the initial test conditions for a stress relaxation test, while Fig. 10B illustrates a general mechanical response.

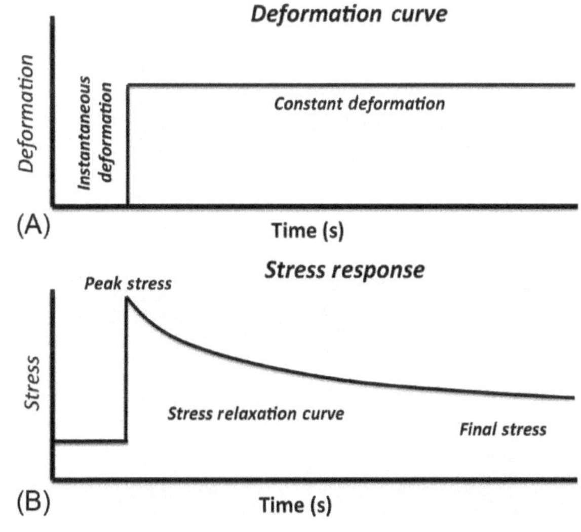

Figure 10 (A) Deformation profile applied for the stress relaxation test. (B) General stress response.

Given enough time for the material to recover following the removal of load, this test should be able to be repeated with the same response from the specimen. If not, then there is likely to have been a permanent change in the specimen as a result of the test (assuming recovery has not been compromised by the environmental conditions surrounding the specimen, eg, dehydration).

Creep

Creep is the time-dependent deformation of a material subjected to a level of stress lower than the yield point of the material, that is, when plastic deformation begins to occur (Fig. 11A). Once a constant load is applied to the specimen and held, strain continues to increases in response. Higher loads will often result in faster creep.

A very simple method to perform creep testing is to suspend a material from a fixed plane and then apply a weight to the opposing end of the material and let it hang. The changes in length of the material can then be measured over an extended period of time. A more formal way of testing creep would additionally require a gauge to measure strain or to use optical tracking of markers on the surface of the specimen, shown in Fig. 11B.

For viscoelastic materials, the removal of the load should allow the specimen to eventually return to its initial length. Unlike a purely elastic material, however, this would not occur instantaneously. If the initial length cannot be recovered, it is likely that the material is viscoplastic. This would be confirmed if the amount of plastic deformation that is observed is related to the time in which the load was applied. It is also possible that some,

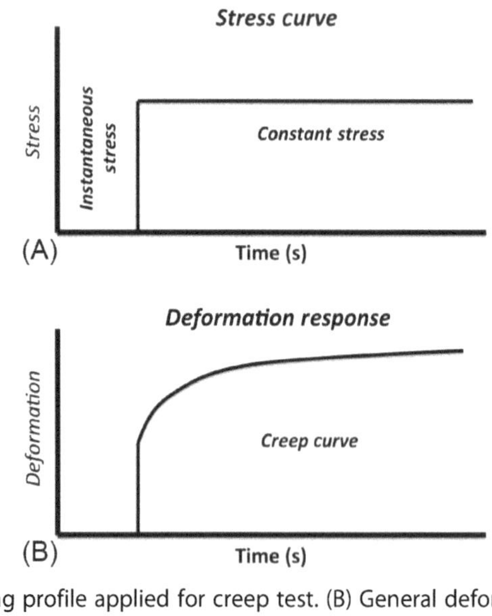

Figure 11 (A) The loading profile applied for creep test. (B) General deformation response.

but not all, of the initial length can be recovered. This would be a characteristic of a specimen that is displaying both viscoelastic and viscoplastic behaviors.

Hysteresis

Hysteresis is the term used to describe the time-dependent dissipation of thermal (heat) energy from a viscoelastic material [26]. It measured as the difference in area under a stress-strain curve as the specimen is loaded versus when it is unloaded. The specific value of hysteresis is a function not only of the applied load, but also all previously applied loads and the rate at which the load was applied. While the following is not completely accurate in all cases, the example is useful in understanding the concept. During loading, a material often experiences internal friction as molecules slide past one another. That internal friction results in heat, some of which is lost to the environment. Thus, the stress-strain curve during loading has a greater area under the curve than it does when unloading, since that area represents the energy in the sample resulting from deformation and a portion of it was given off as heat. The hysteresis loop provides information about the internal state of a material, because the loop changes with each period of cyclic loading [27]. If the peak stress increases, the material is exhibiting cyclic hardening, which indicates the material is becoming stiffer. If the peak stress decreases, the material is exhibiting cyclic softening [27–29]. Hysteresis, and in fact all of the aforementioned viscoelastic behaviors, can be thought of as a feedback loop in which the response to any input is also a function of the prior output. Over time, the effects of hysteresis diminish to an equilibrium value. In fact, cyclic loading of a viscoelastic system can eventually result in a steady state in the viscous response of that system [30], as depicted in Fig. 12. The process of observing

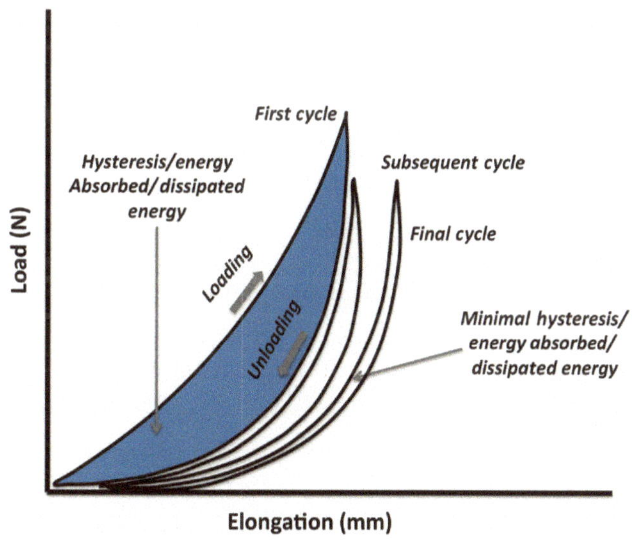

Figure 12 The area of hysteresis reaches a steady state through subsequent cycles.

hysteresis often occurs in conjunction with standard mechanical tests. For tissues, the process of cyclic loading to achieve an equilibrium in terms of hysteresis is known as preconditioning, and is discussed in detail in the following chapter. For most tissues, the hysteresis generally reaches a plateau after 10–20 cycles but can never be eliminated.

CONCLUSION

The mechanical performance of materials can be understood and characterized through mechanical testing. This chapter provided an overview of the fundamental mechanical properties and associated tests. The testing methods described in this chapter pertain specifically to simple materials and will need to be adapted to reliably characterize more complex materials like those seen in the pelvic floor and surrounding organs.

REFERENCES

[1] W.D. Callister, D.G. Rethwisch, Materials Science and Engineering: An Introduction, seventh ed., John Wiley & Sons, New York, 2007.
[2] N.E. Dowling, Mechanical Behavior of Materials: Engineering Methods for Deformation, Fracture, and Fatigue, third ed., Pearson Prentice Hall, New Jersey, 2007.
[3] M.S. Sacks, W. Sun, Multiaxial mechanical behavior of biological materials, Annu. Rev. Biomed. Eng. 5 (2003) 251–284.
[4] F. Guilak, D.L. Butler, S.A. Goldstein, D. Mooney, Functional Tissue Engineering, Springer, New York, 2003.
[5] H. Demiray, Large deformation analysis of some soft biological tissues, J. Biomech. Eng. 103 (2) (1981) 73–78.
[6] Y.C. Fung, Biomechanics: Mechanical Properties of Living Tissues, Springer-Verlag, New York, 1993.
[7] M.R. Hill, et al., A theoretical and non-destructive experimental approach for direct inclusion of measured collagen orientation and recruitment into mechanical models of the artery wall, J. Biomech. 45 (5) (2012) 762–771.
[8] S.C. Cowin, S.B. Doty, Tissue Mechanics, Springer, New York, 2007.
[9] R. Einat, L. Yoram, Recruitment viscoelasticity of the tendon, J. Biomech. Eng. 131 (11) (2009) 111008.
[10] M.F. Hadi, E.A. Sander, V.H. Barocas, Multiscale model predicts tissue-level failure from collagen fiber-level damage, J. Biomech. Eng. 134 (9) (2012) 091005.
[11] C. Hartsuijker, J.W. Welleman, Engineering Mechanics, vol. 2, Springer, Dordrecht, 2007. ISBN 978-1-4020-5763-2.
[12] J.A. Weiss, J.C. Gardiner, Computational modeling of ligament mechanics, Crit. Rev. Biomed. Eng. 29 (4) (2001) 1–70.
[13] K.M. Quapp, J.A. Weiss, Material characterization of human medial collateral ligament, ASME J. Biomech. Eng. 120 (1998) 757–763.
[14] J.P. Shepherd, et al., Uniaxial biomechanical properties of seven different vaginally implanted meshes for pelvic organ prolapse, Int. Urogynecol. J. 23 (5) (2012) 613–620.
[15] V.C. Mow, R. Huiskes, Basic Orthopaedic Biomechanics and Mechano-Biology, third ed., Lippincott Williams & Wilkins, Philadelphia, 2005.
[16] The Composite Materials Handbook-MIL 17: Guidelines for Characterization of Structural Materials, vol. 1, Technomic Publishing Co., Inc. & Materials Sciences Corporation, 1999. ISBN: 9781566768269.
[17] F.C. Campbell, Structural Composite Materials, ASM International, Ohio, 2010.

[18] G. Silber, C. Then, Preventive Biomechanics: Optimizing Support Systems for the Human Body in the Lying and Sitting Position, Springer-Verlag, Berlin, Heidelberg, 2013.
[19] D. Valtorta, Dynamic torsion test for the mechanical characterization of soft biological tissues, Doctoral Thesis, Cuvillier, Göttingen, 2007. ISBN 978-3-86727-184-4.
[20] J.A. Motherway, P. Verschueren, G. Van der Perre, J. Vander Sloten, M.D. Gilchrist, The mechanical properties of cranial bone, in: C.H. James Goh, Chwee Teck Lim (Eds.), World Congress of Biomechanics, Springer Science & Business Media, Singapore, 2010, pp. 776–779.
[21] J.A. Ashton-Miller, J.O. Delancey, On the biomechanics of vaginal birth and common sequelae, Annu. Rev. Biomed. Eng. 11 (2009) 163–176.
[22] D. Jing, J.A. Ashton-Miller, J.O. DeLancey, A subject-specific anisotropic visco-hyperelastic finite element model of female pelvic floor stress and strain during the second stage of labor, J. Biomech. 45 (3) (2012) 455–460.
[23] S.E. Duenwald, R. Vanderby Jr., R.S. Lakes, Viscoelastic relaxation and recovery of tendon, Ann. Biomed. Eng. 37 (6) (2009) 1131–1140.
[24] R.N. Miftahof, H.G. Nam, Biomechanics of the Human Urinary Bladder, Springer-Verlag, Heidelberg, 2013. ISBN: 978-3-642-36145-6.
[25] J.J.M. Pel, E. van Asselt, R. van Mastrigt, Contractile properties of the proximal urethra and bladder in female pig: morphology and function, Neurourol. Urodyn. 25 (2006) 70–77.
[26] N. Özkaya, M. Nordin, Fundamentals of Biomechanics: Equilibrium, Motion, and Deformation, second ed., Springer, New York, 1999.
[27] G. Frunza, E.N. Diaconescu, Hysteresis and mechanical fatigue, Ann. Univ. "Dunarea de jos" galati 8 (7) (2006) 62–66.
[28] E.N. Diaconescu, M. Glovnea, G.H. Frunza, Can metals subjected to mechanical fatigue exhibit a smart feature, in: The Seconds Seminar on Smart Materials and Structures, Suceava, 1996, pp. 105–116.
[29] G.H. Frunza, Elemente de Plasticitate şi Mecanica Ruperii, EDP, Bucuresti, 2005. pp. 256–270.
[30] M.A. Meyers, K.K. Chawla, Mechanical Behavior of Materials, Cambridge University Press, Cambridge, England, 2009. ISBN: 978-0521866750.

CHAPTER FIVE

Biomechanical Characterization of Native Pelvic Floor Organs and Tissues

S. Abramowitch, D. Easley
University of Pittsburgh, Pittsburgh, PA, United States of America

INTRODUCTION

The field of biomechanics refers to the mechanical characterization of objects that are biologic in nature, and includes anything ranging from individual proteins or groups of amino acids, to cells, tissues, organs, and even entire organisms [1]. While the testing and characterization of many basic structures and engineering materials has been standardized, objects of a biological origin often require deeper scrutiny due to their unique composition. In some cases, the developed standards can either be directly applied or slightly modified to provide useful information about biological materials. In other cases, entirely new methodologies, theories, and analytical approaches must be developed.

This chapter will focus on the biomechanical characterization of soft tissues as related to female pelvic organs and tissues. It will describe their structure and composition as related to their known mechanical function. However, there is limited mechanical data available on the organs and tissues in the female pelvis due to the relative infancy of urogynecology; thus, a major focus of this chapter will be to provide an overview of some of the typical mechanical testing protocols and considerations that are relevant to the female pelvis. Where available, data related to female pelvic organs and pelvic organ support will be reviewed. Specific examples of both structural and mechanical testing will be provided for the vagina and pelvic floor. The chapter will conclude by highlighting some directions for future work.

PELVIC FLOOR TISSUE CONSTITUENTS AND CHARACTERISTICS

Tissues are comprised of passive and active components. Components that are unable to generate force are described as passive, such as the extracellular matrix (ECM), which provides a network of mechanical and biochemical support for resident

The authors have no potential conflicts of interests to disclose.

cells [2]. These components only experience force if they are pulled or pushed by something else. Active components, such as muscles, are internally equipped with the ability to independently generate force. Two examples of active components of the female pelvic floor are the striated muscles of the levator ani and the smooth muscle cells that are found in many pelvic organs [3].

The primary components of tissue are collagen, elastin, water, and proteoglycans. The relative abundance of each component (and often their subtypes) and its organization allows for a wide array of load-bearing capabilities for each specific tissue. In general, these are "finely tuned" to allow a tissue or organ to meet its specific functional demands [4]. Like muscle, passive tissues have the ability to adapt to small alterations in mechanical demand through changes in their composition and/or organization, known as remodeling [1]. However, the rate and degree of change that they can accommodate is generally not as great as muscle tissue [1].

Of the primary constituents, collagen, which makes up 84% of pelvic floor tissues, such as the arcus tendineus fasciae pelvis, is responsible for load bearing [5,6]. The high concentration of collagen within vaginal tissue allows the vagina to bear loads and provide support to other pelvic organs, with the help of the connective tissue attachments[7,8]. Vaginal tissues are also comprised of about 16% elastin, the component that gives tissues the ability to recover from deformations. This is analogous to the recoil ability of rubber. This property, along with the likely action of smooth muscle, allows the vagina to recover its shape after the application of large deformations, like those seen during childbirth [8,9]. The elastin within the female reproductive system is considered unique in that it remodels significantly during gestation to allow for extreme expansions of the uterus, growth of the fetus, degradation of the cervix, and softening of the vagina and pelvic floor all in preparation for delivery [10,11]. The remaining 3% of elements are comprised predominantly of proteoglycans and water. Proteoglycans are highly hydrophilic proteins that attract water into tissues, play a significant role in collagen packing and organization, and are believed to contribute to the significant mechanical changes observed during gestation [7,12]. Water serves a number of mechanical functions including hydration, lubrication, and shock absorption, which are essential to the performance and longevity of the tissue [8,13,14].

The two muscle types present in the female pelvis are skeletal and smooth muscle. Skeletal muscles are voluntary muscles, or muscles that require conscious thought or reflex action to activate. For example, if a person wants to smile, they have to consciously move the muscles in their face in order to create that expression. With respect to the pelvic floor, there are skeletal muscles that aid in controlling urination, such as the pubococcygeus muscle and the external urethral sphincter, and defecation, such as the anal sphincter. These muscles can be consciously contracted or relaxed, but also maintain some degree of autonomic or reflex control (ie, activated without thought).

Smooth muscle is involuntary and completely autonomic. For example, there is no thought required to activate the muscles responsible for churning your stomach in order to digest food — the process is involuntary. Hollow organs such as the bladder, vagina, and rectum consist primarily of smooth muscle. Contractions of the smooth muscle are similar to those of skeletal muscles, but they are generally much slower, as skeletal muscle is able to contract and relax about 30 times faster than smooth muscle due to the presence of more actin and myosin filaments [15,16]. Additionally, smooth muscle can exhibit coordinated contractions, due to the gap junctions that allow action potentials to be sent between fibers. Both skeletal and smooth muscles are comprised of actin and myosin filaments. When stimulated, the actin and myosin filaments slide against one another, resulting in contraction of the muscle [15].

These basic constituents and the cells that manufacture them are the reason why biological tissues behave very differently from traditional engineering materials (eg, metals, rubbers, ceramics, or polymers). First and foremost, biological tissues have the ability to grow, shrink, or otherwise adapt over time depending on the mechanical demands that are placed upon them. In the field of mechanics, time-dependent behavior is often referred to as viscoelastic or viscoplastic, as was discussed in the previous chapter. However, the ability of tissues to adapt over time goes beyond this type of behavior. Tissues can physically remodel, changing both their structure and composition in response to changes in their surrounding environment [17].

Second, tissues have the ability to heal in response to damage [18]. Thus, the idea of performing fatigue testing (ie, cyclically loading a tissue over and over until it fails/is unable to sustain an imposed force) to measure its durability either in vivo or ex vivo cannot be studied in the same way that it can for traditional engineering materials. The same tests can be physically performed according to published standards, but the two aforementioned considerations (ie, remodeling and healing) that are unique to biological tissue makes the interpretation of the results from these tests very different, or potentially altogether invalid. After all, a human heart beats an estimated 2.5 billion times over the average life span without undergoing a catastrophic mechanical failure. There are currently no nonliving materials that have that degree of durability. This underscores the importance of having a fundamental understanding of the material that is being tested and how "standards" of mechanical testing and analysis are often not "standard" when it comes to biologic materials. Furthermore, this highlights a major distinction between biomechanical engineers and traditional mechanical engineers. The former are trained to understand the special considerations for biologic materials.

The fundamental components of tissue result in mechanical behaviors that are generally distinct from most standard engineering materials. The material assumptions made in the previous chapter with regard to many common engineering materials, that is, being linear, elastic, homogenous, and isotropic, often do not apply to tissues. Tissues are usually elasto-viscoplastic, nonlinear, inhomogenous, and anisotropic so, among other properties, their ability to resist being stretched changes depending on how much they are stretched (nonlinearity) [19,20]. Tissues tend to become more resistant to stretch the more that they are stretched. In addition, tissues are generally nonhomogenous in that they often vary internally in terms of composition and structure. For example, the vagina is comprised of four distinct sublayers the epithelium, subepithelium, muscularis, and adventitia [8]. While all of these layers serve different functions and display different characteristics, they are all part of the same organ. Furthermore, tissues are often anisotropic — meaning that the mechanical behavior of a tissue is directionally dependent. For example, if vaginal tissue is pulled along its longitudinal axis, it would display different behavior than if it were pulled along its circumferential axis [21]. All of these qualities make fully characterizing the mechanical behavior of tissues very challenging.

FUNCTIONS OF PELVIC FLOOR ORGANS

The female reproductive organs are arguably the most mechanically adaptable organs in humans. For example, the vagina serves many different mechanical roles that shift dramatically during pregnancy and then, in most cases, it is able to return to near prepregnancy levels of function. Throughout a woman's lifetime, one of the primary roles of the vagina is to aid in the mechanical support of the rectum, urethra, uterus, and bladder. It achieves this support directly through its own passive mechanical integrity and smooth muscle tone, and indirectly through its attachments to the pelvis and levator ani muscles [7]. The vagina also serves as a sexual organ where smooth muscle relaxation allows for an expansion in both diameter and length to accommodate intercourse. Then, for what equates to just a few hours or days out of an entire woman's life, it is able to stretch over eight times its initial diameter to allow for passage of a fetus; returning very quickly to its aforementioned functional shape [22].

The cervix and uterus are just as mechanically intriguing. The primary role of the cervix is to serve as a semifirm attachment point for the uterosacral and cardinal ligaments, and as a conduit to allow for fluid (eg, semen, menses) transfer between the vagina and uterus. During pregnancy, its mechanical integrity serves to hold the fetus in the uterus until the time of delivery; at which point, it loses nearly all of its mechanical integrity to dilate more than five times its normal luminal diameter [10,23,24]. Following delivery, the cervix rapidly reconstitutes itself to resume its aforementioned roles. There are no other tissues in the human body that can undergo that type of mechanical transition, especially over such short durations of time. A primary function of the uterus is to host the implantation of a blastocyst following conception. Its smooth muscle composition allows it to stretch and grow significantly throughout pregnancy to accommodate the growing fetus. Unlike skeletal muscle that loses the ability to generate tension when stretched, smooth muscle's ability to generate tension does not change significantly. Its mechanical plasticity, along with its phenotypic plasticity, allows the uterus to stretch and become thin, while maintaining its ability to generate enormous contractile forces at the time of delivery [11,25]. Like the vagina and cervix, it then undergoes a significant transformation to nearly its prepregnancy shape.

The smooth muscle of the rectum and bladder provides them with the ability to expand significantly to store feces and urine, respectively. Contraction of the bladder serves to eliminate urine, while contraction of the rectum aids in defecation. The urethra is primarily responsible for providing a path for the evacuation of urine. In conjunction with the striated external sphincter, the smooth muscle of the urethra (inner urethral sphincter) maintains a constant tone to keep this tube-like structure closed in response to increases in intraabdominal pressure, allowing urine to remain in the bladder until it is time to void [26]. Interestingly, despite being relatively collocated, the smooth muscle of the urethra must undergo relaxation while the smooth muscle of the bladder contracts for successful voiding [1,27].

The rectum and anus are the final portions of the intestines, serving as storage for solid waste accumulated after traveling through the digestive tract. Rectal smooth muscle is unique due to its ability to distend in response to varying degrees of "intraluminal contents" [28]. As the rectum fills with feces, the urge to defecate becomes present. Increased intrarectal pressure in conjunction with the contraction of the rectal muscle, in peristaltic waves, causes waste to pass through.

COMMON EXPERIMENTAL CONDITIONS AND CONSIDERATIONS

While established mechanical testing protocols are generally useful for biological tissues, their unique characteristics, a fundamental understanding of their in vivo function, and an appreciation of their organization and composition must be considered in order to adapt the testing protocols to reliably test tissues. This section will review the conditions and considerations essential to obtaining repeatable and reproducible results when working with biologic materials.

Preconditioning

Preconditioning is the process of performing cyclic loading within the toe region of the stress-strain curve. This technique is critical for testing viscoelastic materials because these materials are history-dependent. This means that the tissue's response to external forces changes due to micro-structural alterations in the tissue is based upon the previous loading condition or treatment of tissues [1]. In other words, the handling and preparation of tissue samples directly impacts the structural properties measured during testing. This sensitivity can make it difficult to successfully reproduce tests. The process of preconditioning minimizes the effects of prior stressors, breaks transient hydrogen bonds that may have formed while having the tissue in an unloaded configuration, and helps to redistribute fluid; thus, it allows engineers to reliably create common baselines between samples through the minimization of hysteresis. This is analogous to repeatedly stretching

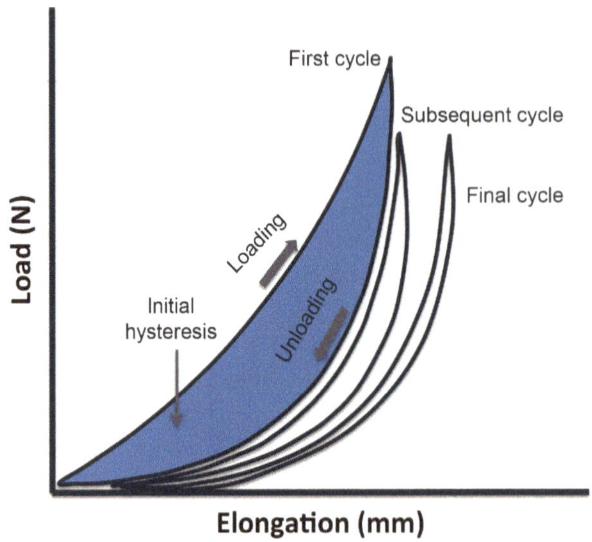

Figure 1 Representative loading and unloading curves for a biological specimen. During preconditioning, note that with each subsequent cycle, the area of hysteresis decreases.

a balloon prior to inflating it. Given enough cycles of preconditioning, the amount of hysteresis eventually reaches a steady-state response (Fig. 1). This provides researchers with an easily reproducible tissue condition from which to initiate testing [29]. While recommended preconditioning protocols exist for a wide array of biomaterials [1,30–32], it is considered best practice to conduct preliminary testing in order to confirm the number of cycles necessary to reach this steady-state response for each specific tissue.

The specific parameters of a preconditioning protocol should be determined through careful examination of the tissue's native physiological environment and its load-elongation profile in response to cyclic loading. Given that preconditioning occurs prior to mechanical testing, it is imperative that loads be restricted to the toe region or just beyond the toe region of the load-elongation curve so as not to risk loading the tissue beyond its physiological limits and thus compromising subsequent mechanical testing [4].

Hydration

The human body is comprised of approximately 75% water. Hydration is the process by which the cells and tissues in our body absorb and retain this water. This process is directly related to tissue mechanics, as it is through hydration that cells and ECM regulate their inter- and intracellular pressures, which in turn form the foundation for the behavior of the bulk tissue structure [33]. The proper amount of absorption and retention allows for dissipation of energy and aids in the overall viscoelastic nature of the tissue. Over- or underhydration (hypotonic or hypertonic states) can have a significant effect on the tissue's mechanical behavior. For example, in order to walk normally, the knee requires the proper amount of lubrication to reduce friction between the tibia and femur. If the pressure in the knee joint is increased as a result of inflammation, it can stiffen the joint and prevent normal ranges of motion. However, underhydration of the cartilage in the knee joint resulting from degeneration and a loss of proteoglycans can lead to increased friction and direct contact between boney surfaces, causing restrictions of motion [1,34].

At the tissue level, excessive or depleted levels of water change the properties of the ECM. For example, when a tissue is dehydrated, the elastin becomes stiff and behaves in a brittle manner rather than a ductile manner, collagen packing and sliding is altered, and the overall viscoelastic response of the tissue is altered [33]. As another example, when cartilage is overhydrated in a hypotonic solution, the relative proportions of ECM structures change with respect to the wet weight of the tissue [4,34–37]. Additionally, many studies have confirmed that mechanical properties are negatively impacted when tissues are immersed in fluids with improper solute concentrations [34–36,38,39]. Therefore, tissues should be kept appropriately hydrated during mechanical testing to maintain their in vivo mechanical properties [16]. Typical hydration methods include pretest saturation, for short duration testing, and complete immersion for extended testing. Saline solutions are the most commonly used hydration medium for passive mechanical testing. Rubod et al. found that saline not only provides hydration for vaginal tissues, but protects against variations caused by humidity [40]. The concentration of sodium chloride can be manipulated to best match in vivo levels during mechanical testing and 0.9% saline solutions are typical for mechanical testing of most tissues [41]. When investigating active mechanical properties, Krebs solution is often used while bubbling with CO_2 and O_2 to help keep the muscle tissue viable and able to generate contractile forces.

Temperature

In order to maintain normal metabolic function, most people maintain a body core temperature of 37°C. However, the body's temperature is also critical for maintaining the body's natural tissue mechanics. Viscoelastic materials, such as tissues, are temperature sensitive. Even a slight elevation of 1 or 2°C of the body's internal temperature can cause tissues to become significantly softer. In fact, temperatures more than 5°C above normal can cause proteins to denature, permanently affecting the structure and mechanical behavior of tissues [42,43]. Of course, decreases in temperature on the same order can result in a stiffer tissue response. As was the case with hydration, temperature must be precisely controlled in order to maintain mechanical properties at in vivo levels. This is most often accomplished by heating the saline bath in which the tissue is being hydrated. However, Rubod studied vaginal tissue of the sheep and found minimal effects due to temperatures from 18 to 37°C and the tissues also suffered fewer effects from 24 hour storage in saline [40].

Tissue Storage

It would be ideal to test tissues immediately following explantation or after euthanasia, but this option is often unavailable. Therefore, tissues must be carefully stored in order to preserve their mechanical properties until testing can be performed. Freezing is the universal method for tissue preservation, as it has been shown to have little to no impact on the passive mechanical properties of the tissue after it is thawed [40,44,45]. Samples needing to be preserved overnight should be stored or thawed at 4°C, known as "fresh storage" [34]. Otherwise, tissues should be wrapped in saline soaked gauze, thereby maintaining proper hydration, and frozen at −20°C to prevent tissue degradation [34]. The saline soaked gauze helps increase the longevity of the sample by protecting it from freezer burn [40]. Some studies have indicated that the viscoelastic behavior of tissues may be compromised by freezing, thus caution should be used when freezing specimens that may undergo stress relaxation or creep tests [8,44–49]. Rubod et al. studied the effects of freezing on vaginal tissues and found that the mechanical properties of vaginal tissues were unaffected by freezing compared to fresh tissues [40].

CROSS-SECTIONAL AREA AND STRAIN MEASUREMENTS

The accurate determination of mechanical properties is dependent on accurate measurements of cross-sectional area and strain. Both of these can be challenging for soft tissues because they generally have irregular shapes and deform easily when contacted. Thus, it is common to classify methods to measure cross-sectional area and strain into contact and noncontact approaches.

Contact approaches for cross-sectional area measurements include the use of calipers, area micrometers, forcing a specimen into a geometry that can be measured, and creating molds of specimens. All of these have been shown to result in significant errors due to the deformability of soft tissue. For tissues that hold their shape well and are relatively stiff, cross-sectional area measurements using calipers can be reasonably accurate and repeatable if the sample being measured has been cut into basic geometric shapes, such as a square or rectangle [50]. Noncontact approaches include shadow amplitude techniques, the profile method, the use of calibrated imaging, and measurements using lasers [50]. The latter has been shown to be highly accurate and repeatable. There are two main types of lasers that are used in these approaches. A collimated laser is used to measure the width of a sample, and is illustrated in Fig. 2. This device emits a wide, thin band of laser light. The specimen is placed within the width of this band, blocking a portion of it from reaching a receiver. The portion of blocked laser light (ie, the shadow) represents the width of the sample. As the laser is rotated around the specimen, the width is recorded at regular intervals such that the shape of the specimen can be reconstructed and the area can be calculated. The major limitation to using this type of laser is that concavities in the specimen cannot be detected. Thus, it is best to use with samples that have a convex geometry.

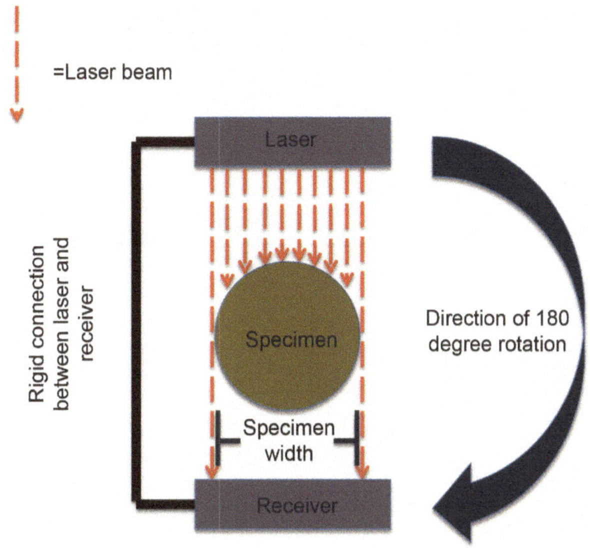

Figure 2 Aerial schematic of a collimated laser. The laser emits a wide band of laser light, and the specimen blocks a portion of the light from reaching the receiver. The portion of blocked laser light represents the width of the sample. The laser rotates around the specimen, collecting information at regular intervals in order to calculate the area of the specimen.

Bounce back laser micrometers are relatively flexible and can be used to measure samples with concave geometries. A laser is reflected off of the surface of a specimen and the distance to the specimen is recorded. As the laser is rotated around the specimen, a surface profile is generated from which the area can be computed. These lasers are also useful for thickness measurements of samples that have a planar geometry [50]. The sample can lay flat on a surface and either the surface can move with respect to the laser or vice versa. As the laser travels over the specimen, the distance measured will decrease as the laser transitions from reflecting off of the surface to the sample revealing the thickness of the specimen. This process is depicted in Fig. 3. A potential limitation of this technology is that the laser may slightly penetrate into the specimen before reflecting back to the laser. This is dependent on a number of factors including the type of tissue and its hydration state. However, this can be overcome by placing India ink on the specimen at the location of the measurement.

Similar to contact measurements of the cross-sectional area, contact methods of measuring strain (ie, placement of strain transducers on or into the specimen) also raise concern because they alter the tissue's ability to deform and/or they create a local defect in the sample. Thus, optical noncontact methods prove to be the most popular and accurate method of strain measurement for tissues [50]. Video recordings can be used to track markers on a sample. The markers define the initial length of the specimen, and the strain on the surface of the specimen can be calculated as the distance between the markers

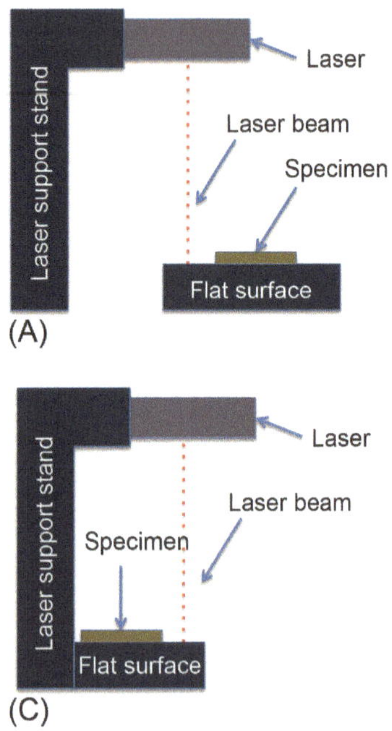

Figure 3 Side view of a bounce back laser configuration. (A) Starting position of the specimen. (B) The specimen is moved into range of the laser, changing the distance between the laser and the surface. (C) The specimen after the range of the laser. The thickness of the specimen is determined by the difference between the initial distance and the distances recorded from (B).

changes during the test. Motion tracking software is then used to track the motion of the markers. Of course, care must be taken in how markers are attached to tissue specimens as the surface strain measurements of the thin films of connective tissue could be altered by glue, etc., from marker placement. Additionally, the markers must provide contrast with respect to the tissue being tested, to allow for optimal visual tracking, displayed in Fig. 4. This method allows for 2D measurements of strain as well as a way to more accurately quantify inhomogeneity in strain by tracking multiple markers [50]. With multiple cameras, it can be extended to 3D measurements. More recently, investigators have developed marker-less tracking algorithms, which track natural features in the surface texture of the specimen [51]. As the scientific community gains confidence in these methods, it is likely that strain tracking will move away from the use of artificial markers.

One method of estimating tissue strain during testing that is commonly reported in the literature is the "clamp-to-clamp" method. Since strain is defined as a change in length divided by an initial length, some investigators choose to measure the initial clamp-to-clamp distance in the sample once a preload (ie, a small amount of tension in the specimen to remove slack) has been applied to the specimen. At this point, the elongation measurement of the testing machine is set to zero. As the specimen is

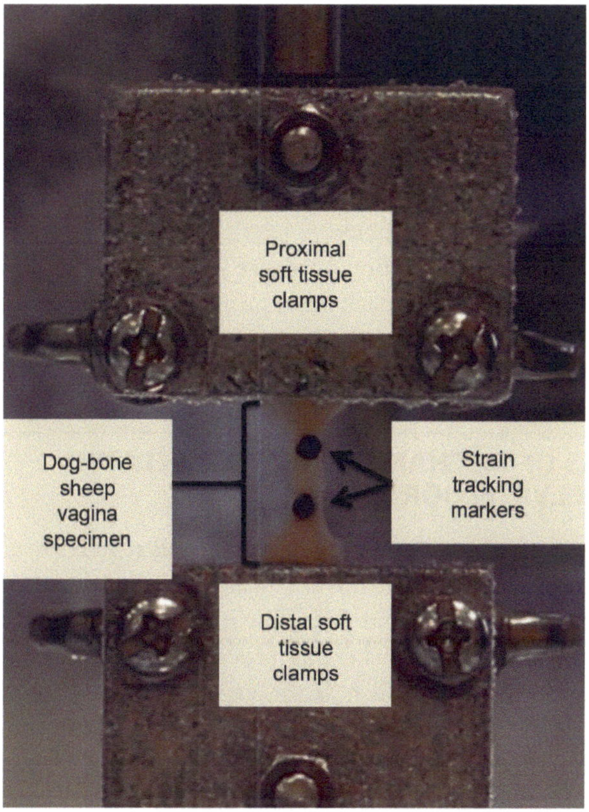

Figure 4 Marker placement on a dog-bone sheep vagina specimen. Markers provide visual contrast to allow for optimal visual tracking.

elongated during testing, the testing machine measures the movement of one clamp with respect to the other, thereby providing the change in length [50,52]. By dividing the change in length as measured by the testing machine by the initial clamp-to-clamp distance, an estimate of strain can be obtained.

While the clamp-to-clamp method of measuring tissue strain is often seen in the literature, it generally provides a significant overestimate of the actual strain that should be reported, that is, the strain originating near the middle of the specimen. The reason for this is that tissue strain for a uniaxial tensile test is nonuniform throughout the length of the specimen [50,53]. The clamps impose an artificial boundary condition on the tissue, preventing it from decreasing its width as it is stretched along its length. In addition, the clamp itself creates a local area of damage within the tissue sample. These two factors combine to result in much higher strains occurring near the clamps that are not reflective of the actual tissue's behavior. Near the middle of the sample, away from the clamps, strains are more uniform and better reflect the response of the sample to uniaxial tension.

In fact, clamp-to-clamp measures of strain can overestimate the strains measured using markers in the mid-substance of the sample by more than 200% [50]. Since calculations of tangent modulus, ultimate strain, and strain energy density are all based on accurate measurement of strain, these values will also reflect that error. In addition, clamp-to-clamp measures of strain are much more likely to be impacted by the specimen slowly slipping within the clamp during testing. Unless the slip is obvious, it can go unnoticed and be a significant source of variability.

Measures of strain and cross-sectional area are just two reasons why different laboratories will report widely different mechanical properties for the same tissue [50,54–56]. It is, therefore, critical for investigators to read the methodology of a paper closely and not take reported values of mechanical properties at face value.

EXAMPLES OF MECHANICAL TESTING RELATED TO THE PELVIC FLOOR

Effective mechanical testing of biological materials requires many more considerations than standard materials. While this is a challenge in general, the unique characteristics of the pelvic floor, and its organs that allow it to adapt and serve multiple functions throughout a woman's lifetime make this process even more complex. Sometimes researchers isolate specific organs such as the vagina or bladder to further understand its properties. Other times, researchers want to gather information about the pelvic floor structure in its entirety, to understand how all of the components contribute to overall support. This section highlights a few structural and mechanical tests performed in the field of urogynecology.

Structural Testing

Clinically, women with prolapse often have compromised connective tissue attachments to the vagina [7,57,58]. However, to test this hypothesis, researchers needed to find an appropriate animal model and testing protocol to understand the structural properties of the pelvic floor. This is primarily due to the difficulty in controlling for confounding variables that would be present in humans as well as collecting a large enough set of cadaver samples for statistical significance. Moalli et al. proved that a rat model could be used to study the structural properties of the vaginal supportive tissue complex, or VSTC [59]. The experimental setup for the structural test is shown in Fig. 5. During dissection, the pelvic organs and muscles of the rat were isolated and left attached to the lumbar spine. The spine was then potted in polymethylmethacrylate to preserve alignment of the structure. Special clamps were used to secure the tissue sample within the uniaxial testing apparatus, with the potted end secured below, and the vagina was secured in the top clamp. This allowed for load to failure tests to be conducted to determine the structural properties of rat pelvic floor.

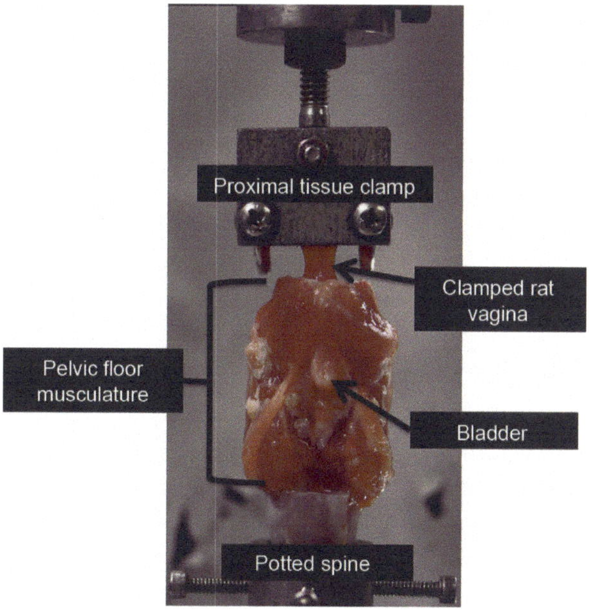

Figure 5 Experimental setup for a structural test of the rat pelvic floor (VSTC).

The rat model was particularly beneficial for this application because it allowed researchers to obtain repeatable results by controlling for primary risk factors such as age, parity, and weight. Structural testing of the rat model showed that, upon failure, level II connective tissue support is disrupted first, later followed by levels I and III [59]. This result agreed with commonly observed paravaginal defects in women suffering from prolapse. The establishment of this repeatable method allowed for subsequent studies to be conducted to further understand the complexity of the vagina-supportive tissue complex.

Lowder et al. utilized this newly established structural test to understand structural changes during gestation, delivery, and recovery and found that stiffness and ultimate load decrease during pregnancy and recover by 4 weeks postpartum [60]. These findings sparked interest in mimicking birth injury in rats, in order to help pinpoint how the rats were recovering from birth so efficiently [8,61,62]. This also further indicates the unique complexity of pelvic floor tissues, as they undergo drastic changes at the cellular level to accommodate the growth and passage of the fetus.

In a second study, Moalli et al. utilized the same structural test to study the effect of hormone deficiency, a hallmark of menopause, on the mechanics of pelvic floor tissues [63]. This study was motivated due to the onset of menopause being a risk factor for developing prolapse. Age was found to be another confounding variable that dictates the properties of the pelvic floor within the rat model [63].

Many more questions still need answers, including how components change and why those changes yield drastically different biomechanical properties. Structural tests play an important role in understanding the response of the pelvic floor to a number of conditions. Although the findings above do not directly translate to humans and cannot tell us specific information about which tissue is undergoing the most significant changes, they provide valuable information toward the formation of a more comprehensive understanding of pelvic anatomy and function and can provide a good starting point for generating additional hypotheses.

Tests to Determine the Passive Mechanical Properties of Tissues

It has been observed clinically that significant biomechanical changes take place within female pelvic tissues during pregnancy to accommodate the growth of the fetus, delivery, and postpartum recovery [8,12,60]. Passive components are believed to have a role in this remodeling because they are typically responsible for the expansion and load-bearing properties of tissues [8]. Feola et al. performed uniaxial tensile testing in an effort to explore the passive mechanical properties of pregnancy on the rat vagina [64]. While it is clinically understood that changes are taking place, this study provided experimental evidence of the dynamic changes in mechanical properties as well as the unique reorganization of elements that may be taking place to prepare for vaginal delivery. They found that the tangent modulus decreases and ultimate strain increases during pregnancy; however, the strain energy density, or toughness of vaginal tissue, remained the same [64]. These changes are believed to increase the distensibility of the vagina in order to reduce maternal birth injuries and fetal birth injuries.

Maternal birth injury is another identified risk factor for developing pelvic organ prolapse [8,59,65–67]. Thus, a method for mitigating maternal birth injury could aid in preventing the onset of prolapse. A study by Alperin et al. investigated the use of a collagen scaffold as a treatment for maternal birth injury in rats [68]. Applying scaffolds to enhance the healing of injuries during delivery could aid in the reduction of scarring as well as inferior mechanical properties that increase the likelihood of developing prolapse. This study found that untreated injured rats displayed grossly inferior mechanical properties. However, rats that were treated with the collagen scaffold maintained mechanical properties comparable to that of the uninjured controls, indicating that the collagen scaffold improved tissue quality through changes in the healing response [68]. While these results are not directly translatable to human patients, it offers an opportunity to reapply techniques to the vagina and pelvic floor that have shown promise in other fields, for example, orthopedics.

While rats are acceptable animal models, nonhuman primates are considered to be a gold-standard animal model for humans due to their semiupright posture, similar female pelvic anatomy of physiology, lengths of gestation, and degree of vaginal stretch resulting from delivery [8,69–71]. Therefore, investigating primates may help elucidate differences between nulliparous women and parous women with respect to the onset and development of prolapse. To this effect, Feola et al. investigated the effect of parity on the mechanical properties of the rhesus macaque vagina [72]. This study found that the tangent modulus was significantly higher for nulliparous animals compared to parous animals. Additionally, they used trichrome staining to determine collagen alignment decreases with parity [72]. These findings provide significant insight through correlating collagen alignment and mechanical properties.

Rubod et al. conducted numerous uniaxial tensile studies using the sheep vagina to explore various experimental conditions [40,73]. In addition to many of the conditions mentioned earlier in this chapter, this research team explored the effect of orientation on mechanical properties. They found that vaginal tissues are anisotropic and display different mechanical properties when loaded in the circumferential or longitudinal directions, and loading in the longitudinal direction proved to be significantly stiffer than the circumferential direction [40,73]. These findings are believed to correlate with the process of delivery to aid in the passage of the fetus. The anisotropic nature of vaginal tissues highlights a major limitation when using uniaxial testing.

For most tissues, especially those related to pelvic organs and pelvic organ support, a simple uniaxial tensile test is often insufficient to fully describe the mechanical properties of a sample. It would only be sufficient for materials that are linear, elastic, isotropic, and homogenous. For example, if a longitudinal strip of vaginal tissue is removed from a patient and tested in uniaxial tension, it is only providing information about the behavior in that direction. We can assume that those properties would be the same in the circumferential direction, but likely they are not. In fact, to make things more complicated, the longitudinal uniaxial test may not be providing a fully accurate representation of the mechanical properties in the longitudinal direction. This is because removing it from the patient has artificially cut the connections between collagen fibers, making them "freer" to rotate along the axis of loading than they would have been otherwise. That is not to say that these data are not useful. They represent a reasonable first approximation of the true mechanical properties of the sample. Nevertheless, they are an approximation and must be interpreted in that context.

Tests to Determine the Active Mechanical Properties of Tissues

Smooth muscle contractions dictate the active mechanics of many organs in the female pelvis and its response is regulated by a number of factors [74]. To test the active properties of vaginal tissues throughout gestation, Feola et al. used a Krebs, oxygen, and CO_2 bath to keep the rat tissue viable and capable of contracting [64]. Notable findings from this study include that significant changes occur during gestation with respect to active mechanics [64]. More specifically, a significant percentage of smooth muscle cells alter their phenotype during gestation becoming less contractile throughout gestation and more contractile just prior to delivery (presumably in preparation for recovery following delivery) [64]. This supports the idea that active and passive mechanics both contribute throughout pregnancy and delivery, but their role may be most critical during the postpartum period.

Further, morphological differences have been identified between the smooth muscle tissue in women with and without prolapse [73,75–79]. This sparked interest in understanding how the tissues from these distinct groups of women respond to caldesmon, a protein responsible for smooth muscle relaxation. It was found that caldesmon expression was lower in women with prolapse. This, in turn, could increase muscle contractility, in an effort to compensate for having fewer cells. Northington et al. went further and investigated differences in anterior vaginal muscularis contraction for women with and without prolapse. It was found that prolapsed and nonprolapsed anterior vaginal muscularis respond similarly to KCl, indicating that the muscularis is able to compensate to some degree. However, women with prolapse did not respond to alpha-1 adrenergic receptor, indicating a decrease in expression of the receptor [74]. These findings were impactful, because they illustrate how the body can adapt to abnormal conditions, while maintaining overall function. Understanding how the body compensates could lead researchers toward solutions to accommodate women suffering from defects in their pelvic floor.

BIAXIAL TENSILE TESTING

Biaxial tests provide a more comprehensive description of the mechanical characteristics of anisotropic tissues because they are more capable of replicating in vivo loading conditions [80]. However, biaxial testing of tissues is particularly complex. The challenges surrounding biaxial testing of tissues include: extraction of a homogenous sample, control of the sample size and thickness, fixation of the tissue boundaries without imparting damage or causing nonphysiologic loading conditions, application of constant force along the tissue boundary, and minimization of the induced stress caused by the fixation of the sample, appropriately identifying and accounting for tissue material axes [80]. Given the complexity of biaxial testing of tissues, it comes as no surprise that high variability has been reported between specimens, making it difficult to draw reliable conclusions from biaxial data. Nevertheless, the models that can be developed based on these data are generally more accurate, robust, and meaningful than those from uniaxial tests, especially when these tests are simultaneously coupled with techniques to measure the orientation of fibers within a sample [80,81].

Boundary conditions present a major challenge for planar biaxial testing (Fig. 6). Researchers must ensure that stress and strain is uniform in the region of interest, which is typically in the center of the sample. To this end, the edges of the sample must be able to

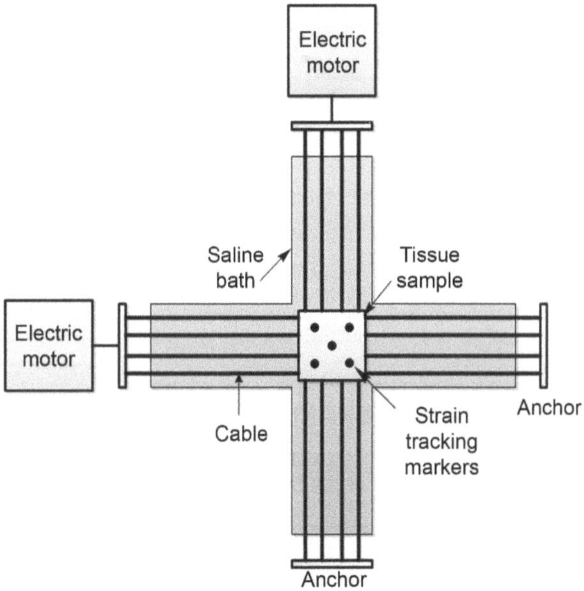

Figure 6 Biaxial test schematic. Edges of the sample are able to move laterally through the anchors. Four or more hooks are attached to the anchors through cables. These not only aid in equally distributing loading forces around the specimen, but allow for lateral motion and in-plane shear. Strain tracking markers allow for noncontact strain measurements to be recorded.

move laterally, and the restraints must be placed far enough away from the region of interest to minimize their effects on the measured stress and strain within said region. This is not a trivial problem and a number of tissue geometries and fixation techniques have been explored to minimize these effects. Many now use four or more hooks along the edges to equally distribute loading forces around the specimen and allow for lateral motion and to allow for in-plane shear [1,80,82].

Planar biaxial testing configurations are more physiologically accurate for tissues that are relatively flat and thin [80,83]. Tubular biaxial testing is often performed on blood vessels and other tissues that can be approximated as thin-walled tubes [84–89]. Unlike planar biaxial testing, in which tension is varied along each of the two axes of the sample, tubular biaxial testing uses changes in luminal pressure and longitudinal tension to explore tissue anisotropy. Of course, any tubular structure can be cut longitudinally and folded open to assume a planar configuration. The concern is that doing so creates an unnatural edge in the sample and causes an artificial stress state as a result of forcing a tissue, which "wants" to be in the shape of tube, to be planar. Thus, it appears that tubular biaxial testing of the entire organ is more appropriate for vaginal tissue. However, it should be noted that the urethra is attached to the anterior wall for much of the length of the vagina. It is difficult to separate it from the anterior vaginal wall without compromising the mechanical integrity of the tissue in that region. The alternative is to cut along the urethra and open the vagina into a planar configuration, which is unnatural. This is a tradeoff that needs to be reconciled for the vagina. In addition, it should be noted that the distal one-third of the vagina has a different embryologic origin than the upper two-thirds, resulting in a slightly different composition and microstructure [90].

A recent study conducted by Becker et al. examined uterosacral and cardinal ligaments of swine [91]. Becker found that these ligaments were orthotropic, and the physiological loading direction displayed higher stiffness values. It was also noted that the ligaments were able to withstand large deformations, which correlates well with their function of supporting pelvic organs. The stress relaxation tests yielded the fact that an increase in stretch causes the ligaments to become stiffer, with the primary causes being the reorganization of fibers as well as a decrease in water in the ECM.

CONCLUSIONS

This chapter discussed the importance of biomechanical characterization of pelvic floor organs and tissues, and shares important findings that help explain the physiological observations seen in human patients. While current studies have expanded the field tremendously over the past decade, there are still many questions that need to be answered, and deeper investigations that need to be pursued. With respect to structural mechanics, urogynecological research is still in its infancy when compared to the fields of orthopedics and cardiovascular research, which emphasizes the need to deepen our understanding of pelvic floor tissues. Pursuing this path is a difficult feat, as tissues of the pelvic floor are some of the most unique tissues in the human body due to their ability to rapidly adapt and recover from pregnancy and birthing. However, continuing to pursue research and testing of the pelvic floor will enhance our ability to model, predict, and better treat pelvic floor disorders.

REFERENCES

[1] Y.C. Fung, Biomechanics: Mechanical Properties of Living Tissues, Springer-Verlag, New York, 1993.
[2] S.C. Cowin, S.B. Doty, Tissue Mechanics, Springer, New York, 2007.
[3] S. Herschorn, Female pelvic floor anatomy the pelvic floor, supporting structures, and pelvic organs, Rev. Urol. 6 (5) (2004).
[4] F. Guilak, D.L. Butler, S.A. Goldstein, D. Mooney, Functional Tissue Engineering, Springer, New York, 2003.
[5] I. Sridharan, et al., Structural and mechanical profiles of native collagen fibers in vaginal wall connective tissues, Biomaterials 33 (5) (2012) 1520–1527.
[6] P.A. Moalli, et al., Impact of menopause on collagen subtypes in the arcus tendineous fasciae pelvis, Am. J. Obstet. Gynecol. 190 (3) (2004) 620–627.
[7] J.O. DeLancey, Anatomy and biomechanics of genital prolapse, Clin. Obstet. Gynecol. 36 (4) (1993) 897–909.

[8] S.D. Abramowitch, et al., Tissue mechanics, animal models, and pelvic organ prolapse: a review, Eur. J. Obstet. Gynecol. Reprod. Biol. 144 (Suppl. 1) (2009) S146–S158.
[9] T. Mizuno, D.A.G. Mickle, C.G. Kiani, R. Li, Overexpression of elastin fragments in infarcted myocardium attenuates scar expansion and heart dysfunction, Am. J. Physiol. Heart Circ. Physiol. 288 (6) (2005).
[10] W.R. Barone, et al., The effect of pregnancy and postpartum recovery on the viscoelastic behavior of the rat cervix, J. Mech. Med. Biol. 12 (1) (2012) 12500091–125000917.
[11] L. Zhao, Collagen studies in late pregnant relaxin null mice, Biol. Reprod. 63 (3) (2000) 697–703.
[12] M. Yanagishita, Proteoglycans and hyaluronan in female reproductive organs, EXS 70 (1994) 14–23.
[13] U. Ulmsten, C. Falconer, Connective tissue in female urinary incontinence, Curr. Opin. Obstet. Gynecol. 11 (5) (1999) 509–515.
[14] Y. Nakashima, T.N. Wight, K. Sueishi, Early atherosclerosis in humans: role of diffuse intimal thickening and extracellular matrix proteoglycans, Cardiovasc. Res. 79 (1) (2008) 14–23.
[15] E.M. Marieb, K. Hoehn, Human Anatomy & Physiology, eighth ed., Pearson, San Francisco, CA, 2010.
[16] D. Shier, J.L. Butler, R. Lewis, Hole's Human Anatomy and Physiology, tenth ed., McGraw-Hill Higher Education, Boston, MA, 2009.
[17] L. Sandra, C. Obbink-Huizer, F.T. Baaijens, A physically motivated constitutive model for cell-mediated compaction and collagen remodeling in soft tissues, Biomech. Model. Mechanobiol. 13 (5) (2014) 985–1001.
[18] W. Prentice, Principles of Athletic Training: A Contemporary-Based Approach, fifteenth ed., McGraw Hill, New York, 2013.
[19] G.A. Holzapfel, R. Eberlein, P. Wriggers, H.W. Weizacker, A new axisymmetrical membrane element for anisotropic, finite strain analysis of arteries, Commun. Numer. Meth. Eng. 12 (8) (1996) 507–517.
[20] R. Akhtar, M.J. Sherratt, J.K. Cruickshank, B. Derby, Characterizing the elastic properties of tissues, Mater. Today 14 (3) (2011) 96–105.
[21] A.J. Feola, Impact of vaginal synthetic prolapse meshes on the mechanics of the host tissue response, Doctoral dissertation, University of Pittsburgh, 2011.
[22] J.A. Ashton-Miller, J.O. Delancey, On the biomechanics of vaginal birth and common sequelae, Annu. Rev. Biomed. Eng. 11 (2009) 163–176.
[23] P.C. Leppert, Anatomy and physiology of cervical ripening, Clin. Obstet. Gynecol. 38 (2) (1995) 267–279.
[24] B.S. Oxlund, et al., Collagen concentration and biomechanical properties of samples from the lower uterine cervix in relation to age and parity in non-pregnant women, Reprod. Biol. Endocrinol. 8 (2010) 82.
[25] R.N. Miftahof, H.G. Nam, Biomechanics of the Gravid Human Uterus, Springer-Verlag, New York, 2011.
[26] J.J.M. Pel, E. van Asselt, R. van Mastrigt, Contractile properties of the proximal urethra and bladder in female pig: morphology and function, Neurourol. Urodyn. 25 (2006) 70–77.
[27] R.N. Miftahof, H.G. Nam, Biomechanics of the Human Urinary Bladder, Springer, 2013. ISBN: 978-3-642-36145-6.
[28] F.H. Dall, C.S. Jorgensen, J.C. Djurhuus, H. Gregersen, Biomechanical wall properties of the porcine rectum: a study using impedance planimetry, Dig. Dis. 9 (6) (1991).
[29] N. Conza, Part 3: tissue preconditioning, Exp. Tech. 29 (2005) 43–46.
[30] K.S. Miller, et al., Effect of preconditioning and stress relaxation on local collagen fiber re-alignment: inhomogeneous properties of rat supraspinatus tendon, J. Biomech. Eng. 134 (3) (2012) 031007.
[31] T.T. Tower, M.R. Neidert, R.T. Tranquillo, Fiber alignment imaging during mechanical testing of soft tissues, Ann. Biomed. Eng. 30 (10) (2002) 1221–1233.
[32] E.O. Carew, J.E. Barber, I. Vesely, Role of preconditioning and recovery time in repeated testing of aortic valve tissues: validation through quasilinear viscoelastic theory, Ann. Biomed. Eng. 28 (9) (2000) 1093–1100.
[33] D. Shahmirzadi, A.H. Hsieh, An efficient technique for adjusting and maintaining specific hydration levels in soft biological tissues in vitro, Med. Eng. Phys. 32 (7) (2010) 795–801.

[34] V.C. Mow, R. Huiskes, Basic Orthopaedic Biomechanics and Mechano-Biology, third ed., Lippincott Williams & Wilkins, Philadelphia, PA, 2005.
[35] W.M. Lai, V.C. Mow, A triphasic theory for the swelling and deformation behaviors of articular cartilage, J. Biomech. Eng. 113 (1991) 245–258.
[36] W.M. Lai, V.C. Mow, D.D. Sun, et al., On the electric potentials inside a charged soft hydrated biological tissue: streaming potential vs. diffusion potential, J. Biomech. Eng. 122 (2000) 336–346.
[37] A.I. Maroudas, Balance between swelling pressure and collagen tension in normal and degenerate cartilage, Nature 260 (1976) 808–809.
[38] V.C. Mow, A. Ratcliffe, M.P. Rosenwasser, et al., Experimental studies on the repair of large osteochondral defects at a high weight bearing area of the knee joint: a tissue engineering study, Trans. Am. Soc. Mech. Eng. 113 (1991) 198–207.
[39] V.C. Mow, X.E. Guo, Mechano-electrochemical properties of articular cartilage: their inhomogeneities and anisotropics, Annu. Rev. Biomed. Eng. 4 (2002) 175–209.
[40] C. Rubod, et al., Biomechanical properties of vaginal tissue. Part 1: new experimental protocol, J. Urol. 178 (1) (2007) 320–325, discussion 325.
[41] R.B. Martin, D.B. Burr, N.A. Sharkey, Skeletal Tissue Mechanics, Springer, New York, 1998.
[42] G.A. Holzapfel, Biomechanics of soft tissue, Comput. Biomech. 7 (2000).
[43] A. Wineman, Nonlinear viscoelastic solids — a review, Math. Mech. Solids 14 (3) (2009) 300–366.
[44] D.K. Moon, et al., The effects of refreezing on the viscoelastic and tensile properties of ligaments, J. Biomech. 39 (6) (2006) 1153–1157.
[45] S.L. Woo, C.A. Orlando, J.F. Camp, W.H. Akeson, Effects of postmortem storage by freezing on ligament tensile behavior, J. Biomech. 19 (1986) 399–404.
[46] A. Viidik, L. Sandqvist, M. Mägi, Influence of postmortal storage on tensile strength characteristics and histology of rabbit ligaments, Acta Orthop. 36 (S79) (1965) 3–38.
[47] P. Clavert, J.F. Kempf, F. Bonnomet, P. Boutemy, L. Marcelin, J.L. Kahn, Effects of freezing thawing on the biomechanical properties of human tendons, Surg. Radiol. Anat. 23 (4) (2001) 259–262.
[48] P.H. Leitschuh, T.J. Doherty, D.C. Taylor, D.E. Brooks, J.B. Ryan, Effects of postmortem freezing on tensile failure properties of rabbit extensor digitorum longus muscle tendon complex, J. Orthop. Res. 14 (5) (1996) 830–833.
[49] C. Rubod, et al., Biomechanical properties of vaginal tissue: preliminary results, Int. Urogynecol. J. Pelvic Floor Dysfunct. 19 (6) (2008) 811–816.
[50] J.A. Weiss, J.C. Gardiner, Computational modeling of ligament mechanics, Crit. Rev. Biomed. Eng. 29 (4) (2001) 1–70.
[51] T.C. Doehring, M. Kahelin, I. Vesely, Direct measurement of nonuniform large deformations in soft tissues during uniaxial extension, J. Biomech. Eng. 131 (6) (2009) 061001.
[52] D.L. Butler, M.D. Kay, D.C. Stouffer, Comparison of material properties in fascicle-bone units from human patellar tendon and knee ligaments, J. Biomech. 19 (1986) 425–432.
[53] K.M. Quapp, J.A. Weiss, Material characterization of human medial collateral ligament, J. Biomech. Eng. 120 (1998) 757–763.
[54] P. Allard, P.S. Thirty, A. Bourgault, G. Drouin, Pressure dependence of the area micrometer method in evaluation of cruciate ligament cross-section, J. Orthop. Res. 1 (1979) 265–267.
[55] T.Q. Lee, S.L.-Y. Woo, A new method for determining cross-sectional shape and area of soft tissues, J. Biomech. Eng. 110 (1988) 110–114.
[56] F. Iaconis, R. Steindler, G. Marinozzi, Measurements of cross-sectional area of collagen structures (knee ligaments) by means of an optical method, J. Biomech. 20 (10) (1987) 1003–1010.
[57] J.O. DeLancey, Anatomic aspects of vaginal eversion after hysterectomy, Am. J. Obstet. Gynecol. 166 (1992) 1717–1724.
[58] P.A. Norton, Pelvic floor disorders: the role of fascia and ligaments, Clin. Obstet. Gynecol. 36 (4) (1993) 926–938.
[59] P.A. Moalli, et al., A rat model to study the structural properties of the vagina and its supportive tissues, Am. J. Obstet. Gynecol. 192 (1) (2005) 80–88.
[60] J.L. Lowder, K. Debes, D.K. Moon, N. Howden, S.D. Abramowitch, P.A. Moalli, Biomechanical adaptations of the rat vagina and supportive tissues in pregnancy to accommodate delivery, Obstet. Gynecol. 109 (1) (2007) 136–143.

[61] M.S. Damaser, C. Whitbeck, P. Chichester, R.M. Levin, Effect of vaginal distension on blood flow and hypoxia of urogenital organs of the female rat, J. Appl. Physiol. 98 (2005) 1884–1890.
[62] Y.H. Lin, G. Liu, F. Daneshgari, A mouse model of simulated birth trauma induced stress urinary incontinence, Neurourol. Urodyn. 27 (2008) 353–358.
[63] P.A. Moalli, et al., Hormones restore biomechanical properties of the vagina and supportive tissues after surgical menopause in young rats, Am. J. Obstet. Gynecol. 199 (2) (2008) 161.e1–161.e8.
[64] A. Feola, et al., Impact of pregnancy and vaginal delivery on the passive and active mechanics of the rat vagina, Ann. Biomed. Eng. 39 (1) (2011) 549–558.
[65] H. Gurel, S.A. Gurel, Pelvic relaxation and associated risk factors the results of logistic regression analysis, Acta Obstet. Gynecol. Scand. 78 (4) (1999) 290–293.
[66] S.L. Hendrix, S.A. Gürel, et al., Pelvic organ prolapse in the women's health initiative: gravity and gravidity, Am. J. Obstet. Gynecol. 186 (6) (2002) 1160–1166.
[67] E.S. Lukacz, J. Lawrence, R. Contreras, C.W. Nager, K.M. Luber, Parity, mode of delivery, and pelvic floor disorders, Obstet. Gynecol. 107 (6) (2006) 1253–1260.
[68] M. Alperin, et al., Collagen scaffold: a treatment for simulated maternal birth injury in the rat model, Am. J. Obstet. Gynecol. 202 (6) (2010) 589.e1–589.e8.
[69] M. Schimpf, P. Tulikangas, Evolution of the female pelvis and relationships to pelvic organ prolapse, Int. Urogynecol. J. Pelvic Floor Dysfunct. 16 (4) (2005) 315–320.
[70] L.N. Otto, et al., The rhesus macaque as an animal model for pelvic organ prolapse, Am. J. Obstet. Gynecol. 186 (3) (2002) 416–421.
[71] K.W. Coates, H.L. Galan, B.L. Shull, T.J. Kuehl, The squirrel monkey: an animal model of pelvic relaxation, Am. J. Obstet. Gynecol. 103 (1995) 31–40.
[72] A. Feola, et al., Parity negatively impacts vaginal mechanical properties and collagen structure in rhesus macaques, Am. J. Obstet. Gynecol. 203 (6) (2010) 595.e1–595.e8.
[73] P. Takacs, et al., Caldesmon expression is decreased in women with anterior vaginal wall prolapse: a pilot study, Int. Urogynecol. J. Pelvic Floor Dysfunct. 20 (8) (2009) 985–990.
[74] G.M. Northington, et al., Contractile response of human anterior vaginal muscularis in women with and without pelvic organ prolapse, Reprod. Sci. 18 (3) (2011) 296–303.
[75] M.K. Boreham, R.T. Miller, J.I. Schaffer, R.A. Word, Smooth muscle myosin heavy chain and caldesmon expression in the anterior vaginal wall of women with and without pelvic organ prolapse, Am. J. Obstet. Gynecol. 185 (2001) 944–952.
[76] M.K. Boreham, C.Y. Wai, R.T. Miller, J.I. Schaffer, R.A. Word, Morphometric analysis of smooth muscle in the anterior vaginal wall of women with pelvic organ prolapse, Am. J. Obstet. Gynecol. 187 (2002) 56–63.
[77] M.K. Boreham, C.Y. Wai, R.T. Miller, J.I. Schaffer, R.A. Word, Morphometric properties of the posterior vaginal wall in women with pelvic organ prolapse, Am. J. Obstet. Gynecol. 187 (2002) 1501–1509.
[78] B. Gabriel, et al., Uterosacral ligament in postmenopausal women with or without pelvic organ prolapse, Int. Urogynecol. J. Pelvic Floor Dysfunct. 16 (2005) 475–479.
[79] O. Ozdegirmenci, et al., Smooth muscle fraction of the round ligament in women with pelvic organ prolapse: a computer-based morphometric analysis, Int. Urogynecol. J. Pelvic Floor Dysfunct. 16 (2005) 39–43.
[80] M.S. Sacks, Biaxial mechanical evaluation of planar biological materials, J. Elast. 61 (2000) 1990246.
[81] M.S. Sacks, D.B. Smith, E.D. Hiester, A small angle light scattering device for planar connective tissue microstructural analysis, Ann. Biomed. Eng. 25 (4) (1997) 678–689.
[82] Y. Lanir, Structure-strength relations in mammalian tendon, Biophys. J. 24 (1978) 541–554.
[83] M.S. Sacks, W. Sun, Multiaxial mechanical behavior of biological materials, Annu. Rev. Biomed. Eng. 5 (2003) 251–284.
[84] J.T. Keyes, et al., Comparisons of planar and tubular biaxial tensile testing protocols of the same porcine coronary arteries, Ann. Biomed. Eng. 41 (7) (2013) 1579–1591.
[85] J. Kim, S. Baek, Circumferential variations of mechanical behavior of the porcine thoracic aorta during the inflation test, J. Biomech. 44 (10) (2011) 1941–1947.
[86] J.P. Vande Geest, M.S. Sacks, D.A. Vorp, The effects of aneurysm on the biaxial mechanical behavior of human abdominal aorta, J. Biomech. 39 (7) (2006) 1324–1334.

[87] W. Wan, H. Yanagisawa, R.L. Gleason Jr., Biomechanical and microstructural properties of common carotid arteries from fibulin-5 null mice, Ann. Biomed. Eng. 38 (12) (2010) 3605–3617.
[88] L. Hansen, W. Wan, R.L. Gleason, Microstructurally motivated constitutive modeling of mouse arteries cultured under altered axial stretch, J. Biomech. 131 (10) (2009) 101015.
[89] J.T. Keyes, et al., Adaptation of a planar microbiaxial optomechanical device for the tubular biaxial microstructural and macroscopic characterization of small vascular tissues, J. Biomech. Eng. 133 (7) (2011) 075001.
[90] Y. Cai, Revisiting old vaginal topics: conversion of the Mullerian vagina and origin of the 'sinus' vagina, Int. J. Dev. Biol. 53 (2009) 925–934.
[91] W.R. Becker, R. De Vita, Biaxial mechanical properties of swine uterosacral and cardinal ligaments, Biomech. Model. Mechanobiol. 14 (3) (2015) 549–560.

CHAPTER SIX

Pelvic Floor Biomechanics From Animal Models

S.S. Patnaik*, A. Borazjani†, B. Brazile*, B. Weed*, D. Christiansen*, P.L. Ryan*, C.H. van der Vaart‡, M. Damaser†,§,¶, J. Liao*

*Mississippi State University, Starkville, MS, United States of America
†Lerner Research Institute, Cleveland Clinic, Cleveland, OH, United States of America
‡University Medical Center Utrecht, Utrecht, Netherlands
§Advanced Platform Technology Center, Cleveland VA Medical Center, Cleveland, OH, United States of America
¶Professor of Molecular Medicine, Cleveland Clinic Lerner College of Medicine, Case Western Reserve University, Cleveland, OH, United States of America

CURRENT ANIMAL MODELS AND THEIR APPLICATIONS

Erect posture is one of the significant landmark developments in the evolutionary history of human beings, which enabled people to walk upright, but also makes childbirth difficult [1]. Pelvic floor anatomical structures of humans, including pelvic floor muscles and connective tissues, have evolved to support the pelvic organs [2] and play a vital role in the childbirth process [3]. Due to the adaptation of erect posture by our ancestors, there was a considerable reduction in the distance between the hip joint and the sacroiliac joint, leading to a shortening of anteroposterior diameter, and widening of the transverse diameter at the pelvic inlet [4]. As a result, the overall sacral structure moved toward the pubic symphysis in humans. This shift in anatomy brings considerable change in the birthing process [5] compared to quadruped animals. Humans have a larger fetus with narrow pelvic cavity, whereas quadrupeds have a smaller fetus and a wider pelvic cavity [3], making humans vulnerable to certain pathologies (eg, obstetric fistula) that are relatively uncommon in quadrupeds. As a result, many scientists believed that quadrupeds were not prone to pelvic floor disorders (PFDs). However, vaginal prolapse or pelvic organ prolapse (POP) is one condition that is innate to both quadrupeds and humans.

PFDs, including POP, result from the loss or damage of structures that support the pelvic organs (ie, rectum, bowel, bladder, etc.) with vaginal wall prolapse (anterior and posterior) being the most common presentation [6]. POP is multifactorial in origin but most commonly follows childbirth [7]. POP is sufficiently common among women, where it has been estimated that almost 13% of women will undergo some form of surgery for POP in their lifetime [8]. Female pelvic musculature can stretch up to three times its size to accommodate the delivery process [9], a remarkable feature of the female body. However, certain risk factors, including genetic and evolutionary predispositions, can result in damage to pelvic floor support structures during childbirth [3,10]. This leads to further weakening of the levator ani muscles and other connective tissue structures which not only control mechanical function, but also help support neurological and anatomical functions [11].

Animal models have been instrumental in furthering our understanding of the mechanisms of pathophysiology of PFDs. It is likely that the risk for POP in humans is increased due to humans' uniquely upright posture; nonetheless, many quadrupeds including sheep, cows, pigs, dogs, deer, and even horses also develop POP [12]. Moreover, the characteristics and risk factors for POP in animals are remarkably similar to those in humans [12]. This finding points toward a molecular-based mechanism of disease in contrast to the current thinking of strictly anatomical or biomechanical etiologies, since development of POP in quadruped animals is not solely dependent on the specific action of gravity on the pelvic floor.

SMALL ANIMAL MODELS

Small animals, including rats, mice, and rabbits, have advantages of ease of handling, short lifespan, and relatively low costs compared to larger animals. These advantages allow for high sample sizes to be attained and enable investigations with population-based outcomes.

Rodent Models

Rat and mouse models have been used for biocompatibility testing (ie, preclinical evaluation of mesh or other biomaterials) as well as in other POP-related research. These models are advantageous due to their availability, cost effectiveness, varied genetics, including knock-out and knock-in strains, as well as for aging-related studies. The disadvantages of these models are their ability to quickly repair an injury, their differences in pelvic floor support structure (horizontally oriented as compared to humans), muscles located in a dorsal orientation to control the tail, and the lower fetal head-to-birth canal size with several small pups per litter [13]. Nonetheless, rodents are perhaps the most reliable and feasible models for investigating molecular mechanisms of disease due to the relative ease with which variables, including genetics, can be controlled [12]. Rodents are therefore the most commonly used animal models for studying POP [12]. In addition, rodents have been extensively used to study urinary incontinence and physiological changes in the pelvic floor [14].

The pelvic musculature of rodents is similar to that of humans in anatomy but significantly different in function, since pelvic skeletal muscles (eg, levator ani) in rodents are not involved in pelvic support. The muscles in rodents homologous to the levator ani in humans are the pubocaudal, iliocaudal, and coccygeal muscles. Whereas in humans these muscles work to support the pelvic organs, in rodents their primary function is to move the tail. They attach to the medial surface of the bony pelvis and are connected via tendons to vertebrae in the tail [15]. In male mice, the levator ani muscles are also thought to be involved in sexual function by increasing penile bulb pressure [16]. In female rats, it has been suggested that the levator ani involutes during development [17]. In one study, this involution was prevented by the administration of testosterone [18]. Although the authors were unable to visualize these muscles with the naked eye, they were able to confirm their existence [19].

Biomechanical findings suggest that the primary structures of pelvic organ support in rodents are the connective tissue attachments to the pelvic side wall and pelvic muscles [20,21]. Given these findings, and that the primary function of the pelvic floor muscles in rodents is tail movement, rodents provide a unique animal model that allows for the targeted study of the pelvic connective tissue support structures [21]. Since the muscular components of pelvic anatomy are not geared for supporting the pelvic organs, it can be deduced that connective tissue structures play a more prominent role.

Rat Models

Bremer et al. found that anatomy of the female rat pelvic musculature is similar to the human pelvic musculature with clearly identifiable levator ani muscles [22]. The primary differences found between rat and human pelvic musculature included the point of origin of the iliocaudalis and coccygeus muscles and the point of insertion of the pubocaudalis and iliocaudalis muscles. The iliocaudalis muscle originates from the ilium, and the analogous muscle in women (iliococcygeus) originates from the arcus tendineus levator ani. The coccygeus muscle in rats originates from the pubic bone, but in women, the coccygeus muscle originates from the ischial spine and sacrospinous ligament. The pubocaudalis and iliocaudalis muscles in rats insert into the vertebrae of the tail, and the analogous muscles in women (pubococcygeus and iliococcygeus) connect to the surrounding fascia, which ultimately attach to the coccyx. Additionally, they found that innervation of the pelvic musculature is similar in rats and humans [22].

Although some similarities in the muscular component of pelvic anatomy have been noted between rodents and humans, there are greater similarities in the connective tissue components of pelvic support. Moalli et al. found analogous connective tissue support structures to humans in female rats [20]. Specifically, they found uterosacral attachments between the upper vagina and lower spine (level I), paravaginal attachments from the vagina to the lateral bony pelvic (level II), and attachments between the distal vagina and ischiopubic rami (level III). Upon histological evaluation, a densely organized condensation of the paravaginal attachment was observed analogous to the arcus tendineus fascia pelvis in humans [20].

For ethical reasons, studies investigating pelvic floor biomechanics and connective tissue components during pregnancy are difficult to perform in humans. Nonetheless, animal models can provide valuable insight. Daucher et al. investigated the effects of pregnancy using Long Evans rats grouped according to gestational age and postpartum period [23]. They performed quantitative histological analyses of collagen fiber area fractions and smooth muscle characterization (synthetic, intermediate, or contractile phenotype). They found that the area of collagen fibers decreases during pregnancy and returns to normal levels in the late postpartum period (3 weeks after delivery). They also found that smooth muscle cells shift from a contractile to a synthetic phenotype during pregnancy and shift back to a contractile phenotype in the late postpartum period [23]. The reports of collagen density from this study are similar to one of the few publications, by Manabe and Yoshida, investigating tissue changes in partial-thickness vaginal tissue biopsies from women in three different groups: nonpregnant, pregnant, and immediately following vaginal delivery [24]. They found that during pregnancy and immediately after delivery, collagen fibers are less dense than in the nonpregnant state [24], similar to the results from Daucher et al. in rats. These results can explain changes in biomechanical properties of vaginal tissues across the peripartum period, such as decreased stiffness or increased laxity during pregnancy, and may help us understand pathophysiology of POP and other PFDs.

Lowder et al. conducted a study using Long Evans rats grouped into virgin, mid- and late pregnant, vaginal delivery (immediate and 4-week postpartum), and C-section (immediate and 4-week postpartum) [25]. They found that the biomechanical properties of the vagina and its supportive tissues are dynamic during the peripartum period, confirming the study by Daucher et al. [23]. Pregnancy resulted in decreased stiffness and decreased failure load in tissues, although the route of delivery (C-section or vaginal) was not associated with the changes in biomechanical properties. They also found that after delivery (C-section or vaginal), the tissues are more distensible than in the virgin or pregnant state. Most importantly, they found that the biomechanical properties recover to the nonpregnant state by 4 weeks postpartum [25].

The studies reviewed above describe an adaptation in rats which allows for non-traumatic delivery of offspring in a manner that accommodates favorable neonatal and maternal outcomes [23,25]. The study by Lowder et al. was novel in that it illustrated an adaptation in biomechanical properties of the pelvic tissues that occurs in an animal that does not develop POP and generally does not have difficult or traumatic births. As discussed by the authors, it may be the case that in some women, adaptations of the pelvic tissues to pregnancy and parturition may be deficient, incomplete, or exceeded [25]. This may help account for both the high incidence of obstetric trauma and development of POP in parous women.

Mouse Models

Mice are the only rodents that have been documented to spontaneously develop POP [12]. Four strains of mice are known to develop POP, but only two of these strains are commonly utilized in pelvic floor research. Mice lacking uterosacral ligaments do not develop POP [26].Therefore, it is likely that POP in mice results from weakening of the vaginal and paravaginal connective tissue components of pelvic support. Thus, the commonly utilized mouse models of POP both involve knockouts (KOs) of genes important for connective tissue homeostasis.

The two most commonly used mice for the POP research are the *Fbln5* KO and the *Loxl1* KO mouse models [12]. *Fbln5* KO mice develop prolapse early in life (+12 weeks), and the majority of animals will develop POP even if nulliparous [27]. It has been shown that a conserved motif on the *Fbln5* gene regulates expression of matrix metalloproteases (MMPs) [28]. Interestingly, when this motif is disrupted, mice display altered MMP activity, suggesting that the *Fbln5* gene regulates pelvic connective tissues both by regulation of elastic fiber organization and by MMP inhibition [28]. Mice with deficiency in both *Fbln5* and Mmp9 (*Fbln5* and Mmp9 double KO) have significantly reduced POP development, indicating that deficiency in Mmp9 could possibly prevent POP in up to 60% of *Fbln5* KO mice. Although the *Fbln5* model has been useful for studying the pathophysiology of POP, a major limitation in this model is that POP is not reliant on parity, which is the most important risk factor in humans [27].

Unlike *Fbln5* KO mice, nulliparous *Loxl1* KO mice rarely develop POP. POP in these animals is highly correlated to parity [29]. In addition to developing POP primarily after vaginal delivery, *Loxl1* KO mice also display measures correlated to stress urinary incontinence (SUI) [29]. Similarly, women with POP often present with SUI [30]. Elastin von Gieson staining of pelvic tissues of *Loxl1* KO mice with POP shows increased elastic fiber clusters, which are suggested to be associated with abnormal elastic fiber repair. As in humans, the bladder is often the organ contained within the prolapse in mice and significant changes are apparent in the vagina (Fig. 1) [29].

POP in mice can be quantified using the validated mouse organ prolapse quantification system (Fig. 2), which has been used for both *Fbln5* KO mice and *Loxl1* KO mice [29,31,32]. Mice and humans have slightly different presentations of POP. In mice, POP manifests itself as an external perineal bulge, whereas in humans, POP is observed as a bulge or descent of one of the vaginal compartments.

It is important to note that POP in mice is more analogous to cystocele rather than perineocele in humans. Perineoceles, which are relatively rare, occur as a result of an isolated defect often due to surgery (eg, abdominoperineal resection or pelvic exenteration). Moreover, patients with perineoceles typically do not present with POP and the vagina is structurally normal [33]. However, cystoceles are more common and occur as a result of loss of structural support to the vagina.

Rabbit Models

Rabbit models have been used extensively for testing abdominal, hernia, and gynecological meshes [34,35]. The advantages of rabbit models include their ease of handling, their relative low cost, and their larger size, which allows for more surgical flexibility and

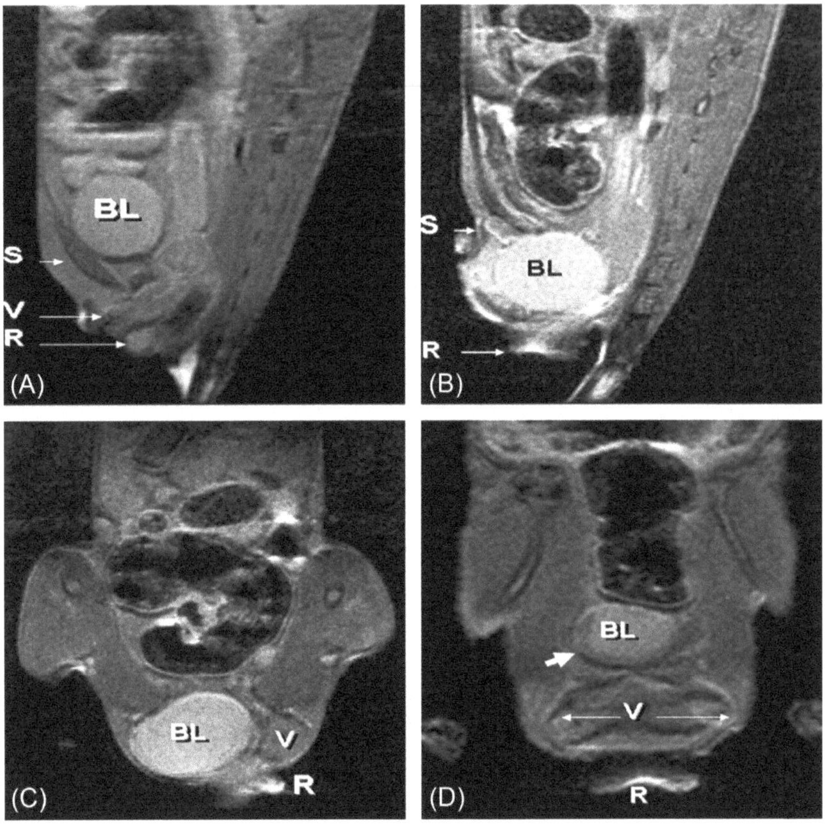

Figure 1 *Loxl1* KO mouse without POP. Bladder (BL), vagina (V), rectum (R), and pubic symphysis (S) are labeled. (A) Sagittal image of LOXL KO mouse without POP. (B-D) Sagittal and coronal images of mice with POP. In a mouse (A), the bladder and rectum are not prolapsed, and the vagina is easily visualized. In mice with POP (B-D), the bladder and rectum are both prolapsed. (A) The bladder is below the pubic symphysis. (D) The vagina is widened in appearance. In the mouse in C, the prolapsed bladder resulted in a displacement of the vagina laterally. The vaginal volume as calculated using 3D reconstruction of the images was found to be 0.013 mL mice without POP and 0.161 mL in mice with POP. *Courtesy: U.J. Lee, et al., Lower urogenital tract anatomical and functional phenotype in lysyl oxidase like-1 knockout mice resembles female pelvic floor dysfunction in humans, Am. J. Physiol. Renal. Physiol. 295 (2) (2008) F545–F555.*

higher precision imaging than in rodent models. The disadvantages of rabbit models include location variation in the reproductive tract, a different collagen metabolism system, a very different anatomy compared to humans, the presence of two different part of vaginas (internal and external portions), the poorly understood immune system, the presence of external connective tissue layer in the anterior vaginal wall, and the high proteolytic activity of the reproductive tract [35,36]. Moreover, spontaneous vaginal prolapse in rabbits is rare, but has been reported by Van Herck et al. as isolated incidents [37].

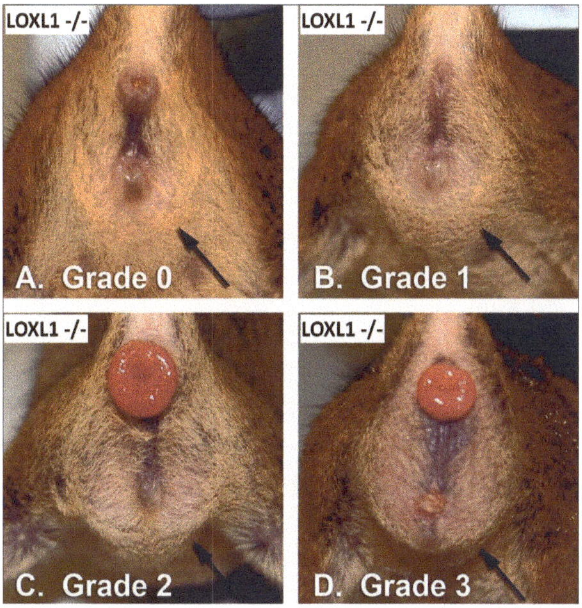

Figure 2 (A) Grade 0 — no prolapse and normal perineum. (B) Grade 1 — mild prolapse and perineal bulge. (C) Grade 2 — moderate prolapse and perineal bulge. (D) Grade 3 — severe prolapse evident by large perineal bulge. Mice in C and D also present with rectal prolapse, which is commonly observed in *Loxl1* KO mice. *Courtesy from: U.J. Lee, et al., Lower urogenital tract anatomical and functional phenotype in lysyl oxidase like-1 knockout mice resembles female pelvic floor dysfunction in humans, Am. J. Physiol. Renal. Physiol. 295 (2) (2008) F545–F555.*

LARGE ANIMAL MODELS

Most explant studies from the above-mentioned small animal models are relatively small in size, which neither replicates the surgical conditions nor provides sufficient tissue to be used for further biomechanical testing. Therefore, larger animal models are beneficial for surgical and imaging purposes, enable testing of a wide variety of meshes, and facilitate the availability of substantially larger explanted specimens for biomechanical testing. Large animals, such as sheep, may be beneficial for studying biomechanics and effects of hormones on the pelvic floor. However, these models are not well-suited for performing large trials, and it is difficult to control for genetic factors.

Humans are the only strictly bipedal mammals, and the species closest to humans are nonhuman primates. In addition to some differences in pelvic anatomy, nonhuman primates differ from humans in that they spend a significant portion of their time in a quadrupedal posture [38]. The functional differences in pelvic anatomy are particularly apparent during parturition. Nonhuman primate neonates emerge from the vagina in mentum anterior position (face toward the mother), whereas in human neonates the optimal position is occiput anterior (face away from the mother) [38–40].

Primate models also have hormonal systems similar to humans and the fetomaternal proportion is most similar to humans. Due to their larger body size, primate models are also convenient for transvaginal surgery [40,41]. Nonetheless, no animal, even nonhuman primates give birth in as traumatic a fashion as humans [42]. Primates are also more costly than other animal models (cost of individual primate is $7000–9000 or more) and have more legal and ethical issues associated. Moreover, as discussed earlier, primates are not firmly bipedal and spend the majority of their time in a quadruped orientation [12,40]. Other disadvantages of primate models include a shortened reproductive tract and the microbial environment (microbiome) of the vaginal tract is very different from humans [40]. However, current studies have primarily focused on nonprolapsed animals, likely due to the fact that it takes many years for these animals to develop POP [12].

With the given issues and compromises between numerous animal models, we believe that sheep may represent an excellent candidate for investigating the etiology behind failure of gynecological mesh implants, as well as developing and assessing novel biomaterials and reconstructive strategies. Recently, scientists have projected sheep as an ideal large animal model for biomedical research due to their ease of handling, economical advantage, similar fetal size, comparative anatomy, and being best suited for surgical procedures [13,43–45].

Chronologically, sheep have been utilized for agricultural research, nutritional studies, wool production, pharmacological research, surgical research (orthopedic, catheterization, etc.) [46], circulatory studies (heart transplantation, hematology, vascular graft, etc.) [47–50], immunological (immunoassay, etc.), and pathological studies (diabetes, asthma, etc.) [51–54]. Pregnant sheep have also been used to investigate fetal development, pregnancy, and other reproductive studies [55–65]. Sheep have been used as a large animal model for studying fetal implantation [66], fetoplacental physiology [63], invasive or noninvasive procedures at the time of pregnancy [63], the role of the immune system in early stages of pregnancy [61], placentation related studies [63,67], as well as for studying potential stem cell- and biomaterial-based therapeutics [68–71].

Sheep have similar fetal size and relatively comparable anatomy to humans, although they are quadrupeds [13,43–45]. In addition, they are more affordable than nonhuman primates, giving them an economical advantage [13]. Notably, sheep exhibit spontaneous vaginal prolapse and have similar vaginal length, diameter, and microstructural characteristics as humans [12,35,72]. Fetomaternal disproportions in sheep lead to dystocia (difficulty in giving birth) [73] and vaginal prolapse soon after delivery [74]. Additionally, due to their relatively large size, sheep can be used for preclinical evaluation of gynecological meshes [35,75–78]. Pregnant sheep are also used for fetal development, pregnancy, and other reproductive studies [55,63]. It is hence essential to explore anatomical similarities and understand biomechanical characteristics of sheep pelvic tissue/organs and their human pelvic analogues.

Bipedal Pelvic Mechanics Versus Quadruped Pelvic Mechanics

In humans, along with evolutionary changes from plantigrade to erect posture, increased intra-abdominal pressure resulted in changes to pelvic floor anatomy compared to quadruped animals [79,80]. The pelvic floor has evolved from quadruped ambulation to modern bipedalism, and the muscles that once supported the tail are supporting the pelvic viscera in humans [81]. The muscles and ligaments of the pelvic floor in humans are connected to the perimeter of the bony pelvic, with openings for the urethra, vagina, and rectum [80]. The human pelvis is a slightly concave bowl-shaped structure with the bottom portion shaped as a funnel.

Moreover, proportions of bones in humans are different compared to quadrupeds. The lumbosacral curve of the spine in humans pushes the pelvic viscera forward and the resultant pressure from the pelvic organs is absorbed by the pubic bones and abdominal wall [80]. Any residual pressure is further concentrated on the sacral portion of the spine and the pelvic floor organs (levator ani muscle, urethra, vagina, and anal canal) [79]. The urethral opening is the smallest and hence most of the pathological presentation of pelvic organ descent is through the other two openings — vagina and anus. The vaginal opening (or genital hiatus), being the largest opening of the pelvic floor, is the most critical anatomical factor in the onset of POP in humans.

In the supine position, the human vagina is aligned with the long axis of the human body; however, in an upright position the lower third of the vagina remains constant, whereas the rest of the vagina and uterus become almost perpendicular to the long axis of the body [80]. This change in position is due to the structural supports of the vagina, that is anterior support, posterior support, apical support, and lateral support [80]. Anterior support of the vagina is provided by muscular and fascial urethral supports (pubourethral and urethropelvic ligaments), whereas uterosacral ligaments, along with the cardinal ligaments, provide posterior support to the vagina. The arcus tendineus, along with uterosacral ligament, constitute the apical support of the vagina, and the arcus tendineus along with cardinal ligament constitutes the lateral support.

The weight of the pelvic floor organs exerts a downward force on the pelvic outlets owing to gravity, whereas the pelvic floor muscles along with ligaments (including fascia) counteract this feat [82]. Hence, from a structural and functional point of view, pelvic floor muscles are well developed in humans and nonhuman primate models, compared to quadrupeds. Furthermore, the abdominal contents of quadrupeds are parallel to the ground and therefore, human pelvic organs are under greater influence of gravity.

Even though the pelvic floor musculature, pelvic loading conditions, and birthing process of humans are more similar to nonhuman primates than sheep [83], the anatomical organization of the vagina and other supporting structures in sheep (quadruped position) are similar to humans [12,35]. Like humans, sheep have prolonged labors, develop pelvic relaxation [84], and exhibit spontaneous prolapse [74] in conjunction with pregnancy. Nonetheless, due to the natural position of quadrupeds, the intra-abdominal movement of the pelvic viscera is greatest toward the abdominal cavity, whereas in humans the movement is more toward the vaginal orifice [82]. The human pelvic diaphragm, consisting of levator ani and coccygeus muscle, bears the loads and pressures from the pelvic cavity [85]. With the absence of the sacrospinous ligament in quadrupeds, the coccygeus muscle is responsible for the tail movements [85]. However, tail docking in sheep, which leads to damage to this coccygeal support of the pelvic girdle, has been associated with development of vaginal prolapse [86]. Moreover, owing to the four-footed position of quadrupeds, thickening of pelvic connective tissue in relation to the vascular pedicle is not observed in sheep [79].

Attachment of the vagina to the pelvic girdle in normal sheep prevents eversion of the vagina, while in the case of vaginal prolapse, the sacroiliac and sacrolumbar joints are usually over stretched [87]. Stretching of the sacroilliac joints (which joins pelvic girdle to the spine) and pubic symphysis (which joins the bones of the pelvic floor canal) allows the birth canal to become wider for successful passage of the fetus during childbirth [88]. As mentioned previously, vaginal prolapse is most often observed in pregnant sheep due to reduction in expanse of the pelvic cavity. With progression of pregnancy in sheep, the length and width of the uterus increases, along with an increased fibroblast and smooth muscle cell content, whereas there is marked reduction in collagen and fibrocytes concentrations [89]. Similar to humans, relaxation of pelvic ligaments, as well as the synovial membrane and fascia, is observed in sheep, as is the resulting increased mobility of pelvic organs [84,88,90]. During delivery or parturition, the pubic symphysis or ischiopubic joint in sheep do not loosen [88], unlike in rodent and murine models [91] and humans [92].

In humans, the bladder is a widely distal pelvic organ and moves proximally during pregnancy, whereas in quadrupeds the bladder is not distal but ascends proximally with progression of pregnancy [93]. Sheep bladders are highly extensible and exert greater pressure on the vagina with bladder filling. This pressure is further compounded in the presence of a growing fetus (or an enlarged uterus). In normal conditions, sheep have negative abdominal pressure (about −5 cm of water) at the pelvic inlet due to the weight of pelvic viscera falling forward and its dependence on abdominal wall for support [94]. Due to overeating, there is an increasing respiratory requirement, and sheep usually lie on the ground or a sloping surface, which relaxes the pressure on the abdomen (about +30 cm of water). Those behaviors, along with other risk factors (nutritional, hormonal, and dietary), can greatly affect the distribution of intra-abdominal pressure. It is well known that the balance of intra-abdominal pressure in the pelvic cavity is critical in both animals and humans. Alteration of intra-abdominal pressure, along with softening of pelvic support in pregnancy (in combination with other confounding risk factors), increases the chance of prolapse in sheep [79].

Arrangement of Pelvic Floor Organs in Sheep

The embryological origins of the human vagina can be traced to the fusion of Müllerian ducts and extension of urogenital sinus [95,96]. Similarly, the sheep vagina derives its dorso-lateral segments from the Müllerian ducts and the lower vaginal segment is formed by a superior projection of the urogenital sinus [97]. The reproductive tract of sheep starts with the vaginal orifice at the posterior/caudal side and ends with uterine horns at the anterior/cranial side (Fig. 3A). The structure of the vaginal tract is also similar to humans, and the narrowest portion is in the vaginal orifice, while the widest section is also located in the proximal/cranial vagina [72].

Moving from the vaginal orifice toward the cranial end of the sheep reproductive tract, the vulva is the first anatomical feature followed by the vagina, cervix, uterus, uterine horns, oviduct, and finally the ovaries [98]. Cranially toward the uterus, the reproductive tract takes almost an *S*-shaped curve and dives down to the cervix. As the uterus is bipartite or two-horned, the tract is bifurcated in shape, and at the end of each horn the oviducts connect to the ovaries [99]. In the region closest to the vaginal

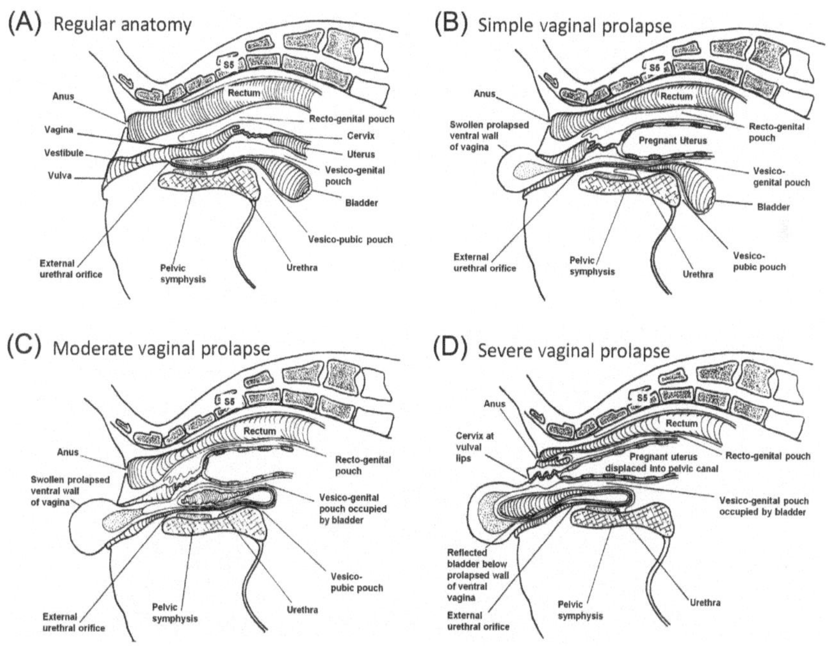

Figure 3 Sheep vaginal prolapse. The figure shows regular sheep anatomy (A), followed by simple vaginal prolapse (B), moderate vaginal prolapse (C), and severe vaginal prolapse (D). *Modified from: J.E. Cox, Surgery of the Reproductive Tract in Large Animals, Liverpool University Press, Liverpool, 1982.*

orifice, the tail section is ventral to the anus followed by the vulva. Similarly, in the middle section of the pelvic cavity, the rectum is ventral to the recto-genital pouch, vaginal canal, urethra (including urethral orifice), and pelvic symphysis. Finally, in the cranial region, bladder and ovaries (including oviduct) are ventral to the vesico-genital pouch, body of uterus, recto-genital pouch, and rectum.

Supporting Structures in the Sheep Pelvic Floor

Sheep have three levels of pelvic floor support, similar to humans [2,13]. The perineal and pelvic region of female sheep are similar to that of humans, with the exception of dorsal and accessory (short and long) sacroiliac ligaments [100]. Similar to humans, the vagina and the uterus are held together by the endopelvic fascia and levator ani muscles. The broad ligament is a thin membranous sheet of connective tissue that covers the pelvic organs (cervix, the body of uterus, and part of the bladder) and holds the ovaries in place [79,101]. Any damage to the fascia, broad ligaments, or muscles will lead to either vaginal or uterine prolapse. The urinary bladder is commonly involved in sheep vaginal prolapse, which may also involve the uterine horns, uterus, small intestine, etc. [102]. Since sheep pelvic floor tissues, like humans, are composed of collagen, elastin, and smooth muscle cells, vaginal prolapse is expected to be similar to that in women.

Sheep Vaginal Prolapse: Characteristics and Risk Factors

Sheep depend on their pelvic viscera to support increased intra-abdominal pressures [103]. With the progression of pregnancy, the pelvic floor tissues of sheep become weaker and the pelvic organs become more mobile inside the pelvic cavity [83,104]. Under the influence of reproductive hormones, there is a marked reduction in collagen content and increase in vaginal wall compliance of pregnant sheep [83,90,104]. Displacement of the pelvic organs and a marked increase in compliance of the vaginal wall has been observed in female sheep during the estrus cycle and pregnancy [84,90,105]. With increasing size of the fetus in the pelvic cavity, the intra-abdominal pressure of pregnant sheep tends to become higher than normal, which further leads to straining of the vaginal wall. A chain of such physiological events, combined with other potential risk factors, can lead to the onset of prolapse of the sheep vaginal wall prior to delivery [106].

Vaginal prolapse in female sheep is characterized by a red ball-shaped protrusion from the birth canal consisting of 8–20 cm of everted vaginal epithelium. Similar to humans, vaginal wall eversion takes place due to the relaxation of the pelvic tissues and a marked reduction in collagen and smooth muscle cells [74,107]. Based on severity of the condition, sheep vaginal prolapse can be categorized as (1) simple, (2) moderate, or (3) severe (Fig. 3) [98]. Simple vaginal prolapse consists of a swollen ventral vaginal wall, while other pelvic organs are not displaced (Fig. 3B). In moderate vaginal prolapse, the bladder, cervix, or intestine can be involved in the vaginal prolapse and edema of the vulva is prominent (Fig. 3C). In severe vaginal prolapse, the uterus and cervix are caudal to their original positions, and protrusion of the cervix is clearly visible (Fig. 3D). The bladder and intestines can also be part of this condition, which can also be complicated by trauma, infection, or necrosis [83]. Uterine prolapse is more sporadic (0.2–0.4%) and less common in sheep compared to vaginal prolapse (5–20%) [86].

As with humans, the risk of vaginal prolapse in female sheep increases with age and parity. Female sheep bearing twins or triplets are at 5 and 12 times increased risk of POP, respectively [108]. Further, female sheep with a history of vaginal prolapse are more likely to have a recurrent prolapse and 10 times more likely to have a cesarean operation to correct the dystocia during subsequent lambing [83].

Fat female sheep with high body condition score (BCS) of 4–5 are highly likely to exhibit vaginal prolapse, and farmers are advised to keep the female sheep BCS at 2 out of 5 [109]. Elevated BCS is equivalent to human obesity; it increases abdominal pressure, which is consequently distributed to the pelvic floor. Other factors, such as less exercise or restricted movement, weight gain during the latter half of pregnancy, increased intake of roughage, excessive straining, vitamin D deficiency, hereditary attributes, and poor quality feed have also been investigated as possible risk factors for sheep vaginal prolapse [108,110].

Current Urogynecological Applications of the Sheep Model

Sheep have been utilized as a large animal model for studying POP and preclinical evaluation of gynecological meshes [12,13]. Rezapour et al. [77] reported one of the first implantation studies comparing the tension free vaginal tape procedure using Prolene (polypropylene) mesh versus a shorter Ethisorb coated polypropylene mesh in sheep. They found the meshes to be similar in biomechanical strength and observed integration with paravaginal tissues, without collagen deposition [77]. De Tayrac et al. compared the effect of collagen coating on polypropylene meshes in sheep, and found that collagen could possibly help delay the mesh erosion process [75]. In a similar study by Krause and Goh [35], Monarc mesh (American Medical Systems) and Gynemesh (Ethicon) materials were implanted in rabbit as well as sheep via vaginal and abdominal routes. Sheep anatomical features were found to be closer to human and facilitated surgical advantages [35]. Alcalay et al. [111] performed a comparative mesh pull-out study of three methods of mesh attachment using sheep, and found that the EndoFast Reliant™ system using fasteners was stronger than trocar-based methods; however, this study was performed using thigh fascia instead of vaginal or paravaginal tissues.

Following De Tayrac et al. [75], Deprest and coworkers have used sheep as large animal models for preclinical evaluation of variety of gynecological meshes (coated, uncoated, or cross-linked) [76,78,112]. They have shown that graft-related complications in humans can be replicated in a sheep model, and this larger model is advantageous for postsurgical ex vivo biomechanical evaluation of mesh-tissue grafts [76]. Tissue biomechanics plays a significant role in POP development and therefore has been characterized in sheep pelvic organs and support structures.

CONCLUSION

In this chapter, we have reviewed the various animal models of POP and provided an indepth analysis of sheep as a robust large animal model for urogynecological research. Each animal model has its own set of advantages and disadvantages. The ease of handling, short lifespan, and relatively low costs are major advantages of rodent models, which have been used extensively to investigate connective tissue physiology and pathophysiology as it relates to POP. Sheep have supporting structures for pelvic organs, as well as structural and mechanical properties more similar to humans. Many of the risk factors for sheep prolapse have close analogues in humans, including high fetal weight, obesity, dystocia, parity, and family history of prolapse. In addition, large animal models, such as sheep, are likely more appropriate for evaluation of novel therapeutic strategies for treatment. Future research using sheep and other animal models will illuminate the pathophysiology of POP, as well as provide essential information on pelvic floor biomechanics and treatment efficacy.

REFERENCES

[1] T. Hogervorst, H.W. Bouma, J. de Vos, Evolution of the hip and pelvis, Acta Orthop. Suppl. 80 (336) (2009) 1–39.
[2] J.O. DeLancey, The anatomy of the pelvic floor, Curr. Opin. Obstet. Gynecol. 6 (4) (1994) 313–316.
[3] M.M. Abitbol, Growth of the fetus in the abdominal cavity, Am. J. Phys. Anthropol. 91 (3) (1993) 367–378.
[4] M.M. Abitbol, Ontogeny and evolution of pelvic diameters in anthropoid primates and in Australopithecus afarensis (AL 288-1), Am. J. Phys. Anthropol. 85 (2) (1991) 135–148.
[5] K.R. Rosenberg, The evolution of modern human childbirth, Am. J. Phys. Anthropol. 35 (S15) (1992) 89–124.
[6] J.E. Jelovsek, C. Maher, M.D. Barber, Pelvic organ prolapse, Lancet 369 (9566) (2007) 1027–1038.
[7] S.L. Hendrix, et al., Pelvic organ prolapse in the Women's Health Initiative: gravity and gravidity, Am. J. Obstet. Gynecol. 186 (6) (2002) 1160–1166.
[8] J.M. Wu, et al., Lifetime risk of stress urinary incontinence or pelvic organ prolapse surgery, Obstet. Gynecol. 123 (6) (2014) 1201–1206.
[9] J.A. Ashton-Miller, J.O.L. DeLancey, Functional anatomy of the female pelvic floor, Ann. N. Y. Acad. Sci. 1101 (1) (2007) 266–296.
[10] C.M. Kim, et al., Risk factors for pelvic organ prolapse, Int. J. Gynecol. Obstet. 98 (3) (2007) 248–251.
[11] J.O. DeLancey, R.A. Starr, Histology of the connection between the vagina and levator ani muscles. Implications for urinary tract function, J. Reprod. Med. 35 (8) (1990) 765–771.
[12] B.M. Couri, et al., Animal models of female pelvic organ prolapse: lessons learned, Expert Rev. Obstet. Gynecol. 7 (3) (2012) 249–260.
[13] S.D. Abramowitch, et al., Tissue mechanics, animal models, and pelvic organ prolapse: a review, Eur. J. Obstet. Gynecol. Reprod. Biol. 144 (Suppl. 1) (2009) S146–S158.
[14] A. Hijaz, et al., Animal models of female stress urinary incontinence, J. Urol. 179 (6) (2008) 2103–2110.
[15] W. Pradidarcheep, et al., Anatomy and histology of the lower urinary tract, in: K.-E. Andersson, M.C. Michel (Eds.), Urinary Tract, Springer, Berlin, 2011, pp. 117–148.

[16] G.M. Holmes, B.D. Sachs, Physiology and mechanics of rat levator ani muscle: evidence for a sexual function, Physiol. Behav. 55 (2) (1994) 255–266.

[17] A. Poortmans, J. Wyndaele, M. levator ani in the rat: does it really lift the anus? Anat. Rec. 251 (1) (1998) 20–27.

[18] R. Cihak, E. Gutmann, V. Hanzlikova, Involution and hormone-induced persistence of the M. sphincter (levator) ani in female rats, J. Anat. 106 (Pt 1) (1970) 93.

[19] Y. Joubert, C. Tobin, Satellite cell proliferation and increase in the number of myonuclei induced by testosterone in the levator ani muscle of the adult female rat, Dev. Biol. 131 (2) (1989) 550–557.

[20] P.A. Moalli, et al., A rat model to study the structural properties of the vagina and its supportive tissues, Am. J. Obstet. Gynecol. 192 (1) (2005) 80–88.

[21] M. Alperin, et al., LOXL1 deficiency negatively impacts the biomechanical properties of the mouse vagina and supportive tissues, Int. Urogynecol. J. 19 (7) (2008) 977–986.

[22] R.E. Bremer, et al., Innervation of the levator ani and coccygeus muscles of the female rat, Anat. Rec. A: Discov. Mol. Cell. Evol. Biol. 275 (1) (2003) 1031–1041.

[23] J.A. Daucher, et al., Adaptations of the rat vagina in pregnancy to accommodate delivery, Obstet. Gynecol. 109 (1) (2007) 128–135.

[24] Y. Manabe, Y. Yoshida, Collagenolysis in human vaginal tissue during pregnancy and delivery: a light and electron microscopic study, Am. J. Obstet. Gynecol. 155 (5) (1986) 1060–1066.

[25] J.L. Lowder, et al., Biomechanical adaptations of the rat vagina and supportive tissues in pregnancy to accommodate delivery, Obstet. Gynecol. 109 (1) (2007) 136–143.

[26] K.A. Connell, et al., HOXA11 is critical for development and maintenance of uterosacral ligaments and deficient in pelvic prolapse, J. Clin. Invest. 118 (3) (2008) 1050.

[27] P.G. Drewes, et al., Pelvic organ prolapse in fibulin-5 knockout mice: pregnancy-induced changes in elastic fiber homeostasis in mouse vagina, Am. J. Pathol. 170 (2) (2007) 578–589.

[28] M. Budatha, et al., Extracellular matrix proteases contribute to progression of pelvic organ prolapse in mice and humans, J. Clin. Invest. 121 (5) (2011) 2048.

[29] U.J. Lee, et al., Lower urogenital tract anatomical and functional phenotype in lysyl oxidase like-1 knockout mice resembles female pelvic floor dysfunction in humans, Am. J. Physiol. Ren. Physiol. 295 (2) (2008) F545–F555.

[30] S. Bai, et al., Relationship between stress urinary incontinence and pelvic organ prolapse, Int. Urogynecol. J. 13 (4) (2002) 256–260.

[31] C.K. Wieslander, et al., Quantification of pelvic organ prolapse in mice: vaginal protease activity precedes increased MOPQ scores in fibulin 5 knockout mice, Biol. Reprod. 80 (3) (2009) 407–414.

[32] A.M. Gustilo-Ashby, et al., The impact of cesarean delivery on pelvic floor dysfunction in lysyl oxidase like-1 knockout mice, Female Pelvic Med. Reconstr. Surg. 16 (1) (2010) 21–30.

[33] K.S. Eilber, et al., Perineocele: symptom complex, description of anatomic defect, and surgical technique for repair, Urology 67 (2) (2006) 265–268.

[34] L.M. Pierce, et al., Biomechanical properties of synthetic and biologic graft materials following long-term implantation in the rabbit abdomen and vagina, Am. J. Obstet. Gynecol. 200 (5) (2009) 549.e1–549.e8.

[35] H. Krause, J. Goh, Sheep and rabbit genital tracts and abdominal wall as an implantation model for the study of surgical mesh, J. Obstet. Gynaecol. Res. 35 (2) (2009) 219–224.

[36] M. Jacques, et al., The normal microflora of the female rabbit's genital tract, Can. J. Vet. Res. 50 (2) (1986) 272–274.

[37] H. Van Herck, et al., Prolapsus vaginae in the IIIVO/JU rabbit, Lab. Anim. 23 (4) (1989) 333–336.

[38] K. Rosenberg, W. Trevathan, Birth, obstetrics and human evolution, BJOG 109 (11) (2002) 1199–1206.

[39] A. Feola, et al., Parity negatively impacts vaginal mechanical properties and collagen structure in rhesus macaques, Am. J. Obstet. Gynecol. 203 (6) (2010) 595.e1–595.e8.

[40] J.A. Mattson, et al., Evaluation of the aged female baboon as a model of pelvic organ prolapse and pelvic reconstructive surgery, Am. J. Obstet. Gynecol. 192 (5) (2005) 1395–1398.

[41] L.N. Otto, et al., The rhesus macaque as an animal model for pelvic organ prolapse, Am. J. Obstet. Gynecol. 186 (3) (2002) 416–421.

[42] W. Trevathan, Human Birth: An Evolutionary Perspective, Aldine De Gruyter, New York, 1987.
[43] W.A. Kues, H. Niemann, The contribution of farm animals to human health, Trends Biotechnol. 22 (6) (2004) 286–294.
[44] A. Mortell, S. Montedonico, P. Puri, Animal models in pediatric surgery, Pediatr. Surg. Int. 22 (2) (2006) 111–128.
[45] A. Bähr, E. Wolf, Domestic animal models for biomedical research, Reprod. Domest. Anim. 47 (s4) (2012) 59–71.
[46] Y.H. An, R.J. Freidman, Animal Models in Orthopaedic Research, CRC Press, Boca Raton, FL, 1998.
[47] P.K. Chow, R.T. Ng, B.E. Ogden, Using Animal Models in Biomedical Research: A Primer for the Investigator, World Scientific, Singapore, 2008.
[48] M. Ali, et al., The sheep as an animal model for heart valve research, Vascular 4 (4) (1996) 543–549.
[49] A. Kónya, et al., Animal models for atherosclerosis, restenosis, and endovascular aneurysm repair, in: P.M. Conn (Ed.), Sourcebook of Models for Biomedical Research, Springer, New York, 2008, pp. 369–384.
[50] A. Trollope, et al., Animal models of abdominal aortic aneurysm and their role in furthering management of human disease, Cardiovasc. Pathol. 20 (2) (2011) 114–123.
[51] K. Srinivasan, P. Ramarao, Animal models in type 2 diabetes research: an overview, Indian J. Med. Res. 125 (3) (2007) 451–472.
[52] J.-P.Y. Scheerlinck, et al., Biomedical applications of sheep models: from asthma to vaccines, Trends Biotechnol. 26 (5) (2008) 259–266.
[53] P.J. Psaltis, et al., An ovine model of toxic, nonischemic cardiomyopathy — assessment by cardiac magnetic resonance imaging, J. Card. Fail. 14 (9) (2008) 785–795.
[54] G. Matute-Bello, C.W. Frevert, T.R. Martin, Animal models of acute lung injury, Am. J. Physiol. Lung Cell. Mol. Physiol. 295 (3) (2008) L379.
[55] M. Dreyfus, et al., The pregnant ewe: an animal model for fetoscopic surgery, Eur. J. Obstet. Gynecol. Reprod. Biol. 71 (1) (1997) 91–94.
[56] B. Mandon-Pépin, et al., Expression profiles and chromosomal localization of genes controlling meiosis and follicular development in the sheep ovary, Biol. Reprod. 68 (3) (2003) 985–995.
[57] S. Fabre, et al., Regulation of ovulation rate in mammals: contribution of sheep genetic models, Reprod. Biol. Endocrinol. 4 (2006) 20.
[58] R.J. Scaramuzzi, et al., A review of the effects of supplementary nutrition in the ewe on the concentrations of reproductive and metabolic hormones and the mechanisms that regulate folliculogenesis and ovulation rate, Reprod. Nutr. Dev. 46 (4) (2006) 339–354.
[59] C. Viñoles, et al., The use of a 'first-wave' model to study the effect of nutrition on ovarian follicular dynamics and ovulation rate in the sheep, Reproduction 140 (6) (2010) 865–874.
[60] G.C. Liggins, Premature delivery of foetal lambs infused with glucocorticoids, J. Endocrinol. 45 (4) (1969) 515–523.
[61] O. Sandra, N. Mansouri-Attia, P. Chavatte-Palmer, Towards a better understanding of immunology of early pregnancy using alternative animal models: the contribution of ruminants, Adv. Neuroimmune Biol. 2 (1) (2011) 125–134.
[62] J.P. Renard, et al., Session 23: building new bridges between animal and human reproduction science, Hum. Reprod. 27 (Suppl. 2) (2012) ii35.
[63] O. Morel, et al., The use of ruminant models in biomedical perinatal research, Theriogenology 78 (8) (2012) 1763–1773.
[64] M. Fransolet, et al., Strategies for using the sheep ovarian cortex as a model in reproductive medicine, PLoS One 9 (3) (2014) e91073.
[65] R. Pearson, D. Mellor, Some physiological changes in pregnant sheep and goats before, during and after surgical insertion of uterine catheters, Res. Vet. Sci. 19 (1) (1975) 102–104.
[66] K.Y. Lee, F.J. DeMayo, Animal models of implantation, Reproduction 128 (6) (2004) 679–695.
[67] A.M. Carter, Animal models of human placentation — a review, Placenta 28 (Suppl. A) (2007) S41–S47.

[68] J. Harding, R. Roberts, O. Mirochnitchenko, Large animal models for stem cell therapy, Stem Cell Res. Ther. 4 (2) (2013) 23.
[69] L.F. Malaver-Ortega, et al., The state of the art for pluripotent stem cells derivation in domestic ungulates, Theriogenology 78 (8) (2012) 1749–1762.
[70] M. Casal, M. Haskins, Large animal models and gene therapy, Eur. J. Hum. Genet. 14 (3) (2005) 266–272.
[71] K. Abi-Nader, et al., Animal models for prenatal gene therapy: the sheep model, in: C. Coutelle, S.N. Waddington (Eds.), Prenatal Gene Therapy, Humana Press, New York, NY, 2012, pp. 219–248.
[72] K.T. Barnhart, et al., Baseline dimensions of the human vagina, Hum. Reprod. 21 (6) (2006) 1618–1622.
[73] K.D. McSporran, E.D. Fielden, Studies on dystocia in sheep. II: Pelvic measurements of ewes with histories of dystocia and eutocia, N. Z. Vet. J. 27 (4) (1979) 75–78.
[74] B.D. Hosie, Prolapse and hernia, in: Diseases of Sheep, Blackwell Publishing Ltd, Oxford, 2008, pp. 94–99.
[75] R. De Tayrac, A. Alves, M. Thérin, Collagen-coated vs noncoated low-weight polypropylene meshes in a sheep model for vaginal surgery. A pilot study, Int. Urogynecol. J. 18 (5) (2007) 513–520.
[76] S. Manodoro, et al., Graft-related complications and biaxial tensiometry following experimental vaginal implantation of flat mesh of variable dimensions, BJOG 120 (2) (2013) 244–250.
[77] M. Rezapour, et al., A 3-month preclinical trial to assess the performance of a new TVT-like mesh (TVTx) in a sheep model, Int. Urogynecol. J. 18 (2) (2007) 183–187.
[78] A. Feola, et al., Host reaction to vaginally inserted collagen containing polypropylene implants in sheep, Am. J. Obstet. Gynecol. 212 (4) (2015) e1–e8.
[79] R. Zacharin, Anatomy of the genital tract supports, in: Pelvic Floor Anatomy and the Surgery of Pulsion Enterocoele, Springer, Vienna, 1985, pp. 7–64.
[80] H.P. Drutz, S. Herschorn, N.E. Diamant, Female Pelvic Medicine and Reconstructive Pelvic Surgery, Springer, London, 2007.
[81] M. Schimpf, P. Tulikangas, Evolution of the female pelvis and relationships to pelvic organ prolapse, Int. Urogynecol. J. Pelvic Floor Dysfunct. 16 (4) (2005) 315–320.
[82] R.W. Kistner, et al., Kistner's Gynecology: Principles and Practice, Mosby, St. Louis, MO, 1995.
[83] D.E. Noakes, et al., Arthur's Veterinary Reproduction and Obstetrics, W.B. Saunders, London, 2001.
[84] E.G. Bassett, D.S.M. Phillips, Pelvic Relaxation in Sheep, Nature 174 (4439) (1954) 1020–1021.
[85] V. Singh, Textbook of Anatomy Abdomen and Lower Limb, Elsevier Health Sciences APAC, London, 2014.
[86] P.R. Scott, Sheep Medicine, CRC Press, Boca Raton, FL, 2006.
[87] RAR Centre, Animal Research in the New Zealand Department of Agriculture, New Zealand Department of Agriculture, Wellington, 1948.
[88] E.G. Bassett, Some effects of endogenous hormones on muscles and connective tissue, with special reference to the ewe, Proc. N. Z. Soc. Anim. Prod. 23 (1963) 107–120.
[89] E. Aughey, et al., Pregnancy-associated changes in the physical and microscopic characteristics of the ovine cervix, J. Anat. 136 (Pt 2) (1983) 389–399.
[90] E.G. Bassett, D.S.M. Phillips, Changes in the pelvic region of the ewe during pregnancy and parturition, N. Z. Vet. J. 3 (1) (1955) 20–25.
[91] H.H. Ortega, et al., Morphological characteristics of the interpubic joint (Symphysis pubica) of rats, guinea pigs and mice in different physiological situations. A comparative study, Cells Tissues Organs 173 (2) (2003) 105–114.
[92] I. Becker, S.J. Woodley, M.D. Stringer, The adult human pubic symphysis: a systematic review, J. Anat. 217 (5) (2010) 475–487.
[93] J.J. Lawrence (Ed.), The Medical Brief: A Monthly Journal of Scientific Medicine and Surgery, Ninth and Olive Streets, St. Louis, MO, 1904.
[94] J.W. McLean, J.H. Claxton, Vaginal prolapse in ewes. Part VII: the measurement and effect of intra-abdominal pressure, N. Z. Vet. J. 8 (3) (1960) 51–61.
[95] H. Ulfelder, S.J. Robboy, The embryologic development of the human vagina, Am. J. Obstet. Gynecol. 126 (7) (1976) 769–776.

[96] S.S. Patnaik, et al., Mayer-Rokitansky-Kuster-Hauser (MRKH) syndrome: a historical perspective, Gene 555 (1) (2015) 33–40.
[97] D. Bulmer, The early stages of vaginal development in the sheep, J. Anat. 90 (Pt 1) (1956) 123–134.
[98] J.E. Cox, Surgery of the Reproductive Tract in Large Animals, Liverpool University Press, Liverpool, 1982.
[99] W.A. Rushworth, The Sheep: A Historical and Statistical Description of Sheep and Their Products. The Fattening of Sheep. Their Diseases, With Prescriptions for Scientific Treatment. The Respective Breeds of Sheep and Their Fine Points. Government Inspection, etc., With Other Valuable Information ... Also an Appendix Containing Sheep Breeders' Directory, Buffalo Review Company, New York, 1899.
[100] E. Bassett, The anatomy of the pelvic and perineal regions of the ewe, Aust. J. Zool. 13 (2) (1965) 201–242.
[101] E.G. Bassett, The comparative anatomy of the pelvic and perineal regions of the cow, goat and sow, N. Z. Vet. J. 19 (12) (1971) 277–290.
[102] P.R. Scott, M.E. Gessert, Ultrasonographic examination of 12 ovine vaginal prolapses, Vet. J. 155 (3) (1998) 323–324.
[103] J.W. McLean, Vaginal prolapse in sheep, N. Z. Vet. J. 4 (2) (1956) 38–55.
[104] E. Ayen, D.E. Noakes, Displacement of the tubular genital tract of the ewe during pregnancy, Vet. Rec. 141 (20) (1997) 509–512.
[105] E. Ayen, D.E. Noakes, S.J. Baker, Changes in the capacity of the vagina and the compliance of the vaginal wall in ovariectomized, normal cyclical and pregnant ewes, before and after treatment with exogenous oestradiol and progesterone, Vet. J. 156 (2) (1998) 133–143.
[106] E. Ayen, Intra-abdominal pressure changes in pregnant ewe, J. Vet. Res. 57 (3) (2002) 97–101.
[107] M.H. Kerkhof, L. Hendriks, H.A.M. Brölmann, Changes in connective tissue in patients with pelvic organ prolapse — a review of the current literature, Int. Urogynecol. J. Pelvic Floor Dysfunct. 20 (4) (2009) 461–474.
[108] R. Jackson, et al., An epidemiological study of vaginal prolapse in ewes, in: International Symposia on Veterinary Epidemiology and Economics Proceedings, Proceedings of the 10th Symposium of the International Society for Veterinary Epidemiology and Economics, Vina del Mar, Chile, International Symposia on Veterinary Epidemiology and Economics, vol. ISVEE 10, 2003, p. 348.
[109] M. Bulgin, Diseases of the periparturient ewe, in: S.Y. Robert, R.T. Walter (Eds.), Current Therapy in Large Animal Theriogenology, second ed., W.B. Saunders, Saint Louis, 2007, pp. 695–700 (Chapter 92).
[110] B.D. Hosie, et al., Nutritional factors associated with vaginal prolapse in ewes, Vet. Rec. 128 (9) (1991) 204–208.
[111] M. Alcalay, et al., Mesh pullout force: comparative study of different deployment techniques in a sheep model, Int. Urogynecol. J. 25 (1) (2014) 103–107.
[112] M. Endo, et al., Cross-linked xenogenic collagen implantation in the sheep model for vaginal surgery, Gynecol. Surg. 12 (2015) 113–122.

CHAPTER SEVEN

Mechanics of Pelvic Floor Prosthetic Devices

K. Knight, P.A. Moalli
University of Pittsburgh, Pittsburgh, PA, United States of America

INTRODUCTION

The use of prosthetic devices in the treatment of pelvic floor disorders (PFDs) began in the 4th century at which time a wide range of devices from fruits to metals were being used to treat pelvic organ prolapse (POP) [1]. Today, pessaries, biologic grafts, and synthetic meshes are the most commonly used prosthetic devices in PFD treatments. These devices aid in treatment by providing structural support to the organs, connective tissues, and muscles within the female pelvis. This is particularly true for synthetic meshes. Given that structural support is one of the primary functions of prosthetic devices, it is imperative that the mechanics of these devices be understood. In terms of prosthetic devices for PFD repairs, mechanics is a broad term that refers to the ex vivo and in vivo mechanical behaviors of these devices. Though the structural properties are a main focus, the specific mechanical properties of the materials used in these devices are also relevant. Additionally, the mechanical behavior is intimately tied to the textile properties (eg, weight, pore size, knit pattern) of these devices. Collectively, this information is crucial as the mechanics and textile properties of prosthetic devices are directly related to biocompatibility and indirectly to patient outcomes.

Mechanics have been widely used to study prosthetic devices in a variety of medical fields (eg, cardiovascular and orthopedics) for many decades. However, the evaluation of PFD devices from a mechanical perspective is relatively new and studies are limited. Therefore, the overall goal of this chapter is to familiarize the reader with the types of tests that have been used to evaluate the mechanics of PFD devices, to discuss the results of these tests, and to explain how their ex vivo mechanics are related to their in vivo mechanical behavior and ultimately to the host response to these devices following implantation. This chapter will begin by briefly introducing pessaries and biologic grafts; however, synthetic meshes will be the primary focus of the chapter. In order to understand and appreciate the mechanics of these meshes, one must be knowledgeable about the mechanical/functional role of synthetic meshes in PFD repairs. Thus, these roles will be discussed with a concentration on synthetic meshes used to repair POP and stress urinary incontinence (SUI) (also referred to as urogynecologic meshes). Next, the types of tests that have been utilized to assess the mechanics of synthetic meshes will be reviewed. An emphasis will be placed on understanding the relevance and methodology of these tests with regard to synthetic meshes. Additionally, the type of data that can be obtained from these tests, and more importantly, the limitations of each method will be explained. The angle of this chapter will then switch to an application perspective in which the results of mechanical testing of synthetic meshes and the biological implications of these results, as observed experimentally and clinically, will be described. Finally, this chapter will conclude with a prospective insight into the future of the mechanical evaluation of prosthetic devices in PFD treatments.

OVERVIEW OF PROSTHETIC DEVICES

Pessary

The use of prosthetics in the female pelvis dates back to the 4th century, to Polybus, at which time uterine prolapse was treated by placing half a pomegranate in the vaginal area or by packing the vagina with a sponge [1]. These "prosthetics" were the earliest forms of devices referred to today as pessaries. One of the first more "modern" pessaries was described in the late 16th century by Ambrose Pare' who designed oval-shaped pessaries made out of brass and waxed cork for uterine prolapse [1]. Essentially, the function of these pessaries was to act as a plug for the vaginal canal, preventing the prolapsed organ(s) from protruding outside of the body. Over the years, the design, materials, and function of pessaries has evolved such that these devices are widely used as a non-surgical treatment option for both POP and SUI.

A pessary is a device that is designed to provide structural support. They are not only used to relieve POP and SUI-related symptoms, but they are also used to treat vaginal wind, neonatal prolapse, prolapse in pregnancy, and voiding dysfunction [2–8]. Modern-day pessaries are made primarily of silicone or an alternative inert plastic and fall into one of two main categories: space-filling and support pessaries (Fig. 1) [8,9]. Similar to the early pessaries, space-filling pessaries fill the vaginal canal by (1) creating suction between the vaginal walls and the pessary and/or by (2) having a diameter that is larger than the genital hiatus. The cube (suction mechanism), donut (large diameter mechanism), and Gellhorn (suction and large diameter mechanism) are all examples of space-filling pessaries. Instead of simply filling a space, support pessaries are placed in the space between the posterior fornix and the pubic symphysis; thus, providing support to the vagina. The ring pessary is one of the most commonly used support pessaries.

Although successful, pessaries have been associated with complications. Common complications include irritation of the vaginal mucosa, bleeding, vaginal excoriation, ulceration, and impaction. Vaginal discharge and vaginal odor are also problems associated with pessary usage [7,8]. From a biomechanical perspective, pessaries that are too

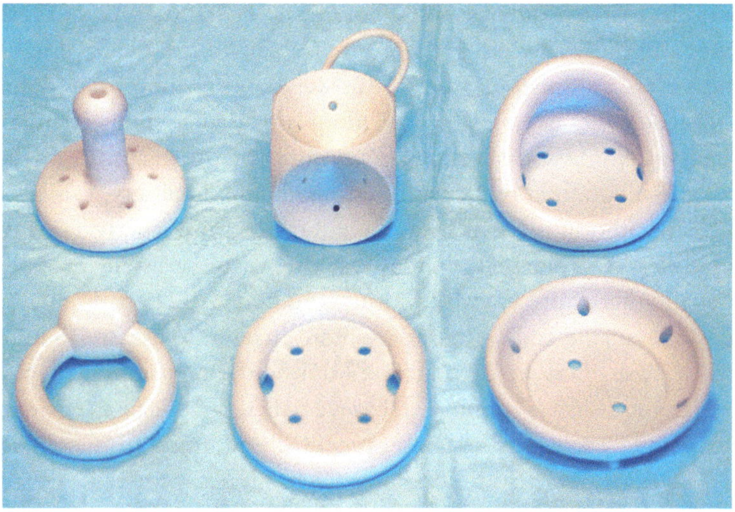

Figure 1 Commonly used pessaries. From left to right: top (1) Gellhorn, (2) cube, (3) Mar-Land and bottom from left to right, (4) ring with knob, (5) oval with platform, (6) dish.

large create friction between the device and the vaginal walls, which essentially wears down the walls of the vagina resulting in abrasion, erosion, and in extreme cases, fistula formation. Thus, the fit of the pessary is a critical component of success. Despite these complications, pessaries are a viable option for patients that decline or are unfit for surgery and for those seeking temporary relief of POP and SUI-related symptoms.

Biologic Grafts and Synthetic Meshes

Overview

Traditionally, surgical repair of POP and SUI used a patient's native connective tissues and muscles to provide support to weakened and/or compromised pelvic structures. Though successful, the concern of failure was great for these repairs given that the integrity of the existing connective tissue and muscles may be compromised [10–15]. For POP and urinary incontinence, the failure rate for native tissue repair requiring reoperation is approximately 40% [16–18]. As a way to provide stronger repairs, biologic grafts and synthetic meshes were introduced.

Biologic grafts are derived from the patient (autologous), another human source excluding the patient (allograft), or an animal source (xenograft). Autologous grafts are advantageous as they are harvested from the patient; thus, the chances of causing a foreign body reaction are minimized. These grafts also incorporate well with native tissues [19]. However, using autologous grafts increases the operation time, is associated with harvest site morbidity, and the strength of the repair, and hence long-term durability, and is dependent upon the quality of the host tissue [19,20]. Allografts, such as cadaveric fascia lata, on the other hand do not impact operation time nor is there a concern for donor site morbidity, as they are often obtained from cadaveric tissue banks. There is, however, an inherent risk of disease transmission, from donor to host. Processing methods (eg, chemical or radiation) are used to reduce immunogenicity and to reduce the transmission of diseases. Depending on the treatment method utilized, the quality of the allograft, in particular the strength and durability, can be significantly impacted [21]. In addition, allografts are often obtained from elderly patients with multiple comorbidities or prolonged illnesses resulting in inconsistent quality. Thus, the number of quality allografts available to surgeons and patients is limited. Xenografts are more readily available and are currently more frequently used relative to the other biologic grafts. These grafts are decellularized during the preparation process; thus, the risk of disease transmission is minimal. However, these grafts fail over time and this is likely due to a host versus graft response, potentially resulting from residual antigenic material [22,23]. Although used in urogynecologic surgeries since the 1950s, synthetic grafts (referred to as synthetic meshes) became popular in the 1990s as concerns regarding the quality, durability, and strength of biologic grafts increased. The primary purpose of synthetic mesh is to provide structural support to the vagina and hence the pelvic organs in an effort to restore the normal anatomy and function of these structures.

To date, the majority of synthetic meshes are knitted, lightweight (<45 g/m^2), wide pore (>1 mm, porosity $>55\%$), and type I polypropylene (monofilament, macroporous). The concept of using synthetic mesh for PFD repairs originates from the repair of abdominal hernias. In the abdominal hernia literature, superior outcomes were demonstrated in patients undergoing a mesh-augmented repair versus a native tissue repair [24,25]. In the 1970s, gynecologists were using meshes indicated for hernia repair to repair POP abdominally. By the 1990s, hernia meshes were being used for the transvaginal repair of POP and for the surgical treatment of SUI. Up until this point, there were no preconfigured meshes for SUI and POP repairs. Thus, surgeons were simply cutting hernia meshes to fit the desired shape needed for a POP or SUI repair. Soon, companies began to make meshes specifically designed for these types of repairs. Surgical fabrics (ProteGen Sling, Boston Scientific Corporation, Marlborough, MA, USA) was the first preconfigured mesh indicated for surgical treatment of SUI in 1996 followed by the tension-free vaginal tape (TVT™, Ethicon, Somerville, NJ, USA) in 1998, which introduced the terms sling and tape for mesh use in SUI repair. Four years later, Gynemesh PS (Ethicon, Somerville, NJ, USA) was cleared as the first preconfigured mesh for POP repair in the form of the Prolift™ transvaginal mesh kit (Ethicon, Somerville, NJ, USA).

Given that existing hernia meshes were already in use prior to the 1976 Medical Device Amendments Act, companies were able to remarket their hernia products for treatment of POP and SUI as 510K devices, thereby avoiding costly premarket testing of an entirely new device. Unfortunately, this meant that these devices were brought to the market without considering the unique composition and loading conditions of the tissues in the female pelvis. Nevertheless, mesh usage escalated following the introduction of the technically easier transvaginal approach and by 2010, meshes were being used in one-third of all cases to repair prolapse and in over 80% of SUI surgeries [26]. However, as vaginal mesh use increased, so did the incidence of mesh-related complications prompting the FDA to issue two public health warnings in 2008 and 2011. Complications cited include infection, mesh shrinkage, mesh erosion into an adjacent structure, mesh exposure into the vaginal lumen, and pain, with the latter two by far being the most commonly reported. Currently, the mechanism(s) responsible for mesh complications is unknown; however, new research in this area is shedding light on plausible mechanisms.

Mechanical Role of Synthetic Meshes

Stress Urinary Incontinence

Hypermobility of the bladder neck is believed to be one of the leading causes of SUI [27]. SUI is defined by the International Continence Society as "the complaint of involuntary loss of urine on effort or physical exertion, or on sneezing or coughing" [28]. As such, it is characterized by leakage that occurs when intra-abdominal pressure exceeds urethral pressure (eg, with coughing, sneezing, and straining) [29]. In continent women, increases in intra-abdominal pressure result in a compressive force that causes the lumen of the urethra to close. The anterior vaginal wall and the endopelvic fascia are key contributors to closure of the lumen. These structures support the urethra in a hammock-like fashion, and they provide a stable platform upon which the urethra sits and presses against with increases in intra-abdominal pressure [30]. A lack of this supportive structure may result in the inability of the urethra to close and consequently, leakage (ie, the development of SUI). Thus, synthetic meshes are implanted to provide this hammock-like supportive structure.

Initially, synthetic meshes were implanted via the pubovaginal sling procedure. In this technique, the sling (ie, the synthetic mesh) is placed underneath the urethra at the bladder neck and the arms are fixed to either the rectus fascia or pubic bone. The sling is then tensioned and the amount of tensioning applied is highly variable among surgeons. Surgical correction of SUI is accomplished with the pubovaginal sling through the increase in structural support to the bladder neck (via repositioning of the bladder neck); however, successful closure of the lumen of the urethra with compression is variable (possibly dictated by the amount of tension applied). Over the years, this implantation technique has evolved and the majority of current synthetic meshes for SUI repair are being implanted using the TVT method with placement of the mesh under the mid-urethra.

Ulmsten et al. first introduced the TVT procedure in 1996, and this procedure is defined by the integral theory of female urinary incontinence. This theory states that the pubourethral ligament supports the urethra by attaching the urethra to the pubic bone [31]. A malfunction or loss of this supportive mechanism will result in the inability of the urethra to adapt to increases in intra-abdominal pressure and the development of incontinence. Thus, the TVT procedure involves implanting the sling at the mid-urethra (these meshes are often referred to as mid-urethral slings) with the arms of the sling being self-fixing. As its name suggests, there is minimal tension applied to the sling, and hence the urethra, as the sling is placed during the procedure. The correction of SUI with the TVT is attributed to the ability of the urethra to kink during stress maneuvers and the ability of the mesh to move such that it compresses the tissue against the pubic symphysis [32].

Pelvic Organ Prolapse

The pathogenesis of POP is multifactorial with advancing age, obesity, and vaginal delivery being cited as the leading risk factors [33–36]. Of these, vaginal delivery is the leading risk factor for the development of POP [33,34,36]. The general consensus for POP development is that support to the pelvic organs has either been compromised or lost due to a loss of support to the vagina. For more information regarding the structural support of the pelvic organs and tissues, the interested reader should refer to the chapter on pelvic anatomy and pathology.

The overall goal of using synthetic mesh in prolapse repair is to restore normal support to the vagina (and ultimately the surrounding pelvic organs) while concomitantly providing a repair that is more or equally durable to autologous tissue [37]. Synthetic meshes are implanted using one of two primary approaches: transabdominally via sacrocolpopexy or transvaginally in the anterior, apical, and/or posterior compartments [38]. An abdominal sacrocolpopexy, often performed laparoscopically or robotically, is considered the "gold-standard" procedure, which provides apical support to the vagina by attaching the vagina through a mesh bridge to the sacrum. During this procedure, two independent mesh straps or the arms of a Y mesh are placed between the bladder and the vagina anteriorly and the rectum and the vagina posteriorly. The arms or straps are then anchored to the sacrum effectively restoring the vagina to its normal anatomical position via a mesh bridge. In a transvaginal procedure, the mesh is placed between the vaginal wall and the underlying viscera (bladder anteriorly or rectum posteriorly), and most often attached to the sacrospinous ligament or the pelvic sidewall via attachment to the arcus tendineous fasciae pelvis. To accomplish this, the vaginal wall must first be opened surgically and dissected off the underlying tissues. After the mesh is placed, the vaginal wall is then closed over it. The aim of these attachments is to correct the prolapse by suspending the vagina and hence, the viscera supported by it. A critical part of this procedure is the requirement of a "full thickness" vaginal dissection prior to placement of the mesh.

Loading Conditions of Synthetic Meshes: Abdominal Hernia Versus POP

Although synthetic meshes used in urogynecologic repairs are simply hernia meshes remarketed for the indication of POP and SUI, the loading conditions in the abdomen and pelvis are distinct. In abdominal hernia repair, synthetic mesh (when implanted flat) is generally uniformly loaded around the boundaries of the mesh. This is not the case for POP and SUI repairs, in which loading is more discrete and uniaxial for both abdominal sacrocolpopexy and transvaginal procedures (Fig. 2). It is therefore important that tests which seek to understand the mechanical behavior and structural properties of PFD synthetic meshes mimic the in vivo loading conditions as much as possible, especially from a uniaxial perspective.

MECHANICAL TESTS USED TO EVALUATE SYNTHETIC MESHES

There are numerous synthetic mesh products available on the market with no guide for clinicians and surgeons to select one mesh over another. Mechanical testing of synthetic meshes has proven to be invaluable for distinguishing the various meshes through the analysis of structural properties and mechanical behavior of these meshes under various boundary (ie, boundaries that restrict and/or limit movement in a given direction(s)) and loading (ie, how deformation or force is applied) conditions. All of these tests have assumptions and limitations that must be understood in order to properly conduct and interpret the results. Additionally, it is critical to note that not all mechanical tests aim to simulate in vivo scenarios. Limitations in sample size and available testing methods often hinder the physiological relevance of some methods. Therefore, understanding the limitations of each test and appropriately interpreting the results based on boundary conditions is critical for describing the mechanical behavior for synthetic meshes that often times could not be obtained otherwise. The next few sections will

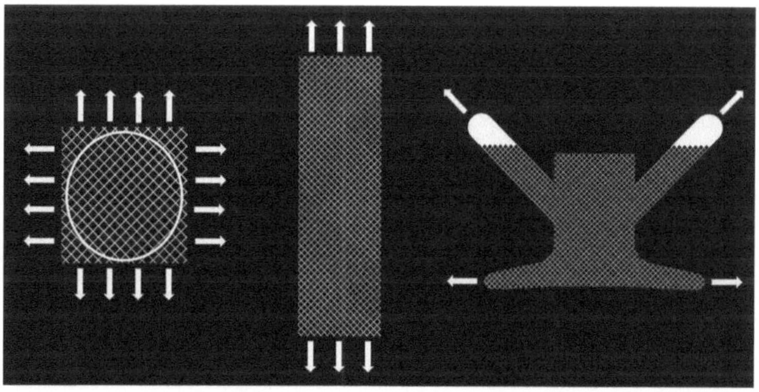

Figure 2 Depiction of the tensile forces *(arrows)* applied to synthetic meshes for abdominal hernia *(left)*, abdominal sacrocolpopexy *(middle)*, and transvaginal *(right)* repairs. Abdominal hernia meshes are loaded in multiple directions, whereas loading is more uniaxial for abdominal sacrocolpopexy and transvaginal repairs. Note that the angle at which the mesh arms are pulled for a transvaginal repair may vary; however, forces along these arms are still largely uniaxial.

discuss the various mechanical tests that have been used to evaluate synthetic meshes and the special considerations that must be made for mesh when using similar testing methods that were discussed in previous chapters for tissue.

Synthetic Mesh Versus Vaginal Tissue

Many of the mechanical tests used to assess the behavior and properties of vaginal tissue are also used in the assessment of synthetic meshes. However, it is important to note that there are some key differences between the two that ultimately impact the applicability of the testing method, the experimental protocol, and the interpretation of the results. The structure/composition of synthetic meshes is drastically different from that of vaginal tissue. Synthetic meshes are made of polymeric fibers that are often knitted together to create a porous, knitted network. From a mechanics perspective, these meshes are considered to be discontinuous materials as a result of their porosity. Vaginal tissue, on the other hand, is composed of cells, proteins, fibers (eg, collagen and elastin), connective tissue, muscle, and water. As a result, vaginal tissue is often assumed to be a continuous material (also referred to as a continuum, ie, there is interconnected material at every location in the sample with no voids) and this level of continuity is assumed at the level of fibers. In contrast, the discontinuity of synthetic meshes restricts the use of analyses that are applicable only to continuous materials. Additionally, the knit pattern of synthetic meshes limits the shape of meshes used for testing. For example, vaginal tissue is often cut into a dog-bone shape for uniaxial tensile testing to achieve a uniform strain field in the center of the sample. Creating a dog-bone shaped mesh disrupts the knit pattern thereby altering the mechanical behavior of the mesh. Moreover, the intertwined network of fibers causes very nonuniform and local deformations to the overall mesh structure.

Reporting Results from the Mechanical Testing of Synthetic Meshes

The interpretation and reporting of results from mechanical tests is largely dictated by the structure/composition of synthetic meshes. Often, the literature inappropriately reports the properties obtained from mechanical tests of mesh as stress, strain, tangent modulus/Young's modulus, and strain energy density. However, because synthetic meshes are filled with voids and cannot be treated as a continuum, only structural properties (eg, load, elongation, stiffness, and energy absorbed) or normalized structural properties (eg, membrane tension) should be reported.

Uniaxial Tensile Test

A uniaxial tensile test (tensile test) is one of the most common analyses used to evaluate synthetic meshes. The methodology utilized in a tensile test stems from standard testing methods for textiles (ASTM D5035-11, Standard Test Method for Breaking Force and Elongation of Textile Fabrics, Strip Method and ISO 13934-1, Textiles — Tensile Properties of Fabrics — Part 1: Determination of Maximum Force and Elongation at Maximum Force Using the Strip Method) that has been slightly modified to accommodate testing of prolapse and incontinence synthetic meshes. In this method, and similar to that described for tissue in previous chapters, a uniaxial load is applied in order to simulate loading conditions similar to those following an abdominal sacrocolpopexy. It involves first securing the opposing ends of the mesh, usually with clamps. While one end of the clamped mesh is rigidly attached to the base of a testing machine, the other end is attached to a force sensor (load cell) that is interfaced with the crosshead of the testing machine (Fig. 3). Slack is then removed from the mesh through the application of a minimal preload. Finally, the mesh is elongated as the crosshead moves upward at a constant rate. Often the mesh is elongated to failure and this is referred to as a "load to failure test." This testing method is a relatively simple but powerful tool that has been used to

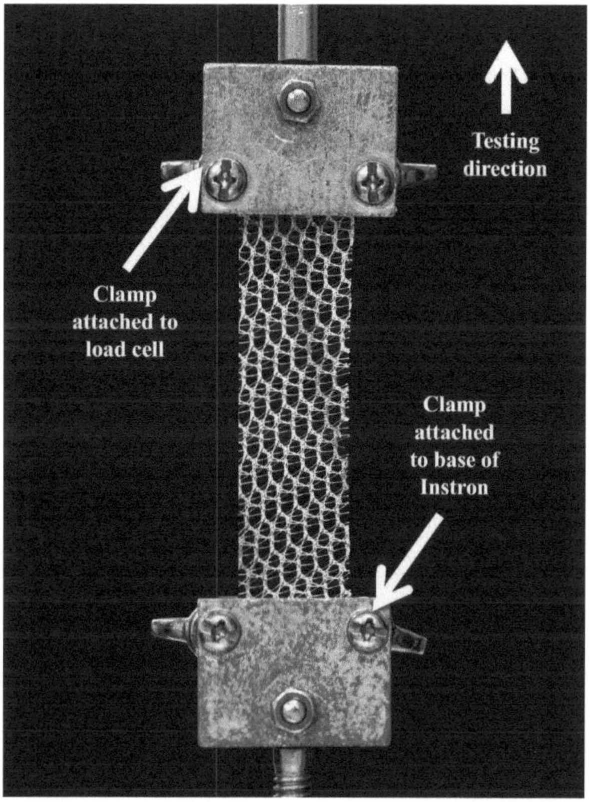

Figure 3 Synthetic mesh clamped and attached to the Instron testing machine (Norwood, MA, USA) for uniaxial tensile testing.

determine the structural properties of synthetic mesh, to predict the behavior of synthetic mesh in response to various loading conditions, and to compare synthetic mesh products.

Caveats to the Uniaxial Tensile Test Protocol for Synthetic Mesh

The protocol for conducting a uniaxial tensile test is relatively straightforward and similar to that used for testing vaginal tissue. However, there are some important caveats to consider, especially when the results from this test are to be compared for different meshes. As noted previously, the creation of a dog-bone shaped geometry for synthetic meshes is inappropriate as this will damage the structure of the mesh and affect the behavior and structural integrity of the mesh. Additionally, given that only structural properties can be reported, it is important that the area of the meshes be kept consistent. This can be accomplished by testing mesh samples that have the same aspect ratio (length to width). Alignment is another key aspect of testing. Two assumptions of a uniaxial tensile test are (1) force is being applied uniformly along one direction and (2) shear is negligible. Once the mesh sample has been mounted in the testing machine, care should be taken to ensure that the mesh is straight and that it is aligned with the plane of testing. Failure to properly align the mesh sample could result in the introduction of shear, nonuniform loading along the testing direction, premature failure (in the case of a load to failure test), and an underestimation of the structural properties of the mesh.

Once the mesh sample is properly aligned, a preload is applied and the distance between the clamps is measured. For consistency, and to provide a common starting point, the same preload should be applied to all mesh samples. The magnitude of the preload is somewhat arbitrary; however, it should be just enough tension/force to remove slack from the mesh sample (eg, the preload can be as small as 0.1 N or smaller). Some recommend this value to be a certain percentage of the failure load, but that can vary significantly from mesh type to mesh type resulting in very different initial reference states. Following the application of a preload, the mesh is then loaded, often to failure, at a constant rate (referred to as an elongation or displacement rate). This is in contrast to uniaxial tensile testing of vaginal tissue in which preconditioning follows the application of a preload. Vaginal tissue specimens are innately different due to biological variability and viscoelastic behavior; this difference is irrespective of whether the specimens come from the same species. Preconditioning is performed to minimize hysteresis; thus, creating a common starting point for all tissue specimens being tested. The synthetic and consistent nature of meshes in terms of composition, the lack of transient hydrogen bond formation during handling, and the absence of water content make the preparation for testing mesh samples with a common baseline less rigorous and the process of preconditioning unnecessary for these materials.

There is no standard elongation rate for synthetic mesh testing. Studies in the literature have used elongation rates as low as 1 mm/min and as high as 1200 mm/min [39–43]. Bazi et al. [39] showed that the displacement rate affects the extension at peak load and stiffness but does not affect the peak load itself [39]. The group recommended that uniaxial tensile tests be conducted at more than one displacement rate and that these rates be within the physiologic range. Currently, the majority of studies in the literature use only one displacement rate that is often not physiologic. This method is limited, as the results of these studies may not be capturing differences in the time-dependent components of the meshes. Regardless of whether one or multiple displacement rates are used, it is important to understand the implications of each method, to justify the use of one method over another, and to report the displacement rate(s) utilized. In addition to using a displacement rate that is physiologically relevant, the use of a heated saline bath is a way to simulate physiologic conditions and to control the testing temperature. Dietz et al. [44] have shown that the mechanical properties of synthetic meshes are independent of testing conditions [44]. However, it is also important to keep in mind that the in vivo behavior of the mesh may differ from the ex vivo behavior, given that in vivo there are environmental factors to consider as well as tissue incorporation into the synthetic mesh which may alter mesh behavior. Thus, measures should be taken to mimic environmental conditions as close as possible.

During a load to failure test, the load and corresponding elongation are recorded and used to generate a load-elongation curve. The shape of this curve is typically nonlinear consisting of three regions: an initial toe region, a linear region (that may be bilinear), and a failure region (Fig. 4). This curve is then used to determine the structural properties of synthetic meshes: the ultimate load, ultimate elongation, stiffness, and energy absorbed as described previously for tissues. For load-elongation curves that can be approximated as bilinear, two values for stiffness are typically reported, low stiffness (the minimum stiffness over a defined interval of elongation) and high stiffness (defined as the highest stiffness over a specified range of elongation). The low stiffness region corresponds to the initial

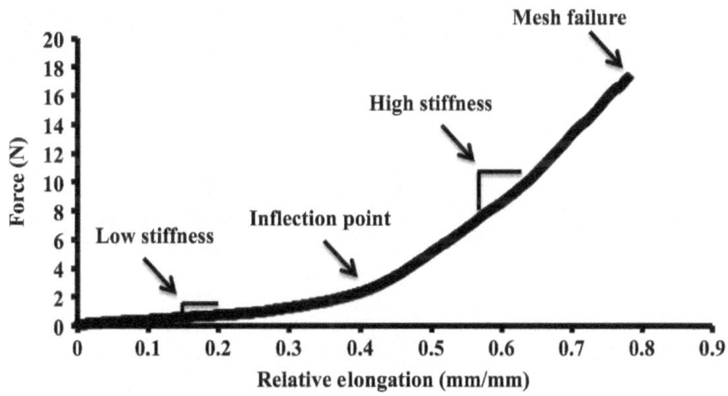

Figure 4 Typical load-elongation curve for synthetic mesh following a uniaxial tensile test to failure.

range of elongation in which the mesh is easily pulled apart, whereas the high stiffness region represents an increased resistance to deformation that occurs after most of the mesh has been reoriented along the direction of loading. It is important when reporting stiffness values that the method used to calculate stiffness is specified and to note whether that value represents an overall stiffness of the load-elongation curve or if low and high stiffness values were calculated. When publishing the results from load to failure tests, load-elongation curves should be presented along with the structural properties.

In addition to the four main structural properties listed above, relative elongation and inflection point are two additional parameters that have been reported in the literature [41–43]. Relative elongation is a normalized version of elongation and is used to account for differences in the initial length of mesh samples. This parameter is calculated by dividing the elongation by the initial clamp-to-clamp distance. Inflection point is used when the load-elongation curve is assumed to be bilinear (as it typically is for synthetic mesh) and it describes the point at which the transition from low stiffness to high stiffness occurs (Fig. 4). The inflection point is calculated by determining the intercept of two tangent lines that have been fit to the low and high stiffness regions of the load-elongation curve. Typically the relative elongation at the inflection is reported.

Membrane tension (or simply tension) is another structural property that is reported for synthetic mesh. The calculation of membrane tension is common in mechanics, and for synthetic meshes, it is a way to normalize the forces obtained during mechanical testing of meshes. It is calculated by dividing these forces by the undeformed width of the mesh, which results in units of force divided by width (ie, N/mm). It is important that tension is not confused with stiffness, which also has the same units (ie, N/mm).

Interpretation and Limitations of Uniaxial Tensile Test for Synthetic Mesh

Structural properties of synthetic meshes are often obtained from uniaxial tests in which the mesh is loaded to failure. Arguably, reporting structural properties at failure is not the most clinically relevant data given that the loads at failure are higher than the loads that meshes will experience under normal physiological conditions. In other words, most synthetic meshes are over-designed in terms of strength. A similar argument can likely be made for high stiffness. The other aforementioned properties, however, provide a better understanding of the clinically relevant behavior of mesh since physiologic stresses most likely occur within the low range of elongation.

As the name suggests, uniaxial tensile testing loads only one axis of the mesh at a time. However, in vivo synthetic meshes are expected to experience loads from multiple directions. In order to evaluate the behavior of the mesh from multiple axes (eg, along the longitudinal and transverse axes) using this technique, samples along each axis are cut and tested separately. Consequently, the interaction between the two axes cannot be examined. Some researchers are using planar biaxial mechanical testing to overcome this limitation [45–47]; however, this method is not entirely appropriate for reasons that will be described in the next section of this chapter.

Planar Biaxial Mechanical Test

Planar biaxial mechanical testing is a technique that involves applying forces along two perpendicular axes simultaneously. This technique has been used to assess the biomechanical behavior of thin soft tissues that are naturally planar or can be made into planar specimens [48,49]. Experimentally, biaxial mechanical testing can be conducted on synthetic meshes; however, there are several important assumptions of planar biaxial mechanical testing that makes this type of testing less appropriate for synthetic meshes.

One major assumption of planar biaxial mechanical testing is that stress and strain are in the x–y plane and zero in the direction perpendicular to this plane. Thus, testing should be restricted to thin specimens that are continuous or the assumption of a continuum can be assumed on some level of scale. This is not the case for synthetic mesh, which is a porous material. In fact, for the reasons mentioned earlier, discussions of stress and strain above and beyond what might be located in an individual fiber are inappropriate. Thus, as with uniaxial testing, this approach is only useful to determine structural behavior.

Additionally, when conducting planar biaxial mechanical tests, the method used to grip the specimen should be chosen such that uniform loading along the edges is achieved and that the local stresses imposed by the gripping method becomes negligible within the central region of the specimen [50]. In some studies that have used biaxial tests to assess the mechanics of synthetic mesh, the edges of the mesh were heat-sealed to allow placement of sutures within this portion of the mesh [47]. However, heat-sealing the mesh alters the mechanical behavior of the mesh by imposing an artificial boundary condition [42]. Heat-sealing restricts movement of the lateral edges of the mesh and this effect is carried through to the rest of the sample. A similar argument can be made if solid grips are used on each of the edges. One could argue that cutting the mesh in a plus sign-shaped configuration would minimize these effects; however, the act of cutting the mesh also changes its behavior and this specific edge pattern may not be at all similar to how the device is implanted in vivo. Thus, biaxial mechanical testing of mesh may be useful for specific questions about fiber interactions at local levels of scale or directionality of mesh; it is generally not useful when evaluating the overall structural performance of mesh as it relates to in vivo behavior, unless such data is used in conjunction with validated computational analyses.

Ball-burst Test

Ball-burst testing (sometimes referred to as biaxial tensiometry) is another mechanical test that is used to characterize the structural properties of synthetic mesh. This test method involves pushing a steel ball head through mesh until it ruptures. It has also been used to assess the structural integrity of mesh-tissue complexes (MTCs), to be discussed later in this chapter. Ball-burst testing originates from the American Society of Testing and Materials (ASTM) D6797-15, the Standard Test Method for Bursting Strength of Fabrics-Constant-Rate-of-Extension Ball Burst Test. In order to apply this testing method to synthetic meshes that are of the size to be implanted in vivo, a scaled version of this ASTM standard is often utilized.

The protocol for conducting ball-burst testing is similar to that for uniaxial tensile testing; however, the testing setup is slightly different. Instead of testing rectangular specimens of synthetic mesh, square or circular specimens are used for testing. The specimens are secured between two flat clamps that have a circular hole in the middle for the steel ball head to pass through. The clamped specimen is then mounted onto a custom stand that is attached to the base of a testing machine (Fig. 5). The steel ball head is connected in series with a load cell that is attached to a moveable crosshead. A small preload is applied (eg, 0.5 N) to ensure that contact is made between the mesh and the ball head with minimum deformation applied to the mesh. The ball head is then pushed through the synthetic mesh at a constant rate until it breaks through the mesh (failure). The resulting load and elongation are recorded and used to create a load-elongation curve (Fig. 6). Similar to uniaxial tensile tests, the ultimate load, maximum elongation, stiffness, and energy absorbed are reported.

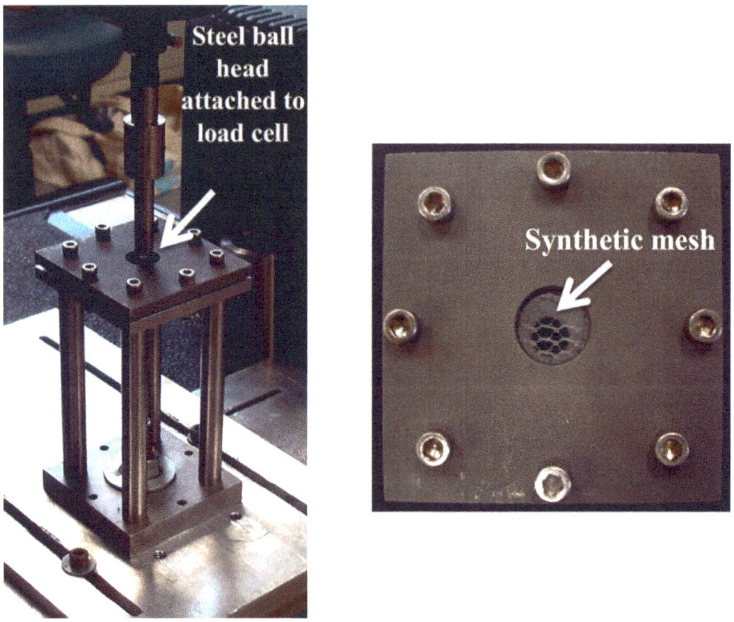

Figure 5 The ball-burst testing apparatus *(left image)* with synthetic mesh centered between two clamps *(right image)*.

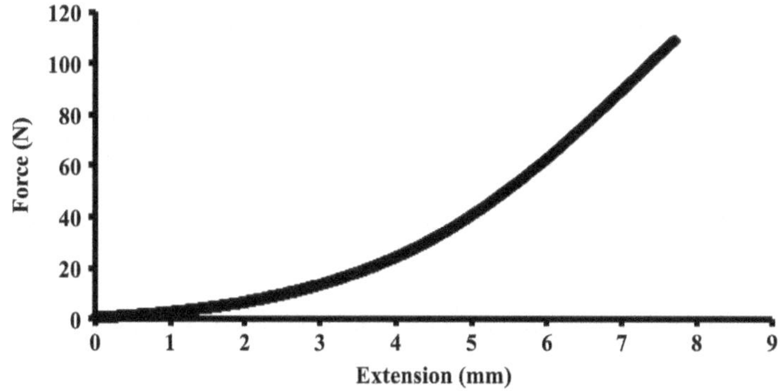

Figure 6 Typical load-elongation curve for synthetic mesh following a ball-burst test to failure.

There are some important caveats to consider when conducting ball-burst testing. Experimentally, it is important that during the preparation process, the mesh is securely clamped along the entire perimeter and that there is enough space between the edges of the mesh and the hole in the middle of the clamps to ensure that the mesh would not slip out of the clamps during testing. Additionally, the steel ball should be placed in the center of the mesh. Finally, the diameter of the ball needs to be significantly larger than the maximum pore size of the mesh. These considerations are critical for uniform loading of the mesh and for the repeatability of the data from sample to sample. A limitation of ball-burst testing is the boundary conditions, which restrict testing to a small area. This "artificially" increases the stiffness of the mesh and it is also not an accurate reflection of the in vivo boundary (and loading) conditions for synthetic urogynecologic meshes. However, the purpose of ball-burst testing is not to simulate in vivo conditions, but it has been used in the literature to assess the structural properties of mesh and to make comparisons between mesh devices. These structural properties can provide useful information and are used in the assessment of MTCs following implantation of synthetic mesh into the abdomen and vagina of animal models. While theoretically it is possible to determine a maximal stress from this test for a continuous material, this is not valid for mesh, which is a discontinuous material and the reporting of mechanical properties is inappropriate.

It is important to note that the ultimate load, maximum elongation, stiffness, and energy absorbed are the same structural properties reported with a uniaxial tensile test. Though these properties are calculated the same way for both tests, the values obtained are not the same (see Tables 1 and 2). These two testing methods, despite being structural tests, have different boundary and loading conditions, which yield different values in structural properties. This demonstrates the importance of maintaining the same testing conditions (ie, the same boundary and loading conditions and mesh dimensions) when comparisons are to be made between meshes.

Table 1 Uniaxial tensile structural properties of synthetic meshes

Synthetic mesh	Low stiffness (N/mm)	High stiffness (N/mm)	Load at mesh failure (N)	Relative elongation at failure (%)	Relative elongation at inflection point (%)
Boston Scientific Polyform™	0.130 ± 0.01	1.42 ± 0.11	53.8 ± 4.8	86.5 ± 2.4	39.9 ± 1.5
Coloplast NovaSilk™	0.072 ± 0.05	0.508 ± 0.09	19.6 ± 4.5	89.4 ± 21.4	44.6 ± 7.5
Gynecare Gynemesh PS™	0.286 ± 0.02	1.37 ± 0.09	46.3 ± 2.6	66.7 ± 4.6	25.0 ± 0.89
Gynecare Ultrapro™ (Artisyn®)	0.009 ± 0.00	0.236 ± 0.02	7.83 ± 0.7	87.9 ± 5.6	46.5 ± 5.2
Mpathy Smartmesh™	0.178 ± 0.03	0.592 ± 0.04	22.7 ± 1.8	68.5 ± 2.5	29.2 ± 1.0

Resulting structural properties of synthetic meshes used for pelvic organ prolapse repair following uniaxial tensile tests to failure. Data represented as mean ± standard deviation.

Data from J.P. Shepherd, A.J. Feola, S.D. Abramowitch, P.A. Moalli, Uniaxial biomechanical properties of seven different vaginally implanted meshes for pelvic organ prolapse, Int. Urogynecol. J. Pelvic Floor Dysfunct. 23 (5) (2012) 613–620.

Table 2 Ball-burst structural properties of synthetic meshes

Synthetic mesh	Stiffness (N/mm)	Failure load (N)	Extension (mm)	Energy absorbed (J)
Boston Scientific Polyform™	28 ± 0.43	108 ± 5.7	7.8 ± 0.05	261 ± 27
Coloplast NovaSilk™	16 ± 5.5	54 ± 19	6.3 ± 0.56	113 ± 43
Gynecare Gynemesh PS™	28 ± 2.7	108 ± 8.6	7.3 ± 0.31	288 ± 37
Gynecare Ultrapro™ (Artisyn®)	22 ± 2.8	76 ± 12	7.3 ± 0.21	170 ± 11
Mpathy Smartmesh™	11 ± 0.89	45 ± 3.8	6.7 ± 0.45	109 ± 11

Resulting structural properties of synthetic meshes used for pelvic organ prolapse repair following ball-burst tests to failure. Data represented as mean ± standard deviation.

Data from A. Feola, W. Barone, P. Moalli, S. Abramowitch, Characterizing the ex vivo textile and structural properties of synthetic prolapse mesh products. Int. Urogynecol. J. Pelvic Floor Dysfunct. 24 (4) (2013) 559–564.

MECHANICAL ANALYSIS OF MESH-TISSUE COMPLEXES

Synthetic meshes are often implanted into the vagina and abdomen of animal models, such as the nonhuman primate and rabbit. When MTCs are explanted, the assessment of the structural properties/integrity of these complexes is often one of the primary outcome measures. In general, large sample dimensions are required to perform a meaningful uniaxial and biaxial structural test. If sample dimensions are too small, inter-sample variability increases significantly as a result of cutting the mesh to achieve a specific size, which changes the transmission of force throughout the mesh. Alternatively, the distribution of pore sizes may vary resulting in increased variability. Yet, in small animal models, longer and bigger MTCs are often impossible due to the size constraints dictated by the anatomy. In addition to size, another concern with MTCs is that they are a composite (heterogeneous) material (mesh and tissue); thus, stress and strain fields are likely to be highly heterogeneous with a significant amount of shearing and force transmission between the mesh and tissue. Though it is physically possible to use uniaxial and biaxial mechanical tests to mechanically test MTCs, these experiments would need to be rigorously controlled in order to extract meaningful structural properties.

Although not the most ideal or physiologically relevant mechanical test, ball-burst testing can be used to assess the structural properties/integrity of MTCs. This test method is advantageous as it can be performed on synthetic mesh, on tissue, and on MTCs. Thus, when MTCs are explanted, the structural properties of the complex can be obtained and the mechanical impact of mesh on the structural integrity of the tissue can be roughly deduced if the mesh has not changed significantly while it was implanted. Also, this test can be performed on relatively small samples with few factors potentially contributing to variability. Nevertheless, as the MTC is a composite, ball-burst testing when applied to MTCs should be interpreted solely as a structural test (only the ultimate load, ultimate elongation, stiffness, and energy absorbed are to be reported).

Ball-burst testing is conducted on MTCs in a similar manner as it is with tissue and synthetic mesh. However, it is important to note that not all MTCs are appropriate for ball-burst testing. MTCs should be large enough such that it can be secured between the two clamps without slipping during testing. The same concerns regarding pore size of the mesh relative to the size of the ball also apply here. Additionally, care should be taken during the clamping process to ensure that the MTC is centered within the hole of the clamps. Also during the clamping process, it is important that the method used to secure the clamps (eg, screws) applies tension equally (or as equally as possible) to the edges of the clamps. For example, the screws on one side of the clamps should not be tighter than the other sides. This can be visually confirmed by checking to make sure that the gap between the two clamps is uniform on all sides. In addition, both mesh and tissue should be secured around the entire perimeter. Hydration of the MTCs should be maintained throughout testing.

APPLICATION OF MECHANICAL TESTS TO THE CHARACTERIZATION OF THE EX VIVO STRUCTURAL PROPERTIES AND MECHANICAL BEHAVIORS OF SYNTHETIC MESHES

Structural Properties

At first glance, most synthetic meshes appear to be very similar. They are lightweight, knitted, have large pores (>1 mm), and are composed primarily of polypropylene. However, upon close examination via microscopy and mechanical testing, these meshes are not the same. In terms of textile properties, the knit patterns, pore geometries, and pore diameters of synthetic meshes are distinct. Uniaxial tensile testing as well as ball-burst testing has shown that the structural properties, particularly stiffness, of synthetic meshes vary considerably, see Tables 1 and 2 [40–44,51]. Gynemesh PS™ (Ethicon) is the first synthetic mesh specifically marketed for POP repair. This mesh is identical to the Prolene soft hernia mesh (Ethicon, Somerville, NJ, USA). Relative to newer generation meshes, Gynemesh PS™ is currently one of the stiffest meshes available on the market. This is not surprising given the current trend of manufacturing meshes with decreasing weight and porosity, which results in a decrease in stiffness. Additionally, from ball-burst testing, textile properties are shown to correlate with structural properties, with stiffness, ultimate load, and energy absorbed being positively correlated to specific weight and porosity [40].

Anisotropy

The structural properties of synthetic meshes are not always independent of the testing direction. This type of behavior is referred to as anisotropy, in which the mechanical behavior is dependent on the direction in which the material, in this case mesh, is evaluated. UltraPro (marketed as Prolift+M™ by Ethicon, Somerville, NJ, USA) is an example of a prolapse synthetic mesh that has directionally dependent structural properties. Using uniaxial tensile testing, the stiffness of UltraPro in the direction that is parallel to the blue orientation lines was found to be orders of magnitude stiffer than in the direction that is perpendicular to the blue orientation lines, 0.258 ± 0.085 and 0.009 ± 0.002 N/mm, respectively [52]. The anisotropic nature of synthetic meshes has also been demonstrated with some abdominal hernia meshes [53]. Understanding and reporting the anisotropic behavior of synthetic meshes for urogynecologic repairs is of utmost importance to surgeons. With this information, surgeons will have a better knowledge of the mesh that they are implanting. Additionally, they will also be able to make a more scientifically driven decision regarding the mesh they should implant and the way in which this mesh should be implanted that will result in good outcomes for patients.

Permanent Deformation With Repetitive Loading

Synthetic meshes experience some degree of permanent deformation (ie, irreversible deformation), and this occurs at loads that are well within the physiologic range, 0.5–15 N [41,43]. At lower loads, this occurs because the structure of the mesh is undergoing reorganization (ie, knots are tightening, fibers are realigning, etc.). Permanent deformation resulting from plastic deformation of the fiber material itself occurs at much higher loads. Combining this knowledge with the current trend of manufacturing meshes with decreasing stiffness has resulted in the use of synthetic meshes that are more susceptible to permanent deformation. This finding may be indicative of adverse clinical outcomes; however, permanent deformation has not been associated clinically or experimentally with mesh complications.

Pore Size Reduction and Loss of Porosity

The pore size and consequently porosity (defined as the amount of open space relative to the total area of the mesh) tends to decrease when synthetic mesh is elongated [54]. Uniaxial loading of synthetic meshes, within a physiologic range, can result in dramatic changes in mesh pore geometry (ie, pore collapse) and in a reduction of porosity [55]. Porosity reduction has been reported as high as 87% when 5 N of force was applied and this reduction increased with 10 N of force. Similarly, Otto et al. [54] demonstrated that a loss of porosity occurred in both the arms and central body of transvaginal synthetic meshes when mesh samples were elongated to relatively small loads (1000 g maximum applied to the arms and 2000 g maximum applied to the body). For some meshes, a decrease in porosity was also associated with a reduction (or complete loss) in the effective porosity. Effective porosity is defined as the percentage of pores with dimensions that are greater than 1000 μm in all directions [56]. Previously, 1000 μm was identified as the minimum distance between two polypropylene filaments required to allow for tissue ingrowth and to prevent bridging fibrosis, the merging of fibrosis from adjacent areas [57–59]. Relating this knowledge to the mechanical behavior of urogynecologic synthetic meshes, decreases in pore size, porosity, and effective porosity potentially reduce the biocompatibility of these meshes. This in turn may lead to an increased risk for poor tissue integration, inflammation, fibrosis, and bridging fibrosis, and it is hypothesized that these parameters will worsen as pore sizes decrease. Collectively, the reduction in mesh biocompatibility may lead to complications and ultimately poor patient outcomes. Future studies are needed to substantiate this claim for urogynecologic meshes.

Boundary Conditions and Mesh Wrinkling

The boundary conditions that synthetic meshes are subjected to impact the response of these materials to loading. Typically the top and bottom edges of synthetic meshes are completely constrained during uniaxial tensile tests. In this configuration, loading and deformation is uniformly distributed and the mesh is allowed to contract in the direction perpendicular to loading. However, this configuration does not allow for out-of-plane deformation nor is it the most physiologically relevant boundary condition. Surgeons use sutures to attach synthetic mesh to the vagina and to anchoring points such as the sacrum and sacrospinous ligament. The number of sutures used and placement of these sutures varies for each surgeon and patient. These sutures essentially become point loads, which result in uneven distribution of loading and deformation. Barone et al. [60] demonstrated that applying point loads to synthetic meshes results in out-of-plane deformation in response to uniaxial loading and the magnitude of this deformation increases with the number of point loads applied (Fig. 7) [60]. Out-of-plane deformation was observed as mesh wrinkling/buckling/folding. It is important to note that this behavior was observed at loads that are within the physiologic range, 1 and 10 N. These results are

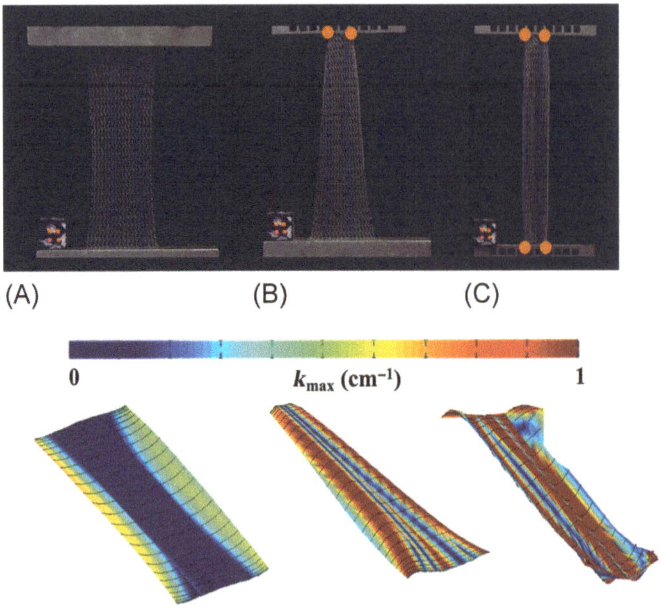

Figure 7 Impact of boundary conditions on mesh deformation in response to 10 N of force. Increasing the number of point loads *(suture)* resulted in increased mesh wrinkling (k_{max}). Images A-C depict uniaxial loading *(above)* and corresponding computational analysis *(below)*. Orange dots represent suture attachment points.

significant, and it is hypothesized that areas of mesh wrinkling results in an increase in mesh density (ie, mesh bunching), which may result in an enhanced host foreign body response. It also likely contributes to the development of heterogeneous stresses within the vagina. Further research is needed to substantiate this claim. Nonetheless, the number of sutures, location of sutures, and the amount of tension applied are important factors that surgeons must consider.

BIOLOGICAL IMPLICATIONS OF MECHANICAL BEHAVIOR, STRUCTURAL, AND TEXTILE PROPERTIES OF SYNTHETIC MESHES

Textile and structural properties of synthetic meshes obtained prior to implantation provide a baseline understanding of the ex vivo behavior of synthetic meshes. However, the physiologically relevant behavior of synthetic meshes and the impact of textile and structural properties on this behavior will only truly be revealed through in vivo studies. The majority of the studies in the literature currently consist of clinical studies that primarily report management and evaluation of various treatment methods for prolapse, as well as prevalence and types of complications. There is currently a lack of in vivo studies regarding how specific textile and structural properties, such as mesh stiffness, impact the biomechanical behavior and remodeling response of the vagina and the soft tissues that support it. This is largely due to synthetic meshes previously being classified as 510K devices; therefore, rigorous scientific testing and clinical data were not required for mesh to be used in urogynecologic surgeries. However, following the Food and Drug Administration Public Health Warnings released in 2008 and 2011 regarding complications associated with synthetic meshes for POP repairs, meshes have been changed to Class III devices and the climate of research has shifted such that more in vivo studies are emerging.

Clinically Observed Mechanical Behaviors

Surgeons and clinicians have reported that in vivo synthetic meshes bunch, contract/shrink, fold, and wrinkle (buckle) (Fig. 8) [61–63]. These behaviors are typically present in patients experiencing complications associated with mesh implantation. Mesh contraction, often reported as "shrinkage or retraction," is a commonly reported, and particularly problematic, mesh complication. Surprisingly it also occurs in women with mesh without complaints. Vaginal pain, dyspareunia (pain with sexual intercourse), and tenderness are all symptoms associated with mesh contraction [62,64]. The mechanisms that are responsible for these behaviors are likely attributed to the mesh pulling on adjacent muscles and connective tissues; however, the implantation method/surgical technique and mesh behavior with loading may all be contributing. Manufacturers recommend that synthetic meshes be implanted as flat as possible. Despite a flat implantation, deformation of the initial mesh geometry still occurs (Fig. 9). Additionally, the way that the mesh is

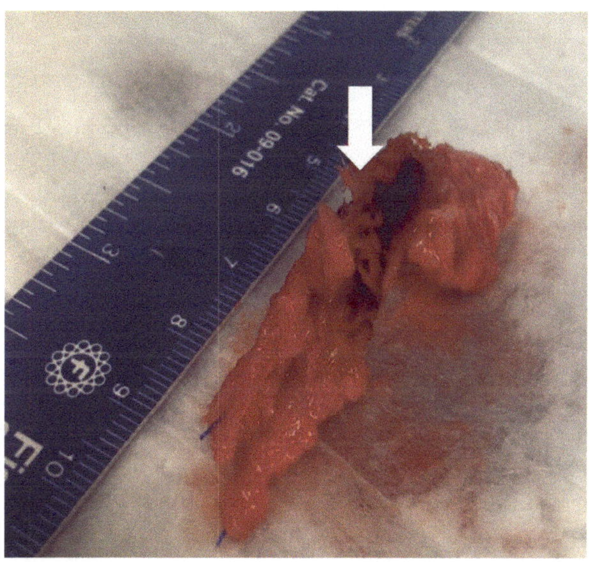

Figure 8 Mesh removed from a woman with a mesh exposure and pain due to a tension band (between two arms) at the apex of her vagina. The area of exposed mesh with little to no tissue incorporation is shown with an arrow. Note that in adjacent areas, there is a thick fibrous capsule.

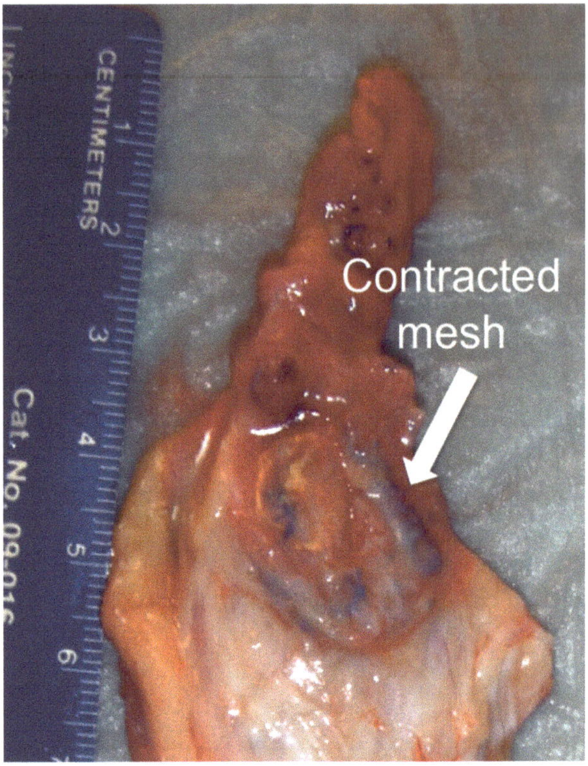

Figure 9 Vagina mesh-tissue complex containing Gynemesh PS explanted from the nonhuman primate demonstrating contraction after implantation.

anchored and the amount of tension applied to the mesh may be contributing factors as well. For example, with transvaginal synthetic meshes that have arms, the angle at which mesh arms are anchored into the pelvic sidewall, and/or unequal tensioning placed on the mesh arms may predispose the mesh to deformation. Biological factors such as tissue incorporation and the quality of the tissue in contact with the mesh may also play a role. Ex vivo and in vivo studies are needed to investigate these mechanical behaviors and to determine their etiology.

Impact of Synthetic Mesh Stiffness on the Vagina

The stiffness of synthetic mesh is a critical structural property and one that can affect the biomechanical integrity and remodeling response of the vagina following mesh implantation. A mesh must be stiff enough to restore normal anatomy to the vagina and prevent recurrent prolapse (or incontinence); however, a mesh that is too stiff can lead to a phenomenon referred to as stress shielding. Stress shielding results when there is a discrepancy in stiffness between adjacent materials such that the stiffer material bears the majority of the load while shielding the less stiff material. A maladaptive remodeling response characterized by degeneration and atrophy is a consequence of stress shielding [65–69]. In studies conducted by Feola et al. [70], Jallah et al. [71], and Liang et al. [72], synthetic meshes with different values of structural stiffness were implanted into nonhuman primates via an abdominal sacrocolpopexy following a hysterectomy [70–72]. The results of these studies suggest that synthetic mesh negatively impacts the structural properties of the underlying and newly incorporated vagina, particularly the smooth muscle, collagen, and elastin. Indeed, in the presence of Gynemesh PS, the contractility, organization, and volume of vaginal smooth muscle were all compromised (Fig. 10) and this was also associated with a decrease in nerve innervation. In addition, collagen and elastin degradation were increased. Interestingly, these degenerative effects of mesh were related to the stiffness of the mesh as Gynemesh PS was the stiffest mesh that was implanted in these studies and it had the greatest negative impact on the vagina. Additionally, it was the mesh associated with the greatest amount of deformation. The negative impact of mesh on the smooth muscle is particularly concerning given that studies have shown that women with POP have a decreased fractional area of smooth muscle, increased smooth muscle apoptosis, and disorganization of the smooth muscle layer [10,73,74]. Thus, implanting current synthetic meshes into an already comprised vagina may increase the chances of inducing a maladaptive remodeling response, which could ultimately lead to complications, such as mesh erosion.

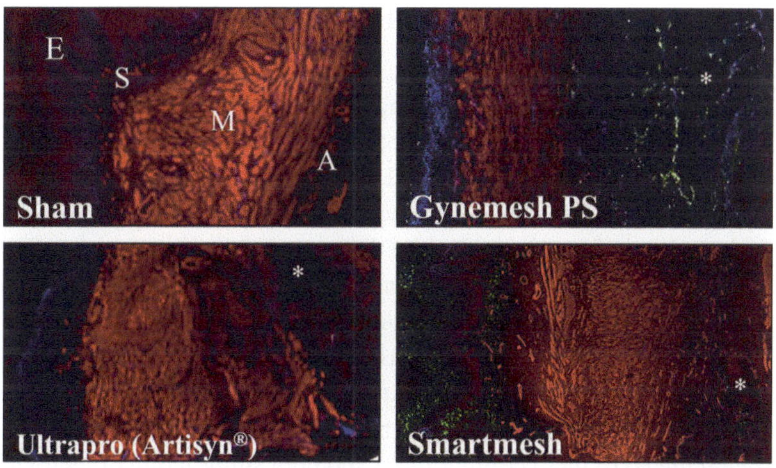

Figure 10 Fluorescently labeled smooth muscle *(red signal)* of tissue sections following implantation without (Sham) and with synthetic mesh in the nonhuman primate. Gynemesh PS had the greatest negative impact on smooth muscle organization and volume. Additionally, there was an increased amount of apoptosis *(green signal)* surrounding the fibers *(*)* of Gynemesh PS. *E*, epithelium; *S*, subepithelium; *M*, muscularis; *A*, adventitia.

Impact of Textile Properties and Synthetic Mesh Behavior on the Host Immune Response: Lessons from the Abdominal Hernia Literature

The impact of textile properties on the host immune response has been extensively studied for synthetic meshes used in abdominal hernia repairs. Unfortunately, this is not the case for POP and incontinence meshes. However, given that the abdominal hernia mesh is the predecessor for urogynecologic meshes, understanding the biological implications of textile properties for hernia meshes may aid with understanding the same for urogynecologic meshes and elicit possible etiologies for the complications that are associated with these meshes.

Pore size and porosity are key textile properties that significantly impact the tissue's reaction to and biocompatibility of synthetic meshes [58,75–77]. Meshes with large pores and high porosity yield better tissue integration with decreased inflammation and fibrosis relative to meshes with small pores and low porosity [58,77]. Additionally, pore size has been shown to be inversely proportional to bridging fibrosis [58,59,77]. As pore size increases, bridging fibrosis between mesh fibers decreases. Along with pore size, the weight of the mesh is also an important parameter for biocompatibility. Studies show that lightweight meshes are associated with less pain, less discomfort, and decreased fibrotic and foreign body responses [77–79]. Similar to urogynecologic meshes, folding, shrinkage, and contraction are also mechanical behaviors observed clinically with abdominal hernia meshes [76,80,81]. However, according to a study by Zogbi et al. [81], lightweight meshes are more prone to shrinkage than heavyweight meshes [81].

MECHANCIS OF PELVIC FLOOR PROSTHETIC DEVICES 10 YEARS FROM NOW

Much of the current research relating to synthetic mesh mechanics has mainly focused on characterizing the structural properties of mesh (eg, ultimate load, ultimate elongation, stiffness, and energy absorbed). However, it is anticipated that over the next 5–10 years, ex vivo mechanical tests and in vivo studies will aim to better understand the physiologically relevant behavior of synthetic meshes. The assessments of permanent deformation following repetitive loading and loss in porosity with mesh elongation are examples of two such studies. These studies were conducted ex vivo and on a limited number of meshes. Given the potential consequence of a synthetic mesh that is susceptible to permanent deformation and porosity loss, it is anticipated that more synthetic meshes will undergo testing to assess these types of behaviors. Additionally, with small pore sizes being linked to mesh complications and poor surgical outcomes for abdominal hernias, in vivo studies are needed to determine the biological ramifications for the decrease/loss in porosity observed with urogynecologic synthetic meshes.

Limited studies have used constitutive and computational analysis to understand the mechanical behavior of synthetic meshes. Feola et al. [52] were able to show that differences in textile geometry could account for the nonlinear mechanical behavior of synthetic meshes using constitutive modeling. An advantage of this technique is that data obtained from ex vivo mechanical tests used to determine structural properties (eg, following load to failure tests) can be used to develop the constitutive model; additional testing is not always necessary. Constitutive models can then be implemented into computational analysis and used to understand how synthetic meshes will behave in vivo. Computational analysis is advantageous as it allows for the simulation of in vivo scenarios without the need for actual in vivo experiments, which can be costly depending on the animal model utilized and the study design. Collectively, constitutive modeling and computational analysis will be of particular importance for understanding how synthetic mesh placement and tensioning impacts mesh mechanics. Preliminary results from experimental and computational analyses suggest that the position and amount of tension placed on the arms of synthetic meshes impact the degree of pore collapse (ie, loss of porosity) [55]. Additionally, these analytical techniques will be instrumental in understanding how textile properties, particularly pore geometry, impact mesh behavior. However, as with ex vivo mechanical testing, in vivo studies are needed to complement information gained from computational analysis.

Increased complication rates and litigation surrounding the use of synthetic mesh for urogynecologic surgery have resulted in a need to understand the pathogenesis of mesh-related complications. In addition to the previously mentioned techniques (eg, ex vivo mechanical testing, computational modeling), advanced imaging techniques such as ultrasound will be crucial in the investigation of mesh complications. These techniques will allow for the visualization of the mesh in vivo without disturbances that occur as a result of explantation. Additionally, given that synthetic meshes were initially designed for abdominal hernias, it is anticipated that as more knowledge is gained on the biomechanics of the vagina, the design of synthetic meshes will be tailored to match this behavior. Ultimately, it is hoped that a better understanding of synthetic mesh mechanics (and vaginal biomechanics) will lead to improved patient treatment and outcomes.

REFERENCES

[1] L.A. Emge, R.B. Durfee, Pelvic organ prolapse: four thousand years of treatment, Clin. Obstet. Gynecol. 9 (4) (1966) 997–1032. Retrieved from, http://www.scopus.com/inward/record.url?eid=2-s2.0-0013991920&partnerID=40&md5=f227a2bbb82b422b63841bf86e309d66.

[2] R.E. Dixon, A.A. Acosta, R.L. Young, Penrose pessary management of neonatal genital prolapse, Am. J. Obstet. Gynecol. 119 (6) (1974) 855–857. Retrieved from, http://www.scopus.com/inward/record.url?eid=2-s2.0-0016161841&partnerID=40&md5=09a104ea62a1450b0bca3623b5c68c6a.

[3] M.J. Donnelly, S. Powell-Morgan, A.L. Olsen, I.E. Nygaard, Vaginal pessaries for the management of stress and mixed urinary incontinence, Int. Urogynecol. J. Pelvic Floor Dysfunct. 15 (5) (2004) 302–307, http://dx.doi.org/10.1007/s00192-004-1163-7.

[4] H. Krissi, C. Medina, S.L. Stanton, Vaginal wind — a new pelvic symptom, Int. Urogynecol. J. Pelvic Floor Dysfunct. 14 (6) (2003) 399–402, http://dx.doi.org/10.1007/s00192-003-1086-8.

[5] B.H.C. Lamers, B.M.W. Broekman, A.L. Milani, Pessary treatment for pelvic organ prolapse and health-related quality of life: a review, Int. Urogynecol. J. Pelvic Floor Dysfunct. 22 (6) (2011) 637–644, http://dx.doi.org/10.1007/s00192-011-1390-7.
[6] J.R. Loret De Mola, S.E. Carpenter, Management of genital prolapse in neonates and young women, Obstet. Gynecol. Surv. 51 (4) (1996) 253–260, http://dx.doi.org/10.1097/00006254-199604000-00022.
[7] R. Oliver, R. Thakar, A.H. Sultan, The history and usage of the vaginal pessary: a review, Eur. J. Obstet. Gynecol. Reprod. Biol. 156 (2) (2011) 125–130, http://dx.doi.org/10.1016/j.ejogrb.2010.12.039.
[8] M.E. Vierhout, The use of pessaries in vaginal prolapse, Eur. J. Obstet. Gynecol. Reprod. Biol. 117 (1) (2004) 4–9, http://dx.doi.org/10.1016/j.ejogrb.2003.10.037.
[9] R. Thakar, S. Stanton, Management of genital prolapse, Br. Med. J. 324 (7348) (2002) 1258–1262. Retrieved from, http://www.scopus.com/inward/record.url?eid=2-s2.0-0037172306&partnerID=40&md5=03118cafe2dbf887075e59c2cb875d96.
[10] M.K. Boreham, C.Y. Wai, R.T. Miller, J.I. Schaffer, R.A. Word, Morphometric analysis of smooth muscle in the anterior vaginal wall of women with pelvic organ prolapse, Am. J. Obstet. Gynecol. 187 (1) (2002) 56–63, http://dx.doi.org/10.1067/mob.2002.124843.
[11] A. Feola, R. Duerr, P. Moalli, S. Abramowitch, Changes in the rheological behavior of the vagina in women with pelvic organ prolapse, Int. Urogynecol. J. Pelvic Floor Dysfunct. 24 (7) (2013) 1221–1227. Retrieved from, http://www.scopus.com/inward/record.url?eid=2-s2.0-84886841156&partnerID=40&md5=d982e3dd82a44d921260a89c5ef609c6.
[12] C. Goepel, L. Hefler, H.D. Methfessel, H. Koelbl, Periurethral connective tissue status of postmenopausal women with genital prolapse with and without stress incontinence, Acta Obstet. Gynecol. Scand. 82 (7) (2003) 659–664, http://dx.doi.org/10.1034/j.1600-0412.2003.00019.x.
[13] S.R. Jackson, N.C. Avery, J.F. Tarlton, S.D. Eckford, P. Abrams, A.J. Bailey, Changes in metabolism of collagen in genitourinary prolapse, Lancet 347 (9016) (1996) 1658–1661, http://dx.doi.org/10.1016/S0140-6736(96)91489-0.
[14] P.A. Moalli, S.H. Shand, H.M. Zyczynski, S.C. Gordy, L.A. Meyn, Remodeling of vaginal connective tissue in patients with prolapse, Obstet. Gynecol. 106 (5 I) (2005) 953–963. Retrieved from, http://www.scopus.com/inward/record.url?eid=2-s2.0-27444438817&partnerID=40&md5=12e512c88a3783fb5edf0210cc5059c0.
[15] W. Zong, S.E. Stein, B. Starcher, L.A. Meyn, P.A. Moalli, Alteration of vaginal elastin metabolism in women with pelvic organ prolapse, Obstet. Gynecol. 115 (5) (2010) 953–961, http://dx.doi.org/10.1097/AOG.0b013e3181da7946.
[16] M.D. Barber, L. Brubaker, K.L. Burgio, H.E. Richter, I. Nygaard, A.C. Weidner, S.A. Menefee, E.S. Lukacz, P. Norton, J. Schaffer, J.N. Nguyen, D. Borello-France, P.S. Goode, S. Jakus-Waldman, C. Spino, L.K. Warren, M.G. Gantz, S.F. Meikle, Eunice Kennedy Shriver National Institute of Child Health and Human Development Pelvic Floor Disorders Network, Comparison of 2 transvaginal surgical approaches and perioperative behavioral therapy for apical vaginal prolapse: the OPTIMAL randomized trial, JAMA 311 (10) (2014) 1023–1034, http://dx.doi.org/10.1001/jama.2014.1719.
[17] D.M. Morgan, M.A.M. Rogers, M. Huebner, J.T. Wei, J.O. DeLancey, Heterogeneity in anatomic outcome of sacrospinous ligament fixation for prolapse: a systematic review, Obstet. Gynecol. 109 (6) (2007) 1424–1433, http://dx.doi.org/10.1097/01.AOG.0000264066.89094.21.
[18] A.L. Olsen, V.J. Smith, J.O. Bergstrom, J.C. Colling, A.L. Clark, Epidemiology of surgically managed pelvic organ prolapse and urinary incontinence, Obstet. Gynecol. 89 (4) (1997) 501–506. Retrieved from, http://www.scopus.com/inward/record.url?eid=2-s2.0-0031127072&partnerID=40&md5=47c8768a1cf8d0738554d61741065585.
[19] S.M. Jakus, A. Shapiro, C.D. Hall, Biologic and synthetic graft use in pelvic surgery: a review, Obstet. Gynecol. Surv. 63 (4) (2008) 253–266, http://dx.doi.org/10.1097/OGX.0b013e318166fb44.
[20] P.A. Moalli, Cadaveric fascia lata, Int. Urogynecol. J. Pelvic Floor Dysfunct. 17 (Suppl. 7) (2006) S48–S50, http://dx.doi.org/10.1007/s00192-006-0106-x.
[21] M.P. FitzGerald, J. Mollenhauer, P. Bitterman, L. Brubaker, Functional failure of fascia lata allografts, Am. J. Obstet. Gynecol. 181 (6) (1999) 1339–1346. Retrieved from, http://www.scopus.com/inward/record.url?eid=2-s2.0-0033371839&partnerID=40&md5=425ca2fabb45a251321512ce6b3b5708.

[22] K.Z. Konakci, B. Bohle, R. Blumer, W. Hoetzenecker, G. Roth, B. Moser, G. Boltz-Nitulescu, M. Gorlitzer, W. Klepetko, E. Wolner, H.J. Ankersmit, Alpha-Gal on bioprostheses: xenograft immune response in cardiac surgery, Eur. J. Clin. Investig. 35 (1) (2005) 17–23, http://dx.doi.org/10.1111/j.1365-2362.2005.01441.x.
[23] L.H. Quiroz, R.E. Gutman, S. Shippey, G.W. Cundiff, T. Sanses, J.L. Blomquist, V.L. Handa, Abdominal sacrocolpopexy: anatomic outcomes and complications with Pelvicol, autologous and synthetic graft materials, Am. J. Obstet. Gynecol. 198 (5) (2008) 557.e551–557.e555, http://dx.doi.org/10.1016/j.ajog.2008.01.050.
[24] J.W.A. Burger, R.W. Luijendijk, W.C.J. Hop, J.A. Halm, E.G.G. Verdaasdonk, J. Jeekel, Long-term follow-up of a randomized controlled trial of suture versus mesh repair of incisional hernia, Ann. Surg. 240 (4) (2004) 578–585, http://dx.doi.org/10.1097/01.sla.0000141193.08524.e7.
[25] R.W. Luijendijk, W.C. Hop, M.P. van den Tol, D.C. de Lange, M.M. Braaksma, J.N. IJzermans, R. U. Boelhouwer, B.C. de Vries, M.K. Salu, J.C. Wereldsma, C.M. Bruijninckx, J. Jeekel, A comparison of suture repair with mesh repair for incisional hernia, N. Engl. J. Med. 343 (6) (2000) 392–398, http://dx.doi.org/10.1056/NEJM200008103430603.
[26] FDA, Surgical mesh for treatment of women with pelvic organ prolapse and stress urinary incontinence: FDA executive summary, (2011). Retrieved from, http://www.fda.gov/downloads/UCM270402.pdf.
[27] H.P. Dietz, B. Clarke, P. Herbison, Bladder neck mobility and urethral closure pressure as predictors of genuine stress incontinence, Int. Urogynecol. J. Pelvic Floor Dysfunct. 13 (5) (2002) 289–293, http://dx.doi.org/10.1007/s001920200063.
[28] B.T. Haylen, D. de Ridder, R.M. Freeman, S.E. Swift, B. Berghmans, J. Lee, A. Monga, E. Petri, D.E. Rizk, P.K. Sand, G.N. Schaer, An international urogynecological association (IUGA)/international continence society (ICS) joint report on the terminology for female pelvic floor dysfunction, Neurourol. Urodyn. 29 (1) (2010) 4–20, http://dx.doi.org/10.1002/nau.20798.
[29] I.E. Nygaard, M. Heit, Stress urinary incontinence, Obstet. Gynecol. 104 (3) (2004) 607–620, http://dx.doi.org/10.1097/01.AOG.0000137874.84862.94.
[30] J.O.L. DeLancey, Structural support of the urethra as it relates to stress urinary incontinence: the hammock hypothesis, Am. J. Obstet. Gynecol. 170 (6) (1994) 1713–1723. Retrieved from, http://www.scopus.com/inward/record.url?eid=2-s2.0-0028356078&partnerID=40&md5=3dfd9e267f3270e52807d0a9aa5a8349.
[31] U. Ulmsten, L. Henriksson, P. Johnson, G. Varhos, An ambulatory surgical procedure under local anesthesia for treatment of female urinary incontinence, Int. Urogynecol. J. Pelvic Floor Dysfunct. 7 (2) (1996) 81–86. Retrieved from, http://www.scopus.com/inward/record.url?eid=2-s2.0-0029786086&partnerID=40&md5=e38256ad9e5dbc9c4223024aafe072b1.
[32] D. Sarlos, M. Kuronen, G.N. Schaer, How does tension-free vaginal tape correct stress incontinence? Investigation by perineal ultrasound, Int. Urogynecol. J. Pelvic Floor Dysfunct. 14 (6) (2003) 395–398, http://dx.doi.org/10.1007/s00192-003-1103-y.
[33] A.H. MacLennan, A.W. Taylor, D.H. Wilson, D. Wilson, The prevalence of pelvic floor disorders and their relationship to gender, age, parity and mode of delivery, Br. J. Obstet. Gynaecol. 107 (12) (2000) 1460–1470. Retrieved from, http://www.scopus.com/inward/record.url?eid=2-s2.0-0034521949&partnerID=40&md5=f638c9f304cb394d15764ec52b6bb25e. http://onlinelibrary.wiley.com/store/10.1111/j.1471-0528.2000.tb11669.x/asset/j.1471-0528.2000.tb11669.x.pdf?v=1&t=i227g1vw&s=b25f18a7ce68516d3adcb804baa84f21f98ad525.
[34] J. Mant, R. Painter, M. Vessey, Epidemiology of genital prolapse: observations from the oxford family planning association study, Br. J. Obstet. Gynaecol. 104 (5) (1997) 579–585. Retrieved from, http://www.scopus.com/inward/record.url?eid=2-s2.0-0030989578&partnerID=40&md5=f63bc5e5feeb23d53247a204deb687fd.
[35] P.A. Moalli, S. Jones Ivy, L.A. Meyn, H.M. Zyczynski, Risk factors associated with pelvic floor disorders in women undergoing surgical repair, Obstet. Gynecol. 101 (5) (2003) 869–874. Retrieved from, http://www.scopus.com/inward/record.url?eid=2-s2.0-0038066291&partnerID=40&md5=f949920552057573dd1387b761bdc86e. http://ac.els-cdn.com/S0029784403000784/1-s2.0-S0029784403000784-main.pdf?_tid=246cd4ea-638e-11e4-bcb0-00000aab0f26&acdnat= 1415042467_63b1c5c2534e9a14f8c3a79d9c8ccb16.

[36] S.E. Swift, The distribution of pelvic organ support in a population of female subjects seen for routine gynecologic health care, Am. J. Obstet. Gynecol. 183 (2) (2000) 277–285. Retrieved from, http://www.scopus.com/inward/record.url?eid=2-s2.0-0033844651&partnerID=40&md5=2b57448a2c363564dace59e426c6bef8. http://ac.els-cdn.com/S000293780053840X/1-s2.0-S000293780053840X-main.pdf?_tid=cf308f02-19a6-11e4-a8f0-00000aacb361&acdnat=1406916676_74f35f359f3edbbaee2b2634a9567cd7.
[37] C.C. Chen, B. Ridgeway, M.F. Paraiso, Biologic grafts and synthetic meshes in pelvic reconstructive surgery, Clin. Obstet. Gynecol. 50 (2) (2007) 383–411, http://dx.doi.org/10.1097/GRF.0b013e31804b184c.
[38] N.Y. Siddiqui, A.L. Edenfield, Clinical challenges in the management of vaginal prolapse, Int. J. Womens Health 6 (2014) 83–94.
[39] T. Bazi, A.H. Ammouri, R.F. Hamade, On the relevance of uniaxial tensile testing of urogynecological prostheses: the effect of displacement rate, Int. Urogynecol. J. Pelvic Floor Dysfunct. 24 (1) (2013) 161–167, http://dx.doi.org/10.1007/s00192-012-1815-y.
[40] A. Feola, W. Barone, P. Moalli, S. Abramowitch, Characterizing the ex vivo textile and structural properties of synthetic prolapse mesh products, Int. Urogynecol. J. Pelvic Floor Dysfunct. 24 (4) (2013) 559–564. Retrieved from, http://www.scopus.com/inward/record.url?eid=2-s2.0-84880552992&partnerID=40&md5=3675a08d724fbeb8d429c21b5817ab07.
[41] K.A. Jones, A. Feola, L. Meyn, S.D. Abramowitch, P.A. Moalli, Tensile properties of commonly used prolapse meshes, Int. Urogynecol. J. Pelvic Floor Dysfunct. 20 (7) (2009) 847–853. Retrieved from, http://www.scopus.com/inward/record.url?eid=2-s2.0-67849121869&partnerID=40&md5=655a244fd25fa1c34868a0ca87acc114.
[42] P.A. Moalli, N. Papas, S. Menefee, M. Albo, L. Meyn, S.D. Abramowitch, Tensile properties of five commonly used mid-urethral slings relative to the TVT™, Int. Urogynecol. J. Pelvic Floor Dysfunct. 19 (5) (2008) 655–663, http://dx.doi.org/10.1007/s00192-007-0499-1.
[43] J.P. Shepherd, A.J. Feola, S.D. Abramowitch, P.A. Moalli, Uniaxial biomechanical properties of seven different vaginally implanted meshes for pelvic organ prolapse, Int. Urogynecol. J. Pelvic Floor Dysfunct. 23 (5) (2012) 613–620. Retrieved from, http://www.scopus.com/inward/record.url?eid=2-s2.0-84863085588&partnerID=40&md5=ee13a1336429600d86a83d3a0345230c.
[44] H.P. Dietz, P. Vancaillie, M. Svehla, W. Walsh, A.B. Steensma, T.G. Vancaillie, Mechanical properties of urogynecologic implant materials, Int. Urogynecol. J. Pelvic Floor Dysfunct. 14 (4) (2003) 239–243, http://dx.doi.org/10.1007/s00192-003-1041-8.
[45] C.R. Deeken, D.M. Thompson, R.M. Castile, S.P. Lake, Biaxial analysis of synthetic scaffolds for hernia repair demonstrates variability in mechanical anisotropy, non-linearity and hysteresis, J. Mech. Behav. Biomed. Mater. 38 (2014) 6–16, http://dx.doi.org/10.1016/j.jmbbm.2014.06.001.
[46] S. Sahoo, K.R. DeLozier, A. Erdemir, K.A. Derwin, Clinically relevant mechanical testing of hernia graft constructs, J. Mech. Behav. Biomed. Mater. 41 (2015) 117–118, http://dx.doi.org/10.1016/j.jmbbm.2014.10.011.
[47] M.T. Wolf, C.A. Carruthers, C.L. Dearth, P.M. Crapo, A. Huber, O.A. Burnsed, R. Londono, S.A. Johnson, K.A. Daly, E.C. Stahl, J.M. Freund, C.J. Medberry, L.E. Carey, A. Nieponice, N.J. Amoroso, S.F. Badylak, Polypropylene surgical mesh coated with extracellular matrix mitigates the host foreign body response, J. Biomed. Mater. Res. A (2013), http://dx.doi.org/10.1002/jbm.a.34671.
[48] M. Sacks, Biaxial mechanical evaluation of planar biological materials, J. Elast. Phys. Sci. Solids 61 (1–3) (2000) 199–246, http://dx.doi.org/10.1023/A:1010917028671.
[49] M.S. Sacks, W. Sun, Multiaxial mechanical behavior of biological materials, Annu. Rev. Biomed. Eng. 5 (2003) 251–284.
[50] W. Sun, M.S. Sacks, M.J. Scott, Effects of boundary conditions on the estimation of the planar biaxial mechanical properties of soft tissues, J. Biomech. Eng. 127 (4) (2005) 709–715. Retrieved from, http://www.scopus.com/inward/record.url?eid=2-s2.0-23644440239&partnerID=40&md5=200ce2968a81ebc390cc98b044fdd927.
[51] J.S. Afonso, P.A.L.S. Martins, M.J.B.C. Girao, R.M. Natal Jorge, A.J.M. Ferreira, T. Mascarenhas, A. Fernandes, J. Bernardes, E. Baracut, G. Rodrigues de Lima, B. Patricio, Mechanical properties of polypropylene mesh used in pelvic floor repair, Int. Urogynecol. J. Pelvic Floor Dysfunct. 19 (3) (2008) 375–380, http://dx.doi.org/10.1007/s00192-007-0446-1.

[52] A. Feola, S. Pal, P. Moalli, S. Maiti, S. Abramowitch, Varying degrees of nonlinear mechanical behavior arising from geometric differences of urogynecological meshes, J. Biomech. 47 (11) (2014) 2584–2589, http://dx.doi.org/10.1016/j.jbiomech.2014.05.027.

[53] E.R. Saberski, S.B. Orenstein, Y.W. Novitsky, Anisotropic evaluation of synthetic surgical meshes, Hernia 15 (1) (2011) 47–52, http://dx.doi.org/10.1007/s10029-010-0731-7.

[54] J. Otto, E. Kaldenhoff, R. Kirschner-Hermanns, T. Mühl, U. Klinge, Elongation of textile pelvic floor implants under load is related to complete loss of effective porosity, thereby favoring incorporation in scar plates, J. Biomed. Mater. Res. A 102 (4) (2014) 1079–1084. Retrieved from, http://www.scopus.com/inward/record.url?eid=2-s2.0-84894476047&partnerID=40&md5=6fee4d1518854386b660c964bbde9a0d.

[55] W.R. Barone, Mechanical characterization of synthetic mesh for pelvic organ prolapse repair, Ph.D. Doctoral Dissertation, University of Pittsburgh, University of Pittsburgh ETD, 2015.

[56] T. Mühl, M. Binnebösel, U. Klinge, T. Goedderz, New objective measurement to characterize the porosity of textile implants, J. Biomed. Mater. Res. B Appl. Biomater. 84 (1) (2008) 176–183. Retrieved from, http://www.scopus.com/inward/record.url?eid=2-s2.0-37549041383&partnerID=40&md5=7a9ea7e73d8907b6753205a37ff5bb92.

[57] J. Conze, K. Junge, C. Weiß, M. Anurov, A. Oettinger, U. Klinge, V. Schumpelick, New polymer for intra-abdominal meshes — PVDF copolymer, J. Biomed. Mater. Res. B Appl. Biomater. 87 (2) (2008) 321–328. Retrieved from, http://www.scopus.com/inward/record.url?eid=2-s2.0-55049131098&partnerID=40&md5=f81c2ad70ad03a5f450c3c2665c270b3.

[58] U. Klinge, B. Klosterhalfen, V. Birkenhauer, K. Junge, J. Conze, V. Schumpelick, Impact of polymer pore size on the interface scar formation in a rat model, J. Surg. Res. 103 (2) (2002) 208–214. Retrieved from, http://www.scopus.com/inward/record.url?eid=2-s2.0-0036349949&partnerID=40&md5=cf0da5a633ec7c896c395f9ad5bdb104.

[59] U. Klinge, B. Klosterhalfen, J. Conze, W. Limberg, B. Obolenski, A.P. Öttinger, V. Schumpelick, Modified mesh for hernia repair that is adapted to the physiology of the abdominal wall, Eur. J. Surg. 164 (12) (1998) 951–960. Retrieved from, http://www.scopus.com/inward/record.url?eid=2-s2.0-0032429949&partnerID=40&md5=42cd391ac43e465a6bf6bcd671256a74.

[60] W.R. Barone, R. Amini, S. Maiti, P.A. Moalli, S.D. Abramowitch, The impact of boundary conditions on surface curvature of polypropylene mesh in response to uniaxial loading, J. Biomech. 48 (9) (2015) 1566–1574, http://dx.doi.org/10.1016/j.jbiomech.2015.02.061.

[61] F. Caquant, P. Collinet, P. Debodinance, J. Berrocal, O. Garbin, C. Rosenthal, H. Clave, R. Villet, B. Jacquetin, M. Cosson, Safety of trans vaginal mesh procedure: retrospective study of 684 patients, J. Obstet. Gynaecol. Res. 34 (4) (2008) 449–456, http://dx.doi.org/10.1111/j.1447-0756.2008.00820.x.

[62] B. Feiner, C. Maher, Vaginal mesh contraction: definition, clinical presentation, and management, Obstet. Gynecol. 115 (2 Part 1) (2010) 325–330. Retrieved from, http://www.scopus.com/inward/record.url?eid=2-s2.0-76549131047&partnerID=40&md5=cfebb3c6fff54d4fc903e9727f7f8263.

[63] K. Svabik, A. Martan, J. Masata, R. El-Haddad, P. Hubka, M. Pavlikova, Ultrasound appearances after mesh implantation — evidence of mesh contraction or folding? Int. Urogynecol. J. 22 (5) (2011) 529–533, http://dx.doi.org/10.1007/s00192-010-1308-9.

[64] A. Rogowski, P. Bienkowski, A. Tosiak, M. Jerzak, P. Mierzejewski, W. Baranowski, Mesh retraction correlates with vaginal pain and overactive bladder symptoms after anterior vaginal mesh repair, Int. Urogynecol. J. Pelvic Floor Dysfunct. 24 (12) (2013) 2087–2092, http://dx.doi.org/10.1007/s00192-013-2131-x.

[65] V.K. Goel, T.H. Lim, J. Gwon, J.Y. Chen, J.M. Winterbottom, J.B. Park, J.N. Weinstein, J.Y. Ahn, Effects of rigidity of an internal fixation device: a comprehensive biomechanical investigation, Spine (Phila Pa 1976) 16 (3 Suppl.) (1991) S155–S161. Retrieved from, http://www.ncbi.nlm.nih.gov/pubmed/2028332.

[66] R. Huiskes, H. Weinans, H.J. Grootenboer, M. Dalstra, B. Fudala, T.J. Slooff, Adaptive bone-remodeling theory applied to prosthetic-design analysis, J. Biomech. 20 (11-12) (1987) 1135–1150. Retrieved from, http://www.ncbi.nlm.nih.gov/pubmed/3429459.

[67] A.P. Rumian, E.R. Draper, A.L. Wallace, A.E. Goodship, The influence of the mechanical environment on remodelling of the patellar tendon, J. Bone Joint Surg. (Br.) 91 (4) (2009) 557–564, http://dx.doi.org/10.1302/0301-620X.91B4.21580.

[68] A.J. Tonino, C.L. Davidson, P.J. Klopper, L.A. Linclau, Protection from stress in bone and its effects: experiments with stainless steel and plastic plates in dogs, J. Bone Joint Surg. B 58 (1) (1976) 107–113. Retrieved from, http://www.scopus.com/inward/record.url?eid=2-s2.0-0017289065&partnerID=40&md5=76e36401420e370792cb9c9533ce1cc8.

[69] N. Yamamoto, K. Ohno, K. Hayashi, H. Kuriyama, K. Yasuda, K. Kaneda, Effects of stress shielding on the mechanical properties of rabbit patellar tendon, J. Biomech. Eng. 115 (1) (1993) 23–28.

[70] A. Feola, S. Abramowitch, Z. Jallah, S. Stein, W. Barone, S. Palcsey, P. Moalli, Deterioration in biomechanical properties of the vagina following implantation of a high-stiffness prolapse mesh, BJOG 120 (2) (2013) 224–232. Retrieved from, http://www.scopus.com/inward/record.url?eid=2-s2.0-84871247182&partnerID=40&md5=544a5a3039b0ace8125e9257ce5d21d8.

[71] Z. Jallah, R. Liang, A. Feola, W. Barone, S. Palcsey, S. Abramowitch, N. Yoshimura, P. Moalli, The impact of prolapse mesh on vaginal smooth muscle structure and function. BJOG (2015), http://dx.doi.org/10.1111/1471-0528.13514.

[72] R. Liang, S. Abramowitch, K. Knight, S. Palcsey, A. Nolfi, A. Feola, S. Stein, P.A. Moalli, Vaginal degeneration following implantation of synthetic mesh with increased stiffness, BJOG 120 (2) (2013) 233–243. Retrieved from, http://www.scopus.com/inward/record.url?eid=2-s2.0-84871217055&partnerID=40&md5=87708d6a0f6daa72151f2dab22a06c27.

[73] M.K. Boreham, C.Y. Wai, R.T. Miller, J.I. Schaffer, R.A. Word, A. Weber, Morphometric properties of the posterior vaginal wall in women with pelvic organ prolapse, Am. J. Obstet. Gynecol. 187 (6) (2002) 1501–1509, http://dx.doi.org/10.1067/mob.2002.130005.

[74] P. Takacs, M. Gualtieri, M. Nassiri, K. Candiotti, C.A. Medina, Vaginal smooth muscle cell apoptosis is increased in women with pelvic organ prolapse, Int. Urogynecol. J. Pelvic Floor Dysfunct. 19 (11) (2008) 1559–1564, http://dx.doi.org/10.1007/s00192-008-0690-z.

[75] J. Conze, R. Rosch, U. Klinge, C. Weiss, M. Anurov, S. Titkowa, A. Oettinger, V. Schumpelick, Polypropylene in the intra-abdominal position: influence of pore size and surface area, Hernia 8 (4) (2004) 365–372. Retrieved from, http://www.scopus.com/inward/record.url?eid=2-s2.0-8844279966&partnerID=40&md5=459f5b397c67e0faff221c7cfe93326b.

[76] U. Klinge, B. Klosterhalfen, M. Muller, A.P. Ottinger, V. Schumpelick, Shrinking of polypropylene mesh in vivo: an experimental study in dogs, Eur. J. Surg. 164 (12) (1998) 965–969, http://dx.doi.org/10.1080/110241598750005156.

[77] S.B. Orenstein, E.R. Saberski, D.L. Kreutzer, Y.W. Novitsky, Comparative analysis of histopathologic effects of synthetic meshes based on material, weight, and pore size in mice, J. Surg. Res. 176 (2) (2012) 423–429. Retrieved from, http://www.scopus.com/inward/record.url?eid=2-s2.0-84863988987&partnerID=40&md5=a0b555f022db32e46a6b3e71ba4d86bc.

[78] S. Schmidbauer, R. Ladurner, K.K. Hallfeldt, T. Mussack, Heavy-weight versus low-weight polypropylene meshes for open sublay mesh repair of incisional hernia, Eur. J. Med. Res. 10 (6) (2005) 247–253. Retrieved from, http://www.scopus.com/inward/record.url?eid=2-s2.0-21344465840&partnerID=40&md5=e633eda7f5a21a3cf7de5de413f0baac.

[79] G. Welty, U. Klinge, B. Klosterhalfen, R. Kasperk, V. Schumpelick, Functional impairment and complaints following incisional hernia repair with different polypropylene meshes, Hernia 5 (3) (2001) 142–147, http://dx.doi.org/10.1007/s100290100017.

[80] M. Endo, A. Feola, N. Sindhwani, S. Manodoro, J. Vlacil, A.C. Engels, F. Claus, J.A. Deprest, Mesh contraction: in vivo documentation of changes in apparent surface area utilizing meshes visible on magnetic resonance imaging in the rabbit abdominal wall model, Int. Urogynecol. J. (2014) 1–7. Retrieved from, http://www.scopus.com/inward/record.url?eid=2-s2.0-84892632232&partnerID=40&md5=0723a2d299e7b1dff007370cd347474a.

[81] L. Zogbi, E.N. Trindade, M.R.M. Trindade, Comparative study of shrinkage, inflammatory response and fibroplasia in heavyweight and lightweight meshes, Hernia 17 (6) (2013) 765–772. Retrieved from, http://www.scopus.com/inward/record.url?eid=2-s2.0-84890568867&partnerID=40&md5=33f92d3a5aea7b59b4844a3992b8099d. http://download.springer.com/static/pdf/605/art%3A10.1007%2Fs10029-013-1046-2.pdf?auth66=1407088103_6c2a6496400b86a2e4d6adcbde0ce838&ext=.pdf.

SECTION 2

Influencers of Pelvic Floor Biomechanics

M. Alperin
University of California, San Diego, CA, United States of America

CHAPTER EIGHT

Biochemistry and Ultrastructure of Pelvic Floor Tissues and Organs

K. Downing
Albert Einstein College of Medicine/Montefiore Medical Center, Bronx, NY, United States of America

INTRODUCTION

The pelvic floor is a complex system designed to support the pelvic organs and allow for their optimal function; this requires the proper integration of bones, muscles, and connective tissue. The application of biomechanics to the pelvic floor, as in other fields, is closely linked to a detailed understanding of normal and pathological anatomy [1]. Rightfully, the pelvic floor surgeon obsesses over anatomic detail, creating 3D mental maps of pelvic floor anatomy, and attempts to intuit the biomechanical properties of these structures. They do so with an end goal in mind: to optimize patient outcomes following the surgical management of conditions such as pelvic organ prolapse (POP) and stress urinary incontinence (SUI). However, they also hope to gain greater insight into the pathogenesis of these conditions. What has been made clear in recent studies with respect to surgical outcomes following POP repair is that enduring surgical success with the focus on anatomical repair is difficult to achieve for far too many patients (and surgeons alike) [2,3]. Nevertheless, what a growing body of literature related to pelvic floor disorders (PFDs) indicates is that to truly understand the support of the pelvic floor, and gain greater insight into the pathogenesis of PFDs — that is, to understand normal function and how changes to components of the system alter its function — the pelvic floor surgeon will need to know more than the arrangement of organs, muscles, fascia, and bony parts that make up the pelvic floor. Having knowledge of the ultrastructure of these key pelvic floor tissues and organs, and an understanding of the biochemistry that modulates and maintains these structures, will be an absolute necessity to design and assess more effective interventions to prevent and treat PFDs.

In light of this understanding, the goal of this chapter will be to describe the major ultrastructural components of the pelvic floor musculature and surrounding connective tissue, including the vaginal wall and those structures that have traditionally been referred to as the endopelvic fascia (here referred to as the pelvic floor connective tissue — PFCT).

The authors have no potential conflicts of interests to disclose.

In addition, we review the biochemical processes that regulate these structures and how these processes can affect biomechanics. Finally, we will consider how pregnancy and parturition, and aging may alter these structures and what influence these exposures may have on their biochemistry. Whenever possible, we will refer to findings in human subjects. In areas where lessons have been gleaned from cell culture or animal models, this will be indicated.

PELVIC FLOOR MUSCULATURE

Striated muscle, through the actions of the levator ani muscle complex (LAMC), coccygeus muscle, external urethral sphincters (EUS), and external anal sphincter (EAS), contributes greatly to the biomechanical support of the pelvic floor and maintenance of urinary and anal continence. Smooth muscle (SM) plays an equally essential role with its presence in the vaginal wall (and other areas of PFCT), the internal urethral sphincters (IUS), and internal anal sphincter (IAS). What follows is a brief review of "the basics" regarding striated and SM architecture and physiology, what is known about their biochemical processes, and their ultrastructural characteristics in general, and in pelvic floor tissues and organs, specifically.

Striated Muscle

The basic architecture of striated muscle and its associated connective tissue has been well characterized in many texts and articles. What the pelvic floor surgeon encounters in the operating room is whole muscle and the bundles of muscle fibers (or fasciculi) that determine its size. Continuing from the macro, to micro, and on to a nanoscale, we see that the bundles of muscle fibers are themselves made up of individual muscle fibers (or single muscle cells) that consist of clusters of myofibrils, while individual myofibrils are made of an array of myofilaments as shown in Fig. 1 [4]. The organizational hierarchy of striated muscle — whole muscle, bundles of muscle fibers, and individual myofibrils — is supported by an enveloping connective tissue, the epimysium, perimysium, and endomysium, respectively. The epimysium and perimysium consist mainly of Type I and III collagen, while the endomysium is predominantly Type IV collagen [5]. This connective tissue has a protective and functional role contributing to the passive tension (or elasticity) of the muscle body and by serving as the primary conduit — via its enthesis (the connective tissue attachment of the muscle to the bone) — through which chemical energy converted to mechanical energy leads to movement of the body, or in this example, stabilization and support of the pelvic floor.

The Sarcomere

Striated muscle myofibrils are multinucleated, postmitotic cells, surrounded by a cell membrane, the sarcolemma. The functional unit of the myofibril is the sarcomere. Sarcomeres align end to end with individual sarcomeres residing between Zwischenscheibe

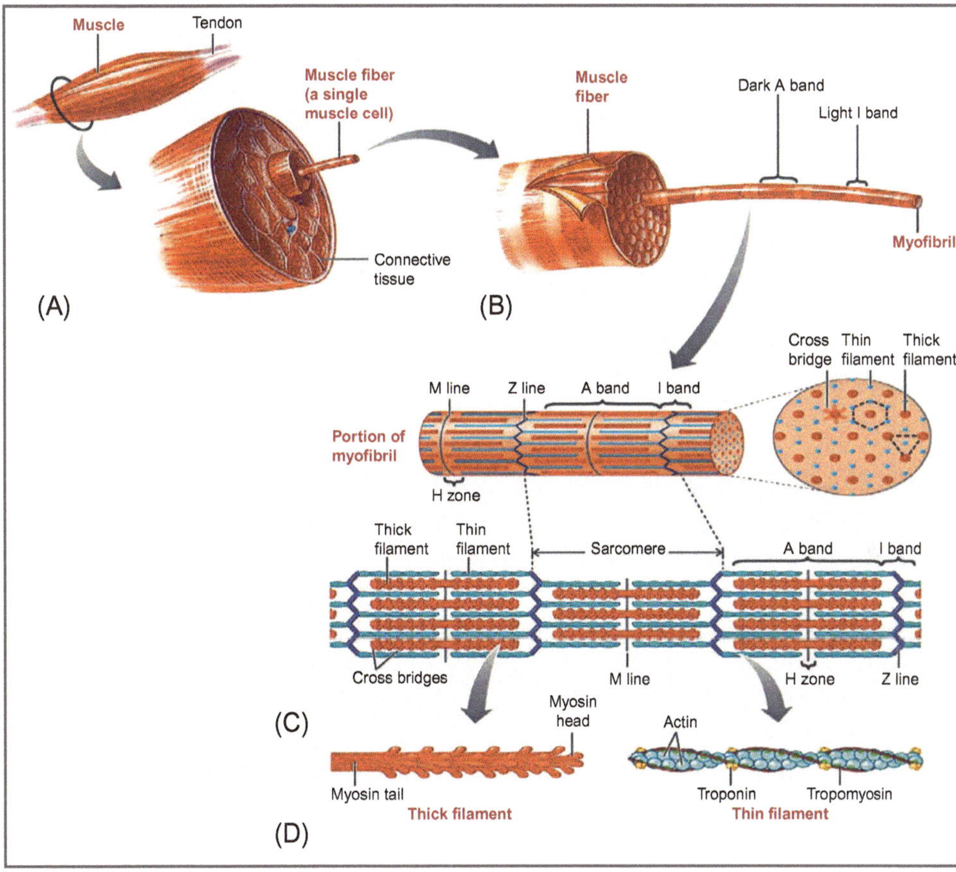

Figure 1 Structure of skeletal muscle. (A) Relationship of a whole muscle and a muscle fiber. (B) Relationship of a muscle fiber and a myofibril. (C) Cytoskeletal components of a myofibril. (D) Protein components of thick and thin filaments. *From W.R. Frontera, J. Ochala, Skeletal muscle: a brief review of structure and function, Calcif. Tissue Int. 96 (3) (2015) 183–195 [Figure 3]; reproduced with permission.*

disks, or Z-disks (or sometimes referred to as disks, lines, or bands), the macromolecular structures that create its borders [6]. Within each sarcomere are 3D arrangements of actin (thin) and myosin (thick) filaments. Actin attaches to the perpendicular Z-disks and projects toward the center of the sarcomere, while myosin is fixed to the central region to a structure called the H-band. It is the organization of these filaments, with their thin and thick components that give rise to the associated bands and lines that give striated muscle its classic "striated" appearance when stained. Titin is an additional protein that attaches to the Z-disks and helps to stabilize and align myosin filaments [7]. It is thought to contribute greatly to the passive tension of muscle due to its elastic properties (likely critical for the LAMC and the stretch it undergoes at the time of vaginal delivery) [8]. Other organelles found in muscle fibers include the transverse tubular system, sarcoplasmic reticulum (SR), mitochondrial network, and satellite cells (SCs) [4].

Excitation-contraction Coupling and Force Generation

Striated muscle activation requires two processes: the transmission of a nerve stimulus to the neuromuscular junction leading to the release of calcium ions (Ca^{2+}) from the cisternae of the SR, and the interaction between actin and myosin that forms cross-bridges. Force generation arises from the interaction of the myosin head with the active site on the actin filament, resulting in a shortening of the sarcomere unit ("the power stroke") as the actin filaments slide over the myosin filaments in what is known as the "Sliding Filament Theory" [9,10]. The force generated by cross-bridge formation and the power stroke is transmitted sequentially from the Z-disk, to serial sarcomere units to the connective tissue coverings of muscle fibers, to the muscle tendon and finally, a bony part or fibrous band (eg, pubic ramus or anococcygeal raphae) [11].

Striated Muscle Biochemical Homeostasis and Growth Pathways

Striated muscle growth or hypertrophy occurs during development and in response to mechanical overload (eg, strength training) or anabolic hormonal stimulation (eg, testosterone or β2-adrenergic agonists). Striated muscle atrophy or decrease in muscle mass or fiber size may result from aging, starvation, cancer, diabetes, bed rest, loss of neural input, or catabolic hormonal stimulation (eg, chronic corticosteroid use). The molecular processes responsible for striated muscle growth and atrophy (and thus, its maintenance) involve various signaling pathways [12].

The two major signaling pathways governing striated muscle growth are the insulin-like growth factor 1-phosphoinositide-3-kinase-Akt/protein kinase B-mammalian target of rapamycin (IGF1-PI3K-Akt/PKB-mTOR) pathway (a positive regulator of muscle growth), and the myostatin-Smad3 pathway (a negative regulator of muscle growth — see Fig. 2). IGF1 acting via the IGF1 receptor activates the PI3K-Akt pathway. Akt stimulates protein synthesis by activating the two main complexes of mTOR, mTORC1, and mTORC2. mTOR activation is thought to stimulate muscle growth via the eukaryotic translation initiation factor S6 kinase 1. However, the exact role of mTOR has not been completely characterized as there is still uncertainty surrounding the downstream effectors of mTOR that regulate transcription [13].

Myostatin, produced by striated muscle, is a member of the transforming growth factor β (TGFβ) superfamily. Myostatin is produced by striated muscle and acts as a negative regulator of muscle growth, as revealed by the finding that myostatin mutations in various mammalian species cause muscle hypertrophy [14]. When myostatin binds to its receptors (activin receptor II A or B and activin-like kinase-4 or 5), intracellular signaling activates the transcription of Smad2 or 3, which translocate to the nucleus and inhibit protein synthesis through mechanisms not clearly defined. It may be that the two pathways cross-talk to regulate muscle growth by direct interaction between Smad3 and Akt, as has been shown in other cell systems [15,16].

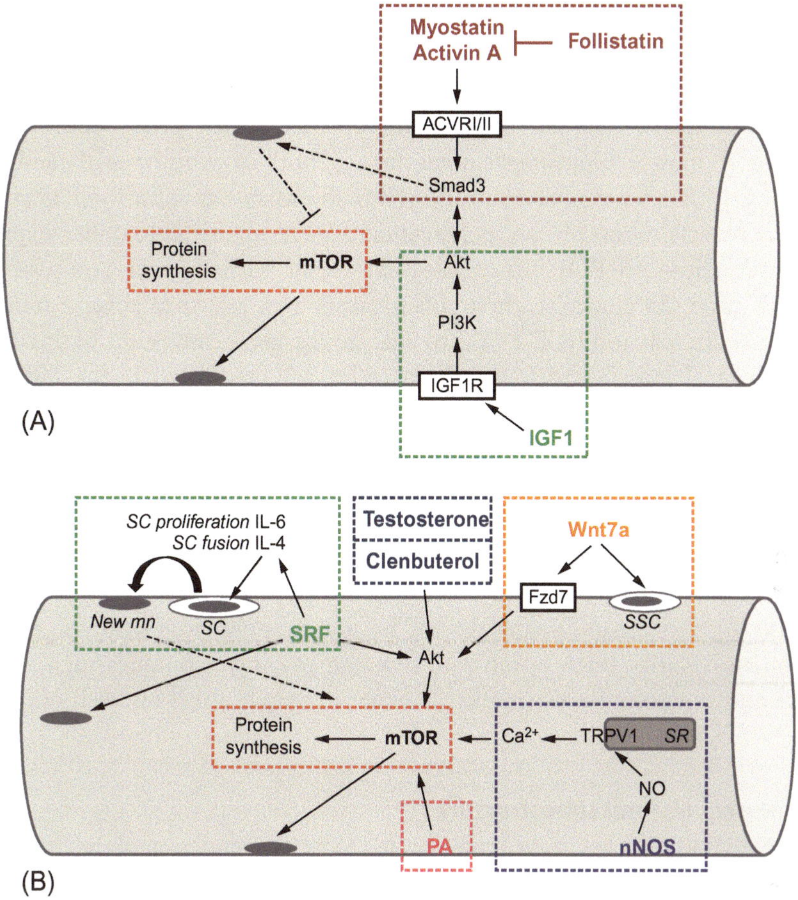

Figure 2 Signaling modules responsible for skeletal muscle growth during development, regeneration, and overload-induced hypertrophy in the adult. (A) Major signaling pathways. IGF1 stimulates mTOR activity and muscle growth via PI3K-Akt. Follistatin induces muscle growth by inhibiting myostatin and activin A. The two pathways cross-talk by direct interaction between Smad3 and Akt. In addition, transcriptional regulation by Smad3/Smad4 heterodimers may repress mTOR and protein synthesis through mechanisms that have not yet been defined. The arrow connecting mTOR with a myonucleus indicates transcriptional roles of mTOR. (B) Additional pathways controlling mTOR activity and protein synthesis. The serum response factor (SRF), phosphatidic acid (PA), and neuronal nitric oxide synthase (nNOS) pathways may be activated by mechanical overload. The dotted arrow connecting newly fused myonuclei (new mn) to the mTOR pathway indicates the postulated increase in protein synthesis and myotube/myofiber growth associated with myoblast/satellite cell fusion. *SC*, satellite cell; *SSC*, satellite stem cell. *From S. Schiaffino, K.A. Dyar, S. Ciciliot, B. Blaauw, M. Sandri, Mechanisms regulating skeletal muscle growth and atrophy, FEBS J. 280 (17) (2013) 4294–4314 [Figure 1]; reproduced with permission.*

Alterations in Striated Muscle Biochemical Pathways

Few studies have evaluated the manner in which risk factors for PFDs influence the molecular processes of striated muscle homeostasis. Visco and Yuan [17] reported on the use of microarray analysis to compare the differential gene expression in the pubococcygeal muscle biopsy specimens in a cohort of women with and without advanced POP (five cases, five controls). They found that myosin-binding protein H and skeletal muscle myosin heavy polypeptide 3 were significantly under-expressed in subjects with advanced POP. In a follow-up study with a larger group of women ($N=40$, 17 cases, 23 controls), these investigators used real-time reverse transcriptase polymerase chain reaction (RT-PCR) to analyze the gene expression of these proteins in pubococcygeal muscle and found them to be downregulated in the subjects with advanced POP [18]. In another small study by Cortes et al. [19], the posterior-lateral portion of the levator ani muscle was biopsied, and the gene expression of IGF1 splice variants analyzed using RT-PCR in a group of 5 nonpregnant premenopausal women and 10 primigravid women at the time of obstetric laceration repair. They found that IGF splice variants were significantly upregulated in the delivery population.

As noted, these studies are few. However, as more is learned about striated muscle homeostasis in other organ systems, we will be able to apply those lessons to pelvic floor striated muscles. Pairing this research with evolving imaging techniques, such as magnetic resonance and ultrasound elastography, is likely to spawn novel research opportunities in PFDs.

Striated Muscle Normal Ultrastructure

Imaging of striated muscle ultrastructure has led to a clear rendering of the lines and disks that give striated muscle its classic appearance (as discussed above). Fig. 3 is an example of a longitudinal electron microscopy view of striated muscle with its lines and disks (or bands) labeled, along with other structures such as the SR, T-tubules (T), and mitochondria (m) [20].

Striated Muscle Ultrastructure Findings Associated With PFDs

Striated muscle of the pelvic floor tissues and organs compose three critical structures: LAMC, EUS, and EAS. Levator ani muscles represent the primary muscular support to the pelvic floor, while the EUS and EAS have essential roles in acting as the "last line of defense" to help maintain urinary and anal incontinence, respectively. Despite these critical functions, few studies have examined the ultrastructural morphology of these muscle bodies. Parks et al. [21] showed an electron micrograph of a LAMC biopsy that had characteristic striated muscle appearance, but also revealed areas of myofibrillar degeneration that were often found in the presence of lipid droplets in patients with anal incontinence. Gosling et al. [22] later showed that striated muscle taken from the EUS and periurethral LAMC consisted only of Type I (slow twitch) fibers in the EUS, but

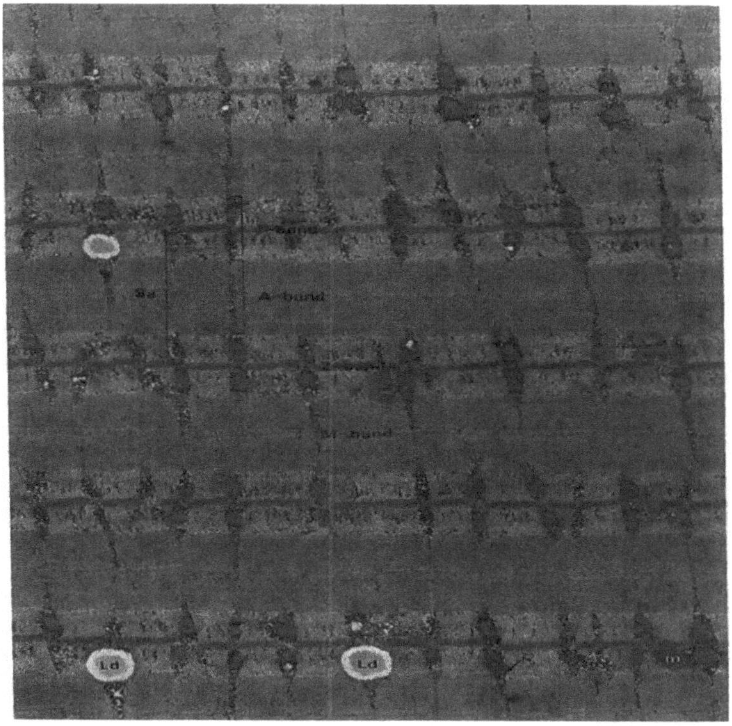

Figure 3 This electron micrograph shows a longitudinal section of a portion of skeletal muscle fiber. Myofibrils show their characteristic pattern of crossbanding with A-, I-, M-, and Z-bands. In the clefts between myofibrils are abundant glycogen particles, lipid droplets (Ld), and sarcoplasmic reticulum. Profiles of mitochondria (m), two to each sarcomere (Sa) occur at the I-band level. *From T. Ebe, S. Kobayashi, Fine Structure of Human Cells and Tissues, Wiley, New York, 1972; reproduced with permission.*

Type I and II (fast twitch) fibers in the LAMC. Notably, slow twitch fibers were found to have a high mitochondrial count, which is thought to be consistent with their role in providing fatigue resistant tonic activity. Hale et al. [23] compared the EUS-striated muscle ultrastructure of women with and without SUI. Like Gosling, they reported the appearance of only Type I fibers in their biopsy samples. Additionally, they noted the endomysial connective tissue made up a greater portion of the biopsy sample in subjects with SUI when compared to controls. However, control subjects were on average 20 years younger than these cases and were less likely to be postmenopausal. Elbadawi [24], in summarizing his work and that of others, commented that both fast and slow twitch-fibers of the EUS had been demonstrated by histochemical and biochemical means, but that Type I (slow-twitch) fibers were the dominant population. Ultrastructural distinctions of Type I fibers included thicker Z-disks, less well-defined M-lines, modest SR, and abundant mitochondria. Similar to the LAMC and EUS, the EAS also consists of Type I and II muscle fibers, providing tonic contraction at rest while also being able to rapidly contract to assist in the maintenance of anal continence [21].

Striated Muscle Ultrastructure Findings Associated With Pregnancy/Parturition

Pregnancy and the events of parturition, particularly vaginal delivery, can have a profound impact on the structure and function of pelvic floor striated muscle. Injuries to the pelvic floor striated muscles, such as the LAMC, EUS, and EAS, are thought to be a major contributing factor to the onset of a PFD [25–27]. Yet, there are no prospective human subject studies assessing ultrastructural alterations in pelvic floor striated muscles following pregnancy and parturition. Animal model studies in this area have been limited as well. Piculo et al. [28] evaluated the urethral striated muscle ultrastructure in (1) diabetic, (2) pregnant, and (3) pregnant, diabetic rodents. They observed increased fiber caliber, normal collagen distribution, and an increased number of mitochondria and vacuoles in the pregnant rodents compared with virgin controls. Pregnant diabetic rodents were found to have increased interstitial collagen, lipid droplets, and numerous subsarcolemmal and intermyofibrillar mitochondria, in addition to larger quantities of glycogen granules, not observed in the pregnant or virgin rodents. Another rodent model study reported on the electron microscopy findings in the striated urethral muscle of virgin, pregnant, 2-day postpartum, and 6-week postpartum rats. Functional testing via rodent cystometry was performed, but direct muscle testing was not, eliminating the possibility of correlating ultrastructural findings with whole tissue and organ function results [29]. Alperin and colleagues used pregnant (mid and late gestation), postpartum (4 and 12 weeks), and age-matched virgin Sprague Dawley rats to assess pregnancy induced adaptations to the architecture and collagen content of pelvic floor skeletal muscle. They observed an increase in skeletal muscle fiber length resulting from the addition of sarcomeres in series and an increase in the collagen content of the skeletal muscle. These findings were not observed in nonpelvic floor muscles (tibialis anterior) [146]. These studies represent limited data, but indicate that changes in the biomechanical properties of genitourinary tract tissue resulting from pregnancy and parturition are likely to result in observable alterations in the ultrastructural morphology, all of which may aid in the understanding of the pathogenesis of PFDs.

Striated Muscle Ultrastructure Findings Associated With Aging

Decline in striated muscle mass (sarcopenia) occurs with aging in both women and men beginning around age 40 and may result in up to a 30–50% loss in striated muscle mass over time [30]. This decline in muscle mass is multifactorial; it is influenced by the preferential loss of Type II myofibers resulting from motor neuron death, physical inactivity, change in hormonal milieu, decreased protein intake, inflammatory mediators, and altered biosynthesis of various proteins [31]. In addition, the declining function of mitochondria — as the source of oxidative metabolism and energy transduction via the Krebs cycle — are thought to play a significant role in muscle aging [32]. In old age, mitochondria suffer a loss of bioenergetics and biochemical activity. In fact, mitochondrial damage has been observed in LAMC tissue of women with POP [33]. These changes in striated muscle result in observable ultrastructure alterations such as decreased muscle fiber size, decreased SCs, segregated calcium stores, and fewer mitochondria [4].

A notable aspect of age-related muscle change is that loss of strength and power exceeds the loss of muscle size and volume [34]. Such age-related changes have been observed in the striated muscle of the EAS and IAS [35–37] and likely explain much of the link between increasing age and the increased prevalence of PFDs [38,39]. Age effects on the LAMC have been observed as well (in human and animal models) [40,41]. In a study of squirrel monkeys — age, weight, and parity matched — with and without POP, gross disruption of the LAMC and age associated changes were observed. However, these alterations to the LAMC were not associated with POP [41]. Nevertheless, Word et al. [42] and Ashton-Miller and DeLancey [25] have posited similar mechanisms by which the loss of LAMC function due to injury and/or aging may lead to POP. Both hypothesized that loss of LAMC support may lead to undue strain on the PFCT, which over long periods of time fail, leading to the clinical presentation of POP.

Smooth Muscle

SM can be distinguished from striated muscle by the absence of the structural proteins that create the appearance of striations. In part, it is the absence of these structural proteins that allows SM to behave in such a dynamic fashion. SM cells are typically smaller than striated muscle and their architecture may vary greatly depending upon the organ/structure in which they reside. SM is surrounded by an extracellular matrix (ECM) that includes collagen, elastic fibers, and other ground matrix. This stroma along with cellular gap junctions likely contributes to the transfer of force generation from individual cells to the target organ. Unlike striated muscle, SM is under autonomic control. However, similar to striated muscle, it may contract in an isometric or isotonic fashion and cross-bridge cycling between actin and myosin resulting from increased intracellular Ca^{2+} serves to initiate contraction [43].

SM Contractile Machinery

SM cells contain actin and myosin filaments. In SM, there are no Z-disks for actin to insert into, but rather dense bodies or bands (as illustrated in Fig. 4) that act as anchors for thin filaments. Actin filaments surround myosin in a lattice-like pattern and force generation is thought to result from cross-bridge formation in the manner proposed by "the sliding filament" theory, similar to striated muscle [1,44].

Biochemistry and Physiology of SM Contraction and Relaxation

Contractile activity of SM is controlled by neural innervation, hormonal input, autocrine/paracrine activity, and load or length changes [45]. Like striated muscle, SM contraction is initiated by increased cytosolic Ca^{2+} (from intracellular and extracellular stores). Ca^{2+} then binds to calmodulin, creating the Ca^{2+}-calmodulin complex which stimulates the phosphorylation of the light chain of myosin. Ca^{2+} sensitization of the

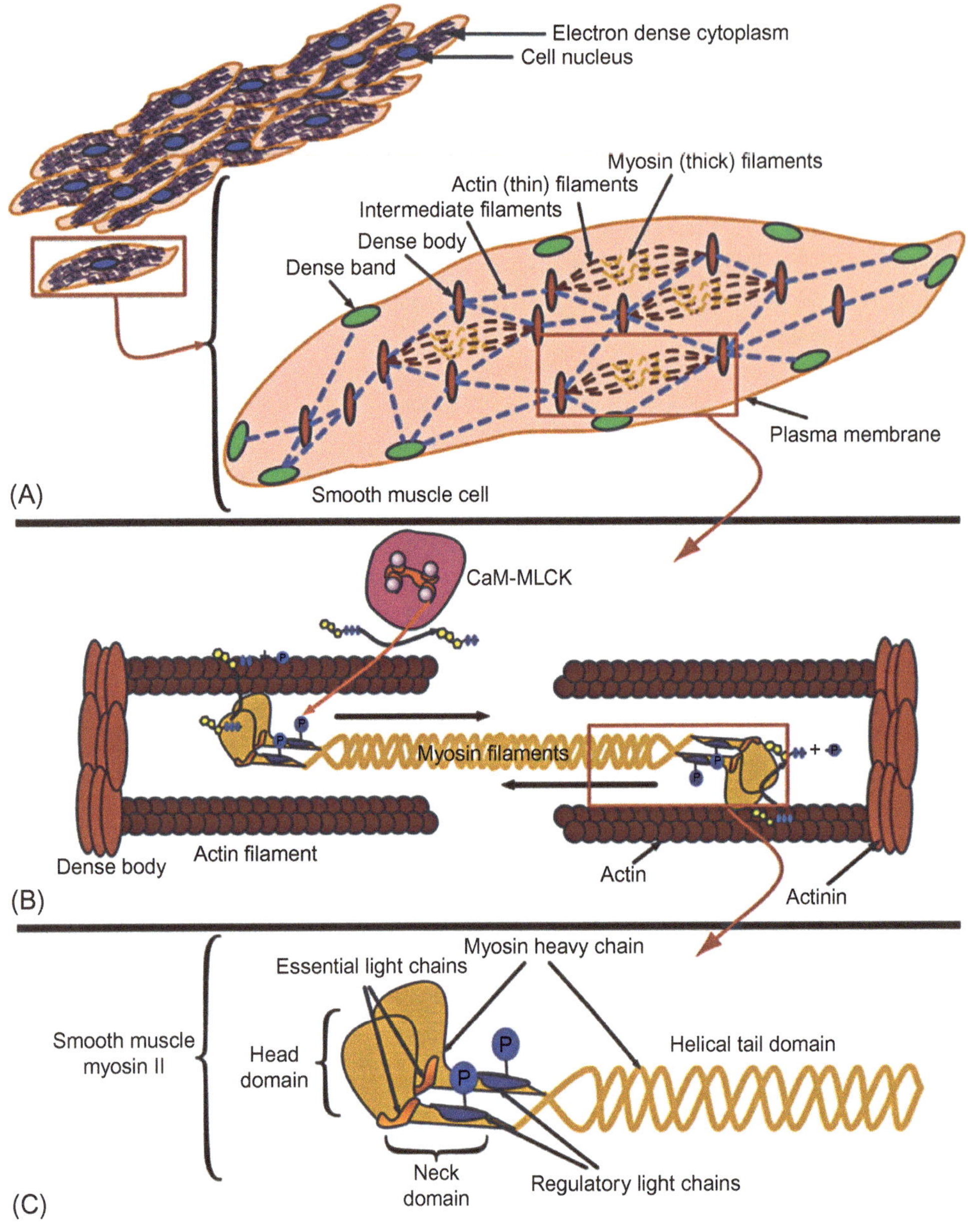

Figure 4 Smooth muscle (SM) contractile machinery. (A) The smooth muscle cell cytoplasm is densely packed with elements of the contractile machinery (thick and thin filaments), and other structural components (dense bodies, dense bands, intermediate filaments). The network formed by the combination of these elements results in force transduction along the longitudinal axis of the cell and cell shortening. (B) The contractile elements are composed of myosin thick filaments and actin thin filaments anchored to dense bodies. The movement of thin filaments caused by phosphorylation of myosin light chains and subsequent ATP hydrolysis by the myosin II ATPase decreases the distance between anchor points. (C) Myosin II is a hexamer composed of two heavy chains, two essential light chains, and two regulatory light chains. Phosphorylation of the two regulatory light chains causes formation of a cross bridge between actin and myosin filaments and also creates a change in the angle of the neck region of myosin II, which causes motion of the actin thin filaments resulting in shortening of the cell. *From H.N. Aguilar, B.F. Mitchell, Physiological pathways and molecular mechanisms regulating uterine contractility, Hum. Reprod. Update. 16 (6) (2010) 725–744 [Figure 2]; reproduced with permission.*

contractile proteins is signaled by the RhoA/Rho kinase pathway to inhibit the dephosphorylation of the light chain by myosin phosphatase, thereby maintaining force generation. Removal of Ca^{2+} from the cytosol and stimulation of myosin phosphatase initiate the process of SM relaxation [46].

As noted above, the interaction of contractile proteins actin and myosin are similar to those found in striated muscle. In SM, actin has four isoforms (α, β, and two forms of γ). The proportion of actin isoforms will vary by organ, but in the human bladder they are 33:25:42%, respectively [47]. SM myosin belongs to the myosin II superfamily of filament forming motors [48]. SM-myosin heavy chain (SM-MHC) forms two isoforms, SM1 and SM2, formed by alternative splicing, while two other MHC isoforms, SM-A and SM-B, are formed by alternative splicing in the amino-terminal region [49]. The velocity of actin shortening can be modulated by the isoform of myosin into which it interacts [50]. SM also expresses two types of light chains, LC_{17a} and LC_{17b} thought to contribute to actin-myosin binding. However, the role of these two proteins on contractile kinetics is not entirely known [51].

Hypertrophic growth of SM of the bladder (detrusor muscle) is well documented in human and animal models. It typically results from urinary obstruction or neurogenic causes and involves multiple cell signaling pathways [52]. Assessing ultrastructural changes that occur during detrusor hypertrophy, Gabella and Uvelius [53] demonstrated that muscle bundles became larger and longer and cell area increased, no mitoses were found, and gap junctions were few or absent. They also noted an increase in intermediate filaments, mitochondria per cell, and interstitial material (see below regarding normal ultrastructure of SM).

Alterations in SM Biochemical Pathways in PFDs

A few studies have attempted to ascertain the biochemical factors that play a role in SM homeostasis in PFDs. Boreham et al. [54] studied a group of women with and without anterior vaginal wall prolapse and found SM-MHC and caldesmon gene expression in vaginal muscularis tissue were increased in women with POP. The authors postulated that caldesmon upregulation may result in the inhibition of vaginal tone, as caldesmon has been shown to decrease myosin ATPase activity [55]. Meanwhile, Takacs et al. [56] measured the messenger RNA expression of caldesmon, SM actin, MHC, and zinc finger protein in the uterosacral ligaments of women with ($n=9$) and without ($n=9$) uterine prolapse and found that women with uterine prolapse had significantly increased caldesmon mRNA and ratio of caldesmon-SM actin expression. Kerkhof et al. [57] biopsied the anterior vaginal cuff in premenopausal women undergoing hysterectomy — 13 without POP and 13 with $\geq$stage 2 anterior vaginal wall prolapse — and found increased SM cells determined by the morphometric analysis of desmin-positive staining cells versus total tissue. They also reported that the percentage of SM cells in the nonprolapsed compartment of the women with POP was similar to controls. In light of these findings, the investigators concluded that the changes observed were the effect of vaginal prolapse rather than its cause. Bortolini et al. [58] assessed the gene expression of SM-MHC (MHY11) and caldesmon (CALD1) from pre- and postmenopausal women with and without POP. They reported the dysregulation of vaginal SM cells finding that MHY11 gene expression was downregulated in premenopausal women with POP compared to premenopausal controls, while CALD1 gene expression was increased six fold in postmenopausal women with POP compared to controls.

Pelvic Floor SM

SM Ultrastructure

SM is found as sheets or layers of cells in various hollow organs of the body affecting organ dimension and mechanical function. In the pelvic floor SM makes up the IUS, IAS and is a component of PFCT (including the vaginal wall). There is great variability in the nature of SM contraction, but there are common features to all SM ultrastructure [53]. SM cells are relatively small in size, uninucleated, and their cytoplasm is packed with contractile filaments. Mitochondria are present and caveoli (ie, invaginations of the cell membrane) can be found at the plasma membrane of SM cells, along with numerous cell-to-cell junctions and cell-to-stroma junctions. The extracellular space is filled with stroma consisting of mostly collagen and elastic fibers [53].

Thin (actin) and thick (myosin) filaments are present in SM cell cytoplasm in abundance. Thin filaments measure 6–8 nm in diameter, are uniform in appearance in various preparations and different muscles, and are readily preserved for electron microscopy after glutaraldehyde fixation. Thick filaments measure 14–16 mm in diameter. They are rather difficult to preserve for microscopy. Also present are strand-like cytoplasmic densities called dense bodies. They represent regions where actin filaments insert or are anchored. Dense bodies are associated with a network of the cytostriated intermediate filaments (10 nm) and the plasma membrane, thus corresponding to the Z-disks of striated muscle fibrils [53].

SM Ultrastructure Findings Associated With PFDs

It has been the rare study that has attempted to correlate observations of pelvic floor SM ultrastructure with a clinical symptom or sign. Hale et al. [23] compared urethral SM ultrastructure in continent and stress urinary incontinent women and found them to be morphologically indistinguishable. Fitzgerald et al. [59], in an analysis of 10 women (5 mixed incontinence, 5 continent), reported smaller electron dense areas of SM sarcolemma, but no other differing morphologic characteristics. Swash et al. [60] compared IAS ultrastructure between a small cohort of women with neurogenic fecal incontinence and another without. In those with fecal incontinence, they observed a loss of SM cells, fewer cell-to-cell junctions, and an increased connective tissue content. In light of our current understanding of the role of fibrillar proteins such as collagen and elastic fibers on pelvic floor support, it may be worth returning to these types of studies (in human subjects and animal models), but with the addition of ultrastructural analysis of PFCT components, and gene and protein expression analysis.

SM Ultrastructure Findings Associated With Pregnancy/Parturition

A notable result in a rodent model study on the effects of pregnancy on bladder and urethra ultrastructure was finding a decrease in subsarcolemmal caveoli [29]. Caveolae are flask-shaped invaginations, 50–100 nm in size that line the SM cell surface [61]. They are thought to contribute to multiple important cellular processes related to the regulation of physiologic bladder function, including contractility. Based on this background, Bakircioglu et al. [29] hypothesized that the decrease in caveolae in pregnancy may impair signal propagation and the coordination of SM cells and thereby negatively impact the continence mechanism.

SM Ultrastructure Findings Associated With Aging

In aging SM, the cell size is slightly larger and they display deep invaginations into the cell membrane and have irregular borders. The amount of stroma increases and is made of large elastic fibers and collagen fibrils up to 105 nm in diameter [53]. As in pregnancy, there is a loss of caveolae that is thought to occur with advancing age. Lowalekar et al. [62] used electron microscopy to assess the bladders of male Sprague-Dawley rats at 10 weeks, 6 months, and 1 year of age to explore this proposition. They noted a consistent age-dependent reduction in the number of caveolae lining cell membranes. These findings along with those noted in Alterations in SM Biochemical Pathways in Pelvic Floor Disorders section give support to the notion expressed by Mei et al. [55] upon their review of the literature on this topic: that abnormalities in PFCT SM may contribute to the pathogenesis of POP, and it is imperative to further this line of research to attain a better understanding of these processes with the goal of improving patient outcomes.

PELVIC FLOOR CONNECTIVE TISSUE

Although the association of PFDs with pregnancy, parturition, and aging has been firmly established, the pathogenesis of PFDs remains incompletely understood [42]. Add to this the conundrum that most women exposed to the risk factors of pregnancy and parturition will not develop a symptomatic PFD; and yet for those who do, the onset will be years after the exposure [63]. Much of the current thinking about the pathogenesis of PFDs relates to the role of PFCT homeostasis and its ability to maintain pelvic floor support [64]. The importance of PFCT to the support of the pelvic floor organs has been an area of interest and debate for some time amongst physicians who care for women with PFDs [65]. The PFCT is the sheath of ECM — composed of collagen, elastic fibers, SM, and other ground matrix — that envelops the pelvic floor muscles and organs, and is thought to provide support in the manner that has been thoughtfully described by DeLancey [66]. The importance of the individual components of the PFCT with respect to PFDs truly came to the fore with the publication of studies in knockout mice that develop POP secondary to null mutations in genes that encode proteins critical to elastic fiber synthesis and assembly [67–69]. In these mice, pregnancy, parturition (vaginal delivery, more so than caesarean) [70], and/or advancing age result in the onset of POP and voiding dysfunction. This work, and that of others which has linked abnormal gene expression with abnormal protein structure or function to abnormal tissue mechanics, has propelled research in the area of PFDs in novel directions in which biomechanics, biomaterials, molecular biology, stem cell therapy, regenerative medicine, and advanced imaging are colliding. What follows is a delineation of the PFCT components that appear to play a dominant role in supporting the pelvic organs with an emphasis on their basic molecular structure, biochemistry, and ultrastructural appearance in PFCT.

Collagen

Collagen Structure

Collagen is the most abundant ECM protein in the human body and plays a vital role in providing structure and tensile strength to tissues. The signature molecular form of collagen consists of three alpha polypeptide chains bound in a triple helix configuration and stabilized by interchain hydrogen bonds. These alpha chains are composed of a series of triplet amino acid pattern: glycine-X-Y, where X and Y can be any amino acid (but are frequently proline and hydroxyproline, respectively) [71]. Collagens exist as either homotrimers or heterotrimers. There have been at least 28 different collagen types described. Collagens I, III, and V have been found in the PFCT where I and III are most abundant. Collagen I fibers provide the majority of tissue resistance to tension. Collagen III confers greater flexibility and distension to tissue, while collagen V appears to be of minor quantitative importance as it forms small fibers of very low tensile strength. The copolymerization of collagen I with III and V forms fibrils that influence the biomechanics of tissue [72]. It is thought that the higher the I/III ratio in a structure the greater its strength (compared to a lower ratio). These fibrillar collagens are thought to be the principal determinants of strength in the vaginal wall. Changes in ratios of these collagens likely contribute to altered vaginal tissue behavior [73].

Collagen Homeostasis (Synthesis and Degradation)

The balance of protein synthesis and degradation is essential to the maintenance of PFCT. The low-level turnover of ECM proteins is part of routine, background homeostasis. Alteration to this level of turnover may occur in physiological processes (development and tissue repair) or pathological processes (tumorigenesis and metastasis). Collagen synthesis begins in the endoplasmic reticulum (ER) where pre-pro-α-chains are synthesized and the triple helix formation occurs along with other posttranslation modifications. Procollagen consists of an amino-terminal propeptide followed by a short, nonhelical N-telopeptide, a central triple helix, a C-telopeptide, and a carboxy-terminal propeptide. Once transported to the extracellular space, proteolytic enzymes (N and C proteinase) remove the extension peptides of procollagen. Collagen maturation (ie, cross-linking) then follows as mature cross-links are produced via the lysyl oxidase pathway as well as nonenzymatic cross-linking referred, to as glycation or Maillard reaction [74]. The latter of these two forms of maturation accumulates with age and is thought to contribute to collagen-rich tissue dysfunction seen in the elderly [75,76].

Collagen degradation is driven by the regulation of matrix metalloproteinases (MMPs). The MMPs are a family of highly conserved, zinc-dependent endopeptidases that participate in the physiologic and pathologic degradation of ECM [77]. Of the two main collagens found in PFCT (collagen I and III), both are preferentially cleaved by MMP-1 (interstitial collagenase), MMP-8 (neutrophil collagenase), and MMP-13 (collagenase 3). The activity of MMPs is regulated by endogenous inhibitors referred to as tissue-derived inhibitors of metalloproteases (TIMPs), in addition to other cell signaling mediators such as TGF-β, and thrombospondin (TSP-1) [78]. There are four TIMPs in humans (TIMP 1–4) that display inhibitory activity against all MMPs [79,80]. The inhibitory ability of TIMPs against MMPs results from the 1:1 formation of enzyme-inhibitor complexes. A balance of MMP activation and TIMP inhibition is essential to the homeostasis of ECM tissues [81].

Collagen Ultrastructure: Overview

Collagen exists in various morphologic forms ranging from long fibrils to complex, interwoven structures. On electron microscopy imaging, collagen fibers consist of bundles of collagenous fibrils held together by inter-fibrillar substance. Fibrils may range in diameter from 10 to 500 nm depending on type and location of the fibril [82]. They are characterized by periodic cross-banding at distinct intervals readily apparent in longitudinal sections. In cross-section, collagen fibers appear as a grouping of punctate marks, each consisting of bundles of fibrils.

Collagen I is about 300 nm long as deduced by electron microscopic analysis and noted to be composed of coarse fibers of many closely packed thick fibrils, whereas collagen III (also referred to as reticular collagen) forms thinner fibers of about 2 μm in diameter that are composed of few loosely disposed thin fibrils [82]. Nevertheless, typing collagen solely by fiber diameter can be misinterpreted [83]. Collagen III forms a delicate network of interwoven fibers that creates a lacework pattern when observed on electron microscopy.

Collagen Ultrastructure in PFCT

Collagen ultrastructure in the pelvic floor has been described in a few studies. In cross-section, collagen fibrils appear as sharply demarcated bundles, evenly spaced, and numerous [84]. When comparing the periurethral collagen content of reproductive age women with and without SUI, Falconer et al. [85] reported larger collagen fibril diameter in the subjects with SUI. However, conducting the same analysis with postmenopausal women, they found no difference in the ultrastructural appearance of collagen [86]. In a study of 30 women, Fitzgerald et al. [84] described differences between the periurethral collagen ultrastructure of those with and without SUI. The authors noted differences in the collagen morphology of subjects with SUI described as obscure (poorly stained, difficult to visualize), dense (collapsed appearance with fiber bunching), and degenerative (rare fibrils surrounded by amorphous material). In a later study in which arcus tendineous fascia pelvis tissue was biopsied from premenopausal women, postmenopausal women on hormone replacement therapy (HRT), and postmenopausal women not on HRT, collagen fibers were noted via scanning electron microscopy to form long parallel bundles (primarily composed of Type III collagen), and no qualitative differences were observed in fiber appearance based on menopause and HRT replacement status [87]. Studies analyzing collagen ultrastructure in the uterosacral-cardinal ligaments of women with and without POP have found collagen fibril diameter to be increased and fibers more loosely packed in women with POP [88,89]. Nevertheless, the small number of subjects in these studies limits the ability to draw firm conclusions regarding ultrastructural collagen morphology and altered pelvic floor tissue biomechanics.

Beyond electron microscopy, the application of atomic force microscopy (AFM) to the study of pelvic floor tissues provides an opportunity to assess ultrastructure and biomechanical properties simultaneously. AFM is unique in its ability to assess tissue elasticity at the microscopic and nanometer scale, distinguishing mechanical properties that can be attributed to the tissue matrix versus individual matrix fibers. The reported use of AFM in analyzing PFCT has been limited, but this is likely to change as the technique is further refined [90,91].

Collagen Biochemistry in PFCT

As the most abundant fibrillar protein found in PFCT, collagen and its metabolism has been of great interest to investigators who want to better understand the pathophysiology of PFDs [92]. Many studies in human subjects have explored this relationship. A common approach has been to compare collagen type (or alteration in the ratio of collagen type) and MMP/TIMP gene and/or protein expression in whole tissue biopsied from PFCT (or fibroblasts cultured and isolated from biopsied tissue, or markers of collagen degradation) in a cohort of women with and without a PFD. Results from these kinds of studies have been collated in excellent review articles by Kerkhof et al. [93], Campeau et al. [94], and most recently, Lim et al. [95]. Nevertheless, it has been difficult to draw firm conclusions about collagen metabolism from these studies due to inconsistent results and the amount of heterogeneity found in this body of literature. The variation found in these studies includes differences in the type and amount of tissue analyzed, biochemical methods employed, and the criteria used to determine cases and controls; a majority of these studies include a small number of subjects. Further, the absence of an animal model of SUI or POP resulting from a primary collagen defect (unlike the lysyl oxidase like-1 (LOXL-1) or fibulin 5 (FBLN 5) KO for elastic fibers) has made it all the more challenging to gain greater insight into the complex regulation of collagen metabolism in PFCT.

In addition to the above, analyzing differences in cases with a PFD versus controls without a PFD does not allow one to make a determination about whether differences observed are the cause, or effect of the PFD (especially true for POP). To overcome this limitation in study design, Kerkhof et al. [57] analyzed histologic (including collagen type) and biochemical (number of collagen cross-links) features of anterior vaginal wall connective tissue in age- and parity-matched subjects with and without POP, but notably included *nonprolapsed* tissue from the anterior vaginal wall of POP subjects. These specimens were taken from the precervical anterior vaginal wall and would allow those subjects with POP to act as a self-control. Thus nonprolapsed tissues in controls (POP-Q $\geq$stage 0–1) were compared to nonprolapsed and prolapsed tissues in cases (POP-Q $\geq$stage 2 at anterior vaginal wall), while the nonprolapsed and prolapsed tissues for each case would be compared as well. These investigators found an increase in collagen cross-links and SM cell density in prolapsed tissue, but no differences in the nonprolapsed tissues between cases and controls. They concluded that their results indicate the differences observed were more likely the effect of, rather than the cause of POP. Using the same study population, Ruiz-Zapata et al. [96] made similar observations and drew similar conclusions regarding the mechano-responsiveness of fibroblast cells cultured and isolated from the same tissue specimens. In their discussion, they note the clear difference observed in nonprolapsed versus prolapsed tissues in subjects with POP, but offer that their findings represent an opportunity for the introduction of autologous based cell therapies (eg, combining biomaterials with unaffected autologous fibroblasts or stem cells).

Pregnancy and parturition have a significant effect on PFCT remodeling and thus can considerably impact collagen metabolism, but it has been the rare study that has investigated collagen biochemistry in pregnancy or during the postpartum period [97]. Much of the early research in this area focused on collagen content and tissue biomechanics, as opposed to the expression or quantification of collagen type or MMP activity [98–100]. More recent studies have explored similar themes, but they, too, have focused on collagen quantification and tissue biomechanics as opposed to biochemical change [101].

With advancing age, collagen synthesis slows while its degradation increases leading to significant quantitative and qualitative changes [76]. Investigators have observed differences in collagen expression in the PFCT of pre- and postmenopausal women [87,102,103]. It is likely that these changes, along with others, explain (in part) the observation of increased prevalence of PFDs with advancing age. Change in collagen expression associated with aging has been shown to be reversible with estrogen therapy in postmenopausal women and ovariectomized animal models [87,104–106]. Nevertheless, a clinical role for estrogen therapy for the prevention or treatment of POP (perhaps as an adjuvant therapy) has not been established.

Elastin and Elastic Fibers

Elastin and Elastic Fibers Structure

Elastin is an amorphous appearing protein that consists of cross-linked tropoelastin (its precursor). Elastic fibers (distinct from elastin) consist of an elastin protein core surrounded by a microfibrillar scaffold. Elastic fibers are found in the connective tissue of various organ systems endowing these tissues with the properties of resistance and recoil [74]. Elastic fiber morphology varies depending on the tissue in which it is found. In ligaments, lung, and skin, elastic fibers are small, rope-like, and of variable length. In arteries they form concentric sheets or lamellae, while in cartilage they form 3D honeycomb arrangements of large anastomosing fibers [107]. The "elasticity" of elastin is thought to arise from its entropic nature. That is, elastin is in a state of maximum disorder when unstretched. When a force is applied and stretch is induced, order is increased, thereby decreasing entropy. Once the force is removed, recoil will occur because the partially ordered particles will return to their initial random state to maximize entropy [108].

Elastic Fiber Homeostasis (Synthesis and Degradation)

Elastin is an insoluble polymer that consists of cross-linked tropoelastin monomers, while microfibrils are made up a complex array of macromolecules. Tropoelastin is encoded by a single human gene and is secreted as an ~60–70 kDa unglycosylated protein by a variety of cells, including fibroblasts, endothelial and SM cells, and chondrocytes. Following transcription, it undergoes extensive alternative splicing before leaving the cell nucleus. Translation occurs on the surface of the rough ER after which intracellular tropoelastin is thought to be chaperoned by elastin-binding protein to the cell surface avoiding self-aggregation and premature degradation. Once there, tropoelastin is secreted into the extracellular space where LOXL-1 is thought to play a role in the cross-linking of tropoelastin monomers to create mature elastin polymers, and FBLN 5 is thought to play a role in the transfer of elastin aggregates to microfibrils, the microfibrils acting as the scaffolding of elastic fibers (see Fig. 5 [109]) [110].

Elastic fiber homeostasis results from a complex interplay between its synthesis and degradation. In healthy adult tissues, elastin turnover is very slow, which in part explains its long half-life. Significant elastic fiber degradation is only likely to occur in certain pathological conditions, as a result of inflammation, during wound healing, and in pregnancy [99,110]. Elastic fiber degradation also occurs with aging, contributing to the appearance of aged skin and onset of cardiovascular disease [111]. Three classes of proteolytic enzymes are capable of degrading elastin: serine proteases (eg, neutrophil elastase), cysteine proteases (eg, cathepsins), and the MMPs (MMP-2, MMP-9, MMP-12) as previously discussed [112].

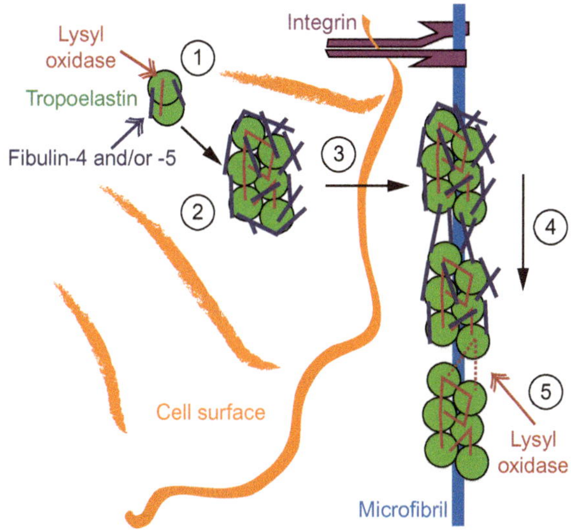

Figure 5 Model of elastic fiber assembly. (1) Tropoelastin is transported to assembly sites on the plasma membrane where it is organized into small aggregates that are cross-linked by a LOX (eg, LOXL-1). Cell surface receptors/binding proteins, such as heparan sulfate proteoglycans, at these sites may serve to assist with the initial assembly steps. Interaction with FBLN 4 and/or 5 may facilitate cross-linking or possibly help limit the size of the aggregates. (2) The aggregates remain on the cell surface while newly secreted elastin is added to increase the size. (3) The aggregates are then transferred to extracellular microfibrils, which interact with the cell through integrins. FBLN 4 and/or 5, or another microfibril-associated protein, may assist the transfer of elastin aggregates to the microfibril. (4) Elastin aggregates on the microfibril coalesce into larger structures. FBLN 4 and/or 5 may facilitate this process. (5) The elastin aggregates are further cross-linked by a LOX to form the complete elastic fiber. *From J.E. Wagenseil, R.P. Mecham, New insights into elastic fiber assembly, Birth Defects Res. C Embryo Today 81 (4) (2007) 229–240 [Figure 7]; reproduced with permission.*

Elastin and Elastic Fibers Ultrastructure: Overview

Ultrastructural examination of elastic fibers reveals two distinct components: an electron-lucent amorphous core component (insoluble elastin) surrounded by 10–12 nm tubular appearing microfibrils. Using cyrofixation and freeze-fracture electron microscopy, the amorphous core that appears to be a homogenous mass has been determined to be a network of thin elastin fibrils bound together in bundles [113]. Elastic fibers are notoriously stable with a half-life close to the life span [114].

Ultrastructure in PFCT

Few studies have analyzed the morphology of elastic fibers in the PFCT in human subjects [115–117] and even fewer have investigated elastic fiber ultrastructure. Goepel and Thomssen [118] used transmission electron microscopy to study periurethral connective tissue in stress incontinent postmenopausal women with normotonic or hypotonic urethras. Women with hypotonic urethras were observed to have irregular, fragmented elastin and loss of the amorphous elastin core, while women with normotonic urethras had normal appearing amorphous elastin in close proximity to collagen fibrils, SM cells, and fibrocytes. Despite the paucity of data in human and animal studies regarding elastic fiber ultrastructure, a number of studies in human and rodent tissues have observed that alterations in elastic fiber morphology directly impact tissue biomechanics [68,117,119,120]. Further, if we draw from the literature of other organ systems, it is likely that histologically abnormal elastic fibers will have detectable ultrastructural abnormalities as well [121,122]. Nevertheless, this warrants further study with regard to PFCT ultrastructure and PFDs.

Elastic fiber degradation is known to occur with advancing age in all organs and tissues, including the pelvic floor [123]. This is thought to result from the constitutive expression of MMPs, the accumulation of oxidative damage, and the accumulation of pathological cross-links [124]. Such age-related changes surely impact the biomechanics of pelvic floor tissues and organs. In older adults, elastic fibers show signs of degradation becoming more tortuous and fragmented. The ultrastructural correlation to these microscopic changes can be observed in the work of Pasquali-Ronchetti and Baccarani-Contri [125] who described the ultrastructural changes to dermal elastic fibers during development and with aging. Their work revealed that with aging elastic fibers gradually become tortuous, frayed, and porous; holes develop in the amorphous component of the elastin protein, and isolated fibers appear with atypical organization and interwoven filaments.

Biochemistry in PFCT

The importance of elastic fibers to the maintenance of pelvic floor support has been revealed in the study of knockout mice that develop POP secondary to null mutations in genes that encode proteins critical to elastic fiber synthesis and assembly: LOXL-1 and FBLN 5 [67–69,126]. These findings are supported by the observed association of urinary incontinence and POP in women with connective tissue disorders such as Marfan syndrome, Cutis Laxa and Ehlers-Danlos syndrome [127–129].

A number of studies have attempted to assess alterations in elastic fiber homeostasis in women with PFDs. Yamamoto et al. [130] were among the first to demonstrate a decrease in elastin gene expression and protein synthesis in cardinal ligaments of women with POP. Others have noted a decreased expression in the lysyl oxidase (LOX) family of genes related to elastic fiber synthesis and/or FBLN 5 in the PFCT of female subjects with POP when compared to those without [131–136]. Chen et al. [137] evaluated the mRNA and protein expression of elastolytic proteins in the vaginal tissue of women with SUI and POP and found that alpha-1 antitrypsin — an elastase inhibitor — was downregulated in women with SUI and POP. Others have shown that elastolytic enzymes are upregulated in the PFCT of women with PFDs and that this may result from the upregulation of enzymes such as MMP-2 or MMP-9, or by the suppression of inhibitors such as TIMPs 1–4 [137–141]. These findings have led to the conclusion that altered elastic fiber homeostasis contributes to PFDs. Nevertheless, the biochemical mechanisms by which this occurs remain obscure; however, the biomechanics of elastin suggests a lack of resilience and inability to restore to normal pelvic anatomy following each increase in abdominal pressure.

The impact of pregnancy and parturition on elastic fiber biochemistry has been best studied in animal models. Drewes et al. [126] studied the regulation of elastic fiber synthesis and assembly in the vaginal wall of mice during pregnancy and after vaginal delivery. They showed that in the immediate postpartum period, a burst in the protein expression of FBLN 5, tropoelastin, and desmosine occurs. Subsequently, Wieslander et al. [142] demonstrated a marked increase in MMP-2 and 9 in the vaginal wall tissue of wild-type mice 12–48 h after vaginal delivery. This led both groups to hypothesize that proper elastic fiber remodeling was critical to the recovery of pelvic floor support. Testing this hypothesis, Drewes injected elastase into the posterior vaginal wall of postpartum wild-type mice. This resulted in the development of posterior wall prolapse in the elastase injected mice, but not in saline or inactivated elastase-treated animals [126]. Rahn et al. [143] tested this same hypothesis by exposing nulliparous FBLN 5 KO, FBLN 5 heterozygotes, and wild-type mice to vaginal balloon distension at 6 weeks of age (prior to the spontaneous onset of POP in FBLN 5 KO mice) and found that only the FBLN 5 KO mice developed POP. They also noted that vaginal distention resulted in elastic fibers appearing fragmented and disrupted. Furthermore, Budatha et al. [144], using a knockin mouse with a mutant arginine/glycine/aspartic acid motif (RGD-a sequence known to mediate binding to cell surface integrin receptors) and a double FBLN 5 KO/MMP-9 mouse, showed that the development of POP was correlated to the presence of aberrant elastic fibers and the upregulation of MMP-9. As a result of these studies and others, leaders in the field have hypothesized that PFDs such as POP and SUI — to a large extent — result from an acquired elastinopathy in which the primary event is abnormal elastic fiber remodeling following parturition and/or that which results from aging [42,64]. This is an intriguing hypothesis as a causal pathway of PFDs, one that deserves further investigation.

CONCLUSION

As with most complex systems, a deeper understanding of its parts is required to gain a greater understanding of the whole. In considering the pelvic floor, we are still at an early stage of understanding, especially with respect to the ultrastructure and biochemistry of its "parts." Nevertheless, since the call to action expressed in the published proceedings of the 2002 National Institute of Health-sponsored meeting on basic and translational research in female PFDs, significant advances in these areas have been made — as evidenced by the work discussed in this chapter [92]. In addition, there is good reason to be hopeful more progress is to come as clinicians, engineers, and basic scientists begin to work collaboratively on disorders that have adversely affected women for millennia [145]. It is these multidisciplinary efforts that will lead to the design and conduct of innovative research to advance knowledge in this area, and provide the foundation from which novel ideas for therapeutic interventions, disease assessment and preventive measures will spring forth.

REFERENCES

[1] Y.C. Fung, Biomechanics: Mechanical Properties of Living Tissues, second ed., Springer, New York, 1993.
[2] I. Nygaard, L. Brubaker, H.M. Zyczynski, et al., Long-term outcomes following abdominal sacrocolpopexy for pelvic organ prolapse, JAMA 309 (19) (2013) 2016–2024.
[3] M.D. Barber, L. Brubaker, K.L. Burgio, et al., Comparison of 2 transvaginal surgical approaches and perioperative behavioral therapy for apical vaginal prolapse: the OPTIMAL randomized trial, JAMA 311 (10) (2014) 1023–1034.
[4] W.R. Frontera, J. Ochala, Skeletal muscle: a brief review of structure and function, Calcif. Tissue Int. 96 (3) (2015) 183–195.
[5] N. Light, A.E. Champion, Characterization of muscle epimysium, perimysium and endomysium collagens, Biochem. J. 219 (3) (1984) 1017–1026.
[6] R. Knoll, B. Buyandelger, M. Lab, The sarcomeric Z-disc and Z-discopathies, J. Biomed. Biotechnol. 2011 (2011) 569628.
[7] K. Wang, R. McCarter, J. Wright, J. Beverly, R. Ramirez-Mitchell, Regulation of skeletal muscle stiffness and elasticity by titin isoforms: a test of the segmental extension model of resting tension, Proc. Natl. Acad. Sci. U. S. A. 88 (16) (1991) 7101–7105.
[8] T. Funatsu, H. Higuchi, S. Ishiwata, Elastic filaments in skeletal muscle revealed by selective removal of thin filaments with plasma gelsolin, J. Cell Biol. 110 (1) (1990) 53–62.
[9] A.F. Huxley, R. Niedergerke, Structural changes in muscle during contraction: interference microscopy of living muscle fibres, Nature 173 (4412) (1954) 971–973.
[10] H. Huxley, J. Hanson, Changes in the cross-striations of muscle during contraction and stretch and their structural interpretation, Nature 173 (4412) (1954) 973–976.
[11] S.M. Greising, H.M. Gransee, C.B. Mantilla, G.C. Sieck, Systems biology of skeletal muscle: fiber type as an organizing principle, Wiley Interdiscip. Rev. Syst. Biol. Med. 4 (5) (2012) 457–473.
[12] S. Cohen, J.A. Nathan, A.L. Goldberg, Muscle wasting in disease: molecular mechanisms and promising therapies, Nat. Rev. Drug Discov. 14 (1) (2015) 58–74.
[13] S. Schiaffino, K.A. Dyar, S. Ciciliot, B. Blaauw, M. Sandri, Mechanisms regulating skeletal muscle growth and atrophy, FEBS J. 280 (17) (2013) 4294–4314.
[14] S.J. Lee, Regulation of muscle mass by myostatin, Annu. Rev. Cell Dev. Biol. 20 (2004) 61–86.
[15] I. Remy, A. Montmarquette, S.W. Michnick, PKB/Akt modulates TGF-beta signalling through a direct interaction with Smad3, Nat. Cell Biol. 6 (4) (2004) 358–365.
[16] A.R. Conery, Y. Cao, E.A. Thompson, C.M. Townsend Jr., T.C. Ko, K. Luo, Akt interacts directly with Smad3 to regulate the sensitivity to TGF-beta induced apoptosis, Nat. Cell Biol. 6 (4) (2004) 366–372.
[17] A.G. Visco, L. Yuan, Differential gene expression in pubococcygeus muscle from patients with pelvic organ prolapse, Am. J. Obstet. Gynecol. 189 (1) (2003) 102–112.
[18] A.F. Hundley, L. Yuan, A.G. Visco, Skeletal muscle heavy-chain polypeptide 3 and myosin binding protein H in the pubococcygeus muscle in patients with and without pelvic organ prolapse, Am. J. Obstet. Gynecol. 194 (5) (2006) 1404–1410.
[19] E. Cortes, L.F. te Fong, M. Hameed, et al., Insulin-like growth factor-1 gene splice variants as markers of muscle damage in levator ani muscle after the first vaginal delivery, Am. J. Obstet. Gynecol. 193 (1) (2005) 64–70.
[20] T. Ebe, S. Kobayashi, Fine Structure of Human Cells and Tissues, Wiley, New York, 1972.

[21] A.G. Parks, M. Swash, H. Urich, Sphincter denervation in anorectal incontinence and rectal prolapse, Gut 18 (8) (1977) 656–665.
[22] J.A. Gosling, J.S. Dixon, H.O. Critchley, S.A. Thompson, A comparative study of the human external sphincter and periurethral levator ani muscles, Br. J. Urol. 53 (1) (1981) 35–41.
[23] D.S. Hale, J.T. Benson, L. Brubaker, M.C. Heidkamp, B. Russell, Histologic analysis of needle biopsy of urethral sphincter from women with normal and stress incontinence with comparison of electromyographic findings, Am. J. Obstet. Gynecol. 180 (2 Pt 1) (1999) 342–348.
[24] A. Elbadawi, Functional anatomy of the organs of micturition, Urol. Clin. N. Am. 23 (2) (1996) 177–210.
[25] J.A. Ashton-Miller, J.O. Delancey, On the biomechanics of vaginal birth and common sequelae, Annu. Rev. Biomed. Eng. 11 (2009) 163–176.
[26] K.P. Sajadi, B.C. Gill, M.S. Damaser, Neurogenic aspects of stress urinary incontinence, Curr. Opin. Obstet. Gynecol. 22 (5) (2010) 425–429.
[27] A.H. Sultan, M.A. Kamm, C.N. Hudson, J.M. Thomas, C.I. Bartram, Anal-sphincter disruption during vaginal delivery, N. Engl. J. Med. 329 (26) (1993) 1905–1911.
[28] F. Piculo, G. Marini, A.M. Barbosa, et al., Urethral striated muscle and extracellular matrix morphological characteristics among mildly diabetic pregnant rats: translational approach, Int. Urogynecol. J. 25 (3) (2014) 403–415.
[29] M.E. Bakircioglu, K.D. Sievert, A. Lau, C.S. Lin, T.F. Lue, The effect of pregnancy and delivery on the function and ultrastructure of the rat bladder and urethra, BJU Int. 85 (3) (2000) 350–361.
[30] J.A. Faulkner, L.M. Larkin, D.R. Claflin, S.V. Brooks, Age-related changes in the structure and function of skeletal muscles, Clin. Exp. Pharmacol. Physiol. 34 (11) (2007) 1091–1096.
[31] I.G. Ko, J.W. Jeong, Y.H. Kim, et al., Aerobic exercise affects myostatin expression in aged rat skeletal muscles: a possibility of antiaging effects of aerobic exercise related with pelvic floor muscle and urethral rhabdosphincter, Int. Neurourol. J. 18 (2) (2014) 77–85.
[32] C.M. Peterson, D.L. Johannsen, E. Ravussin, Skeletal muscle mitochondria and aging: a review, J. Aging Res. 2012 (2012) 194821.
[33] R. Yiou, F.J. Authier, R. Gherardi, C. Abbou, Evidence of mitochondrial damage in the levator ani muscle of women with pelvic organ prolapse, Eur. Urol. 55 (5) (2009) 1241–1243.
[34] M.V. Narici, N. Maffulli, Sarcopenia: characteristics, mechanisms and functional significance, Br. Med. Bull. 95 (2010) 139–159.
[35] H. Strasser, M. Tiefenthaler, M. Steinlechner, I. Eder, G. Bartsch, G. Konwalinka, Age dependent apoptosis and loss of rhabdosphincter cells, J. Urol. 164 (5) (2000) 1781–1785.
[36] D. Perucchini, J.O. DeLancey, J.A. Ashton-Miller, A. Galecki, G.N. Schaer, Age effects on urethral striated muscle. II. Anatomic location of muscle loss, Am. J. Obstet. Gynecol. 186 (3) (2002) 356–360.
[37] D.J. Boyle, C.H. Knowles, J. Murphy, et al., The effects of age and childbirth on anal sphincter function and morphology in 999 symptomatic female patients with colorectal dysfunction, Dis. Colon Rectum 55 (3) (2012) 286–293.
[38] I. Nygaard, M.D. Barber, K.L. Burgio, et al., Prevalence of symptomatic pelvic floor disorders in US women, JAMA 300 (11) (2008) 1311–1316.
[39] J.M. Wu, C.A. Matthews, C.P. Vaughan, A.D. Markland, Urinary, fecal, and dual incontinence in older U.S. adults, J. Am. Geriatr. Soc. 63 (5) (2015) 947–953.
[40] T. Dimpfl, C. Jaeger, W. Mueller-Felber, et al., Myogenic changes of the levator ani muscle in premenopausal women: the impact of vaginal delivery and age, Neurourol. Urodyn. 17 (3) (1998) 197–205.
[41] L.M. Pierce, S. Baumann, M.R. Rankin, et al., Levator ani muscle and connective tissue changes associated with pelvic organ prolapse, parity, and aging in the squirrel monkey: a histologic study, Am. J. Obstet. Gynecol. 197 (1) (2007) 60. e61–69.
[42] R.A. Word, S. Pathi, J.I. Schaffer, Pathophysiology of pelvic organ prolapse, Obstet. Gynecol. Clin. N. Am. 36 (3) (2009) 521–539.
[43] G. Gabella, Ultrastructure of Smooth Muscle, first ed., Kluwer, Boston, MA, 1990.
[44] H.N. Aguilar, B.F. Mitchell, Physiological pathways and molecular mechanisms regulating uterine contractility, Hum. Reprod. Update 16 (6) (2010) 725–744.

[45] M.S. Barany, Biochemistry of Smooth Muscle Contraction, Elsevier, San Diego, CA, 1996.
[46] R.C. Webb, Smooth muscle contraction and relaxation, Adv. Physiol. Educ. 27 (1–4) (2003) 201–206.
[47] U. Malmqvist, A. Arner, B. Uvelius, Cytoskeletal and contractile proteins in detrusor smooth muscle from bladders with outlet obstruction — a comparative study in rat and man, Scand. J. Urol. Nephrol. 25 (4) (1991) 261–267.
[48] J.R. Sellers, Myosins: a diverse superfamily, Biochim. Biophys. Acta 1496 (1) (2000) 3–22.
[49] R. Nagai, M. Kuro-o, P. Babij, M. Periasamy, Identification of two types of smooth muscle myosin heavy chain isoforms by cDNA cloning and immunoblot analysis, J. Biol. Chem. 264 (17) (1989) 9734–9737.
[50] T.J. Eddinger, D.P. Meer, Single rabbit stomach smooth muscle cell myosin heavy chain SMB expression and shortening velocity, Am. J. Physiol. Cell Physiol. 280 (2) (2001) C309–316.
[51] A.P. Somlyo, Myosin isoforms in smooth muscle: how may they affect function and structure? J. Muscle Res. Cell Motil. 14 (6) (1993) 557–563.
[52] K.E. Andersson, A. Arner, Urinary bladder contraction and relaxation: physiology and pathophysiology, Physiol. Rev. 84 (3) (2004) 935–986.
[53] G. Gabella, B. Uvelius, Urinary bladder of rat: fine structure of normal and hypertrophic musculature, Cell Tissue Res. 262 (1) (1990) 67–79.
[54] M.K. Boreham, R.T. Miller, J.I. Schaffer, R.A. Word, Smooth muscle myosin heavy chain and caldesmon expression in the anterior vaginal wall of women with and without pelvic organ prolapse, Am. J. Obstet. Gynecol. 185 (4) (2001) 944–952.
[55] S. Mei, M. Ye, L. Gil, et al., The role of smooth muscle cells in the pathophysiology of pelvic organ prolapse, Female Pelvic Med. Reconstr. Surg. 19 (5) (2013) 254–259.
[56] P. Takacs, M. Gualtieri, M. Nassiri, K. Candiotti, A. Fornoni, C.A. Medina, Differential expression of smooth muscle regulatory proteins in the uterosacral ligaments of women with uterine prolapse, Am. J. Obstet. Gynecol. 202 (6) (2010) 620. e621–625.
[57] M.H. Kerkhof, A.M. Ruiz-Zapata, H. Bril, et al., Changes in tissue composition of the vaginal wall of premenopausal women with prolapse, Am. J. Obstet. Gynecol. 210 (2) (2014) 168. e161–169.
[58] M.A. Bortolini, O. Shynlova, H.P. Drutz, et al., Expression of genes encoding smooth muscle contractile proteins in vaginal tissue of women with and without pelvic organ prolapse, Neurourol. Urodyn. 31 (1) (2012) 109–114.
[59] M.P. FitzGerald, B. Russell, D. Hale, J.T. Benson, L. Brubaker, Ultrastructure of detrusor and urethral smooth muscle in women with urinary incontinence, Am. J. Obstet. Gynecol. 182 (4) (2000) 879–884.
[60] M. Swash, A. Gray, D.Z. Lubowski, R.J. Nicholls, Ultrastructural changes in internal anal sphincter in neurogenic faecal incontinence, Gut 29 (12) (1988) 1692–1698.
[61] M. Murakumo, T. Ushiki, T. Koyanagi, K. Abe, Scanning electron microscopic studies of smooth muscle cells and their collagen fibrillar sheaths in empty, distended and contracted urinary bladders of the guinea pig, Arch. Histol. Cytol. 56 (4) (1993) 441–449.
[62] S.K. Lowalekar, V. Cristofaro, Z.M. Radisavljevic, S.V. Yalla, M.P. Sullivan, Loss of bladder smooth muscle caveolae in the aging bladder, Neurourol. Urodyn. 31 (4) (2012) 586–592.
[63] R.C. Bump, P.A. Norton, Epidemiology and natural history of pelvic floor dysfunction, Obstet. Gynecol. Clin. N. Am. 25 (4) (1998) 723–746.
[64] I. Nygaard, Pelvic floor recovery after childbirth, Obstet. Gynecol. 125 (3) (2015) 529–530.
[65] R.F. Porges, J.C. Porges, G. Blinick, Mechanisms of uterine support and the pathogenesis of uterine prolapse, Obstet. Gynecol. 15 (1960) 711–726.
[66] J.O. DeLancey, Anatomy and biomechanics of genital prolapse, Clin. Obstet. Gynecol. 36 (4) (1993) 897–909.
[67] X. Liu, Y. Zhao, J. Gao, et al., Elastic fiber homeostasis requires lysyl oxidase-like 1 protein, Nat. Genet. 36 (2) (2004) 178–182.
[68] X. Liu, Y. Zhao, B. Pawlyk, M. Damaser, T. Li, Failure of elastic fiber homeostasis leads to pelvic floor disorders, Am. J. Pathol. 168 (2) (2006) 519–528.

[69] H. Yanagisawa, E.C. Davis, B.C. Starcher, et al., Fibulin-5 is an elastin-binding protein essential for elastic fibre development in vivo, Nature 415 (6868) (2002) 168–171.
[70] A.M. Gustilo-Ashby, U. Lee, D. Vurbic, et al., The impact of cesarean delivery on pelvic floor dysfunction in lysyl oxidase like-1 knockout mice, Female Pelvic Med. Reconstr. Surg. 16 (1) (2010) 21–30.
[71] R.O. Hynes, A. Naba, Overview of the matrisome — an inventory of extracellular matrix constituents and functions, Cold Spring Harb. Perspect. Biol. 4 (1) (2012) a004903.
[72] D.E. Birk, Type V collagen: heterotypic type I/V collagen interactions in the regulation of fibril assembly, Micron 32 (3) (2001) 223–237.
[73] M. Alperin, A. Feola, L. Meyn, R. Duerr, S. Abramowitch, P. Moalli, Collagen scaffold: a treatment for simulated maternal birth injury in the rat model, Am. J. Obstet. Gynecol. 202 (6) (2010) 589. e581–588.
[74] R.P. Mecham, Overview of extracellular matrix, Curr. Protoc. Cell Biol. (2012), http://dx.doi.org/10.1002/0471143030.cb1001s57.
[75] R.G. Paul, A.J. Bailey, Glycation of collagen: the basis of its central role in the late complications of ageing and diabetes, Int. J. Biochem. Cell Biol. 28 (12) (1996) 1297–1310.
[76] A.J. Bailey, R.G. Paul, L. Knott, Mechanisms of maturation and ageing of collagen, Mech. Ageing Dev. 106 (1–2) (1998) 1–56.
[77] S. Ricard-Blum, The collagen family, Cold Spring Harb. Perspect. Biol. 3 (1) (2011) a004978.
[78] A. Page-McCaw, A.J. Ewald, Z. Werb, Matrix metalloproteinases and the regulation of tissue remodelling, Nat. Rev. Mol. Cell Biol. 8 (3) (2007) 221–233.
[79] G. Murphy, H. Nagase, Progress in matrix metalloproteinase research, Mol. Asp. Med. 29 (5) (2008) 290–308.
[80] A.H. Baker, D.R. Edwards, G. Murphy, Metalloproteinase inhibitors: biological actions and therapeutic opportunities, J. Cell Sci. 115 (Pt 19) (2002) 3719–3727.
[81] H. Birkedal-Hansen, Proteolytic remodeling of extracellular matrix, Curr. Opin. Cell Biol. 7 (5) (1995) 728–735.
[82] D.A.D. Parry, A.S. Craig, Growth and development of collagen fibrils in connective tissue, in: A. Ruggeri, P.M. Motta (Eds.), Ultrastructure of the Connective Tissue Matrix, vol. 3, Springer, Boston, MA, 1984, pp. 34–64.
[83] D.R. Keene, L.Y. Sakai, H.P. Bachinger, R.E. Burgeson, Type III collagen can be present on banded collagen fibrils regardless of fibril diameter, J. Cell Biol. 105 (5) (1987) 2393–2402.
[84] M.P. Fitzgerald, J. Mollenhauer, D.S. Hale, J.T. Benson, L. Brubaker, Urethral collagen morphologic characteristics among women with genuine stress incontinence, Am. J. Obstet. Gynecol. 182 (6) (2000) 1565–1574.
[85] C. Falconer, B. Blomgren, O. Johansson, et al., Different organization of collagen fibrils in stress-incontinent women of fertile age, Acta Obstet. Gynecol. Scand. 77 (1) (1998) 87–94.
[86] C. Falconer, G. Ekman-Ordeberg, B. Blomgren, et al., Paraurethral connective tissue in stress-incontinent women after menopause, Acta Obstet. Gynecol. Scand. 77 (1) (1998) 95–100.
[87] P.A. Moalli, L.C. Talarico, V.W. Sung, et al., Impact of menopause on collagen subtypes in the arcus tendineous fasciae pelvis, Am. J. Obstet. Gynecol. 190 (3) (2004) 620–627.
[88] M.C. Salman, O. Ozyuncu, M.F. Sargon, T. Kucukali, T. Durukan, Light and electron microscopic evaluation of cardinal ligaments in women with or without uterine prolapse, Int. Urogynecol. J. 21 (2) (2010) 235–239.
[89] L. Han, L. Wang, Q. Wang, H. Li, H. Zang, Association between pelvic organ prolapse and stress urinary incontinence with collagen, Exp. Ther. Med. 7 (5) (2014) 1337–1341.
[90] H.K. Graham, N.W. Hodson, J.A. Hoyland, et al., Tissue section AFM: in situ ultrastructural imaging of native biomolecules, Matrix Biol. 29 (4) (2010) 254–260.
[91] I. Sridharan, Y. Ma, T. Kim, W. Kobak, J. Rotmensch, R. Wang, Structural and mechanical profiles of native collagen fibers in vaginal wall connective tissues, Biomaterials 33 (5) (2012) 1520–1527.
[92] A.M. Weber, G.M. Buchsbaum, B. Chen, et al., Basic science and translational research in female pelvic floor disorders: proceedings of an NIH-sponsored meeting, Neurourol. Urodyn. 23 (4) (2004) 288–301.

[93] M.H. Kerkhof, L. Hendriks, H.A. Brolmann, Changes in connective tissue in patients with pelvic organ prolapse — a review of the current literature, Int. Urogynecol. J. Pelvic Floor Dysfunct. 20 (4) (2009) 461–474.
[94] L. Campeau, I. Gorbachinsky, G.H. Badlani, K.E. Andersson, Pelvic floor disorders: linking genetic risk factors to biochemical changes, BJU Int. 108 (8) (2011) 1240–1247.
[95] V.F. Lim, J.K. Khoo, V. Wong, K.H. Moore, Recent studies of genetic dysfunction in pelvic organ prolapse: the role of collagen defects, Aust. N. Z. J. Obstet. Gynaecol. 54 (3) (2014) 198–205.
[96] A.M. Ruiz-Zapata, M.H. Kerkhof, B. Zandieh-Doulabi, H.A. Brolmann, T.H. Smit, M.N. Helder, Functional characteristics of vaginal fibroblastic cells from premenopausal women with pelvic organ prolapse, Mol. Hum. Reprod. 20 (11) (2014) 1135–1143.
[97] L. Zhao, C.S. Samuel, G.W. Tregear, F. Beck, E.M. Wintour, Collagen studies in late pregnant relaxin null mice, Biol. Reprod. 63 (3) (2000) 697–703.
[98] M.L. Harkness, R.D. Harkness, The collagen content of the reproductive tract of the rat during pregnancy and lactation, J. Physiol. 123 (3) (1954) 492–500.
[99] J.F. Woessner, T.H. Brewer, Formation and breakdown of collagen and elastin in the human uterus during pregnancy and post-partum involution, Biochem. J. 89 (1963) 75–82.
[100] A. Rundgren, Physical properties of connective tissue as influenced by single and repeated pregnancies in the rat, Acta Physiol. Scand. Suppl. 417 (1974) 1–138.
[101] K. Yoshida, C. Reeves, J. Vink, et al., Cervical collagen network remodeling in normal pregnancy and disrupted parturition in Antxr2 deficient mice, J. Biomech. Eng. 136 (2) (2014) 021017.
[102] C. Falconer, G. Ekman-Ordeberg, U. Ulmsten, G. Westergren-Thorsson, K. Barchan, A. Malmstrom, Changes in paraurethral connective tissue at menopause are counteracted by estrogen, Maturitas 24 (3) (1996) 197–204.
[103] E. Trabucco, M. Soderberg, L. Cobellis, et al., Role of proteoglycans in the organization of periurethral connective tissue in women with stress urinary incontinence, Maturitas 58 (4) (2007) 395–405.
[104] A.L. Clark, O.D. Slayden, K. Hettrich, R.M. Brenner, Estrogen increases collagen I and III mRNA expression in the pelvic support tissues of the rhesus macaque, Am. J. Obstet. Gynecol. 192 (5) (2005) 1523–1529.
[105] P.A. Moalli, K.M. Debes, L.A. Meyn, N.S. Howden, S.D. Abramowitch, Hormones restore biomechanical properties of the vagina and supportive tissues after surgical menopause in young rats, Am. J. Obstet. Gynecol. 199 (2) (2008) 161. e161–168.
[106] T.I. Montoya, P.A. Maldonado, J.F. Acevedo, R.A. Word, Effect of vaginal or systemic estrogen on dynamics of collagen assembly in the rat vaginal wall, Biol. Reprod. 92 (2) (2015) 43.
[107] C.M. Kielty, M.J. Sherratt, C.A. Shuttleworth, Elastic fibres, J. Cell Sci. 115 (Pt 14) (2002) 2817–2828.
[108] C.A. Hoeve, P.J. Flory, The elastic properties of elastin, Biopolymers 13 (4) (1974) 677–686.
[109] J.E. Wagenseil, R.P. Mecham, New insights into elastic fiber assembly, Birth Defects Res. C Embryo Today 81 (4) (2007) 229–240.
[110] S.M. Mithieux, A.S. Weiss, Elastin, Adv. Protein Chem. 70 (2005) 437–461.
[111] C.M. Kielty, Elastic fibres in health and disease, Expert Rev. Mol. Med. 8 (19) (2006) 1–23.
[112] Z. Werb, M.J. Banda, J.H. McKerrow, R.A. Sandhaus, Elastases and elastin degradation, J. Investig. Dermatol. 79 (Suppl. 1) (1982) 154s–159s.
[113] I. Pasquali Ronchetti, C. Fornieri, M. Baccarani-Contri, D. Volpin, The ultrastructure of elastin revealed by freeze-fracture electron microscopy, Micron 10 (2) (1979) 89–99.
[114] S.D. Shapiro, S.K. Endicott, M.A. Province, J.A. Pierce, E.J. Campbell, Marked longevity of human lung parenchymal elastic fibers deduced from prevalence of D-aspartate and nuclear weapons-related radiocarbon, J. Clin. Invest. 87 (5) (1991) 1828–1834.
[115] G.Y. el-Kholi, S.N. Mina, Elastic tissue of the vagina in genital prolapse. A morphological study, J. Egypt. Med. Assoc. 58 (3–4) (1975) 196–204.
[116] J.A. Karam, D.V. Vazquez, V.K. Lin, P.E. Zimmern, Elastin expression and elastic fibre width in the anterior vaginal wall of postmenopausal women with and without prolapse, BJU Int. 100 (2) (2007) 346–350.

[117] C. Goepel, Differential elastin and tenascin immunolabeling in the uterosacral ligaments in postmenopausal women with and without pelvic organ prolapse, Acta Histochem. 110 (3) (2008) 204–209.
[118] C. Goepel, C. Thomssen, Changes in the extracellular matrix in periurethral tissue of women with stress urinary incontinence, Acta Histochem. 108 (6) (2006) 441–445.
[119] D.D. Rahn, M.D. Ruff, S.A. Brown, H.F. Tibbals, R.A. Word, Biomechanical properties of the vaginal wall: effect of pregnancy, elastic fiber deficiency, and pelvic organ prolapse, Am. J. Obstet. Gynecol. 198 (5) (2008) 590. e591–596.
[120] K.T. Downing, M. Billah, E. Raparia, et al., The role of mode of delivery on elastic fiber architecture and vaginal vault elasticity: a rodent model study, J. Mech. Behav. Biomed. Mater. 29 (2014) 190–198.
[121] K. Tamura, Y. Fukuda, M. Ishizaki, Y. Masuda, N. Yamanaka, V.J. Ferrans, Abnormalities in elastic fibers and other connective-tissue components of floppy mitral valve, Am. Heart J. 129 (6) (1995) 1149–1158.
[122] Y. Fukuda, M. Kawamoto, A. Yamamoto, M. Ishizaki, F. Basset, Y. Masugi, Role of elastic fiber degradation in emphysema-like lesions of pulmonary lymphangiomyomatosis, Hum. Pathol. 21 (12) (1990) 1252–1261.
[123] Y. Jiang, W. Zong, H. Luan, et al., Decreased expression of elastin and lysyl oxidase family genes in urogenital tissues of aging mice, J. Obstet. Gynaecol. Res. 40 (8) (2014) 1998–2004.
[124] M.J. Sherratt, Tissue elasticity and the ageing elastic fibre, Age 31 (4) (2009) 305–325.
[125] I. Pasquali-Ronchetti, M. Baccarani-Contri, Elastic fiber during development and aging, Microsc. Res. Tech. 38 (4) (1997) 428–435.
[126] P.G. Drewes, H. Yanagisawa, B. Starcher, et al., Pelvic organ prolapse in fibulin-5 knockout mice: pregnancy-induced changes in elastic fiber homeostasis in mouse vagina, Am. J. Pathol. 170 (2) (2007) 578–589.
[127] M.E. Carley, J. Schaffer, Urinary incontinence and pelvic organ prolapse in women with Marfan or Ehlers Danlos syndrome, Am. J. Obstet. Gynecol. 182 (5) (2000) 1021–1023.
[128] S.V. Choudhary, S. Bisati, S. Koley, Congenital cutis laxa with rectal and uterovaginal prolapse, Indian J. Dermatol. Venereol. Leprol. 77 (3) (2011) 321–324.
[129] M. Castori, Ehlers-danlos syndrome, hypermobility type: an underdiagnosed hereditary connective tissue disorder with mucocutaneous, articular, and systemic manifestations, ISRN Dermatol. 2012 (2012) 751768.
[130] K. Yamamoto, M. Yamamoto, K. Akazawa, S. Tajima, H. Wakimoto, M. Aoyagi, Decrease in elastin gene expression and protein synthesis in fibroblasts derived from cardinal ligaments of patients with prolapsus uteri, Cell Biol. Int. 21 (9) (1997) 605–611.
[131] W. Kobak, J. Lu, A. Hardart, C. Zhang, F.Z. Stanczyk, J.C. Felix, Expression of lysyl oxidase and transforming growth factor beta2 in women with severe pelvic organ prolapse, J. Reprod. Med. 50 (11) (2005) 827–831.
[132] J. Klutke, Q. Ji, J. Campeau, et al., Decreased endopelvic fascia elastin content in uterine prolapse, Acta Obstet. Gynecol. Scand. 87 (1) (2008) 111–115.
[133] H.J. Jung, M.J. Jeon, G.W. Yim, S.K. Kim, J.R. Choi, S.W. Bai, Changes in expression of fibulin-5 and lysyl oxidase-like 1 associated with pelvic organ prolapse, Eur. J. Obstet. Gynecol. Reprod. Biol. 145 (1) (2009) 117–122.
[134] P. Takacs, M. Nassiri, K. Candiotti, J. Yang, S. Yavagal, C.A. Medina, Differential expression of fibulins in the uterosacral ligaments of women with uterine prolapse, Arch. Gynecol. Obstet. 282 (4) (2010) 389–394.
[135] B.H. Zhao, J.H. Zhou, Decreased expression of elastin, fibulin-5 and lysyl oxidase-like 1 in the uterosacral ligaments of postmenopausal women with pelvic organ prolapse, J. Obstet. Gynaecol. Res. 38 (6) (2012) 925–931.
[136] Y. Zhou, O. Ling, L. Bo, Expression and significance of lysyl oxidase-like 1 and fibulin-5 in the cardinal ligament tissue of patients with pelvic floor dysfunction, J. Biomed. Res. 27 (1) (2013) 23–28.
[137] B. Chen, Y. Wen, M.L. Polan, Elastolytic activity in women with stress urinary incontinence and pelvic organ prolapse, Neurourol. Urodyn. 23 (2) (2004) 119–126.

[138] M. Dviri, E. Leron, J. Dreiher, M. Mazor, R. Shaco-Levy, Increased matrix metalloproteinases-1,-9 in the uterosacral ligaments and vaginal tissue from women with pelvic organ prolapse, Eur. J. Obstet. Gynecol. Reprod. Biol. 156 (1) (2011) 113–117.
[139] Y.J. Moon, J.R. Choi, M.J. Jeon, S.K. Kim, S.W. Bai, Alteration of elastin metabolism in women with pelvic organ prolapse, J. Urol. 185 (5) (2011) 1786–1792.
[140] M. Budatha, S. Silva, T.I. Montoya, et al., Dysregulation of protease and protease inhibitors in a mouse model of human pelvic organ prolapse, PLoS ONE 8 (2) (2013) e56376.
[141] M. Alarab, H. Kufaishi, S. Lye, H. Drutz, O. Shynlova, Expression of extracellular matrix-remodeling proteins is altered in vaginal tissue of premenopausal women with severe pelvic organ prolapse, Reprod. Sci. 21 (6) (2014) 704–715.
[142] C.K. Wieslander, S.I. Marinis, P.G. Drewes, P.W. Keller, J.F. Acevedo, R.A. Word, Regulation of elastolytic proteases in the mouse vagina during pregnancy, parturition, and puerperium, Biol. Reprod. 78 (3) (2008) 521–528.
[143] D.D. Rahn, J.F. Acevedo, R.A. Word, Effect of vaginal distention on elastic fiber synthesis and matrix degradation in the vaginal wall: potential role in the pathogenesis of pelvic organ prolapse, Am. J. Physiol. Regul. Integr. Comp. Physiol. 295 (4) (2008) R1351–1358.
[144] M. Budatha, S. Roshanravan, Q. Zheng, et al., Extracellular matrix proteases contribute to progression of pelvic organ prolapse in mice and humans, J. Clin. Invest. 121 (5) (2011) 2048–2059.
[145] K.T. Downing, Uterine prolapse: from antiquity to today, Obstet. Gynecol. Int. 2012 (2012) 649459.
[146] M. Alperin, D.M. Lawley, M.C. Esparza, R.L. Lieber, Pregnancy-induced adaptations in the intrinsic structure of rat pelvic floor muscles. Am. J Obstet. Gynecol. 213 (2) (2015) 191 e191–197.

CHAPTER NINE

Impact of Genetics on Pelvic Floor Biomechanics

M.A.T. Bortolini
Federal University of São Paulo, São Paulo, Brazil

INTRODUCTION

Pelvic floor function depends on the integrity and coordination of the structures that support pelvic organs: connective tissues in the form of ligaments and endopelvic fascia, and pelvic floor muscles [1]. Dysregulation in the molecular components of these integral structures may be associated with pelvic floor biomechanical imbalance in women with, or predisposed to, pelvic dysfunction such as pelvic organ prolapse (POP) and stress urinary incontinence (SUI), either as cause or effect [2–4].

The molecular alterations observed in the damaged pelvic floor tissues may result from conditions in which they are overstretched by excessive mechanical loading such as pregnancy, vaginal childbirth, chronic coughing, or obesity [1,3]. Proper tissue remodeling, wound healing, and repair after damage are fundamental to maintain mechanical stability of the pelvic floor support structures. If healing is ineffective, pelvic tissues would not regain their strength and POP and SUI could eventually occur [3,5]. On the other hand, women may have inherited susceptible tissues that predispose them to develop pelvic dysfunctions [6,7].

In this chapter, we present evidence that links molecular content in the pelvic floor tissues and pelvic dysfunction, and the influence of genetic background, to the constitution and biomechanics of the pelvic floor, and the development of pelvic floor disorders.

MOLECULAR AND BIOCHEMICAL CHANGES IN PELVIC FLOOR

The majority of studies that investigate molecular changes associated with pelvic floor dysfunctions are focused on the connective tissue components of the pelvic floor support apparatus. Collagen and elastin are two main proteins composing the

No conflicts to declare.

extracellular matrix (ECM) of connective tissues. Collagen is responsible for the tensile strength and integrity, while elastin provides elasticity and resilience to the tissues, affording the tissues the ability to extend and recoil. ECM homeostasis relies upon coordinated biosynthesis and biodegradation. Several proteins are involved in those processes including (1) bone morphogenetic protein 1 (BMP1), lysyl oxidases family of proteins (LOXs), fibulins (FIB), laminins (LAM), and fibrillins as responsible for the ECM biogenesis; and (2) matrix metalloproteinases (MMPs) and tissue inhibitors of MMPs (TIMPs), neutrophil elastases, and their inhibitors alpha-1 antitrypsin, and cathepsins that play a role in the ECM breakdown and remodeling [8–15].

The amount of collagen and elastin and variations in their metabolism are important factors influencing maternal tissue damage during parturition, the degree of resistance to abdominal pressure transmission to the pelvic floor, the propensity to repair injured tissues and consequent pelvic floor relaxation [3,5,16]. There is substantial evidence that a deficient synthesis and/or degradation of collagen and elastic fibers may be associated with POP and SUI, further described in this chapter (Figs. 1 and 2). It is not known definitively at this time if these pathological changes in the ECM stem from genetic causes, from environmental influences, or life events, but it is likely all three play a role in proteolytic disorder contribution to POP and SUI.

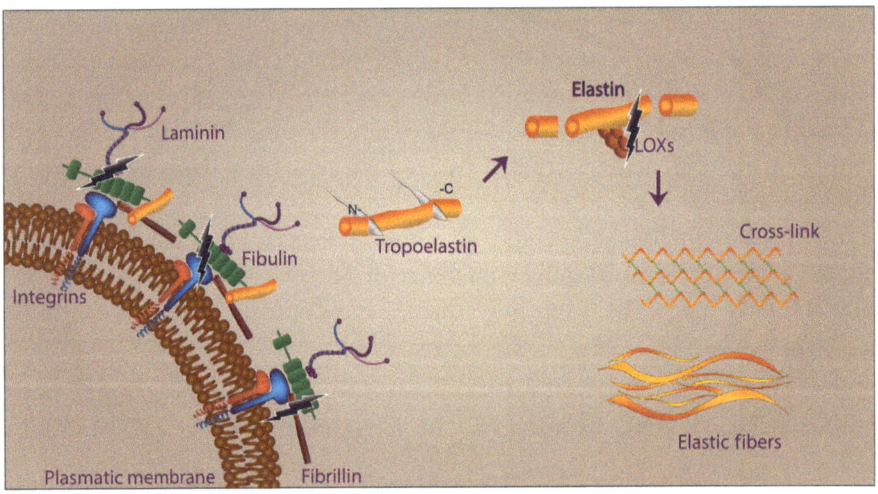

Figure 1 Schematic representation of elastogenesis pathway. LOXs act to crosslink microfibrils to create mature elastin fibers. Fibulin proteins stabilize and organize elastic fibers. Laminins are basement membrane proteins that contribute to the structure of ECM. Fibrillins provide a scaffold for deposition of elastin and are responsible for different cell–matrix interactions. Lightning symbols represent breakdown points in the elastin pathway that may interfere in the ECM homeostasis and may be related to the development of pelvic floor dysfunctions.

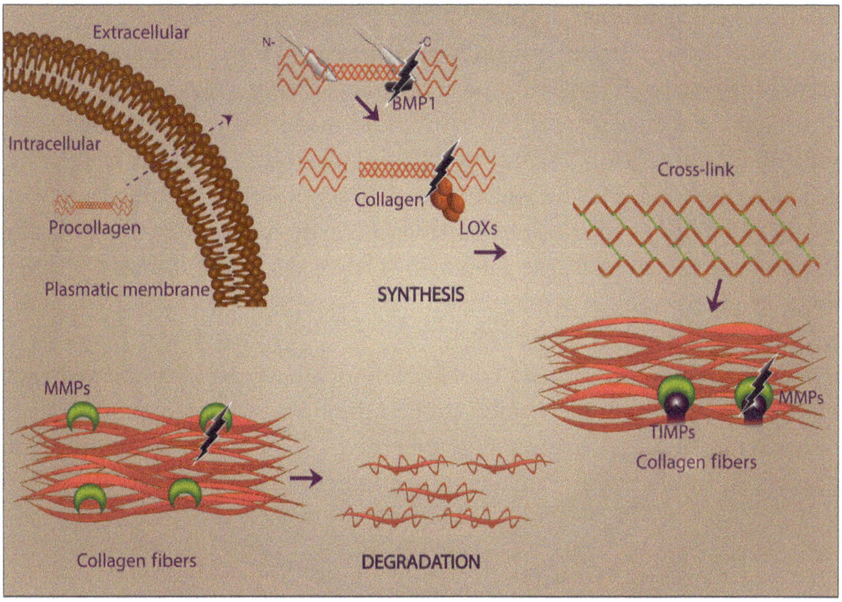

Figure 2 Schematic representation of collagen pathway. Laminins are basement membrane proteins that contribute to the structure of ECM. BMP1 acts in the maturation of procollagen chains and in the activation of LOXs. LOXs act to crosslink collagen microfibrils to create mature collagen fibers. MMPs are proteolytic enzymes while TIMPs inhibit the MMPs action. Lightning symbols represent breakdown points in the collagen pathway that may interfere in the ECM homeostasis and may be related to the development of pelvic floor dysfunctions.

Experimental Studies: Genetic Knockout Models

In animal studies involving genetic knockout or null models, specific genes are blocked, and their role in the ECM metabolism, pelvic floor support, and in POP and SUI development can be investigated.

LOX enzymes act primarily to crosslink collagen and elastin microfibrils to create mature elastin and collagen fibers [13,14]. Deficiency in lysyl oxidase-like-1 (LOXL1), a protein essential for the postnatal deposition of elastic fibers and their repair after injury, leads to severe POP in mice, but only after vaginal delivery due to a failure to replenish mature crosslinked elastin fibers, and results in weakness of the vaginal wall, paraurethral pathology, and lower urinary tract dysfunction [5,17,18]. In a genetic knockout model, Alperin et al. have demonstrated that nulliparous mice deficient in the LOXL1 protein presented altered biomechanical behavior of the vagina and supportive tissues, represented by a 31% decrease in ultimate load at failure when compared to age-matched wild-type animals. Those findings suggest mechanically weaker tissues and the role of LOXL1 in ECM function [18].

Fibulin-5 (FIB-5) protein stabilizes and organizes elastic fibers [12]. Drewes and colleagues have shown that FIB-5 null mice develop POP, suggesting that synthesis and assembly of elastic fibers are crucial for recovery of pelvic organ support after vaginal delivery, and that disordered elastic fiber homeostasis is associated with POP development in mice [16]. Vaginas of FIB-5 knockout mice that developed POP exhibited decreased maximal stress, increased distensibility and strain, and decreased stiffness [19], phenotypes that resemble pelvic floor dysfunction in humans. Similarly, animals with null mutation in the Fibulin-3 (FIB-3) gene develop a POP phenotype comprising bladder, vaginal vault, descended uterine horns associated with rectal prolapse, flank, and inguinal hernias [20].

Therefore, experimental genetic knockout models have demonstrated an association between ECM proteins, biomechanical properties of pelvic floor supportive tissues and pelvic dysfunctions. Whether individual inherited DNA mutations or variations that would cause similar alterations remains an open question.

Human Case-Control Studies

Based on the possible target proteins identified in animal studies, several researchers have investigated similar molecular alterations in the pelvic floor of women with pelvic floor disorders.

Despite controversial findings, it seems that patients with POP and SUI present with decreased total collagen content in their pelvic floor [21–23]. In addition to the amount, the quality of the collagen and its subtypes has also been investigated in pelvic tissues. Collagen type I fibers are resistant to tension, while collagen type III offers flexibility and distention, influencing the biomechanical characteristics of a given tissue. Qualitative analysis of collagen has found reduced ratio of collagen subtypes I/III in the arcus tendineous fasciae pelvis and vagina [24,25] of postmenopausal women with POP compared to matched women with stage I or no POP. Barbiero et al. observed that the parametrium of women with uterine prolapse was shorter, thinner, and with more disorderly arranged collagen fibers compared to women without POP [26]. Therefore, those findings suggest more laxity of tissues in women with POP. In regards to maturity, Jackson et al. found that patients with advanced apical prolapse presented vaginal tissues with relatively high content of immature collagen crosslinks compared to non-POP patients, and are hence more susceptible to rupture [21].

Similar to collagen, overall elastin content seems diminished in pelvic floor tissues of women with POP and SUI [27–30]. Reduction of LOX, LOXL1, and LOXL3 gene and protein levels were detected in vaginal tissues of premenopausal women with advanced POP compared to women without POP [31]. By evaluating uterosacral ligaments, Klutke and coworkers observed decreased mRNA expression of LOX, LOXL1 and LOXL2, and decreased desmosine content in women with uterine prolapse compared to women without POP [32]. Desmosine concentration reflects the crosslink and elastin maturation processes, suggesting that reduced LOX family of proteins and desmosine may be implicated in high levels of immature collagen and elastin and consequently, deficient support function.

Other authors reported decreased FIB-5 in pelvic supportive tissues in POP. FIB-5 mRNA was diminished in the vagina of women with anterior vaginal wall prolapse compared to age-matched women without POP. However, the evaluation was not controlled by menopausal status [33]. Jung et al. observed lower expression levels of the FIB-5 gene and protein in uterosacral ligaments of postmenopausal women with advanced POP compared to postmenopausal women with no pelvic floor defect [34]. Those findings corroborate deficient elastin synthesis in the genesis or progression of POP.

BMP1, also known as procollagen C proteinase, acts in the maturation of procollagen chains and in the activation of LOXs [14,15]. BMP1 gene expression is decreased in pre- and postmenopausal women with severe POP compared to asymptomatic women, suggesting that dysregulation of BMP1 may contribute to deficient vaginal connective tissue and pelvic floor support [35].

Disturbances in collagen and elastin metabolism may lead to tissues with impaired mechanical strength, elasticity, and resilience. These previous data suggest that alterations in the biogenesis of both proteins may play a role in the development of pelvic floor dysfunctions.

On the other hand, activated collagenolysis and enhanced elastolytic activity might be also connected to pelvic floor dysfunction. Collagen amount and turnover depend on the critical balance between the proteolytic enzymes, MMPs and TIMPs [9,10,36]. Jackson et al. reported an increase in MMP activity (pro and active MMP-2 and active MMP-9) in vaginal tissue of premenopausal women with POP compared to women without POP [21]. Chen et al. reported an increase in proMMP-1 mRNA, decrease in TIMP-1 gene expression, and total collagen content in periurethral vaginal tissues of premenopausal women with SUI and POP, but not in matched controls, and no difference in gene expression of proMMP-2, proMMP-9, TIMP-2, or TIMP-3 [37]. Moalli et al. found an increased active form of MMP-9 in vaginal tissues of women with advanced POP compared to women with <stage II prolapse, while Gabriel et al. found that active MMP-2 was increased in uterosacral ligaments of patients with POP controlled by age and parity [25,38]. Alarab et al. observed that the MMP-2 gelatinase activity was upregulated in premenopausal patients with advanced POP, irrespective of menstrual phase status, compared to matched controls.

TIMP-1–4 gene and TIMP-1 protein expression were significantly reduced, whereas protein expression of MMP-12 (pro and active forms) was significantly increased in vaginal biopsies of patients with POP in the proliferative phase of the menstrual cycle compared to corresponding controls [39]. Despite controversies due to different study designs, the data suggest that remodeling of connective tissue is increased in the vagina and supportive tissues of women with POP. The increased metabolic turnover may result in impaired pelvic floor support.

Alpha-1 antitrypsin is a secretory glycoprotein that neutralizes the effects of proteases, by binding and inhibiting neutrophil elastase (a protease that degrades a wide range of ECM proteins), thereby preventing deterioration of tissue and ECM [8,11]. Chen and colleagues demonstrated a significant decrease in alpha-1 antitrypsin in periurethral vaginal wall tissues from women with SUI and/or POP compared to women with no pelvic floor dysfunctions [29]. Those findings suggest dysregulation of the finely tuned balance of ECM production/degradation in women with POP, which may compromise the quality of connective tissue in these women and reduce the strength of pelvic floor support. Those observational case-control studies immensely contribute to identify potential markers and molecular mechanisms behind POP and SUI, but they do not clarify innumerous shortcomings in the literature.

In this section, we have didactically mentioned some of the studies that support the most acceptable theory of imbalance between synthesis and degradation of ECM proteins in the genesis and/or progression of pelvic dysfunctions. Nevertheless, inconsistent findings are commonly noted and can be explained by interstudies differences in the population enrolled, site and tissue sampling chosen to represent the pelvic floor, and techniques applied for molecular and biochemical analysis which can detect and quantify the ECM contents via mRNA level, precursor or mature molecules. In addition, the majority of studies were biased for demographic characteristics of the study population regarding menopause status, age, ethnicity, POP stage and lifestyle, all of which play a role in the pathophysiology of this condition. The cross-section study design does not allow researchers to draw conclusions regarding the association of each risk factor and the biochemical changes observed in the tissues of women with pelvic dysfunction.

There is a general agreement that dysregulation in the components of the pelvic floor connective tissue in POP and SUI may be the cause or effect of these dysfunctions. Women may have inherited "deficient tissues" that predispose them to develop the dysfunction spontaneously or initiated by events such as pregnancy, vaginal childbirth, menopause, or aging. On the other hand, pregnancy and childbirth trauma may directly influence the biochemical content, or the molecular changes can result from POP progression per se due to stretch that the pelvic floor tissues undergo.

Studies of Association Between Biochemical Changes and Biomechamical Properties of Pelvic Floor

Some investigations addressed whether biomechanical properties are correlated to biochemical changes in pelvic tissues or associated with independent risk factors for pelvic dysfunctions. In 2012, Zhou et al. reported that anterior and posterior vaginal walls of premenopausal and postmenopausal women with POP were significantly stiffer (higher elasticity constant) than those of controls and this was correlated to the increase in the collagen type III protein expression in postmenopausal affected women compared to asymptomatic controls [40]. Epstein et al. detected that extensibility was related to POP quantification stage in a linear fashion [2].

An in vitro study by Zong et al. demonstrated that cultured human vaginal fibroblasts are mechano-sensitive after cycles of biaxial stretch. In the presence of stretch, culture media analysis has shown increased endogenous collagenase activity compared to media from cultured vaginas that had not been stretched. Collagenase activity reflects the balance between active collagenases and their inhibitors, and its increase was linearly correlated with the magnitude of the mechanical elongation. This suggests that exposing vaginal tissue to repetitive mechanical stretch, such as conditions that may increase intra-abdominal pressure, may contribute to POP causality and progression [41].

Recently, Ruiz-Zapata reported in vitro functional characteristics of vaginal fibroblasts from premenopausal women with cystocele and non-POP women matched for age and parity, and correlated them with ECM proteolytic enzymes [42]. To rule out an acquired or intrinsic defect in patients with cystocele, and the influence of the aging process, samples from the anterior vaginal wall and the precervical region (non-POP site) were collected from the same POP patient. A contractility assay was used to test cultured fibroblast-mediated collagen contraction. Cellular mechano-responses were evaluated by measuring MMP and TIMP gene and protein expression. The authors observed that fibroblastic cells from the prolapsed site showed delayed fibroblastic-mediated collagen contraction and decreased production of MMP-2, but not TIMP-2, on collagen coated plates compared to non-POP site in the same POP patient [42]. No difference was observed in the contraction and in the total levels of released MMP-2 between controls without POP and in the non-POP site of affected women. These findings suggest an acquired, rather than an intrinsic deficiency for patients with cystocele, an effect rather than the cause of the disease [42]. The different proteolytic profile observed by Zong et al. may be explained by their small sample size and the inter-variability between POP patients and controls used in their study. Additional investigations are needed to explore this topic.

In an experimental study, Rahn et al. reported that pregnancy per se conferred remarkable changes in the mid-vaginal segments of mice that include increased distensibility and decreased stiffness and maximal stress [19]. Using a model of uniaxial loading of sections of vaginal tissues of rats, Feola et al. demonstrated significant changes in the passive mechanical properties that reflect ECM content (decreased tangent modulus and tensile strength, and increased ultimate strain) and in active contractile properties of smooth muscle (decrease in the contractile force) in the vagina of pregnant rats compared to virgin animals [43]. Those findings suggest that tissue remodeling occurs to maximize the ability to distend during vaginal delivery to facilitate the passage of the fetus with less injury [19,43].

More recently, Ulrich et al. reported results in an ovine model [44]. By evaluating vaginal tissues, the authors observed that pregnant sheep had higher elastin and lower collagen content, and a thicker smooth muscle layer compared with tissues from virgin and parous animals. The results were associated with low maximum stress, a biomechanical property compatible with tissue remodeling as demonstrated in other species [44]. Failure in the remodeling process during pregnancy may corroborate to pelvic damage.

Parity also impacts vaginal mechanical properties and may contribute to POP development. Feola et al. evaluated the vagina of rhesus macaques using uniaxial tensile tests, and observed that mechanical properties (tangent modulus, tensile strength, and strain energy density) were inferior in tissues from parous versus nulliparous animals. Collagen ratios I/(III + IV) were similar between the groups, but parity was associated with decreased collagen alignment [45].

Even though extrinsic factors may contribute to biochemical and biomechanical alterations in POP, the theory of intrinsic defect has been defended by many authors, as described further. Following the trend to investigate the genetic role in pelvic floor dysfunctions, previously described animal and human studies involving proteins and mRNA analyses identified potential targets for genetic investigations.

GENETICS OF PELVIC FLOOR DISORDERS

Familial and Twin-Sisters Studies

Heredity is the genetic transmission of characteristics from parent to child. The heritable contribution to POP and SUI is strongly suggested by epidemiological studies. In a large population-based study involving 1292 women with POP, Norton and coworkers observed that the relative risk of POP is significantly elevated in first- and third-degree female relatives (RR 4.15; 1.24, respectively) [46].

The risk of POP among siblings of women with severe prolapse who are younger than 55 years old is five times higher than in the general population [47] and the prevalence of symptomatic prolapse is threefold higher in women whose mothers or sisters have undergone surgery for prolapse compared with women with no such family history [48]. However, it is a common misunderstanding that familial aggregation invariably is a result of genetic factors [49]. Familial environmental influences such as smoking habits, exercise, diet and drinking habits, socioeconomic status, etc. may have a direct effect on the transmission of risk for POP. In a large study of female twin pairs (3376 monozygotic and 5067 dizygotic), there was greater similarity among monozygotic twins, suggesting a genetic component for SUI and POP, while in dizygotic twin sisters, genetic and nonshared environmental factors contributed equally to the development of pelvic floor dysfunction (both 40%) [50].

In some cases, POP develops as part of a generalized connective tissue disease affecting joints, skin, and pelvic floor support. History of varicose veins, hernias, hemorrhoids, joint hypermobility, and abdominal striae was found to be significantly and positively associated with symptomatic POP [4,8,51,52]. Women with connective tissue disorders such as Marfan's (mutation in fibulin gene), cutis laxa (mutation in elastin and fibulin-5 genes), and Ehlers-Danlos (mutation in collagen genes) syndromes have significantly higher rates of POP [53]. Dietz et al. showed that the mobility of the bladder is a heritable trait with 50% of the variability due to genetic factors in nulligravid women [54]. Moreover, bladder mobility appears to correlate with joint hyperextensibility, mostly at the elbow, with 14% of the variation in bladder neck mobility accounted for by genes influencing elbow mobility [55].

Genetic Markers

Given previous findings in animal and human studies that evaluated protein and mRNA in pelvic floor tissues, it is not surprising that candidate gene studies have focused on those related to ECM metabolism. To identify potential genetic markers for POP, DNA sequencing, and searches for DNA variations have been performed in the affected population. Genotyping of candidate genes participating in ECM metabolism may elucidate the missing link between the manifestation of the pelvic floor dysfunction and the biochemical changes observed in pelvic floor tissues. The most common genetic investigations are performed by genome linkage, genome association studies, and polymorphisms analysis. These investigations are still in their infancy, but they seem to be promising. We present the currently available literature data addressing genetic studies in POP.

Genome-Wide Linkage Studies

Genetic linkage is the tendency of certain loci or alleles to be inherited together. Genetic loci that are physically close to each other on the same chromosome tend to stay together during meiosis, and are thus genetically linked. Linkage analysis is a technique to identify genetic effects and relies on the fact that if family members affected by the disease share a specific area of a chromosome, not shared by unaffected members, then the genes predisposing to the disease are likely to be on or near that area [56,57]. The results of the analysis are reported in a logarithm of the odd scores (LOD) that measures the likelihood that two loci are near each other in the same chromosome. This approach has the advantage of being an unbiased, comprehensive search across the genome for susceptibility alleles, and has been successfully applied to find the genes for many single gene disorders. However, linkage analysis has been less successful for polygenic diseases and quantitative traits [57].

A single three-generational Filipino family with six affected individuals presenting with early onset POP has shown that, although not significant, a polymorphism C/T (rs10911193), in which the nucleotide cytosine is substituted by a thymine in the promoter of laminin 1 (LAMC1), is present at a frequency of 4.9% in the general population and 22% among probands (the first affected family member recruited to the study) from this cohort of families. The variant affects the binding site for NFIL3 (Nuclear Factor Interleukin 3 Regulated Protein), a transcription factor co-expressed in vaginal tissues, and it may account for the increased predisposition to POP in this family, but likely is not the sole cause of POP in all women [58].

Allen-Brady and coworkers conducted a genome-wide linkage study to identify POP predisposition genes using a resource of 99 familial POP cases in 83 high-risk pedigrees (with at least two female relatives with POP) [59]. A total of 25,436 markers were evaluated. The study showed significant linkage on chromosomes 10q24-26 and 17q25. Candidate genes in the 10q24-26 region include genes such as lysyl oxidase-like-4 (LOXL-4), a member of LOX family that plays a role in the biogenesis of ECM as previously described, and therefore may predispose to POP. Other candidates near this region are: ankyrin repeat domain 2 (involved in slow type 1 muscle fibers), heparanase 2 (HPSE2) (protein that degrades heparin sulfate proteoglycans in the ECM), collagen type 17A1 (COL17A1), and carbohydrate sulfotransferase 15 (CHST15) (important structural component of ECM and assists in proteoglycans formation) [59]. In regards to the candidate genes that may be linked to POP genetic predisposition near the 17q25 region, the most important is TIMP-2, a member of the TIMP family that acts in the ECM remodeling. Hence, those genes are involved in ECM homeostasis and muscle function, major components of pelvic floor support.

Familial case analysis suggests that the most likely mode of transmission of POP is an autosomal dominant pattern with high penetrance through both maternal and paternal relatives [47,58]. It is possible that POP, as with other multifactorial complex and common disorders, is caused by the combined effects of multigenes and nongenetic environmental factors (Fig. 3). Therefore, it is likely that sequence variation alone is not sufficient to predict the risk of disease susceptibility, particularly in homeostatic organisms like humans. Nevertheless, these variants may provide a starting point for further inquiry [57].

Case-Control Studies: Polymorphisms

When any two human genomes are compared side by side, they are 99.9% identical [60]. The remaining 0.1% of DNA contains sequence variations. These variations are associated with diversity in the population, individuality, susceptibility to diseases, and individual responses to medicine [56]. Single nucleotide polymorphisms (SNPs) are the most abundant type of DNA sequence variation in the human genome [61], in which a single base-pair varies from person to person, being the basis for the physical diversity found among humans [62]. As genetic markers, SNPs can be used to follow the inheritance patterns of chromosomal regions. SNPs may result in altered amino acids sequences and altered protein function, playing a direct or indirect role in the resultant phenotype [56,63,64].

Some of the most relevant investigated potential variations are located in the chromosome regions that encode for collagens. The substitution of a guanine for a tyrosine (G/T) in the transcription factor Sp1-binding site of COL1A1 (rs1800012) leads to abnormal production of α-1 collagen chain in comparison to α-2 collagen chain. This

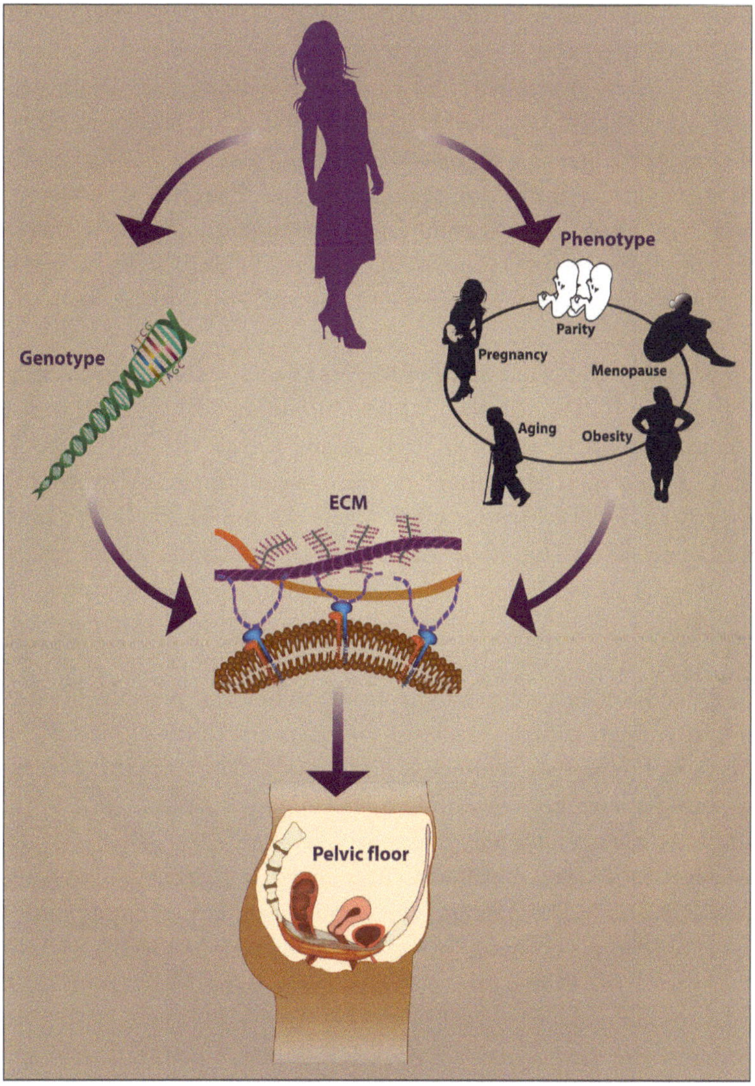

Figure 3 Schematic representation of the combined effects of the genotypic and phenotypic factors that may influence the ECM composition, structure, and function of the pelvic floor support in women.

polymorphism was investigated in Polish, Brazilian, Israeli, and Italian populations; however, no association between GT allele and POP was observed in those studies [65–68]. Despite each individual study being underpowered and of risk for population stratification bias, the pooled effect size was significant (OR, 1.33; 95% CI, 1.02–1.73), showing an association between the polymorphism and POP, according to the meta-analysis by Cartwright et al. [7].

On the other hand, an association between POP and the mutation G/A in the position +2092 of the exon 31 in the sequence that encodes for collagen type III (COL3A1) (rs111929073) was reported in Korean women with POP, evidence that was not observed in the Brazilian mixed population [69,70]. Similar to other investigations, small sample size was of a major concern in both studies. Moreover, the different ethnicity of the studied populations may explain the discrepancy between both POP studies. With that, when data was combined, a nonsignificant pooled effect was noted [7]. Another candidate, the mutation rs1800255 in the coding region of type III collagen (COL3A1 2209G>A) was also not associated with POP in both Taiwanese and Dutch populations [71,72].

Khadzhieva et al. investigated whether there is an association between common SNPs of the FBLN5 gene and POP. A total of 11 tag SNPs of the FBLN5 gene were genotyped using the polymerase chain reaction in 210 Russian patients with POP (stages III–IV) and 292 non-POP controls. The authors revealed significant associations of tag SNPs rs2018736 (protective effect for the minor allele A) and rs12589592 with POP. The top association signal was found for SNP rs2018736 in the entire set: $p=0.0026$, OR=0.42, 95% CI: 0.24–0.75; in the stratum with pelvic floor trauma: $p=0.0018$, OR=0.27, 95% CI: 0.11–0.64; and in the stratum with fetal macrosomia: $p=0.013$, OR=0.14, 95% CI: 0.03–0.71. In the strata without perineal trauma and fetal macrosomia, the effects were not statistically significant, possibly due to the smaller effect sizes. Current data provide evidence that common SNPs of the FBLN5 gene are associated with POP especially after pelvic floor injury, demonstrating the importance of the FBLN5 gene that encodes a key protein of elastic fiber matrix assembly and function that contributes to maintaining pelvic support [73].

Other diverse DNA polymorphisms have been investigated, involving LOXL1, LAMC1, MMP-1, and MMP-3 in different populations. In the unique study evaluating the presence of SNP in the promoter of the LOXL1 gene (45008784A>C, rs16958477), Ferrel et al. analyzed 137 POP patients (POP stage ≥ 2) and 130 controls (POP < 2) in a case-control study involving Caucasians and Afro-American women [74]. No association was observed between the polymorphism and POP phenotype, as well as no allelic frequency was significant linked to ethnicity [74]. The well-known role of LOXL1 protein in the ECM integrity of the pelvic floor justifies further investigations addressing DNA variations in the corresponding chromosome region.

Considering the important gene LAMC1, Wu et al. investigated several polymorphisms in 239 patients with advanced stage III–IV POP and compared them with 197 patients presenting prolapse stage < 2 by POP-Q classification [75]. The authors found no association between 14 different SNPs and POP in a North American white population [75]. It is interesting to note that inclusion criteria vary among the studies which may affect the analysis of combined studies addressing the same DNA variation. Other authors have also not observed significant results after the investigation of rs10911193

LAMC1 polymorphism (alleles CT) and risk of POP [76,77], findings that do not replicate the increased allele frequency in familial cases of early onset POP described by Nikolova et al. [58].

DNA sequence variations that code for the MMP and TIMP family of proteins (MMP-1, MMP-3, MMP-9, TIMP-1, and TIMP-3) have also been searched in POP patients. However, pooled analysis did not show significant effects of genotypes on the POP phenotype [68,78–81]. The major criticism and pitfalls regarding these primary genetic markers studies is the small sample size used for the analysis. Larger cohort studies are needed to identify potential genetic markers for POP and to identify to which populations each genetic marker is relevant.

Genome-Wide Association Studies

Initial discoveries of genetic-based diseases were Mendelian in nature, in which a disease was transmitted by a single gene autosomal, dominant or recessive, in a simple relationship between genotype and phenotype. More recently, the polygenic model of genetic risk to more complex diseases has changed the paradigm of Mendel's law, launched by advances in genotyping and sequencing techniques [82].

Genome-wide association studies (GWAS) are based on a case-control association design in which the whole genome is genotyped with DNA markers in a nonrelated population presenting the same phenotype [49]. The new technology of computerized chip array containing up to millions of SNPs makes GWAS possible to detect genetic predisposition for common polygenic disorders. However, GWAS are not without their own problems, such as confounders arising from population stratification when genetically heterogeneous populations are analyzed together without further adjustment; the need for large sample sizes to detect minor effect alleles; and inflated false-positive association rates arising from the several thousand tests that are an integral part of any such study [83].

An important study is this field was published by Allen-Brady et al. in which the authors analyzed 115 Caucasian patients presenting moderate or severe POP with at least one sibling who also had advanced dysfunction, and compared them with 2976 genetic-matched white North European descendent controls without POP [84]. Both groups were also matched for age of diagnosis (mean between 48 and 49 years old), BMI (mean between 25 and 26.6), but were different with regard to parity (cases: 4.3 ± 2.4; controls: 2.4 ± 0.7) [84]. Genotyping showed that six SNPs were significantly associated with POP, localized to chromosomes 4q21 (rs1455311), 8q24 (rs1036819), 9q22 (rs430794), 15q11 (rs8027714), 20p13 (rs1810636), and 21q22 (rs2236479). The chromosome 21q region codes for collagen type XVIII (COL18A1). None of the single candidate markers for POP reported in linkage and case-control polymorphism investigations were identified in this GWAS study. Although not replicated yet, this study confirmed the genetic influence on POP as a polygenic disorder [84].

INFLUENCE OF GENETICS ON PELVIC FLOOR BIOMECHANICS

Even though current evidence points toward an association between inheritance and pelvic dysfunctions, genomic markers that potentially reflect the biomechanical properties of pelvic tissues are still not well explored. Knowledge of the role of ECM proteins in pelvic floor support and possible DNA mutations that may predispose women to develop POP suggests that genetic variations lead to impaired biomechanical properties of the pelvic floor.

Investigations of biomechanics of different tissues and DNA variations have been performed in the field of cardiology, orthopedics, dermatology, and rheumatology, attempting to identify candidate markers for the diseases. These can be used to consider similar relationships between genetics and biomechanics of the pelvic floor. Some examples are briefly described.

Marfan syndrome is caused by mutations in the fibrillin-1 gene [85] that lead to alterations in the ECM of the aortic wall, producing a stiffer aorta and elevated pulse pressure [86,87]. Clinical manifestations include aortic dilation and dissection. Teixido-Tura et al. studied aortic biomechanics by magnetic resonance imaging in 80 Marfan patients without advanced aortic disease and 36 age- and sex-matched controls [88]. The authors observed that aortic distensibility was lower in the entire aorta in Marfan compared to healthy patients. Therefore, Marfan patients with the DNA mutation that encodes for fibrillin protein showed clear impairment of aortic biomechanics [88]. In addition, the T allele of the polymorphism (2046G/T) located in the promoter region of collagen type I (COL1A1) was associated with higher aortic pulse wave velocity and mean arterial pressure. This SNP alters Sp1 binding and therefore affects transcription rates of the COL1A1 gene, suggesting that altered collagen deposition is directly related to the effects on arterial stiffening [89].

Ehlers-Danlos syndrome (EDS) represents a heterogeneous group of connective tissue disorders caused by mutations in collagen genes. In most cases, it is characterized by hyperextensible (cutis laxa), soft and easily bruising skin associated with joint laxity and vascular fragility. Vascular EDS is a rare autosomal dominant collagen vascular disorder caused by a 2q31 COL3A1 gene mutation encoding pro-alpha1 chain of type III collagen (in contrast to classic Ehlers-Danlos, caused by a COL5A1 mutation) [90].

Flagothier et al. measured the mechanical properties of skin using a computerized suction device and took skin biopsies of the forearms of nine members of four generations of a family who suffered from large vessel dissections and cardiac infarct at a young age, with cutaneous changes consistent with EDS [91]. In the experiments, the authors observed variations in the mechanical parameters of patients affected by EDS. The group with the highest maximum distention values exhibited the most severe ultrastructural disorganization of the skin assessed by biopsies, characterized by a loose network of thin collagen bundles and clumsy elastic fibers compared to control patients [91]. The most prominent changes showed the most striking biomechanical alterations characterized by increased biologic elasticity without any excess in skin extensibility, below the normal range of values. A positive linear correlation between maximum distension and biological elasticity of the skin was found [91]. With that, the maximum distensibility of skin is rapidly reached. The absence of reserve for further deformation under increasing mechanical stress leads to higher risk for rupture of stress-sustaining fibers. Therefore, EDS patients with the DNA mutation that encodes for collagen proteins showed clear impairment of skin biomechanics and possible aortic biomechanical dysfunction [91].

Recent research indicates that heredity may be largely responsible for the degeneration of intervertebral discs. Annunen et al. pointed out that Trp2 amino acid (tryptophan) allele of *COL9A2* has a significant role in the pathogenesis of disc disease [92]. Confined compression experiments were conducted on nucleus pulposus of human intervertebral discs excised from young patients with thoracolumbar and lumbar scoliosis during anterior scoliosis surgery to study the impact of the presence of the Trp2 allele in *COL9A2* on the compressive mechanical properties. The presence of Trp2 is associated with the reduction of both swelling pressure and compressive modulus properties compared to tissues of patients that did not present with the Trp2 allele. This association is evident even at a very young age, suggesting genetic influences on disc biomechanics [93].

Normally, there are no Trp residues in the collagenous domains of collagen IX in humans [94]. Trp may disrupt the collagen triple helix, as well as interfere with the interaction between collagens IX and II or prevent the action of lysyl oxidase, which catalyzes crosslink formation. Then, lower mechanical stresses cause a drastic fall in the swelling pressure, the primary load bearing mechanism, in patients with Trp+, reflecting disc diseases.

Eighty-four recreationally active, Caucasian men and women, aged 18–39, with no history of injuries to the knee, and a body mass index between 18.5 and 30 were recruited to examine the association between three polymorphisms of the MMP-3 gene (rs679620, rs591058, and rs650108) and one of COL5A1 (rs12722) and patellar tendon dimensional and mechanical properties in vivo. Patellar tendon dimensions (volume) and functional (elastic modulus) properties were assessed in vivo using geometric modeling, isokinetic dynamometry, electromyography, and ultrasonography. Polymorphisms were evaluated using real-time polymerase chain reaction. No association between the presence of such SNPs and tendon biomechanical parameters were detected [95,96].

The search for SNPs as a genetic marker of a disease should be ideally evaluated at the level of effects in the function of the tissues and organs. This is due to the fact that only a minority of DNA variations has functional significance. The majority of them are of no biological consequence [62].

Future studies addressing the genetic basis of pelvic floor dysfunction are required and investigations on biomechanical properties of pelvic floor support should be performed together with identification of candidate genes for pelvic dysfunctions to determine the relationship between genetics and biomechanics in the pelvic floor.

REFERENCES

[1] J.O.L. DeLancey, Anatomy and biomechanics of genital prolapse, Clin. Obstet. Gynecol. 36 (1993) 897–909.
[2] L.B. Epstein, C.A. Graham, M.H. Heit, Systemic and vaginal biomechanical properties of women with normal vaginal support and pelvic organ prolapse, Am. J. Obstet. Gynecol. 197 (2) (2007) 165e1–165e6.
[3] R.A. Word, S. Pathi, J.I. Schaffer, Pathophysiology of pelvic organ prolapse, Obstet. Gynecol. Clin. North Am. 36 (2009) 521–539.
[4] M.H. Kerkhof, L. Hendriks, H.A. Brölmann, Changes in connective tissue in patients with pelvic organ prolapse: a review of the current literature, Int. Urogynecol. J. Pelvic Floor Dysfunct. 20 (4) (2009) 461–474.
[5] X. Liu, Y. Zhao, B. Pawlyk, M. Damaser, T. Li, Failure of elastic fiber homeostasis leads to pelvic floor disorders, Am. J. Pathol. 168 (2) (2006) 519–528.
[6] M.A. Bortolini, D.E. Rizk, Genetics of pelvic organ prolapse: crossing the bridge between bench and bedside in urogynecological research, Int. Urogynecol. J. 22 (10) (2011) 1211–1219.
[7] R. Cartwright, A.C. Kirby, K.A. Tikkinen, et al., Systematic review and metaanalysis of genetic association studies of urinary symptoms and prolapse in women, Am. J. Obstet. Gynecol. 212 (2) (2015) 199e1–199e24.
[8] L. Paul, Alpha1-antitrypsin deficiency, Pulm. Crit. Care Update 5 (1989) 2–8.
[9] D.E. Gomez, D.F. Alonso, H. Yoshiji, U.P. Thorgeirsson, Tissue inhibitors of metalloproteinases: structure, regulation and biological functions, Eur. J. Cell Biol. 74 (1997) 111–122.
[10] S.D. Shapiro, Matrix metalloproteinase degradation of extracellular matrix: biological consequences, Curr. Opin. Cell Biol. 10 (1998) 602–608.
[11] K. Kawabata, T. Hagio, S. Matsuoka, The role of neutrophil elastase in acute lung injury, Eur. J. Pharmacol. 451 (2002) 1–10.
[12] T. Nakamura, P.R. Lozano, Y. Ikeda, Fibulin-5/DANCE is essential for elastogenesis in vivo, Nature 415 (6868) (2002) 171–175.
[13] H.M. Kagan, W. Li, Lysyl oxidase: properties, specificity, and biological roles inside and outside the cell, J. Cell. Biochem. 88 (2003) 660–672.
[14] P.C. Trackman, Diverse biological functions of extracellular collagen processing enzymes, J. Cell. Biochem. 96 (5) (2005) 927–937.
[15] G. Ge, D.S. Greenspan, Developmental roles of the BMP1/TLD metalloproteinases, Birth Defects Res., Part C 78 (2006) 47–68.
[16] P.G. Drewes, H. Yanagisawa, B. Starcher, et al., Pelvic organ prolapse in fibulin-5 knockout mice: pregnancy-induced changes in elastic fiber homeostasis in mouse vagina, Am. J. Pathol. 170 (2) (2007) 578–589.
[17] G. Liu, F. Daneshgari, M. Li, et al., Bladder and urethral function in pelvic organ prolapsed lysyl oxidase like-1 knockout mice, BJU Int. 100 (2) (2007) 414–418.
[18] M. Alperin, K. Debes, S. Abramowitch, L. Meyn, P.A. Moalli, LOXL1 deficiency negatively impacts the biomechanical properties of the mouse vagina and supportive tissues, Int. Urogynecol. J. Pelvic Floor Dysfunct. 19 (7) (2008) 977–986.
[19] D.D. Rahn, M.D. Ruff, S.A. Brown, H.F. Tibbals, R.A. Word, Biomechanical properties of the vaginal wall: effect of pregnancy, elastic fiber deficiency, and pelvic organ prolapse, Am. J. Obstet. Gynecol. 198 (5) (2008) 590e1–590e6.
[20] D.D. Rahn, J.F. Acevedo, S. Roshanravan, et al., Failure of pelvic organ support in mice deficient in fibulin-3, Am. J. Pathol. 174 (1) (2009) 206–215.
[21] S.R. Jackson, N.C. Avery, J.F. Tarlton, S.D. Eckford, P. Abrams, A.J. Bailey, Changes in metabolism of collagen in genitourinary prolapse, Lancet 347 (9016) (1996) 1658–1661.

[22] C.C. Takano, M.J. Girão, M.G. Sartori, et al., Analysis of collagen in parametrium and vaginal apex of women with and without uterine prolapse, Int. Urogynecol. J. Pelvic Floor Dysfunct. 13 (6) (2002) 342–345.
[23] M.W. Soderberg, C. Falconer, B. Bystrom, A. Malmstrom, G. Ekman, Young women with genital prolapse have a low collagen concentration, Acta Obstet. Gynecol. Scand. 83 (2004) 1193–1198.
[24] P.A. Moalli, L.C. Talarico, V.W. Sung, et al., Impact of menopause on collagen subtypes in the arcus tendineous fasciae pelvis, Am. J. Obstet. Gynecol. 190 (2004) 620–627.
[25] P.A. Moalli, S.H. Shand, H.M. Zyczynski, S.C. Gordy, L.A. Meyn, Remodeling of vaginal connective tissue in patients with prolapse, Obstet. Gynecol. 106 (2005) 620–627.
[26] E.C. Barbiero, M.G. Sartori, M.J. Girão, E.C. Baracat, G.R. Lima de, Analysis of type I collagen in the parametrium of women with and without uterine prolapse, according to hormonal status, Int. Urogynecol. J. 14 (2003) 331–334.
[27] K. Yamamoto, M. Yamamoto, K. Akazawa, S. Tajima, H. Wakimoto, M. Aoyagi, Decrease in elastin gene expression and protein synthesis in fibroblasts derived from cardinal ligaments of patients with prolapsus uteri, Cell Biol. Int. 21 (1997) 605–611.
[28] A.A. Ewies, F. Al-Azzawi, J. Thompson, Changes in extracellular matrix proteins in the cardinal ligaments of post-menopausal women with or without prolapse: a computerized immunohistomorphometric analysis, Hum. Reprod. 18 (2003) 2189–2195.
[29] B. Chen, Y. Wen, M.L. Polan, Elastolytic activity in women with stress urinary incontinence and pelvic organ prolapse, Neurourol. Urodyn. 23 (2) (2004) 119–126.
[30] J.A. Karam, D.V. Vazquez, V.K. Lin, P.E. Zimmern, Elastin expression and elastic fibre width in the anterior vaginal wall of postmenopausal women with and without prolapse, BJU Int. 100 (2007) 346–350.
[31] M. Alarab, M.A. Bortolini, H. Drutz, S. Lye, O. Shynlova, LOX family enzymes expression in vaginal tissue of premenopausal women with severe pelvic organ prolapse, Int. Urogynecol. J. Pelvic Floor Dysfunct. 21 (11) (2010) 1397–1404.
[32] J. Klutke, Q. Ji, J. Campeau, et al., Decreased endopelvic fascia elastin content in uterine prolapse, Acta Obstet. Gynecol. Scand. 87 (2008) 111–115.
[33] P. Takacs, M. Nassiri, A. Viciana, K. Candiotti, A. Fornoni, C.A. Medina, Fibulin-5 expression is decreased in women with anterior vaginal wall prolapse, Int. Urogynecol. J. Pelvic Floor Dysfunct. 20 (2) (2009) 207–211.
[34] H.J. Jung, M.J. Jeon, G.W. Yim, S.K. Kim, J.R. Choi, S.W. Bai, Changes in expression of fibulin-5 and lysyl oxidase-like 1 associated with pelvic organ prolapse, Eur. J. Obstet. Gynecol. Reprod. Biol. 145 (1) (2009) 117–122.
[35] M.A.T. Bortolini, O. Shynlova, H.P. Drutz, et al., Expression of bone morphogenetic protein-1 in vaginal tissue of women with severe pelvic organ prolapse, Am. J. Obstet. Gynecol. 204 (6) (2011) 544e1–544e8.
[36] T. Strinic, M. Vulic, S. Tomic, V. Capkun, I. Stipic, I. Alujevic, Matrix metalloproteinases-1, -2 expression in uterosacral ligaments from women with pelvic organ prolapse, Maturitas 64 (2009) 132–135.
[37] B.H. Chen, Y. Wen, H. Li, M.L. Polan, Collagen metabolism and turnover in women with stress urinary incontinence and pelvic prolapse, Int. Urogynecol. J. Pelvic Floor Dysfunct. 13 (2) (2002) 80–87.
[38] B. Gabriel, D. Watermann, K. Hancke, et al., Increased expression of matrix metalloproteinase 2 in uterosacral ligaments is associated with pelvic organ prolapse, Int. Urogynecol. J. Pelvic Floor Dysfunct. 17 (5) (2006) 478–482.
[39] M. Alarab, H. Kufaishi, S. Lye, H. Drutz, O. Shynlova, Expression of extracellular matrix-remodeling proteins is altered in vaginal tissue of premenopausal women with severe pelvic organ prolapse, Reprod. Sci. 21 (6) (2014) 704–715.
[40] L. Zhou, J.H. Lee, Y. Wen, et al., Biomechanical properties and associated collagen composition in vaginal tissue of women with pelvic organ prolapse, J. Urol. 188 (3) (2012) 875–880.
[41] W. Zong, Z.C. Jallah, S.E. Stein, S.D. Abramowitch, P.A. Moalli, Repetitive mechanical stretch increases extracellular collagenase activity in vaginal fibroblasts, Female Pelvic Med. Reconstr. Surg. 16 (5) (2010) 257–262.

[42] A.M. Ruiz-Zapata, M.H. Kerkhof, B. Zandieh-Doulabi, et al., Functional characteristics of vaginal fibroblastic cells from premenopausal women with pelvic organ prolapse, Mol. Hum. Reprod. 20 (11) (2014) 1135–1143.
[43] A. Feola, P. Moalli, M. Alperin, R. Duerr, R.E. Gandley, S. Abramowitch, Impact of pregnancy and vaginal delivery on the passive and active mechanics of the rat vagina, Ann. Biomed. Eng. 39 (1) (2011) 549–558.
[44] D. Ulrich, S.L. Edwards, K. Su, et al., Influence of reproductive status on tissue composition and biomechanical properties of ovine vagina, PLoS One 9 (4) (2014) e93172.
[45] A. Feola, S. Abramowitch, K. Jones, S. Stein, P. Moalli, Parity negatively impacts vaginal mechanical properties and collagen structure in rhesus macaques, Am. J. Obstet. Gynecol. 203 (6) (2010) 595e1–595e8.
[46] P.A. Norton, K. Allen-Brady, L.A. Cannon-Albright, The familiality of pelvic organ prolapse in the Utah population database, Int. Urogynecol. J. 24 (2013) 413–418.
[47] G.S. Jack, G. Nikolova, E. Vilain, S. Raz, L.V. Rodríguez, Familial transmission of genitovaginal prolapse, Int. Urogynecol. J. Pelvic Floor Dysfunct. 17 (5) (2006) 498–501.
[48] A. Miedel, G. Tegerstedt, M. Maehle-Schmidt, O. Nyrén, M. Hammarström, Nonobstetric risk factors for symptomatic pelvic organ prolapse, Obstet. Gynecol. 113 (5) (2009) 1089–1097.
[49] P. Norton, I. Milsom, Genetics and the lower urinary tract, Neurourol. Urodyn. 29 (2010) 609–611.
[50] D. Altman, M. Forsman, C. Falconer, P. Lichtenstein, Genetic influence on stress urinary incontinence and pelvic organ prolapse, Eur. Urol. 54 (4) (2008) 918–922.
[51] P.A. Norton, J.E. Baker, H.C. Sharp, J.C. Warenski, Genitourinary prolapse and joint hypermobility in women, Obstet. Gynecol. 85 (2) (1995) 225–228.
[52] T.R. Sayer, J.S. Dixon, G.L. Hosker, D.W. Warrel, A study of paraurethral connective tissue in women with stress incontinence of urine, Neurourol. Urodyn. 9 (1990) 319–320.
[53] M.E. Carley, J. Schaffer, Urinary incontinence and pelvic organ prolapse in women with Marfan or Ehlers Danlos syndrome, Am. J. Obstet. Gynecol. 182 (5) (2000) 1021–1023.
[54] H.P. Dietz, N.K. Hansell, M.E. Grace, A.M. Eldridge, B. Clarke, N.G. Martin, Bladder neck mobility is a heritable trait, BJOG 112 (3) (2005) 334–339.
[55] N.K. Hansell, H.P. Dietz, S.A. Treloar, B. Clarke, N.G. Martin, Genetic covariation of pelvic organ and elbow mobility in twins and their sisters, Twin Res. 7 (3) (2004) 254–260.
[56] B.S. Shastry, SNP alleles in human disease and evolution, J. Hum. Genet. 47 (2002) 561–566.
[57] J.N. Hirschhorn, Genetic approaches to studying common diseases and complex traits, Pediatr. Res. 57 (5 Pt 2) (2005) 74R–77R.
[58] G. Nikolova, H. Lee, S. Berkovitz, et al., Sequence variation in laminin gamma 1 (LAMC1) gene associated with familial pelvic organ prolapse, Hum. Genet. 120 (2007) 847–856.
[59] K. Allen-Brady, L.A. Cannon-Albright, J.M. Farnham, P.A. Norton, Evidence for pelvic organ prolapse predisposition genes on chromosomes 10 and 17, Am. J. Obstet. Gynecol. 212 (6) (2015) 771e1–771e7.
[60] D.N. Cooper, B.A. Smith, H.J. Cooke, S. Niemann, J. Schmidtke, An estimate of unique DNA sequence heterozygosity in the human genome, Hum. Genet. 69 (1985) 201–205.
[61] A.J. Brookes, The essence of SNPs, Gene 234 (1999) 177–186.
[62] F.S. Collins, M.S. Guyer, A. Charkravarti, Variations on a theme: cataloging human DNA sequence variation, Science 278 (1997) 1580–1581.
[63] H.D. Lohrer, U. Tangen, Investigations into the molecular effects of single nucleotide polymorphism, Pathobiology 68 (2000) 283–290.
[64] P.Y. Kwok, X. Chen, Detection of single nucleotide polymorphisms, Curr. Issues Mol. Biol. 5 (2003) 43–60.
[65] P. Skorupski, P. Miotła, K. Jankiewicz, T. Rechberger, Polymorphism of the gene encoding alpha-1 chain of collagen type I and a risk of pelvic organ prolapse: a preliminary study, Ginekol. Pol. 78 (11) (2007) 852–855.
[66] A.M. Rodrigues, M.J. Girão, I.D. da Silva, M.G. Sartori, F. Martins K de, A. Castro R de, COL1A1 Sp1-binding site polymorphism as a risk factor for genital prolapse, Int. Urogynecol. J. Pelvic Floor Dysfunct. 19 (11) (2008) 1471–1475.

[67] B. Feiner, F. Fares, N. Azam, R. Auslender, M. David, Y. Abramov, Does COL1A1 SP1-binding site polymorphism predispose women to pelvic organ prolapse? Int. Urogynecol. J. 20 (2009) 1061–1065.
[68] M.M. Ferrari, G. Rossi, M.L. Biondi, P. Vigano, C. Dell'Utri, M. Meschia, Type I collagen and matrix metalloproteinase 1, 3 and 9 gene polymorphisms in the predisposition to pelvic organ prolapse, Arch. Gynecol. Obstet. 285 (2012) 1581–1586.
[69] M.J. Jeon, S.M. Chung, J.R. Choi, H.J. Jung, S.K. Kim, S.W. Bai, The relationship between COL3A1 exon 31 polymorphism and pelvic organ prolapse, J. Urol. 181 (3) (2009) 1213–1216.
[70] K. Martins, Z.I.K. de Jarmy-DiBella, A. Fonseca, et al., Evaluation of demographic, clinical characteristics, and genetic polymorphism as risk factors for pelvic organ prolapse in Brazilian women, Neurourol. Urodyn. 30 (7) (2011) 1325–1328.
[71] H.Y. Chen, Y.W. Chung, W.Y. Lin, J.C. Wang, F.J. Tsai, C.H. Tsai, Collagen type 3 alpha 1 polymorphism and risk of pelvic organ prolapse, Int. J. Gynaecol. Obstet. 103 (1) (2008) 55–58.
[72] S.L. Lince, L.C. van Kempen, J.R. Djikstra, J. IntHout, M.E. Vierhout, K.B. Kluivers, Collagen type III alpha 1 polymorphism (rs1800255, COL3A1 2209G>a) assessed with high-resolution melting analysis is not associated with pelvic organ prolapse in the Dutch population, Int. Urogynecol. J. 25 (2014) 1237–1242.
[73] M.B. Khadzhieva, S.V. Kamoeva, A.G. Chumachenko, et al., Fibulin-5 (FBLN5) gene polymorphism is associated with pelvic organ prolapse, Maturitas 78 (2014) 287–292.
[74] G. Ferrell, M. Lu, P. Stoddard, et al., A single nucleotide polymorphism in the promoter of the LOXL1 gene and its relationship to pelvic organ prolapse and preterm premature rupture of membranes, Reprod. Sci. 16 (5) (2009) 438–446.
[75] J.M. Wu, A.G. Visco, E.A. Grass, et al., Comprehensive analysis of LAMC1 genetic variants in advanced pelvic organ prolapse, Am. J. Obstet. Gynecol. 206 (2012) 447e1–447e6.
[76] C. Chen, L.D. Hill, C.M. Schubert, J.F. Strauss 3rd., C.A. Matthews, Is laminin gamma-1 a candidate gene for advanced pelvic organ prolapse? Am. J. Obstet. Gynecol. 202 (5) (2010) 505e1–505e5.
[77] R. Neupane, Z. Sadeghi, R. Fu, S.A. Hagstrom, C.K. Moore, F. Daneshgari, Mutation screen of LOXL1 in patients with female pelvic organ prolapse, Female Pelvic Med. Reconstr. Surg. 20 (6) (2014) 316–321.
[78] H.Y. Chen, W.Y. Lin, Y.H. Chen, W.C. Chen, F.J. Tsai, C.H. Tsai, Matrix metalloproteinase-9 polymorphism and risk of pelvic organ prolapse in Taiwanese women, Eur. J. Obstet. Gynecol. Reprod. Biol. 149 (2) (2010) 222–224.
[79] P. Skorupski, P. Miotła, K. Jankiewicz, T. Rechberger, MMP-1 and MMP-3 gene encoding polymorphism and the risk of the development of pelvic organ prolapse and stress urinary incontinence, Ginekol. Pol. 81 (8) (2010) 594–599.
[80] J.M.J. Wu, A.G.A. Visco, E.A.E. Grass, et al., Matrix metalloproteinase-9 polymorphisms and the risk of pelvic organ prolapse, Obstet. Gynecol. 120 (2012) 587–593.
[81] A. Romero, M. Jamison, Are single nucleotide polymorphisms associated with pelvic organ prolapse? J. Pelvic Med. Surg. 14 (2008) 37–43.
[82] N. Risch, K. Merikangas, The future of genetic studies of complex human diseases, Science 273 (1996) 1516–1517.
[83] B.J. Traynor, The era of genomic epidemiology, Neuroepidemiology 33 (2009) 276–279.
[84] K. Allen Brady, L. Cannon-Albright, J.M. Farnham, et al., Identification of six loci associated with pelvic organ prolapse using genome-wide association analysis, Obstet. Gynecol. 118 (6) (2011) 1345–1353.
[85] H.C. Dietz, G.R. Cutting, R.E. Pyeritz, et al., Marfan syndrome caused by a recurrent de novo missense mutation in the fibrillin gene, Nature 352 (6333) (1991) 337–339.
[86] R.W. Jeremy, H. Huang, J. Hwa, H. McCarron, C.F. Hughes, J.G. Richards, Relation between age, arterial distensibility, and aortic dilatation in the Marfan syndrome, Am. J. Cardiol. 74 (4) (1994) 369–373.
[87] G. Jondeau, P. Boutouyrie, P. Lacolley, et al., Central pulse pressure is a major determinant of ascending aorta dilation in Marfan syndrome, Circulation 99 (20) (1999) 2677–2678.
[88] G. Teixido-Tura, A. Redheuil, J. Rodríguez-Palomares, et al., Aortic biomechanics by magnetic resonance: early markers of aortic disease in Marfan syndrome regardless of aortic dilatation? Int. J. Cardiol. 171 (1) (2014) 56–61.

[89] D.J. Brull, L.J. Murray, C.A. Boreham, et al., Effect of a COL1A1 Sp1 binding site polymorphism on arterial pulse wave velocity: an index of compliance, Hypertension 38 (3) (2001) 444–448.
[90] A. Superti-Furga, E. Gugler, R. Gitzelmann, B. Steinman, Ehlers-Danlos syndrome type IV: a multi-exon deletion in one of the two COL3A1 alleles affecting structures, stability, and processing of type III pro-collagen, J. Biol. Chem. 263 (2008) 6226–6232.
[91] C. Flagothier, V. Goffin, T. Hermanns-Lê, G.E. Piérard, P. Quatresooz, A four-generation Ehlers-Danlos syndrome with vascular dissections. Skin ultrastructure and biomechanical properties, J. Med. Eng. Technol. 31 (3) (2007) 175–180.
[92] S. Annunen, P. Paassilta, J. Lohiniva, et al., An allele of COL9A2 associated with intervertebral disc disease, Science 285 (5426) (1999) 409–412.
[93] D.M. Aladin, K.M. Cheung, D. Chan, et al., Expression of the Trp2 allele of COL9A2 is associated with alterations in the mechanical properties of human intervertebral discs, Spine (Phila Pa 1976) 32 (25) (2007) 2820–2826.
[94] Y. Muragaki, T. Kimura, Y. Ninomiya, B.R. Olsen, The complete primary structure of two distinct forms of human alpha 1 (IX) collagen chains, Eur. J. Biochem. 192 (1990) 703–708.
[95] B.P. Foster, C.I. Morse, G.L. Onambele, A.G. Williams, Human COL5A1 rs12722 gene polymorphism and tendon properties in vivo in an asymptomatic population, Eur. J. Appl. Physiol. 114 (7) (2014) 1393–1402.
[96] B.P. Foster, C.I. Morse, G.L. Onambele, A.G. Williams, Variants within the MMP3 gene and patellar tendon properties in vivo in an asymptomatic population, Eur. J. Appl. Physiol. 114 (12) (2014) 2625–2634.

CHAPTER TEN

Impact of Pregnancy and Delivery on Pelvic Floor Biomechanics

M. Alperin
University of California, San Diego, CA, United States of America

IMPACT OF PREGNANCY ON BIOMECHANICS OF PELVIC FLOOR

There are probably no greater phenotypic transformations in healthy human adults than the ones that occur in female pelvis during pregnancy. Elucidation of pregnancy-induced changes in the mechanical properties of pelvic floor components is, therefore, an absolute prerequisite for designing scientifically sound biomechanical studies and reliable models of vaginal deliveries. Given that investigations in pregnant women are arduous and are associated with obvious ethical constraints, biomechanical studies conducted in animals and computational models are invaluable for advancing our understanding of the effects of pregnancy and delivery on the pelvic floor.

The upright posture associated with human obligate bipedalism resulted in the pelvis becoming an essential load-bearing and supportive structure, with body weight transferred from the spine through the sacroiliac joints to the pelvic girdle [1]. In pregnancy, the center of gravity is shifted forward with a mechanical axis located in front of the femoral heads, leading to compensatory retroversion of the pelvis [2]. In addition to these and other anatomic modifications incurred in pregnancy, the altered hormonal milieu of pregnancy also affects the musculoskeletal system, resulting in decreased stiffness of pelvic and spinal ligaments [3]. These changes, together with increased weight of the gravid uterus, lead to lumbar lordosis. In turn, lumbar hyperlordosis lowers the contractility of the pelvic floor skeletal muscles in a static posture and during dynamic increases in the intraabdominal pressure [4].

One of the main pregnancy-associated hormones that accounts for the remodeling of ligaments and pelvic connective tissues is a member of the insulin family of polypeptide hormones, relaxin. The upsurge of relaxin in pregnancy leads to collagen degradation through upregulation of endogenous proteolytic collagenases — metalloproteinases [3,5–7]. Studies in rodent models demonstrate that breakdown of collagen in the rat pubic symphysis during pregnancy coincides with a peak in relaxin levels, which are highest in the second half of gestation [8,9]. In mice, relaxin reduces content and alters organization of fibrillar collagens in the cervix in late pregnancy [10]. Even though serum levels of relaxin in humans are significantly lower than those in rodents, a similar decrease in human cervical stiffness occurs during gestation [11].

Relaxin receptors have also been identified in human and animal vaginal tissue, which also undergoes significant remodeling in pregnancy [12,13]. The fibromuscular layer of the vagina is mainly composed of collagen and elastin, which comprise the subepithelium, and smooth muscle, which forms the muscularis. Similar to the adaptations seen in the cervix, the connective tissue of the vagina and its supportive structures exhibit biochemical and structural changes during pregnancy [1,13–16]. In pregnant rats treated with antirelaxin antibody that blocks endogenous relaxin, vaginal weight, length, and diameter do not undergo the increase observed in untreated pregnant controls [17]. In a recent study examining changes in the composition of vaginal fibromuscular layers in a sheep model, significantly less collagen and more elastin was identified in pregnant animals, compared to nonpregnant controls [18]. Investigations in a mouse model also demonstrate upregulation of elastin turnover in pregnancy [19]. In parallel with animal data, human studies also demonstrate significant biochemical changes in vaginal fibrillar extracellular matrix (ECM) in response to the hormonal milieu associated with pregnancy [13]. In addition to the alterations in the ECM, pregnancy has been shown to induce changes in vaginal smooth muscle in a rat model [14]. Synthetic smooth muscle cells secrete matrix and have no contractile tone, whereas the contractile smooth muscle cells function as contractile entities. In pregnancy, smooth muscle cells in rat vagina transform their phenotype from contractile to synthetic, likely resulting in the increased production of ECM components such as collagen, elastin, and proteoglycans.

PREGNANCY-INDUCED ALTERATIONS IN VAGINAL BIOMECHANICAL PROPERTIES

The biochemical alterations and structural remodeling of the ECM and smooth muscle components of the vagina, summarized above, are thought to account for the dramatic changes in vaginal passive and active biomechanical properties in preparation for delivery of the fetus [20]. Changes in vaginal mechanical properties are evident on clinical examinations, with pregnant women demonstrating increased vaginal laxity and progressive displacement of the anterior and posterior compartments distally across trimesters relative to nonpregnant controls [13,16,21].

The rat vagina is analogous to the human vagina with respect to the histological layers and the organization of apical and paravaginal levels of support [22]. Decreased stiffness of the vagina-supportive tissue complex and increased vaginal distensibility in pregnancy has been consistently demonstrated in ex vivo and in vivo experiments conducted in a rat model [15,20,23]. The majority of ex vivo biomechanical studies utilize the uniaxial load to failure test along the longitudinal vaginal axis, simulating downward distension of the vagina-supportive tissue complex. The main parameters that describe vaginal passive mechanical properties are determined from the resultant load-elongation curves and the corresponding stress-strain curves. These include the tangent modulus, the ultimate tensile strength, the ultimate tensile strain, and the strain energy density. Invariably, the tangent modulus was significantly lower and the ultimate strain substantially higher in pregnant animals relative to virgin controls [20,23]. Similar results were obtained in studies conducted in mouse and sheep models [18,24]. Consistent with the results of the in vitro experiments, in vivo studies in rat models demonstrate a dramatic increase in vaginal distensibility during pregnancy compared to virgin controls [15].

As opposed to the uniform results seen with respect to distensibility, data regarding the effect of pregnancy on vaginal tensile strength need to be interpreted in the framework of the specific experimental testing and calculations employed. The studies that utilize uniaxial load to failure tests in the longitudinal direction (ie, parallel to the long axis of vagina) and represent the tensile strength either by ultimate force or ultimate stress, derived from the force divided by the *initial* cross-sectional area, demonstrated decreased vaginal tensile strength in pregnancy [20,23]. However, when the same test was performed in the direction perpendicular to the long axis of vagina and the tensile strength was derived from Cauchy stress (ie, force divided by *deformed* cross-sectional area), vaginal tissue was found to have a higher tensile strength in pregnant animals compared to nonpregnant controls [25].

In addition to the changes in the passive mechanical properties, the contractile force of vaginal muscularis decreases in pregnancy [20], a change consistent with the phenotypic shift from contractile to synthetic smooth muscle that occurs in pregnancy [14]. Taken together, the above findings characterize an important maternal adaptation, as increased vaginal compliance is likely to facilitate fetal delivery while minimizing pressure exerted on the vaginal walls during parturition.

PREGNANCY-INDUCED ALTERATIONS IN BIOMECHANICAL PROPERTIES OF PELVIC FLOOR MUSCLES

Pelvic floor striated muscles (PFM) consist of coccygeus and the muscles of the levator ani complex, which include iliococcygeus, pubococcygeus, and puborectalis. PFM provide integral structural support for proper pelvic floor function. Trauma to these muscles related to vaginal childbirth appears to play an important role in the pathogenesis of pelvic floor disorders (PFD). As evident from the above, dramatic biomechanical and biochemical changes occur in the pelvic organs in preparation for delivery [1,13–16]. It is reasonable to expect that pregnancy would also induce substantial alterations in the PFM.

Imaging of the levator hiatus has been used to assess changes in the PFM in pregnancy. Increased hiatal area at rest and with straining has been reported in women at the end of their first pregnancy compared to nonpregnant controls [26,27]. Furthermore, a recent study showed that larger levator hiatus at rest in late pregnancy is associated with a shorter second stage of labor and successful spontaneous vaginal delivery [28]. However, the levator hiatus area during straining was not associated with delivery outcome [28]. The explanation proposed by the authors is based on the findings that the parameters that had significant correlations with delivery outcomes, such as hiatal transverse diameter, are directly related to the bony pelvis. PFM parameters that are less dependent on the dimensions of the bony pelvis, such as anterior-posterior diameter, on the other hand, did not correlate with the delivery outcomes. Furthermore, levator hiatus dimension provides very limited information relevant to PFM structure and function. First, only the most medial portion of the levator ani muscle complex is accounted for when levator hiatus is measured. Second, levator hiatus area during straining is substantially smaller than the area required for the delivery of the fetus [29]. Third, levator hiatus area does not correlate with strength or endurance of PFM contraction [30]. Thus, our current understanding of the impact of pregnancy on the structure of the PFM is severely limited.

In a recent study conducted in our lab, we demonstrated, for the first time, that pregnancy induces unique adaptations in the intrinsic structure of the PFM in a rat model [31]. Specifically, we showed that, under functional conditions of pregnancy, PFM increased their fiber length and consequently their excursion by adding sarcomeres in series. Muscle injury is primarily caused by excessive sarcomere strain [32,33], which is likely to occur during parturition. Increased fiber length is likely protective against such damage as large mechanical deformations are distributed across a greater number of sarcomeres. The importance of determining the impact of pregnancy on the PFM biomechanics is two-fold: (1) to improve our understanding of the pathogenesis of PFM injury during delivery and (2) to identify factors that segregate women who do not sustain PFM birth trauma from those who do.

IMPACT OF VAGINAL DELIVERY ON BIOMECHANICS OF PELVIC FLOOR

Vaginal delivery is the greatest epidemiological risk factor for the development of PFD [34–36]. Although cesarean delivery does not completely protect the pelvic floor from the untoward effects of pregnancy, currently published data clearly indicate that the mode of delivery confers a higher risk for subsequent PFD compared to pregnancy alone [37–41]. Further support for this position is provided by the data indicating that prevalence of pelvic organ prolapse following childbearing does not differ between women who underwent abdominal delivery prior to the onset of labor and women who had cesarean section after reaching the second stage of labor, and is lower in both groups relative to women who gave birth vaginally [37]. Despite high prevalence of the inciting event (vaginal delivery) and PFD that follow, high-resolution studies of the pathological alterations in the supportive pelvic floor structures and the mechanisms that lead to their dysfunction are not well delineated. Determining the biomechanical impact of parturition on the pelvic floor is important for understanding the pathophysiology of its postpartum dysfunction. The above, in turn, will allow to mechanistically link vaginal delivery and PFD.

It is evident from the studies discussed above and other similar investigations that dramatic biochemical changes in tissues take place during pregnancy. These pregnancy-induced adaptations, in turn, have a profound effect on the mechanical function of the main pelvic floor supportive structures during parturition. The main biomechanical impact of pregnancy is manifested in significantly decreased stiffness and a shift toward more viscous material properties of pelvic tissues [25]. These alterations allow tissue to become more extensible and even more tolerant of higher stress applied in the cross-sectional direction, facilitating the delivery of the fetus while protecting against maternal birth injury. Human studies demonstrate correlation between higher vaginal elastase activity and uncomplicated spontaneous vaginal delivery, which, in turn, protects women from pelvic floor trauma associated with operative vaginal deliveries [13]. Ex vivo animal studies conducted in various models show a full recovery of the mechanical function of the vagina and its supportive structures after spontaneous delivery [18,20,23,42]. Furthermore, biomechanical studies of sheep vagina showed a pronounced Mullins effect, which is present when tissues mechanical response depends on the prior applied maximum load. The mechanical response of tissue subsequent to the initial loading was dependent on the previously attained maximum stress and was characterized by the increased vaginal tissue strain in pregnancy [18]. Throughout parturition, pelvic tissues are subjected to loading during the expulsive efforts, followed by tissue "unloading" between the contractions. Therefore, these material properties become an important factor for understanding the mechanical function of pelvic tissue during vaginal delivery.

The adaptive effect of significantly lower tissue stiffness in pregnancy is accompanied by the decrease in tensile strength, as indicated by multiple studies that employ longitudinal load to failure testing protocols. This, in turn, renders tissues more susceptible to trauma in response to the applied load. Thus, the demands of vaginal delivery can easily exceed the load-bearing capacity of pelvic tissues altered by pregnancy, leading to trauma associated with childbirth. In addition, the collagenase activity is upregulated in vaginal tissues subjected to mechanical stresses and strains, with proteolytic activity proportional to the magnitude of the applied strain [43]. Consequently, when substantial stresses and strains occur during parturition, metabolism of ECM shifts further toward degradation. Studies in mouse models have demonstrated that in addition to collagen degradation, increased mechanical loading plays an important role in turnover of vaginal elastin [19]. Interestingly, expression of fibulin-5 and tropoelastin, two proteins essential for production of mature elastin, was augmented in pregnant animals concomitant with activation of vaginal elastases. These data suggest that reassembly of elastin fibers likely plays an important role in restoration of mechanical function of the vagina after delivery [44]. Placing excessive mechanical demands on the vagina during parturition potentially alters the balance between proteolytic activity of elastases and synthesis and assembly necessary for tissue recovery, setting the stage for subsequent pelvic organ prolapse.

Divergent from ex vivo findings, in vivo experiments in a rat model demonstrate that spontaneous vaginal delivery leads to persistently altered mechanical properties postpartum [15]. One possible reason behind this discrepancy is a more complex tissue loading achieved in the in vivo testing, which includes circumferential and longitudinal directions. Secondly, the in vivo experiments are able to assess the collective effect of parturition on the vagina and all the tissues that support and maintain its structure, not captured by ex vivo testing. From experiments in animal models, we know that when vaginal tissue is subjected to large stresses and strains through simulated birth injury, ex vivo testing protocols also demonstrate incomplete recovery of biochemical and biomechanical parameters postpartum [45,46]. Taken together, these persistently untoward changes provide potential mechanistic clues that can link such clinical risk factors as large fetal size, prolonged second stage of labor, and operative vaginal delivery with maternal birth injury and subsequent pelvic floor dysfunction [38,47].

IMPACT OF VAGINAL DELIVERY ON PELVIC FLOOR MUSCLES

Maternal birth trauma to the striated pelvic floor muscles appears to play an important role in the pathogenesis of PFD. A recent study highlighted the long-term negative impact of childbirth on the PFM function, which can last a decade after vaginal delivery [48]. Defects in the levator ani complex are prevalent and have been observed by imaging in a third of women after spontaneous vaginal delivery and in 60% after forceps assisted delivery [49]. Moreover, the risk of developing PFD increases dramatically in women with radiologically abnormal PFM compared to women with intact PFM [35]. Numerous studies suggest that demands placed on the PFM during vaginal childbirth exceed the physiologic limits of striated muscles [50–52]. By the end of pregnancy, the median levator hiatus area that women achieve while straining is four to five times smaller than the cross-sectional area needed for the delivery of a fetal head [28]. To achieve the cross-sectional area required for vaginal childbirth, the PFM are estimated to undergo strain as high as 300% [50,51,53]. Such dramatic strains are presumed to cause radiologically detected abnormalities in the PFM, including macroscopic trauma (referred to as avulsion injury) and microscopic trauma from irreversible overdistension [54–56]. These conclusions are consistent with mechanical studies of limb striated muscles, which demonstrate that the magnitude of muscle injury is a function of strain [33,57,58].

Due to the difficulties and obvious ethical constraints associated with directly probing PFM, the majority of investigations exploring the mechanics of vaginal delivery and its effect on the PFM employ computational models (see Chapter 22). The growing body of literature, exploring the mechanics of childbirth through computational modeling, has significantly improved our understanding of the impact that vaginal delivery has on the PFM. Unfortunately, the existing models are significantly limited by paucity of experimental data available to characterize the material properties of the PFM in general and during pregnancy in particular, and to validate the models. The importance of the material properties was highlighted by a constitutive modeling study, which demonstrated that varying muscle stiffness has a profound effect on the mechanical behavior of the PFM during parturition [59]. Recent models ascribe the following material properties to the PFM: anisotropy and visco-hyperelastic behavior [25,66]. Coinciding with clinical findings, the models predict that PFM located closest to the levator hiatus experience the greatest stresses and strains during parturition. Specifically, the maximum stress and the largest stretch were calculated at the pubococcygeus enthesis or the connective tissue at the site of muscle attachment to the pubis [66].

PFM are a composite material consisting of contractile myofibers embedded within the large connective tissue network of the intramuscular ECM [60]. Activation of PFM myofibers in models has been found to oppose the expulsive forces and result in higher calculated stresses exerted on the pelvic floor structures [61]. While myofibers are responsible for muscle active mechanics, the skeletal muscle ECM bears the majority of muscle's passive load. Thus, skeletal muscle stiffness primarily reflects its ECM properties [60]. Much effort has been devoted to studying the connective tissues of the vagina and its supportive structures [62–65]. However, passive mechanical properties of PFM ECM have not been elucidated to date. Furthermore, the complex anatomy of these muscles and technical restrictions of magnetic resonance imaging inhibit accurate identification of individual muscle borders and direction of muscle fascicles. Quantitative data on pregnancy-induced alterations in tissue biomechanics is essential for the development of models that accurately represent the impact of vaginal delivery on the main constituents of the female pelvic floor.

CONCLUSIONS

- Pregnancy induces dramatic adaptations in the biochemical and biomechanical properties of pelvic connective tissue, and smooth and skeletal muscles.
- Pregnancy-induced adaptations in pelvic tissues mechanical properties appear to facilitate the delivery of the fetus, while protecting against maternal birth injury.
- Computational models must account for changes in the biomechanical tissue properties to make accurate predictions regarding the response of pelvic structures to stresses and strains associated with pregnancy and delivery.

- The destabilizing changes resultant from vaginal delivery need to be defined at the tissue level to establish a mechanistic link between childbirth and dysfunction of pelvic floor supportive structures.
- Elucidating the mechanisms that guide pregnancy-induced adaptations are needed to enable the development of preventive and therapeutic strategies to minimize maternal morbidity related to vaginal birth.

REFERENCES

[1] E.H. Sze, G.B. Sherard 3rd., J.M. Dolezal, Pregnancy, labor, delivery, and pelvic organ prolapse, Obstet. Gynecol. 100 (5 Pt 1) (2002) 981–986.
[2] J.C. Le Huec, et al., Equilibrium of the human body and the gravity line: the basics, Eur. Spine J. 20 (Suppl. 5) (2011) 558–563.
[3] C.S. Samuel, J.P. Coghlan, J.F. Bateman, Effects of relaxin, pregnancy and parturition on collagen metabolism in the rat pubic symphysis, J. Endocrinol. 159 (1) (1998) 117–125.
[4] A.C. Capson, J. Nashed, L. McLean, The role of lumbopelvic posture in pelvic floor muscle activation in continent women, J. Electromyogr. Kinesiol. 21 (1) (2011) 166–177.
[5] E.N. Unemori, et al., Relaxin induces an extracellular matrix-degrading phenotype in human lung fibroblasts in vitro and inhibits lung fibrosis in a murine model in vivo, J. Clin. Invest. 98 (12) (1996) 2739–2745.
[6] S. Palejwala, et al., Relaxin positively regulates matrix metalloproteinase expression in human lower uterine segment fibroblasts using a tyrosine kinase signaling pathway, Endocrinology 142 (8) (2001) 3405–3413.
[7] C.K. Too, G.D. Bryant-Greenwood, F.C. Greenwood, Relaxin increases the release of plasminogen activator, collagenase, and proteoglycanase from rat granulosa cells in vitro, Endocrinology 115 (3) (1984) 1043–1050.
[8] O.D. Sherwood, V.E. Crnekovic, Development of a homologous radioimmunoassay for rat relaxin, Endocrinology 104 (4) (1979) 893–897.
[9] O.D. Sherwood, et al., Radioimmunoassay of relaxin throughout pregnancy and during parturition in the rat, Endocrinology 107 (3) (1980) 691–698.
[10] Y.M. Soh, et al., Relaxin regulates hyaluronan synthesis and aquaporins in the cervix of late pregnant mice, Endocrinology 153 (12) (2012) 6054–6064.
[11] O.D. Sherwood, Relaxin's physiological roles and other diverse actions, Endocr. Rev. 25 (2) (2004) 205–234.
[12] M.A. Harvey, S.L. Johnston, G.A. Davies, Mid-trimester serum relaxin concentrations and post-partum pelvic floor dysfunction, Acta Obstet. Gynecol. Scand. 87 (12) (2008) 1315–1321.
[13] S.S. Oliphant, et al., Maternal adaptations in preparation for parturition predict uncomplicated spontaneous delivery outcome, Am. J. Obstet. Gynecol. 211 (6) (2014) 630.e1–630.e7.
[14] J.A. Daucher, et al., Adaptations of the rat vagina in pregnancy to accommodate delivery, Obstet. Gynecol. 109 (1) (2007) 128–135.
[15] M. Alperin, et al., Pregnancy- and delivery-induced biomechanical changes in rat vagina persist post-partum, Int. Urogynecol. J. 21 (9) (2010) 1169–1174.
[16] A.L. O'Boyle, et al., Pelvic organ support in pregnancy and postpartum, Int. Urogynecol. J. Pelvic Floor Dysfunct. 16 (1) (2005) 69–72, discussion 72.
[17] S. Zhao, O.D. Sherwood, Monoclonal antibodies specific for rat relaxin. X. Endogenous relaxin induces changes in the histological characteristics of the rat vagina during the second half of pregnancy, Endocrinology 139 (11) (1998) 4726–4734.
[18] D. Ulrich, et al., Influence of reproductive status on tissue composition and biomechanical properties of ovine vagina, PLoS ONE 9 (4) (2014) e93172.
[19] D.D. Rahn, J.F. Acevedo, R.A. Word, Effect of vaginal distention on elastic fiber synthesis and matrix degradation in the vaginal wall: potential role in the pathogenesis of pelvic organ prolapse, Am. J. Physiol. Regul. Integr. Comp. Physiol. 295 (4) (2008) R1351–R1358.

[20] A. Feola, et al., Impact of pregnancy and vaginal delivery on the passive and active mechanics of the rat vagina, Ann. Biomed. Eng. 39 (1) (2011) 549–558.
[21] H.P. Dietz, A.B. Steensma, Which women are most affected by delivery-related changes in pelvic organ mobility? Eur. J. Obstet. Gynecol. Reprod. Biol. 111 (1) (2003) 15–18.
[22] P.A. Moalli, et al., A rat model to study the structural properties of the vagina and its supportive tissues, Am. J. Obstet. Gynecol. 192 (1) (2005) 80–88.
[23] J.L. Lowder, et al., Biomechanical adaptations of the rat vagina and supportive tissues in pregnancy to accommodate delivery, Obstet. Gynecol. 109 (1) (2007) 136–143.
[24] X. Liu, et al., Failure of elastic fiber homeostasis leads to pelvic floor disorders, Am. J. Pathol. 168 (2) (2006) 519–528.
[25] D. Jing, Experimental and Theoretical Biomechanical Analyses of the Second Stage of Labor, Department of Bioengineering, University of Michigan, Ann Arbor, MI, 2010.
[26] K.L. Shek, J. Kruger, H.P. Dietz, The effect of pregnancy on hiatal dimensions and urethral mobility: an observational study, Int. Urogynecol. J. 23 (11) (2012) 1561–1567.
[27] J. Staer-Jensen, et al., Ultrasonographic evaluation of pelvic organ support during pregnancy, Obstet. Gynecol. 122 (2 Pt 1) (2013) 329–336.
[28] F. Siafarikas, et al., Levator hiatus dimensions in late pregnancy and the process of labor: a 3- and 4-dimensional transperineal ultrasound study, Am. J. Obstet. Gynecol. 210 (5) (2014) 484.e1–484.e7.
[29] V. Lanzarone, H.P. Dietz, Three-dimensional ultrasound imaging of the levator hiatus in late pregnancy and associations with delivery outcomes, Aust. N. Z. J. Obstet. Gynaecol. 47 (3) (2007) 176–180.
[30] K. Bo, et al., Pelvic floor muscle variables and levator hiatus dimensions: a 3/4D transperineal ultrasound cross-sectional study on 300 nulliparous pregnant women, Int. Urogynecol. J. 25 (10) (2014) 1357–1361.
[31] M. Alperin, D.M. Lawley, M.C. Esparza, R.L. Lieber, Pregnancy induced adaptations in the intrinsic structure of rat pelvic floor muscles, Am. J. Obstet. Gynecol. 213 (2015) 191.e1–191.e7.
[32] R.L. Lieber, J. Friden, Muscle damage is not a function of muscle force but active muscle strain, J. Appl. Physiol. (1985) 74 (2) (1993) 520–526.
[33] G.L. Warren, et al., Mechanical factors in the initiation of eccentric contraction-induced injury in rat soleus muscle, J. Physiol. 464 (1993) 457–475.
[34] N.A. Clark, et al., Levator defects affect perineal position independently of prolapse status, Am. J. Obstet. Gynecol. 203 (6) (2010) 595.e17–595.e22.
[35] J.O.L. DeLancey, et al., Comparison of levator ani muscle defects and function in women with and without pelvic organ prolapse. Obstet. Gynecol. 109 (2 Pt 1) (2007) 295–302, http://dx.doi.org/10.1097/01.AOG.0000250901.57095.ba.
[36] J.A. Ashton-Miller, J.O. Delancey, On the biomechanics of vaginal birth and common sequelae, Annu. Rev. Biomed. Eng. 11 (2009) 163–176.
[37] V.L. Handa, et al., Pelvic floor disorders 5–10 years after vaginal or cesarean childbirth, Obstet. Gynecol. 118 (4) (2011) 777–784.
[38] L.M. Dolan, P. Hilton, Obstetric risk factors and pelvic floor dysfunction 20 years after first delivery, Int. Urogynecol. J. 21 (5) (2010) 535–544.
[39] C. Larsson, K. Kallen, E. Andolf, Cesarean section and risk of pelvic organ prolapse: a nested case–control study, Am. J. Obstet. Gynecol. 200 (3) (2009) 243.e1–243.e4.
[40] H.Y. Chin, et al., Postpartum urinary incontinence: a comparison of vaginal delivery, elective, and emergent cesarean section, Int. Urogynecol. J. Pelvic Floor Dysfunct. 17 (6) (2006) 631–635.
[41] V.L. Handa, et al., Progression and remission of pelvic organ prolapse: a longitudinal study of menopausal women, Am. J. Obstet. Gynecol. 190 (1) (2004) 27–32.
[42] D.D. Rahn, et al., Biomechanical properties of the vaginal wall: effect of pregnancy, elastic fiber deficiency, and pelvic organ prolapse, Am. J. Obstet. Gynecol. 198 (5) (2008) 590.e1–590.e6.
[43] C.M. Yang, et al., Mechanical strain induces collagenase-3 (MMP-13) expression in MC3T3-E1 osteoblastic cells, J. Biol. Chem. 279 (21) (2004) 22158–22165.
[44] P.G. Drewes, et al., Pelvic organ prolapse in fibulin-5 knockout mice: pregnancy-induced changes in elastic fiber homeostasis in mouse vagina, Am. J. Pathol. 170 (2) (2007) 578–589.
[45] M. Alperin, et al., Collagen scaffold: a treatment for simulated maternal birth injury in the rat model, Am. J. Obstet. Gynecol. 202 (6) (2010) 589.e1–589.e8.

[46] K.D. Sievert, et al., The effect of simulated birth trauma and/or ovariectomy on rodent continence mechanism. Part I: functional and structural change, J. Urol. 166 (1) (2001) 311–317.
[47] V.L. Handa, et al., Pelvic floor disorders after vaginal birth: effect of episiotomy, perineal laceration, and operative birth, Obstet. Gynecol. 119 (2 Pt 1) (2012) 233–239.
[48] H.U. Memon, V.L. Handa, Vaginal childbirth and pelvic floor disorders, Womens Health (Lond. Engl.) 9 (3) (2013) 265–277, quiz 276–277.
[49] R. Kearney, et al., Levator ani injury in primiparous women with forceps delivery for fetal distress, forceps for second stage arrest, and spontaneous delivery, Int. J. Gynaecol. Obstet. 111 (1) (2010) 19–22.
[50] K.C. Lien, et al., Levator ani muscle stretch induced by simulated vaginal birth, Obstet. Gynecol. 103 (1) (2004) 31–40.
[51] L. Hoyte, et al., Quantity and distribution of levator ani stretch during simulated vaginal childbirth, Am. J. Obstet. Gynecol. 199 (2) (2008) 198.e1–198.e5.
[52] L. Krofta, et al., Pubococcygeus-puborectalis trauma after forceps delivery: evaluation of the levator ani muscle with 3D/4D ultrasound, Int. Urogynecol. J. Pelvic Floor Dysfunct. 20 (10) (2009) 1175–1181.
[53] K. Svabik, K.L. Shek, H.P. Dietz, How much does the levator hiatus have to stretch during childbirth? BJOG 116 (12) (2009) 1657–1662.
[54] K.L. Shek, H.P. Dietz, Intrapartum risk factors for levator trauma, BJOG 117 (12) (2010) 1485–1492.
[55] H.P. Dietz, V. Lanzarone, Levator trauma after vaginal delivery, Obstet. Gynecol. 106 (4) (2005) 707–712.
[56] J. Cassado Garriga, et al., Tridimensional sonographic anatomical changes on pelvic floor muscle according to the type of delivery, Int. Urogynecol. J. 22 (8) (2011) 1011–1018.
[57] J.A. Faulkner, S.V. Brooks, J.A. Opiteck, Injury to skeletal muscle fibers during contractions: conditions of occurrence and prevention, Phys. Ther. 73 (12) (1993) 911–921.
[58] R.L. Lieber, J. Friden, Mechanisms of muscle injury gleaned from animal models, Am. J. Phys. Med. Rehabil. 81 (11 Suppl.) (2002) S70–S79.
[59] M.P. Parente, et al., The influence of the material properties on the biomechanical behavior of the pelvic floor muscles during vaginal delivery, J. Biomech. 42 (9) (2009) 1301–1306.
[60] A.R. Gillies, R.L. Lieber, Structure and function of the skeletal muscle extracellular matrix, Muscle Nerve 44 (3) (2011) 318–331.
[61] M.P. Parente, et al., The influence of pelvic muscle activation during vaginal delivery, Obstet. Gynecol. 115 (4) (2010) 804–808.
[62] P.A. Norton, Pelvic floor disorders: the role of fascia and ligaments, Clin. Obstet. Gynecol. 36 (4) (1993) 926–938.
[63] E. Hirata, et al., Comparative histological study of levels 1–3 supportive tissues using pelvic floor semiserial sections from elderly nulliparous and multiparous women, J. Obstet. Gynaecol. Res. 37 (1) (2011) 13–23.
[64] B. Gabriel, et al., Uterosacral ligament in postmenopausal women with or without pelvic organ prolapse, Int. Urogynecol. J. 16 (6) (2005) 475–479.
[65] K. Yamamoto, et al., Decrease in elastin gene expression and protein synthesis in fibroblasts derived from cardinal ligaments of patients with prolapsus uteri, Cell Biol. Int. 21 (1997) 605–611.
[66] D. Jing, J.A. Ashton-Miller, J.O. DeLancey, A subject-specific anisotropic visco-hyperelastic finite element model of female pelvic floor stress and strain during the second stage of labor, J. Biomech. 45 (3) (2012) 455–460.

CHAPTER ELEVEN

Biomechanical Environment of the Pelvic Floor

S. Madill
School of Physical Therapy, University of Saskatchewan, Saskatchewan, Canada

PHYSIOLOGICAL FUNCTION OF THE PELVIC FLOOR

The pelvic floor skeletal muscles (PFMs) are the pubovisceral including the puborectalis and the pubococcygeus muscles, the iliococcygeus, and the coccygeus muscles [1]. At rest the PFMs are tonically active [2–5]. This activity is superseded by phasic activity when demands increase due to changes in posture, movement, and respiration [2–5]. Although each muscle within the PFMs has a different fiber direction, which typically would indicate a different action for each muscle, the muscles can only be contracted collectively and thus have an amalgamated action that lifts the PFMs higher into the pelvic cavity, elevates the pelvic organs, narrows the anorectal angle, and flexes the coccyx on the sacrum [6–15]. The contraction also constricts the urogenital hiatus shortening it anterior-posteriorly and reducing the area of the opening [16]. This squeezes the vaginal canal, exerts a compressive force on the urethra, tensions the anterior vaginal wall, and relieves stress from the fascial support [1,3,8,17–22] (see Fig. 1). The ability of the PFMs to elevate the bladder neck seems to depend on the amount of intraabdominal pressure that is generated simultaneously [23]. During activities that generate high levels of intraabdominal pressure, the bladder neck may not elevate despite the contraction of the PFMs. The tonic activity of the PFMs ceases with micturition [2–5]. Similar activity is observed in the smooth muscle fibers that are integrated with the pelvic fascia; they are active during urine storage and relax with micturition [21,24].

In addition to the compression of the urethra resulting from the contraction of the PFMs, the intrinsic structure of the urethra also facilitates continence. The anterior and posterior walls lie against each other and the mucosa falls in longitudinal folds maintaining a hermetic seal [25]. At rest, urethral closure is maintained by three factors in essentially equal proportions: the striated urethral sphincters, mucosa and submucosal vascular bed, and the circular smooth muscle and connective tissue in the urethral and periurethral tissues [26]. As bladder volume increases, urethral circular smooth muscle, the striated urethral sphincter, and the PFMs increase their activity to maintain urethral closure [27]. The increase in urethral pressure is greater than the bladder pressure increase

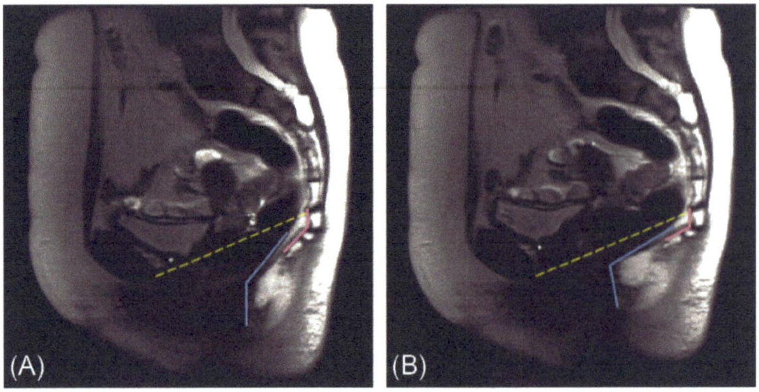

Figure 1 Mid-sagittal images of the female pelvis acquired in supine (A) with the pelvic floor muscles (PFMs) relaxed and (B) with the PFMs maximally contracted. The dashed *yellow line*, from the inferior edge of the pubic symphysis to the anterior sacrococcygeal joint line, is the pubococcygeal line (PCL). Note how the urethrovesical junction (*) and the anorectal angle *(blue)* are higher relative to the PCL in B compared to A. As well, the anorectal angle *(blue)* narrows and coccygeal flexion *(pink)* increases in B compared to A. These differences illustrate the lifting and forward pulling actions of the PFMs.

with filling [28]. The same pattern of urethral pressure increase in response to bladder filling is observed in the sitting position as in the supine position; however, in sitting the maximum urethral pressure is not as high relative to the bladder pressure [28]. Postural changes from a change from lying to upright position increase maximum urethral pressure by an average of 23% [29].

The striated urethral sphincter contracts synergistically with the PFMs during both voluntary and reflex PFM contractions [30–34]. This ability to contract the urethral sphincter concomitantly with a voluntary PFM contraction has been reported to be preserved even in women with weak or damaged PFMs [34,35].

There is also a synergistic relationship between the abdominal muscles and the PFMs [36–41]. Several studies have reported that women are not able to perform a PFM contraction without also activating their abdominal muscles [38,40,42]. Consistently, abdominal muscle contractions are accompanied by an automatic PFM contraction, with the strength of the abdominal muscle contraction and the synergistic PFM contraction positively correlated [36,39,41]. Sapsford and Hodges have reported that a strong transversus abdominis contraction is equally effective as a PFM contraction for stopping urine flow and for increasing urethral pressure in continent women [22,40]. Similarly, Madill and McLean [37] have reported that abdominal muscle activity contributes to the generation of intravaginal pressure, measured at the level where the PFMs act upon the vagina. They found that the initial rise in intravaginal pressure was predominately generated by the PFMs, while the rectus abdominis, internal obliques, and transversus abdominis muscles generated the last 30% of the pressure during a maximum voluntary PFM contraction (Fig. 2). However, Bø et al. [16] found that contracting the transversus abdominis was not effective in reducing the size of the levator hiatus. This, combined with Sapsford and Hodges's findings discussed above, suggests that the synergy between the abdominal muscles and the urethral sphincter may be stronger than the abdominal-PFMs synergy.

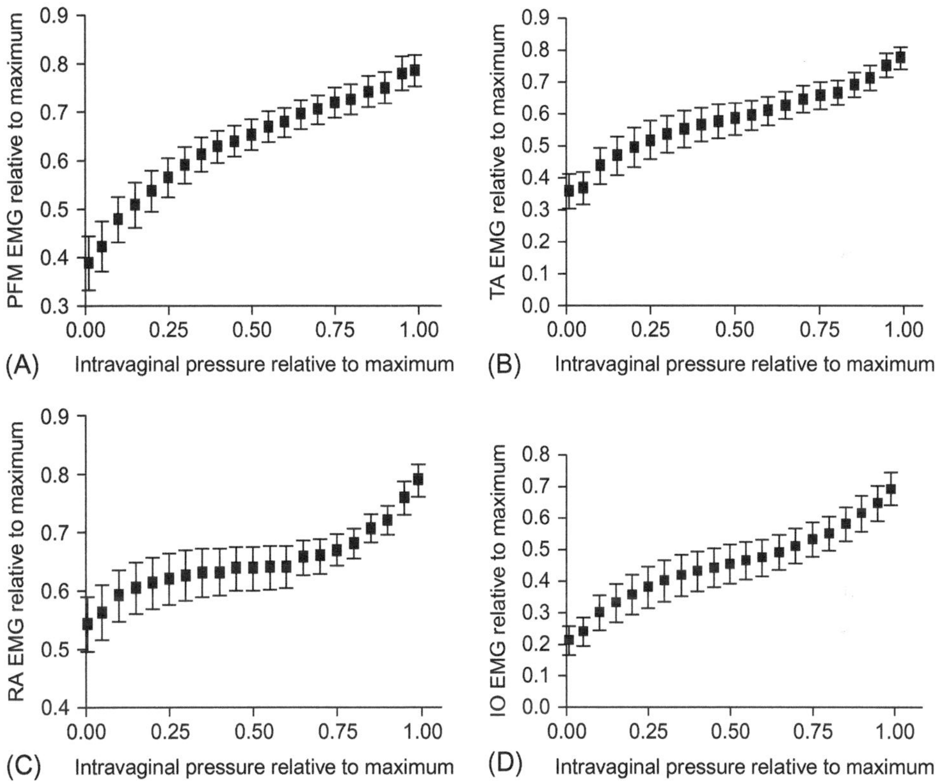

Figure 2 Ensemble average curves of electromyographic (EMG) activation over vaginal pressure generation. (A) Pelvic floor muscle (PFM) EMG versus intravaginal pressure. (B) Transversus abdominus (TA) EMG versus intravaginal pressure. (C) Rectus abdominus (RA) EMG versus intravaginal pressure. (D) Internal obliques (IO) EMG versus intravaginal pressure. Each curve displays that muscle's activation level (percent maximum EMG) for each 5% increase in intravaginal pressure. The *squares* show the mean percent EMG and the *whiskers* show the 95% confidence interval. The EMG does not reach 100% in any case because the muscle activation continued to increase after the pressure reached its maximum. The steep rises in the curves indicate that the muscle was actively contributing to the pressure rise; while the flatter sections of the curves indicate that the muscle was not contributing as much to the pressure generation. *Based on Fig. 5 from S.J. Madill, L. McLean, Relationship between abdominal and pelvic floor muscle activation and intravaginal pressure during pelvic floor muscle contractions in healthy continent women, Neurourol. Urodyn. 25 (7) (2006) 722–730. Published by Wiley Blackwell.*

Vaginal pressure and PFM electromyographic (EMG) activity have been found to increase prior to increases in intraabdominal pressure or abdominal muscle EMG during both voluntary and automatic abdominal muscle contractions, indicating that the PFMs' contraction precedes the increase in intraabdominal pressure [42–48]. This precontraction of the PFMs tenses the anterior vaginal wall in preparation for the downward movement of the abdominal and pelvic contents produced by the increase in intraabdominal pressure. The PFMs resist this downward displacement, limiting bladder neck descent and allowing the urethra to be compressed against the anterior vaginal wall [43,45].

Increases in intraabdominal pressure exert two types of opening forces on the urethra: shear forces produced by unequal separation of the anterior and posterior urethral walls (the anterior wall is restrained by the pubourethral ligaments while the posterior wall moves with the vagina) and expulsive forces caused by increased intraabdominal pressure acting on the bladder and its contents [49–51]. These forces are resisted by contraction of the external urethral sphincter, aided by the PFMs [49]. This is demonstrated by increases in urethral pressure and urethral sphincter EMG amplitudes during coughing and straining maneuvers [30–33,43].

The timing of the PFM and sphincter contractions is vital to preventing urinary incontinence. Coughing, voluntary abdominal contractions, and limb movements all produce preparatory PFM contractions to counteract the associated increases in intra-abdominal pressure [17,42,47,48]. Coughing is accompanied by an increase in urethral pressure that rises approximately 240 ms prior to the rise in bladder pressure in continent women [32,52], and a strong correlation has been found between the forcefulness of a cough and the EMG amplitude of the corresponding external anal sphincter contraction in women without pelvic floor dysfunction [53].

DISRUPTION OF PELVIC FLOOR BIOMECHANICS LEADS TO INCONTINENCE AND PROLAPSE

Stress Urinary Incontinence

The etiology of stress urinary incontinence (SUI) is not well understood, although it has been suggested that the principal mechanism is the failure of the striated urethral sphincter, with the assistance of the PFMs, to adequately resist increases in intraabdominal pressure, which may be compounded by a loss of fascial support [9,32,43,49,54–56]. Damage to the proximal urethra, the external urethral sphincter, the PFMs, the pelvic fascia, the pudendal and levator ani nerves, along with altered reflex PFM and urethral sphincter activity are all implicated in the development of SUI [49,57,58]. Recently, DeLancey et al. [59] reported that decreased urethral closure pressure was more predictive of SUI ($d = 1.47$) than were either measures of urethral support ($d \leq 0.50$) or PFM strength ($d \leq 0.60$) in middle-aged women, and that urethral closure pressure was the single largest predictor of SUI in primiparous women ($d = 0.91$). In another study, the same research group found that urethral closure pressure and urethral support together accounted for 41% of the risk of SUI following vaginal delivery [60].

Several structural differences have been found in women with SUI, including smaller urethral sphincters [60,61], increased likelihood of bladder neck funneling and wider posterior urethrovesical junction angles than continent controls [14]. In the pelvic floor, women with SUI have been found to have smaller and thinner PFMs and wider urogenital hiatuses than continent women using ultrasound and MRI [62–69].

There have been differences in PFM and urethral sphincter activity reported in women with SUI during resting conditions. Tonic PFM activity has been found to be altered in women with SUI during cystometric bladder filling and postural tasks. Women with SUI do not demonstrate the increase in urethral sphincter or PFM EMG amplitudes with cystometric bladder filling that continent controls do [28,70]. Similarly, women with SUI do not demonstrate as large an increase in PFM EMG activity between supported and unsupported sitting postures as continent women do [71]. Conversely, Smith et al. [72] found that women with SUI demonstrated both greater postural sway and higher trunk and PFM EMG amplitudes during static standing compared to continent women. As higher trunk muscle activity is usually associated with a stiffer trunk and decreased postural sway [73], this finding suggests that the PFM and abdominal muscle synergies may be impaired in women with SUI.

Several authors have examined forces in the vagina using dynamometry at rest, that is, passive forces. Passive forces are believed to be produced by a combination of tonic PFM activity and passive tissue properties, including decreased stiffness. Morin et al. [74] reported that postmenopausal women with SUI demonstrated lower passive forces in the vagina, both statically and during a constant velocity stretch, compared to continent women. Conversely, DeLancey et al. [60] found that there was no difference in passive forces in the vagina between continent primiparous women and primiparous women with SUI; however, the passive forces were lower in both groups of primiparous women compared to nulliparous controls. The differences among the groups assessed in these studies may be a product of age: in the younger women the differences in passive force seem to be related more to pregnancy and delivery than continence status, while in the postmenopausal women passive force seems to be more related to continence.

External urethral sphincter activity has been found to be diminished or absent in women with SUI compared to continent controls during voluntary PFM contractions [35,75,76]. The ability to generate urethral pressure with a voluntary PFM contraction has also been found to be decreased in women with SUI [77]. They have also been found to produce less lift with a PFM contraction [78,79].

There is disagreement in the literature about whether the ability to produce a voluntary PFM contraction is impaired in women with SUI. Some authors have found that women with SUI produce lower PFM EMG amplitudes, and less intravaginal pressure or force than continent women when they attempt maximal voluntary PFM contractions

[67,77–82], while others report no differences in the ability to maximally activate the PFMs between women with and without SUI [59,60,72,83–86]. Interestingly, two studies that found no differences in maximum PFM force generation between women with and without SUI did find that women with SUI were less able to generate force quickly, suggesting that fast twitch muscle fibers may be most affected in SUI and that solely measuring maximum pressure, force, or EMG amplitudes may be insufficient to compare PFM function between women with and without SUI [85,86].

Women with SUI also display synergistic abdominal muscle contractions when they perform voluntary PFM contractions; however, they demonstrate lower abdominal muscle EMG activation levels and different abdominal muscle activation patterns compared to continent controls [82,83]. As you can see in Fig. 3, in the women with SUI the abdominal muscles were active to a greater level when intravaginal pressure

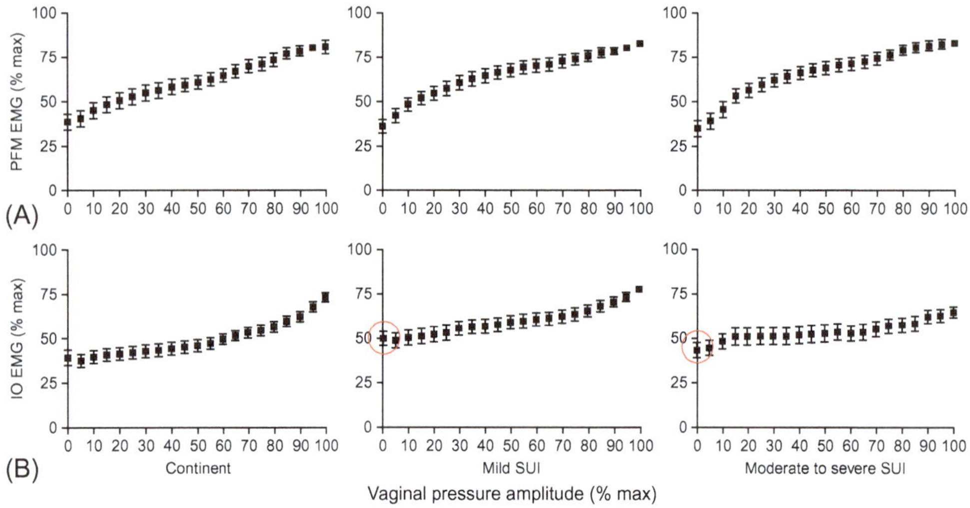

Figure 3 Ensemble average curves of electromyographic (EMG) activation over vaginal pressure generation for continent women compared to women with mild and severe stress urinary incontinence (SUI). (A) Pelvic floor muscle (PFM) EMG versus intravaginal pressure. (B) Internal obliques (IO) EMG versus intravaginal pressure. The curves for IO were typical of the curves for all of the abdominal muscles. Each curve displays that muscle's activation level (percent maximum EMG) for each 5% increase in intravaginal pressure. The *squares* show the mean percent EMG and the *whiskers* show 1 standard deviation. The EMG does not reach 100% in any case because of averaging and because, in some cases, the muscle activation continued to increase after the pressure reached its maximum. Steep rises in the curves indicate that the muscle was actively contributing to the pressure rise; while the flatter sections of the curves indicate that the muscle was not contributing as much to the pressure generation. The *circles* indicate that the IO EMG activation amplitudes (intercepts) were significantly higher at the onset of the pressure rise in the women with SUI compared to the continent women. The slopes of the IO curves were lower in the women with SUI compared to the continent women. The PFM curves were not significantly different among the groups. *Based on Fig. 4 from S.J. Madill, M.A. Harvey, L. McLean, Women with SUI demonstrate motor control differences during voluntary pelvic floor muscle contractions. Int. Urogynecol. J. Pelvic Floor Dysfunct. 20 (4) (2009) 447–459. Published by Springer.*

started to rise and they contributed less to the generation of intravaginal pressure than in women without incontinence. These findings may suggest that altered muscle activation patterns may play a role in SUI or it may suggest that women with SUI avoid activating their abdominal muscles to avoid increasing intraabdominal pressure and provoking leakage.

Automatic PFM contractions have been found to be disrupted in women with SUI. The ability to modulate the strength of the PFM contraction to the forcefulness of a cough is diminished in women with SUI [87]. The rise in urethral pressure is diminished and delayed [52,88–90], and the rise in PFM EMG activity is also delayed [53,91]. Delays have also been found in the time that the intravaginal pressure peaks compared to the time of the PFM EMG peak in women with SUI [84]. This study also found that women with SUI generated intravaginal pressure more slowly and that their PFM EMG was higher at the moment when intravaginal pressure began to rise. These findings suggest that SUI during coughing may be the result of a combination of deficits, including delays in PFM activation and disruptions in force transfer.

External urethral sphincter activity has been found to be diminished or absent in women with SUI compared to continent controls during Valsalva and coughing [92]. In women with SUI, maximum urethral closure pressure decreases with repeated coughs while it remains unchanged in continent women, suggesting that sphincter and/or PFM fatigue may play a role in SUI. Women with SUI have been found to demonstrate more bladder neck movement with coughing than continent women [55], and while continent women's anorectal junctions move ventrally during coughing, in women with SUI the anorectal junction moves dorsally [93]. Contrarily, Bø et al. [94] found that seven nulliparous young women with SUI demonstrated similar synergistic PFM and urethral sphincter contractions in response to coughing to matched continent controls.

During demanding postural tasks, muscle activation patterns are altered in women with SUI. Smith et al. [47] found that PFM EMG activation timing in response to rapid shoulder movements is delayed in women with SUI. They found that continent women activated their PFMs prior to the onset of deltoid activity while women with SUI activated their PFMs following the onset of deltoid activity. They also found that some of the women with SUI demonstrated a period of decreased PFM EMG activity that straddled the onset of deltoid EMG activity. In another study, the same group found that women with SUI displayed higher abdominal muscle EMG amplitudes than continent women when the women caught a 1-kg weight that was dropped unexpectedly [48]. The abdominal muscle EMG amplitudes were even greater when the women performed the task with full bladders.

These apparently contradictory findings in EMG activity, strength, and activation patterns may be evidence that SUI has different etiologies in different populations, and they highlight the challenges to understanding the biomechanical basis for SUI.

Mixed Urinary Incontinence

Much less research has been done in women with mixed urinary incontinence (MUI) compared to women with SUI. An early study found that women with MUI produced lower peak PFM EMG amplitudes than did continent women and that these amplitudes decreased with age in the women with MUI, but not in the continent women [77]. Women with MUI have demonstrated similar difficulties generating urethral pressure with a voluntary PFM contraction as have women with SUI [75]. These difficulties were related to the severity of the incontinence, but not to the type. Older women with MUI have been found to demonstrate differences in pelvic organ support at rest compared to both continent controls and women with SUI using magnetic resonance imaging; their bladder necks and anorectal junctions were both lower relative to a line drawn from the inferior pubic symphysis to the sacrococcygeal joint line [14]. This suggests a greater role for pelvic organ support, both muscular and fascial, in women with MUI compared to those with SUI.

Pelvic Organ Prolapse

Pelvic floor morphology and function are altered in women with pelvic organ prolapse [6]. Several authors have reported that the urogenital hiatus is larger in women with prolapse than it is in controls [95]. Women with prolapse are also more likely to display PFM defects, such as avulsions, PFM weakness, and a loss of fascial support than are controls [96–98]. During straining, women with prolapse demonstrate greater pelvic floor descent than do controls without prolapse [99]. These changes in PFM geometry and function seem to predispose women to prolapse, rather than being an effect of the prolapse [100].

Women with PFM defects are less able to generate PFM force and they are less able to narrow the urogenital hiatus with a PFM contraction than are women with intact PFMs [96,101,102]. The PFM damage is seen primarily in the most inferior-medial portion of the pubovisceral muscle [103,104]. This damage to the muscle is concurrent with damage to the anterior portions of the arcus tendineus fascia pelvis and the arcus tendineus levator ani [105].

PELVIC FLOOR CHANGES WITH REHABILITATION

PFM resting tone, the ability to isolate and correctly perform a PFM contraction, PFM strength, urethral closure pressure, pelvic floor support, and pelvic organ prolapse stage all improve with PFM exercise [106–116]. Surprisingly, continence outcomes may improve before measureable improvements are seen in PFM function [117,118]. This may indicate a learning effect or that the exercises are producing an effect on the striated urethral sphincter. Even women with PFM avulsion have been found to be able to increase urethral pressure with a "PFM" contraction [34], although the magnitude of the increase is smaller in these women as compared to women with intact PFMs [119], supporting both the synergy between the PFMs and the sphincter, and the role of the PFMs in contributing to urethral pressure.

Rest

Improvements in pelvic organ support, as assessed by a narrower anorectal angle and a higher position of the urethrovesical junction, have been found following PFM training programs [13,120]. The increase in urethrovesical junction height was also found during maximum voluntary PFM contractions and straining, suggesting that both the muscular and the fascial support had improved. Pelvic floor muscle training has been shown to increase the thickness and the cross-sectional area of the PFMs on ultrasonography [120,121]. This increase in cross-sectional area is indicative of muscle hypertrophy and is an indirect measure of muscle strength [122].

Recently two groups have reported that PFM muscle training also results in changes to the urethral sphincter, providing another mechanism for how PFM exercise increases urethral pressure and improves continence. First, McLean et al. [123] reported that a 12-week PFM exercise program resulted in a significant increase in the urethral cross-sectional area at the level of the striated urethral sphincter on ultrasound. Subsequently, Madill et al. [124] found that only the striated urethral sphincter, and not other layers of the urethra, increased in thickness following a 12-week PFM exercise intervention using magnetic resonance imaging. The study populations were different between the two studies, with McLean's group including middle-aged women and Madill's only women over 60 years old; this strengthens the findings and suggests that this effect should be seen in women of any age.

PFM Contraction

Women's ability to voluntarily contract their PFMs improves with PFM training. Along with increases in strength and endurance [125], motor control also improves. Madill et al. [13] found that a 12-week PFM exercise program for older women improved the women's ability to rapidly and repeatedly contract their PFMs. They also found that the participants contracted their PFMs sooner before a cough and that they were better able to maintain a PFM contraction between repeated coughs than they had been able to prior to the intervention.

Straining

Braekken et al. [120] reported that the levator hiatus and the PFM length stretched less with maximum straining following PFM training, using ultrasound.

Functional Changes in Women With Pelvic Floor Dysfunction

PFM training has been shown to decrease the volume of urine leaked and the frequency of urinary incontinence episodes [106,125–127]. Skilling and Petros [128] found that women who regained continence with PFM exercise were continent even when they were surprised by increases in intraabdominal pressure. The ability to perform strenuous physical activity without leakage improves, as does overall participation in activity [129–131]. In women with pelvic organ prolapse, PFM training has been reported to produce a reduction of symptoms, a reduction in the prolapse stage, and similar improvements in PFM function as are seen in women with incontinence [132–135].

CONCLUSION

The pelvic floor is a complex biomechanical system of multiple overlapping components. The PFMs are under both voluntary and automatic control and demonstrate finely nuanced reactions that are highly context dependent. Urinary incontinence and pelvic organ prolapse seem to be the product of variable combinations of deficits in this system; the evidence is scanty for a direct one-to-one relationship between defects and symptoms. Overall, PFM exercise has been found to reduce the symptoms of urinary incontinence and prolapse while it addresses many of the observed deficits: it improves urethral closure pressure, urethral sphincter and PFM thickness increase, PFM strength and endurance improve, and PFM and sphincter activation patterns become more like those of continent women. Unfortunately, PFM exercise cannot restore structural damage: such as avulsion or neurological injuries, and, as yet, we are not able to predict how much benefit a woman can expect from PFM rehabilitation.

REFERENCES

[1] J.O.L. DeLancey, The anatomy of the pelvic floor, Curr. Opin. Obstet. Gynecol. 6 (1994) 313–316.
[2] A. Giannantoni, S.M. Di Stasi, A. Cucchi, E. Mearini, V. Bini, M. Porena, Pelvic floor muscle behaviour during Valsalva leak point pressure measurement in males and females affected by stress urinary incontinence, J. Urol. 170 (2003) 485–489.
[3] A. Shafik, S. Doss, S. Asaad, Etiology of the resting myoelectric activity of the levator ani muscle: physioanatomic study with a new theory, World J. Surg. 27 (2003) 309–314.
[4] R.L. Vereecken, J. Derluyn, H. Verduyn, Electromyography of the perineal striated muscles during cystometry, Urol. Int. 30 (1975) 92–98.
[5] R.L. Vereecken, H. Verduyn, The electrical activity of the paraurethral and perineal muscles in normal and pathological conditions, Br. J. Urol. 42 (1970) 457–463.
[6] J.A. Ashton-Miller, J.O. DeLancey, Functional anatomy of the female pelvic floor, Ann. N. Y. Acad. Sci. 1101 (2007) 266–296.
[7] K. Bø, F. Lilleås, T. Talseth, H. Hedland, Dynamic MRI of the pelvic floor muscles in an upright sitting position, Neurourol. Urodyn. 20 (2001) 167–174.
[8] C.E. Constantinou, G. Hvistendahl, A. Ryhammer, L.L. Nagel, J.C. Djurhuus, Determining the displacement of the pelvic floor and pelvic organs during voluntary contractions using magnetic resonance imaging in younger and older women, BJU Int. 90 (2002) 408–424.

[9] J.O.L. DeLancey, Anatomy and physiology of urinary continence, Clin. Obstet. Gynecol. 33 (1990) 298–307.
[10] S. Hjartadottir, J. Nilsson, C. Petersen, G. Lingman, The female pelvic floor: a dome — not a basin, Acta Obstet. Gynecol. Scand. 76 (1997) 567–571.
[11] A.H. Kegel, Progressive resistance exercise in the functional restoration of the perineal muscles, Am. J. Obstet. Gynecol. 56 (1948) 238–248.
[12] S. Madill, A. Tang, S. Pontbriand-Drolet, C. Dumoulin, Comparison of two methods for measuring the pubococcygeal line from sagittal-plane magnetic resonance imaging, Neurourol. Urodyn. 30 (2011) 1613–1619.
[13] S.J. Madill, S. Pontbriand-Drolet, A. Tang, C. Dumoulin, Effects of PFM rehabilitation on PFM function and morphology in older women, Neurourol. Urodyn. 32 (2013) 1086–1095.
[14] S. Pontbriand-Drolet, A. Tang, S.J. Madill, C. Tannenbaum, M.C. Lemieux, J. Corcos, C. Dumoulin, Differences in pelvic floor morphology between continent, stress urinary incontinent, and mixed urinary incontinent elderly women: an MRI study. Neurourol, Urodyn, 2015. http://dx.doi.org/10.1002/nau.22743.
[15] A. Shafik, A new concept of the anatomy of the anal sphincter mechanism and the physiology of defecation: mass contraction of the pelvic floor muscles, Int. Urogynecol. J. Pelvic Floor Dysfunct. 9 (1998) 28–32.
[16] K. Bø, I.H. Braekken, M. Majida, M.E. Engh, Constriction of the levator hiatus during instruction of pelvic floor or transversus abdominis contraction: a 4D ultrasound study, Int. Urogynecol. J. Pelvic Floor Dysfunct. 20 (2009) 27–32.
[17] F.M. Deindl, D.B. Vodusek, U. Hesse, B. Schussler, Activity patterns of pubococcygeal muscles in nulliparous continent women, Br. J. Urol. 72 (1993) 46–51.
[18] J.O.L. DeLancey, Anatomy and mechanics of structures around the vesical neck: how vesical neck position might affect its closure, Neurourol. Urodyn. 7 (1988) 161–162.
[19] J.O.L. DeLancey, Structural support of the urethra as it relates to stress urinary incontinence: the hammock hypothesis, Am. J. Obstet. Gynecol. 170 (1994) 1713–1723.
[20] J.O.L. DeLancey, J.A. Gosling, K. Creed, J.S. Dixon, V. Delmas, D. Landon, P.A. Norton, Gross anatomy and cell biology of the lower urinary tract, in: P. Abrams, L. Cardozo, S. Khoury, A. Wein (Eds.), Incontinence, second ed., Health Publications Ltd., Paris, 2002.
[21] P.E.P. Petros, An integral theory of female urinary incontinence, experimental and clinical considerations, Acta Obstet. Gynecol. Scand. 69 (1990) 7–31.
[22] R.R. Sapsford, B. Clarke, P.W. Hodges, The effect of abdominal and pelvic floor muscle activation patterns on urethral pressure, World J. Urol. 31 (2013) 639–644.
[23] B. Junginger, K. Baessler, R. Sapsford, P.W. Hodges, Effect of abdominal and pelvic floor tasks on muscle activity, abdominal pressure and bladder neck, Int. Urogynecol. J. Pelvic Floor Dysfunct. 21 (2010) 69–77.
[24] V.L. Handa, T.A. Harris, D.R. Ostergard, Protecting the pelvic floor: obstetric management to prevent incontinence and pelvic organ prolapse, Obstet. Gynecol. 88 (1996) 470–478.
[25] D.P. Keane, S. O'Sullivan, Urinary incontinence: anatomy, physiology and pathophysiology, Baillieres Best Pract. Res. Clin. Obstet. Gynaecol. 14 (2000) 207–226.
[26] T. Rud, K.-E. Anderson, M. Asmussen, A. Hunting, U. Ulmsten, Factors maintaining the intraurethral pressure in women, Invest. Urol. 17 (1980) 343–347.
[27] E.A. Tanagho, F.H. Meyers, D.R. Smith, Urethral resistance: its components and implications. I. Smooth muscle component, Invest. Urol. 7 (1969) 136–149.
[28] S. Sørensen, Urethral pressures during bladder filling, Scand. J. Urol. Nephrol. 23 (1989) 45–51.
[29] A. Massey, P. Abrams, Urodynamics of the female lower urinary tract, Urol. Clin. North Am. 12 (1985) 231–246.
[30] K. Baessler, K. Miska, R. Draths, B. Schuessler, Effects of voluntary pelvic floor contraction and relaxation on the urethral closure pressure, Int. Urogynecol. J. Pelvic Floor Dysfunct. 16 (2005) 187–191.
[31] K. Bø, R. Stien, Needle EMG registration of striated urethral wall and pelvic floor muscle activity patterns during cough, Valsalva, abdominal, hip adductor, and gluteal muscle contractions in nulliparous healthy females, Neurourol. Urodyn. 13 (1994) 35–41.

[32] C.E. Constantinou, D.E. Govan, Spatial distribution and timing of transmitted and reflexly generated urethral pressures in healthy women, J. Urol. 127 (1982) 964–969.
[33] G. Lose, H. Colstrup, Urethral pressure and power generation during coughing and voluntary contraction of the pelvic floor in healthy females, Br. J. Urol. 67 (1991) 573–579.
[34] J.M. Miller, W.H. Umek, J.O. DeLancey, J.A. Ashton-Miller, Can women without visible pubococcygeal muscle in MR images still increase urethral closure pressures? Am. J. Obstet. Gynecol. 191 (2004) 171–175.
[35] K. Kenton, L. Brubaker, Relationship between levator ani contraction and motor unit activation in the urethral sphincter, Am. J. Obstet. Gynecol. 187 (2002) 403–406.
[36] K. Bø, B. Kvarstein, R. Hagen, S. Larsen, Pelvic floor muscle exercise for the treatment of female stress urinary incontinence: II. Validity of vaginal pressure measurements of pelvic floor muscle strength and the necessity of supplementary methods for control of correct contraction, Neurourol. Urodyn. 9 (1990) 479–487.
[37] S.J. Madill, L. Mclean, Relationship between abdominal and pelvic floor muscle activation and intravaginal pressure during pelvic floor muscle contractions in healthy continent women, Neurourol. Urodyn. 25 (2006) 722–730.
[38] S.J. Madill, L. Mclean, Quantification of abdominal and pelvic floor muscle synergies in response to voluntary pelvic floor muscle contractions, J. Electromyogr. Kinesiol. 18 (2008) 955–964.
[39] P. Neumann, V. Gill, Pelvic floor and abdominal muscle interaction: EMG activity and intra-abdominal pressure, Int. Urogynecol. J. Pelvic Floor Dysfunct. 13 (2002) 125–132.
[40] R.R. Sapsford, P.W. Hodges, The effect of abdominal and pelvic floor muscle activation on urine flow in women, Int. Urogynecol. J. 23 (2012) 1225–1230.
[41] R.R. Sapsford, P.W. Hodges, C.A. Richardson, D. Cooper, S.J. Markwell, G.A. Jull, Co-activation of the abdominal and pelvic floor muscles during voluntary exercises, Neurourol. Urodyn. 20 (2001) 31–42.
[42] R.R. Sapsford, P.W. Hodges, Contraction of the pelvic floor muscles during abdominal maneuvers, Arch. Phys. Med. Rehabil. 82 (2001) 1081–1088.
[43] J.O.L. DeLancey, J.A. Ashton-Miller, Pathophysiology of adult urinary incontinence, Gastroenterology 126 (2004) S23–S32.
[44] P.W. Hodges, R. Sapsford, L.H.M. Pengel, Postural and respiratory functions of the pelvic floor muscles, Neurourol. Urodyn. 26 (2007) 362–371.
[45] U.M. Peschers, G. Fanger, G.N. Schaer, D.B. Vodusek, J.O. DeLancey, B. Schuessler, Bladder neck mobility in continent nulliparous women, BJOG 108 (2001) 320–324.
[46] M.D. Smith, M.W. Coppieters, P.W. Hodges, Postural response of the pelvic floor and abdominal muscles in women with and without incontinence, in: A. Rainoldi, M.A. Minetto (Eds.), XVI Congress of the International Society of Electrophysiology and Kinesiology, International Society of Electrophysiology and Kinesiology, Turin, Italy, 2006.
[47] M.D. Smith, M.W. Coppieters, P.W. Hodges, Postural activity of the pelvic floor muscles is delayed during rapid arm movements in women with stress urinary incontinence, Int. Urogynecol. J. Pelvic Floor Dysfunct. 18 (2007) 901–911.
[48] M.D. Smith, M.W. Coppieters, P.W. Hodges, Postural response of the pelvic floor and abdominal muscles in women with and without incontinence, Neurourol. Urodyn. 26 (2007) 377–385.
[49] H. Koelbl, V. Nitti, K. Baessler, S. Salvatore, A. Sultan, O. Yamaguchi, Committee 4: pathophysiology of urinary incontinence, faecal incontinence and pelvic organ prolapse, in: P. Abrams, L. Cardozo, S. Khoury, A. Wein (Eds.), Incontinence, fourth ed., Health Publications Ltd., Paris, 2009.
[50] J.L. Mostwin, R. Genadry, R. Saunders, A. Yang, Stress incontinence observed with real time sonography and dynamic fastscan magnetic resonance imaging — insights into pathophysiology, Scand. J. Urol. Nephrol. 35 (2001) 94–99.
[51] R.C. Sanders, R. Genadry, A. Yang, J.L. Mostwin, Imaging the female urethra with ultrasound, Ultrasound Q. 12 (1994) 167–183.
[52] J.B. van der Kooi, P.J.A. van Wanroy, M.C. De Jonge, J.A. Kornelis, Time separation between cough pulses in bladder, rectum and urethra in women, J. Urol. 132 (1984) 1275–1278.

[53] X. Deffieux, P. Raibaut, P. Rene-Corail, R. Katz, M. Perrigot, S.S. Ismael, P. Thoumie, G. Amarenco, External anal sphincter contraction during cough: not a simple spinal reflex, Neurourol. Urodyn. 25 (2006) 782–787.
[54] K. Bø, Bekkenbunnstrening og urininkontinens — tren deg tett! Tidsskr. Nor. Laegeforen. 120 (2000) 3583–3589.
[55] D. Howard, J.M. Miller, J.O.L. DeLancey, J.A. Ashton-Miller, Differential effects of cough, Valsalva, and continence status on vesical neck movement, Obstet. Gynecol. 95 (2000) 535–540.
[56] E.D. James, The behaviour of the bladder during physical activity, Br. J. Urol. 50 (1978) 387–394.
[57] P. Busacchi, T. Perri, R. Paradisi, C. Oliverio, D. Santini, S. Guerrini, G. Barbara, V. Stanghellini, R. Corinaldesi, R. De Giorgio, Abnormalities of somatic peptide-containing nerves supplying the pelvic floor of women with genitourinary prolapse and stress urinary incontinence, Urology 63 (2004) 591–595.
[58] T. Washington Cannon, M.S. Damaser, Pathophysiology of the lower urinary tract: continence and incontinence, Clin. Obstet. Gynecol. 47 (2004) 28–35.
[59] J.O.L. DeLancey, E.R. Trowbridge, J.M. Miller, D.M. Morgan, K. Guire, D.E. Fenner, W.J. Weadock, J.A. Ashton-Miller, Stress urinary incontinence: relative importance of urethral support and urethral closure pressure, J. Urol. 179 (2008) 2286–2290.
[60] J.O. DeLancey, J.M. Miller, R. Kearney, D. Howard, P. Reddy, W. Umek, K.E. Guire, R.U. Margulies, J.A. Ashton-Miller, Vaginal birth and de novo stress incontinence: relative contributions of urethral dysfunction and mobility, Obstet. Gynecol. 110 (2007) 354–362.
[61] R. Kirschner-Hermanns, H.M. Klein, U. Muller, W. Schafer, G. Jakse, Intra-urethra ultrasound in women with stress incontinence, Br. J. Urol. 74 (1994) 315–318.
[62] P. Aukee, J.-P. Usenius, P. Kirkinen, An evaluation of pelvic floor anatomy and function by MRI, Eur. J. Obstet. Gynecol. Reprod. Biol. 112 (2004) 84–88.
[63] I.T. Bernstein, The pelvic floor muscles: muscle thickness in healthy and urinary-incontinent women measured by perineal ultrasonography with reference to the effect of pelvic floor training. Estrogen receptor studies, Neurourol. Urodyn. 16 (1997) 237–275.
[64] L. Hoyte, M. Jakab, S.K. Warfield, S. Shott, G. Flesh, J.R. Fielding, Levator ani thickness in symptomatic and asymptomatic women using magnetic resonance-based 3-dimensional color mapping, Am. J. Obstet. Gynecol. 191 (2004) 856–861.
[65] L. Hoyte, L. Schierlitz, K. Zou, G. Flesh, J.R. Fielding, Two- and 3-dimensional MRI comparison of levator ani structure, volume, and integrity in women with stress incontinence and prolapse, Am. J. Obstet. Gynecol. 185 (2001) 11–19.
[66] R. Kirschner-Hermanns, B. Wein, S. Niehaus, W. Schaefer, G. Jakse, The contribution of magnetic resonance imaging of the pelvic floor to the understanding of urinary incontinence, Br. J. Urol. 72 (1993) 715–718.
[67] S. Mørkved, K.A. Salvesen, K. Bø, S. Eik-Nes, Pelvic floor muscle strength and thickness in continent and incontinent nulliparous pregnant women, Int. Urogynecol. J. Pelvic Floor Dysfunct. 15 (2004) 384–390.
[68] E. Oliveira, R.A. Castro, C.C. Takano, L.R. Bezerra, M.G. Sartori, G.R. Lima, E.C. Baracat, M.J. Girao, Ultrasonographic and Doppler velocimetric evaluation of the levator ani muscle in premenopausal women with and without urinary stress incontinence, Eur. J. Obstet. Gynecol. Reprod. Biol. 133 (2007) 213–217.
[69] J. Stoker, E. Rociu, J.L.H. Ruud Bosch, E.J. Messelink, V.P.M. Van Der Hulst, A.G. Groenendijk, M.J.C. Eijkemans, J.S. Laméris, High-resolution endovaginal MR imaging in stress urinary incontinence, Eur. Radiol. 13 (2003) 2031–2037.
[70] K. Kenton, E. Mueller, L. Brubaker, Continent women have better urethral neuromuscular function than those with stress incontinence, Int. Urogynecol. J. 22 (2011) 1479–1484.
[71] R.R. Sapsford, C.A. Richardson, C.F. Maher, P.W. Hodges, Pelvic floor muscle activity in different sitting postures in continent and incontinent women, Arch. Phys. Med. Rehabil. 89 (2008) 1741–1747.
[72] M.D. Smith, M.W. Coppieters, P.W. Hodges, Is balance different in women with and without stress urinary incontinence, Neurourol. Urodyn. 27 (2008) 71–78.

[73] N.P. Reeves, V. Everding, J. Cholewicki, E. Al, The effects of active and passive trunk stiffness on spine controllability, in: A. Rainoldi, M.A. Minetto (Eds.), Proceedings of the International Society for Electromyography and Kinesiology, Torino, Italy, 2006. p. 67.

[74] M. Morin, D. Gravel, D. Bourbonais, C. Dumoulin, S. Ouellet, Comparing pelvic floor muscle tone in postmenopausal continent and stress urinary incontinent women, Neurourol. Urodyn. 27 (2008) 610–611.

[75] A. Mattiasson, P. Teleman, Abnormal urethral motor function is common in female stress, mixed, and urge incontinence, Neurourol. Urodyn. 25 (2006) 703–708.

[76] P.M. Teleman, A. Mattiasson, Urethral pressure response patterns induced by squeeze in continent and incontinent women, Int. Urogynecol. J. Pelvic Floor Dysfunct. 18 (2007) 1027–1031.

[77] M. Gunnarsson, A. Mattiasson, Female stress, urge, and mixed urinary incontinence are associated with a chronic and progressive pelvic floor/vaginal neuromuscular disorder: an investigation of 317 healthy and incontinent women using vaginal surface electromyography, Neurourol. Urodyn. 18 (1999) 613–621.

[78] P.A. Law, J.C. Danin, D.M. Lamb, L. Regan, A. Darzi, W.M. Gedroyc, Dynamic imaging of the pelvic floor using an open-configuration magnetic resonance scanner, J. Magn. Reson. Imaging 13 (2001) 923–929.

[79] J.A. Thompson, P.B. O'Sullivan, N.K. Briffa, P. Neumann, Assessment of voluntary pelvic floor muscle contraction in continent and incontinent women using transperineal ultrasound, manual muscle testing and vaginal squeeze pressure measurements, Int. Urogynecol. J. Pelvic Floor Dysfunct. 17 (2006) 624–630.

[80] P. Aukee, J. Penttinen, O. Airaksinen, The effect of aging on the electromyographic activity of pelvic floor muscles. A comparative study among stress incontinent patients and asymptomatic women, Maturitas 44 (2003) 253–257.

[81] G. Hilde, J. Staer-Jensen, M. Ellstrom Engh, I.H. Braekken, K. Bo, Continence and pelvic floor status in nulliparous women at midterm pregnancy, Int. Urogynecol. J. 23 (2012) 1257–1263.

[82] J.A. Thompson, P.B. O'Sullivan, K. Briffa, P. Neumann, Altered muscle activation patterns in symptomatic women during pelvic floor muscle contraction and Valsalva manouevre, Neurourol. Urodyn. 25 (2006) 268–276.

[83] S.J. Madill, M.A. Harvey, L. Mclean, Women with SUI demonstrate motor control differences during voluntary pelvic floor muscle contractions, Int. Urogynecol. J. Pelvic Floor Dysfunct. 20 (2009) 447–459.

[84] S.J. Madill, M.A. Harvey, L. Mclean, Women with stress urinary incontinence demonstrate motor control differences during coughing, J. Electromyogr. Kinesiol. 20 (2010) 804–812.

[85] M. Morin, D. Bourbonnais, D. Gravel, C. Dumoulin, M.C. Lemieux, Pelvic floor muscle function in continent and stress urinary incontinent women using dynamometric measurements, Neurourol. Urodyn. 23 (2004) 668–674.

[86] D. Underwood, T. Calteaux, A. Cranston, S. Novotny, J. Hollman, Hip and pelvic floor muscle strength in women with and without stress urinary incontinence: a case-control study, J. Womens Health Phys. Therap. 36 (2012) 55–61.

[87] X. Deffieux, K. Hubeaux, R. Porcher, S.S. Ismael, P. Raibaut, G. Amarenco, Abnormal pelvic response to cough in women with stress urinary incontinence, Neurourol. Urodyn. 27 (2008) 291–296.

[88] H. Heidler, H. Wölk, U. Jonas, Urethral closure mechanism under stress conditions, Eur. Urol. 5 (1979) 110–112.

[89] G. Lose, Urethral pressure and power generation during coughing and voluntary contraction of the pelvic floor in females with genuine stress incontinence, Br. J. Urol. 67 (1991) 580–585.

[90] D. Pieber, F. Zivkovic, K. Tamussino, Timing of urethral pressure pulses before and after continence surgery, Neurourol. Urodyn. 17 (1998) 19–23.

[91] M. Barbic, B. Kralj, A. Cor, Compliance of the bladder neck supporting structures: importance of activity pattern of levator ani muscle and content of elastic fibers of endopelvic fascia, Neurourol. Urodyn. 22 (2003) 269–276.

[92] J.K. Nguyen, G.C. Gunn, N.N. Bhatia, The effect of patient position on leak-point pressure measurements in women with genuine stress incontinence, Int. Urogynecol. J. Pelvic Floor Dysfunct. 13 (2002) 9–14.

[93] Q. Peng, R. Jones, K. Shishido, C.E. Constantinou, Ultrasound evaluation of dynamic responses of female pelvic floor muscles, Ultrasound Med. Biol. 33 (2007) 342–352.

[94] K. Bø, R. Stien, S. Kulseng-Hanssen, M. Kristofferson, Clinical and urodynamic assessment of nulliparous young women with and without stress incontinence symptoms: a case-control study, Obstet. Gynecol. 84 (1994) 1028–1032.

[95] M. Majida, I. Braekken, K. Bo, J. Benth, M. Engh, Anterior but not posterior compartment prolapse is associated with levator hiatus area: a three- and four-dimensional transperineal ultrasound study, BJOG 118 (2011) 329–337.

[96] J.O. DeLancey, D.M. Morgan, D.E. Fenner, R. Kearney, K. Guire, J.M. Miller, H. Hussain, W. Umek, Y. Hsu, J.A. Ashton-Miller, Comparison of levator ani muscle defects and function in women with and without pelvic organ prolapse, Obstet. Gynecol. 109 (2007) 295–302.

[97] R.F. El Sayed, S. El Mashed, A. Farag, M.M. Morsy, M.S. Abdel Azim, Pelvic floor dysfunction: assessment with combined analysis of static and dynamic MR imaging findings, Radiology 248 (2008) 518–530.

[98] M. Huebner, R.U. Margulies, J.O. DeLancey, Pelvic architectural distortion is associated with pelvic organ prolapse, Int. Urogynecol. J. Pelvic Floor Dysfunct. 19 (2008) 863–867.

[99] Y. Hsu, A. Summers, H.K. Hussain, K.E. Guire, J.O. DeLancey, Levator plate angle in women with pelvic organ prolapse compared to women with normal support using dynamic MR imaging, Am. J. Obstet. Gynecol. 194 (2006) 1427–1433.

[100] B.P. Andrew, K.L. Shek, V. Chantarasorn, H.P. Dietz, Enlargement of the levator hiatus in female pelvic organ prolapse: cause or effect? Aust. N. Z. J. Obstet. Gynaecol. 53 (2013) 74–78.

[101] M. Majida, I.H. Braekken, K. Bo, M.E. Engh, Levator hiatus dimensions and pelvic floor function in women with and without major defects of the puboviscceral muscle, Int. Urogynecol. J. 23 (2012) 707–714.

[102] A.B. Steensma, M.L. Konstantinovic, C.W. Burger, D. De Ridder, D. Timmerman, J. Deprest, Prevalence of major levator abnormalities in symptomatic patients with an underactive pelvic floor contraction, Int. Urogynecol. J. 21 (2010) 861–867.

[103] H.P. Dietz, V. Lanzarone, Levator trauma after vaginal delivery, Obstet. Gynecol. 106 (2005) 707–712.

[104] R.U. Margulies, M. Huebner, J.O.L. DeLancey, Origin and insertion points involved in levator ani muscle defects, Am. J. Obstet. Gynecol. 196 (2007) 251.e1–251.e5.

[105] K.A. Larson, J. Luo, A. Yousuf, J.A. Ashton-Miller, J.O. DeLancey, Measurement of the 3D geometry of the fascial arches in women with a unilateral levator defect and "architectural distortion", Int. Urogynecol. J. 23 (2012) 57–63.

[106] J. Balmforth, J. Bidmead, L. Cardozo, A. Hextall, B. Kelvin, J. Mantle, Raising the tone: a prospective observational study evaluating the effect of pelvic floor muscle training on bladder neck mobility and associated improvement in stress urinary incontinence, Neurourol. Urodyn. 23 (2004) 553–554.

[107] K. Bo, Pelvic floor muscle training in treatment of female stress urinary incontinence, pelvic organ prolapse and sexual dysfunction, World J. Urol. 30 (2012) 437–443.

[108] A.M.R. Di Gangi Herms, R. Veit, C. Reisenauer, A. Herms, W. Grodd, P. Enck, A. Stenzl, N. Birbaumer, Functional imaging of stress urinary incontinence, Neuroimage 29 (2006) 267–275.

[109] C. Dumoulin, Q. Peng, K. Stodkilde-Jorgensen, K. Shishido, C. Constantinou, Changes in levator ani anatomical configuration following physiotherapy in women with stress urinary incontinence, J. Urol. 178 (2007) 970–977.

[110] C. Griffin, M.C. Dougherty, H. Yarandi, Pelvic muscles during rest: responses to pelvic muscle exercise, Nurs. Res. 43 (1994) 164–167.

[111] M. Gunnarsson, S. Ahlmann, S. Lindstrom, I. Rosen, A. Mattiasson, Cortical magnetic stimulation in patients with genuine stress incontinence: correlation with results of pelvic floor exercises, Neurourol. Urodyn. 18 (1999) 437–445.

[112] M. Gunnarsson, P. Teleman, A. Mattiasson, J. Lidfeldt, C. Nerbrand, G. Samsioe, Effects of pelvic floor exercises in middle aged women with a history of naive urinary incontinence: a population based study, Eur. Urol. 41 (2002) 556–561.
[113] A. Hirsch, G. Weirauch, B. Steimer, K. Bihler, U.M. Peschers, F. Bergauer, B. Leib, T. Dimpfl, Treatment of female urinary incontinence with EMG-controlled biofeedback home training, Int. Urogynecol. J. Pelvic Floor Dysfunct. 10 (1999) 7–10.
[114] E. Kujansuu, The effect of pelvic floor exercises on urethral function in female stress urinary incontinence: an urodynamic study, Ann. Chir. Gynaecol. 72 (1983) 28–32.
[115] I.H. Pages, S. Jahr, M. Schaufele, E. Conradi, Comparative analysis of biofeedback and physical therapy for treatment of urinary stress incontinence in women, Am. J. Phys. Med. Rehabil. 80 (2001) 494–502.
[116] P.D. Wilson, T. Al Samarrai, M. Deakin, E. Kolbe, A.D.G. Brown, An objective assessment of physiotherapy for female genuine stress incontinence, Br. J. Obstet. Gynaecol. 94 (1987) 575–582.
[117] C. Dumoulin, M.C. Lemieux, D. Bourbonnais, D. Gravel, G. Bravo, M. Morin, Physiotherapy for persistent postnatal urinary incontinence: a randomized controlled trial, Obstet. Gynecol. 104 (2004) 504–510.
[118] M. Sherburn, M. Bird, M. Carey, K. Bo, M.P. Galea, Incontinence improves in older women after intensive pelvic floor muscle training: an assessor-blinded randomized controlled trial, Neurourol. Urodyn. 30 (2011) 317–324.
[119] H.P. Dietz, K.L. Shek, Levator function and voluntary augmentation of maximum urethral closure pressure, Int. Urogynecol. J. 23 (2012) 1035–1040.
[120] I.H. Braekken, M. Majida, M.E. Engh, K. Bo, Morphological changes after pelvic floor muscle training measured by 3-dimensional ultrasonography: a randomized controlled trial, Obstet. Gynecol. 115 (2010) 317–324.
[121] B.T. Bernardes, A.P. Resende, L. Stupp, E. Oliveira, R.A. Castro, Z.I. Bella, M.J. Girao, M.G. Sartori, Efficacy of pelvic floor muscle training and hypopressive exercises for treating pelvic organ prolapse in women: randomized controlled trial, Sao Paulo Med. J. 130 (2012) 5–9.
[122] H.A. Devries, Physiology of Exercise for Physical Education and Athletics, Wm. C. Brown Publishers, Dubuque, IA, 1986.
[123] L. Mclean, K. Varette, E. Gentilcore-Saulnier, M.A. Harvey, K. Baker, E. Sauerbrei, Pelvic floor muscle training in women with stress urinary incontinence causes hypertrophy of the urethral sphincters and reduces bladder neck mobility during coughing, Neurourol. Urodyn. 32 (2013) 1096–1102.
[124] S.J. Madill, S. Pontbriand-Drolet, A. Tang, C. Dumoulin, Changes in urethral sphincter size following rehabilitation in older women with stress urinary incontinence, Int. Urogynecol. J. 26 (2015) 277–283.
[125] E.J.C. Hay-Smith, K. Bø, L.C.M. Berghmans, H.J.M. Hendricks, R.A. De Bie, E.S.C. van Waawijk van Doorn, Pelvic Floor Muscle Training for Urinary Incontinence in Women (Review) [Online], The Cochrane Library, 2002 (accessed 12.11.02).
[126] K. Bø, S. Larsen, O. Oseid, B. Kvarstein, R. Hagen, J. Jørgensen, Knowledge about and ability to correct pelvic floor muscle exercises in women with urinary stress incontinence, Neurourol. Urodyn. 7 (1998) 261–262.
[127] K.L. Burgio, J.C. Robinson, B.T. Engel, The role of biofeedback in Kegel exercise training for stress urinary incontinence, Am. J. Obstet. Gynecol. 154 (1986) 58–64.
[128] P.M. Skilling, P.E.P. Petros, Synergistic non-surgical management of pelvic floor dysfunction: second report, Int. Urogynecol. J. Pelvic Floor Dysfunct. 15 (2004) 106–110.
[129] K. Bø, Urinary incontinence, pelvic floor dysfunction, exercise and sport, Sports Med. 34 (2004) 451–464.
[130] K. Bø, R. Hagen, B. Kvarstein, S. Larsen, Pelvic floor muscle exercise for the treatment of female stress urinary incontinence: III. Effects of two different degrees of pelvic floor muscle exercises, Neurourol. Urodyn. 9 (1990) 489–502.
[131] R.A. Sherman, G.D. Davis, M.F. Wong, Behavioral treatment of exercise-induced urinary incontinence among female soldiers, Mil. Med. 162 (1997) 690–694.

[132] I.H. Braekken, M. Majida, M.E. Engh, K. Bo, Can pelvic floor muscle training reverse pelvic organ prolapse and reduce prolapse symptoms? An assessor-blinded, randomized, controlled trial, Am. J. Obstet. Gynecol. 203 (170) (2010) e1–e7.

[133] S. Hagen, D. Stark, C. Glazener, S. Dickson, S. Barry, A. Elders, H. Frawley, M.P. Galea, J. Logan, A. McDonald, G. Mcpherson, K.H. Moore, J. Norrie, A. Walker, D. Wilson, P.T. Collaborators, Individualised pelvic floor muscle training in women with pelvic organ prolapse (POPPY): a multicentre randomised controlled trial, Lancet 383 (2014) 796–806.

[134] D. Mcclurg, P. Hilton, L. Dolan, A. Monga, S. Hagen, H. Frawley, L. Dickinson, Pelvic floor muscle training as an adjunct to prolapse surgery: a randomised feasibility study, Int. Urogynecol. J. 25 (2014) 883–891.

[135] L. Stüpp, A.P. Resende, E. Oliveira, R.A. Castro, M.J. Girao, M.G. Sartori, Pelvic floor muscle training for treatment of pelvic organ prolapse: an assessor-blinded randomized controlled trial, Int. Urogynecol. J. 22 (2011) 1233–1239.

SECTION 3

Imaging and Segmentation of Pelvic Floor

L. Hoyte
Director, Urogynecology and Female Pelvic Reconstructive Surgery, USF College of Medicine, Tampa, FL, United States of America
Medical Director, Urogynecology and Robotic Surgery, Tampa General Hospital, Tampa, FL, United States of America

CHAPTER TWELVE

Dynamic Visualization of the Female Pelvic Floor Function

C.E. Constantinou
Stanford University, Stanford, CA, United States of America

INTRODUCTION

Previous chapters have shown that the pelvic cavity contains several organs that help to fulfill diverse roles, including sexual intercourse, conception, parturition, and solid and liquid storage and evacuation. These multiple organ systems interact and coordinate with each other in performing their normal physiological functions. Under certain conditions, some of these functions are subject to disruption, manifesting in pathologies such as incontinence and prolapse in a large percentage of women as well as any one of a spectrum of complex problems such as pain whose origin may or may not be readily identifiable. In this context, pelvic floor (PF) dysfunction constitutes a global burden affecting the quality of life of the individual, their family, and society in general. In women, because of the magnitude of the problem, a great deal of attention has been placed on the implication of the PF function as it relates to urinary incontinence (UI) in general and stress urinary incontinence (SUI) in particular. SUI is associated with the involuntary leakage of urine on coughing, sneezing, and physical exertion. The severity of SUI increases with age to the extent that it has been termed a hidden epidemic by DeLancey [1]. As a consequence, diagnostic methodologies, collectively called urodynamic testing, evolved to focus on the primary functions of the bladder and urethra as these structures are directly implicated. In cases of SUI, the function of the PF plays a more critical role because of its action on the urethra.

For this reason, it is critical to formalize the evaluation of the function of the PF using a variety of methods ranging from imaging [2–4] to probe development [5–9]. Evidently the technical requirements of PF dynamics instrumentation are more complex and challenging, requiring a contextual approach and the least possible invasive means. In current practice, manual muscle testing per vagina or rectum is the technique used by most clinicians to evaluate the pelvic floor muscles (PFMs). Unfortunately, due to the location of the PFMs defining their normal function in a noninvasive way is clinically challenging because of the complex 3-D arrangement of muscle groups, as well as the biomechanics of connective tissue and their attachments to the bony pelvis. Moreover there are few diagnostic tools to facilitate the outcome of treatment or objective metrics to document its effectiveness. Unlike urodynamics, PF evaluation has a limited range of dedicated technologies to characterize and localize the extent of dysfunction in a quantitative way. These factors also need to be contrasted by the fact that the physiotherapists contribute, as first responders, in many countries to the treatment of UI using PF training exercises [10–13].

In view of the earlier considerations, in this chapter we will describe methods developed to evaluate PF function in as quantitative a way as possible using our novel technology, parameters, and metrics developed specifically for the purpose. To start with, this chapter is focused toward the noninvasive acquisition of PF function using perineal ultrasound imaging and image processing [14]. Clearly, different biomechanical parameters are generated from these measurements and their utility in the appropriate diagnosis of SUI will await more controlled trials.

Because the fundamental clinical approach in the evaluation of PF function examination is also done using vaginal palpation [14], we shall consider in this chapter a description of a vaginal probe constructed to quantify the biomechanical vaginal closure forces [9]. Data generated using this probe are collected using a protocol of voluntary and reflex activation of the PF of normal subjects and also those with SUI. PF function obtained using our probe is typically invasive because it is inserted into the vagina. The inclusion of data using the vaginal probe can be considered as a complementary method to the digital vaginal examination in that it generates biomechanical outcomes regarding PF function in ways that ultrasound visualization alone does not. As seen in the chapter by Egorov, important biomechanical parameters of tissue constituent properties require direct measurements. Thus, as it will be demonstrated in the following section, while ultrasound can provide information about kinematic parameters such as timing, displacement, velocity acceleration, trajectory, and strain in the PF, it cannot measure the forces that are involved in their generation.

ULTRASOUND IMAGING

Perineal ultrasound imaging has become a method of choice to study the biomechanical characteristics of the female PF [5]. In this chapter, we aim to illustrate some important biomechanical parameters of PF function, which can be helpful in understanding the underlying mechanisms of PF behavior. The dynamic function of PF tissues can be derived from short video segments of ultrasound imaging. To quantitatively derive biomechanical parameters from the image sequences, a coordinate system was established. This coordinate system spans the symphysis, urethra, vagina, and the anorectal junction, all of which are best

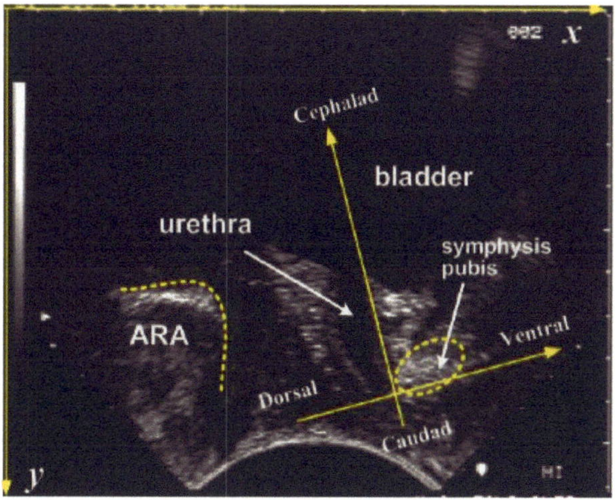

Figure 1 The orthogonal coordinate system fixed on the symphysis pubis. The two orthogonal components (ventral-dorsal and cephalad-caudad components) of the tissues displacements reflect PF functions of squeezing the urethra and supporting the bladder, respectively.

seen together on a sagittal view. For convenience this coordinate system is overlaid on the sagittal view of the ultrasonogram shown in Fig. 1.

During rest, strain, and squeeze maneuvers, displacement of the PF tissues can be defined as occurring in one or more directions: anteroposterior and superior-inferior. A ventral (anterior) component moves the tissues closer to the symphysis, whereas a dorsal (posterior) component moves the tissues away from the symphysis pubis. Furthermore, a cephalad (superior) component moves the tissues cranially, and a caudal (inferior) component moves the tissues toward the feet.

The choice of these directions is supported by other studies that suggest that in a functional PFM contraction, the bladder neck has been shown to move in a ventrocephalad direction [15], increasing the closure pressure within the urethra as it is displaced toward the symphysis pubis [16]. During Valsalva, as the intraabdominal pressure increases, the bladder neck moves in a dorsal-caudal direction. Based on these considerations, the orthogonal coordinate system, fixed on the bony landmark, the symphysis pubis, was established (Fig. 1). The two axes of the coordinate system are parallel and vertical to the urethra at rest, respectively. Using this schema, measurements of displacement and time can be made sonographically. From the time and displacement measures, the biomechanically important outcome, velocity of contraction, can be obtained. Acceleration can be derived by numerically differentiating the velocity curve.

Dynamic Ultrasound Image Analysis

As indicated by the ultrasonogram of Fig. 1, the identified anatomical structure is the angle that the rectal ampulla forms with the anal canal as defined as the anorectal angle (ARA). Movement in the ARA during maneuvers can be used to analyze the PFMs function. This is because the levator ani muscles wrap around the anorectal junction, and changes in the levator ani due to contraction and relaxation can change the position of the anorectal junction. It is well known that displacement of the ARA with respect to the symphysis is closely associated with PFMs activity [17]. As such, tissue displacement over time of the ARA with maneuvers can be defined by their magnitude and direction, with respect to the symphysis.

Such characterization can be accomplished using motion tracking techniques in the cephalocaudad and dorsoventral direction during active voluntary PFMs contractions as well during passive displacement during the Valsalva maneuver. Most importantly, motion tracking can be employed to evaluate the biomechanics of responses during reflex activation such as coughing where the guarding reflex is recruited [18].

Practically speaking, to accurately map the trajectory of the ARA in response to a cough it is necessary to maintain each video frame indexed to the symphysis pubis, which is a stationary, rigid, nondeforming structure. However, the movement of the ultrasound probe in experiments could cause a motion artifact in the image of the symphysis pubis. Therefore, the motion artifact of the symphysis pubis needs to be tracked and subtracted from the motion of the ARA. To accomplish the task of indexing, we developed an adaptive motion tracking algorithm based on a matching template to measure the movement of the symphysis pubis. A similar adaptive matching algorithm is used to track the motion of ARA. However, ARA is a soft tissue structure that deforms, particularly in fast maneuvers like coughing. Therefore, a weight coefficient is introduced to accelerate updating of the template to follow the deformations: to decrease the effect of the size and position of the initial template, manually defined in the first image frame, the tracking procedure is performed four times with different initial templates, and the results are averaged. The relative movement of ARA to the symphysis pubis is then derived by subtracting the motion of symphysis pubis from that of the ARA (Fig. 2).

Rapid continuous image sequences obtained are recorded and stored for analysis. These video sequences contain dynamic information relating to the integrity of the supporting structures. The resting position of the structures and their displacement at the end of specific and incontinence producing events (eg, coughing) were measured. However, these methods do not provide information about the path taken by the structures to reach their final destination. Furthermore, due to the speed inherent in such stress events, the observer is unable to be fully informed about the direction of the overall movement at a given moment in time, the velocity at which it occurs, or the acceleration produced. These parameters are important, both in the field of measurement and rehabilitation of the PFMs.

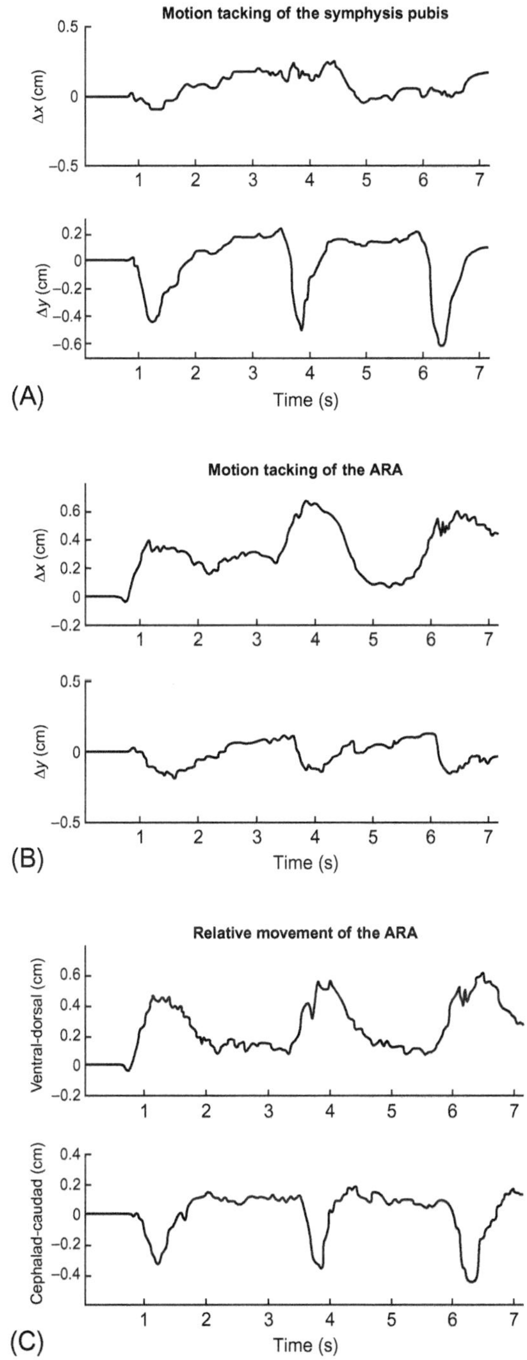

Figure 2 The motion tracking of (A) the symphysis pubis and (B) the ARA. (C) The ventral-dorsal and cephalad-caudad components of the displacement of the ARA.

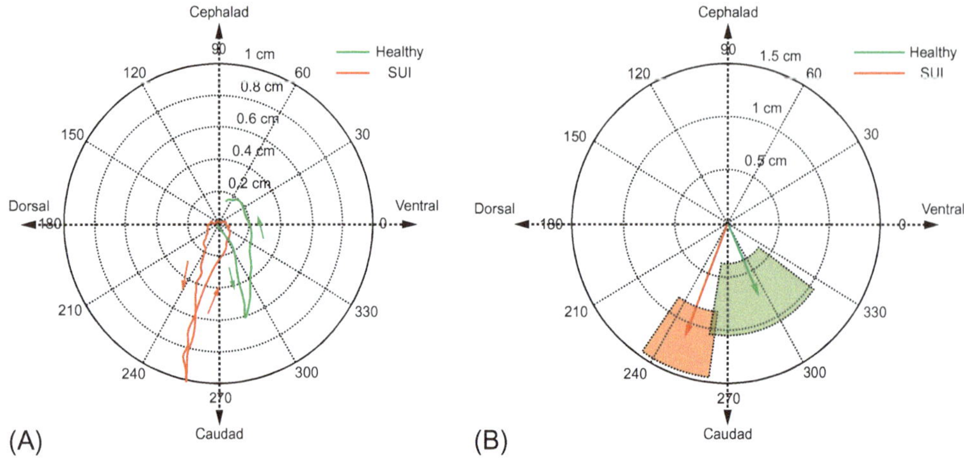

Figure 3 (A) Trajectory maximum displacement vectors of the anorectal junction in coughing. The normal subjects and SUI patients are shown in *green* and *red*, respectively. (B) The standard deviations are marked by the transparent pies.

To better understand these kinematic mechanisms, we developed methods to capture and visualize the sequence of dynamic changes the PFMs produced on the urethra, vagina, and rectum. This is made possible by fixing the coordinate system on the symphysis pubis, which is a recognizable stationary landmark across different maneuvers.

Ultrasound movies of normal subjects and SUI patients in coughing were collected and analyzed using those 2-D motion tracking algorithms. Among the parameters evaluated are the trajectory and the maximum displacement vectors of the anorectal junction in coughing (Fig. 3).

The results of the ultrasound experiments show that, when compared to women with SUI, the PFMs of the continent subjects have two mechanisms to prevent urine leakage during stress maneuvers like coughing.

(1) The well-functioning PFMs contractions give the bladder and the posterior edge of the urethra more support in the cephalad direction (green curves in Fig. 4B), when the abdominal pressure increases and pushes the bladder toward the caudad direction in maneuvers like coughing.

(2) The well-functioning PFMs contraction pushes the urethra against the symphysis pubis in the dorsal direction and therefore increases the closure pressure in the urethra (green curves in Fig. 4A), when the increased abdominal pressure pushes the urethra toward the ventral direction.

Timing of Pelvic Contractions

The appropriate timing control of the PFMs contraction is as important as, if not more important than, the amplitude of the pressures produced by the PFMs contraction in

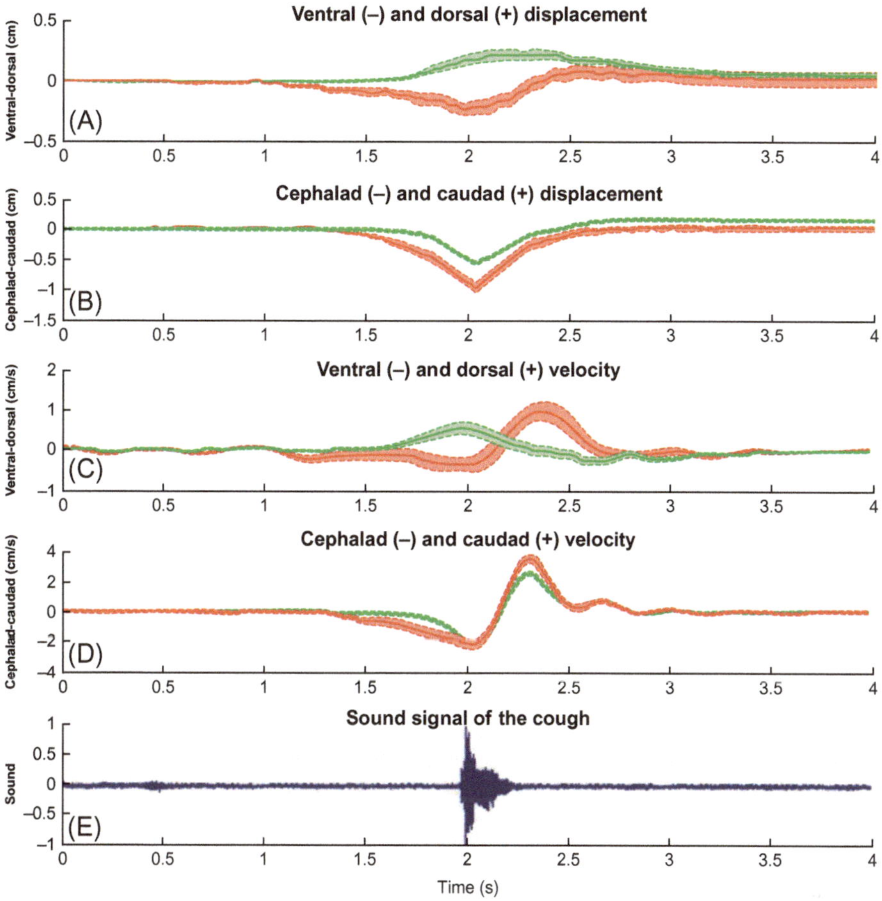

Figure 4 (A, B) The displacement and (C, D) velocity of the anorectal junction in coughing, and (E) the sound signal. The results of normal subjects and SUI patients are shown in *green* and *red*, respectively.

preventing UI. For this reason, measurements were indexed to the sound recording of the cough, taken by a microphone located on the subject's chest. As shown in Fig. 4A and C, the normal subjects' PFMs contraction pushed the anorectal junction to the dorsal direction (toward the urethra) before the increased abdominal pressures were able to push the anorectal junction and urethra to the ventral direction. By contrast, the SUI patients' PFMs contractions were behind the increased abdominal pressures in timing. As a result, the anorectal junction and urethra were pushed to the ventral direction at first and then bounded back to the dorsal direction with higher velocity. Those may lead to the decreasing and vibration of the closure pressures in the urethra.

Therefore, an accurate mechanical model to study mechanics of UI should be able to describe not only the closure pressures in the urethra, as most models do, but also the timing, direction, and amplitude of the PFMs contraction produced pressures and how the pressures affect the internal pressures and the movement of the bladder and urethra.

Visualization of the Pathway of Bladder/Urethra and ARA During a Cough

2-D sagittal imaging is useful to document the movement of structures influenced by the PFMs due to voluntary contraction, involuntary reflex events, coughing, or passive forces. Important information, defining the function and effectiveness of the PFMs, was obtained by examining the mechanics of movement of reliably visible organs such as the bladder, urethra, and rectum and define metrics to parameterize and quantify the results. To do so we scanned subjects capturing and storing digitized video sequences of consecutive frames of dynamic ultrasound images. These methods enabled the measurement of timing of movement of pelvic tissues in fast, stressful maneuvers, and facilitated the understanding of the neuromuscular control continence mechanisms. An example of the resulting postprocessed images is given in Figs. 5–7 where transparent color coding based on the segmentation was used to visualize the pathway of timing, amplitude, and direction of movement.

Fig. 5A demonstrates the sequence of displacement of the bladder/urethra as outlined by segmentation, marking the relative movement produced by the reflex contraction of the PFMs during the first 0.3 s. Fig. 5B and C shows the follow-up of PFMs activity. Total recording period is 1.2 s for Fig. 5A–C. Clearly, the human eye cannot retain the detail or quantify the parameters recorded by visual observation during real-time imaging. This visualization using frame-by-frame segmentation of the outlines of bladder/urethra, symphysis, and ARA show the timing and path of displacement where in (a) fast onset of forward movement is shown, (b) the persistance, and (c) of the recovery during a single cough.

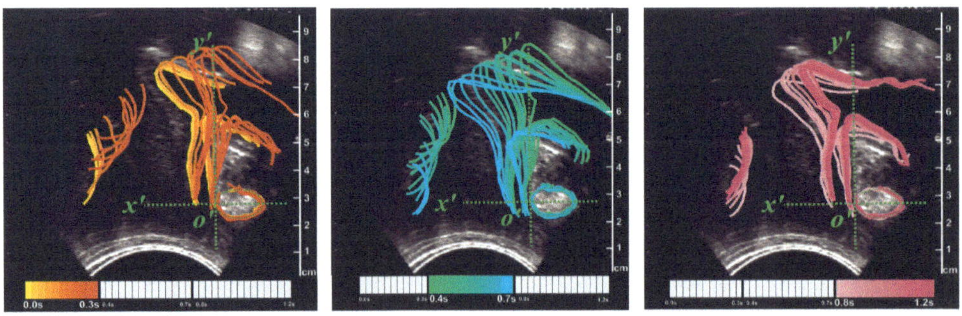

(A) (B) (C)

Figure 5 Frame-by-frame segmentation of outlines of the bladder/urethra, symphysis, and ARA showing the timing and path of displacement. (A) Fast onset; (B) persistence; (C) of the movement of the ARA, bladder, during a single cough.

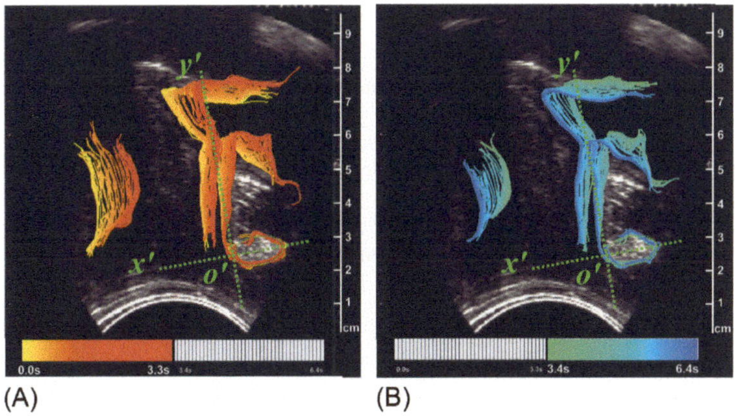

Figure 6 The timing of the movement due to a voluntary PFMs contraction of 6 s duration. (A) Shows the onset phase of the contraction and (B) release and restoration phase.

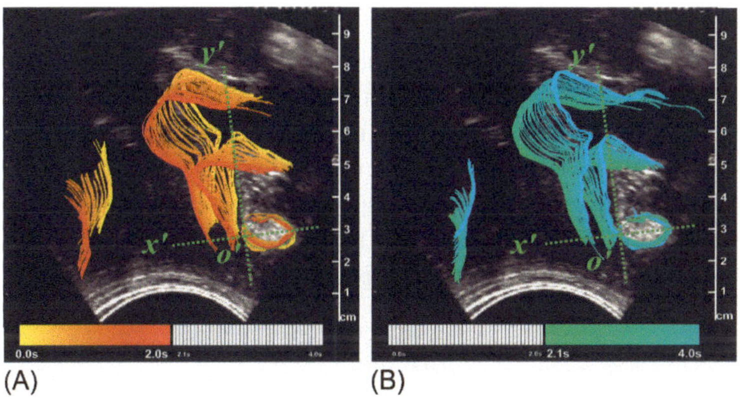

Figure 7 Time and direction of passive movement during a typical Valsalva of 4 s duration. (A) Represents the onset sequence of a downward movement and (B) return to pre-Valsalva position.

Visualization of the Pathway of Bladder/Urethra and ARA During Voluntary PFMs Contraction

Unlike the pattern of fast displacements produced by the reflex action generated by a cough (Fig. 5), voluntary PFMs contraction has a slower time course. As an example, Fig. 6A illustrates the displacement sequence and timing during a typical voluntary PFMs contraction. As the PFMs contraction starts (Fig. 6A), structures move in a ventral-cephalad direction, forward and up and relaxation (Fig. 6B) return to the original resting position occurs. However, during a cough (Fig. 5), the pattern of response of PFMs muscles differ, suggesting the variance of voluntary to reflex contractions.

Comparison between the pattern shown in Fig. 5, showing the reflex response of the PFMs muscles, and Fig. 6, the voluntary activation of PFMs contractions, provides evidence that while the direction is the same, the time course differs significantly.

Visualization of the Pathways of Bladder/Urethra and ARA During Valsalva

While Fig. 6 shows the timing and displacement caused by voluntary PFMs contraction the response during the passive compression produced by voluntarily straining or Valsalva, structures move from their resting position in a dorsal-caudal direction, down and back before returning to their resting position. Clearly, the response of the active contraction produced by PFMs contraction compared to the passive displacement produced by Valsalva also differ in the ways that may produce some insight into the mechanisms involved in their recruitment.

Clearly, as Fig. 7 shows, the functional PFMs in continent females appear to be highly regulated. These observations provide convincing evidence to support the hypothesis that structures prior to and during a cough appear to be acting like a feedback control system, to resist or limit the dorsal-caudal movement that occurs as during a cough. This hypothesis is supported by comparing both the overall direction and magnitude of displacement. On the basis of these imaging observations, evidence suggests that the function of the PFMs can be quantified and mechanistic insight can be found by appropriate analysis. However, there are important limitations using 2-D ultrasound so that it is not possible to incorporate the contribution of all muscle groups. In that sense the potential use of 3-/4-D imaging needs to be explored using similar analytics so that visualizations are not done only in a single plane.

Strain Analysis of 2-D Dynamic Ultrasound Images

In analogy to the use of ultrasound to evaluate cardiac function, measurements of PFMs strain, a dimensionless index reflecting the deformation produced by contraction can be applied [19]. Details of the methodology used for the PF can be found in Ref. [20] where the generated strain was calculated from the displacement due to the dorsal-ventral and cardinal-caudal displacements of the selected landmarks. From these displacement measurements, shown diagrammatically by Fig. 8 where the displacement generated was simplified to a straight line, strain was calculated as follows:

The strain ε at time t is defined as

$$\varepsilon_t = \frac{||\dot{L}_t|| - ||\dot{L}_0||}{||\overset{\text{r}}{L}_0||}, \tag{1}$$

where $||\dot{L}_0||$ is the initial length of the PFM at rest and is measured from the ultrasound images; $||\dot{L}_t||$ is the length of the muscle at time t and is calculated from the equation.

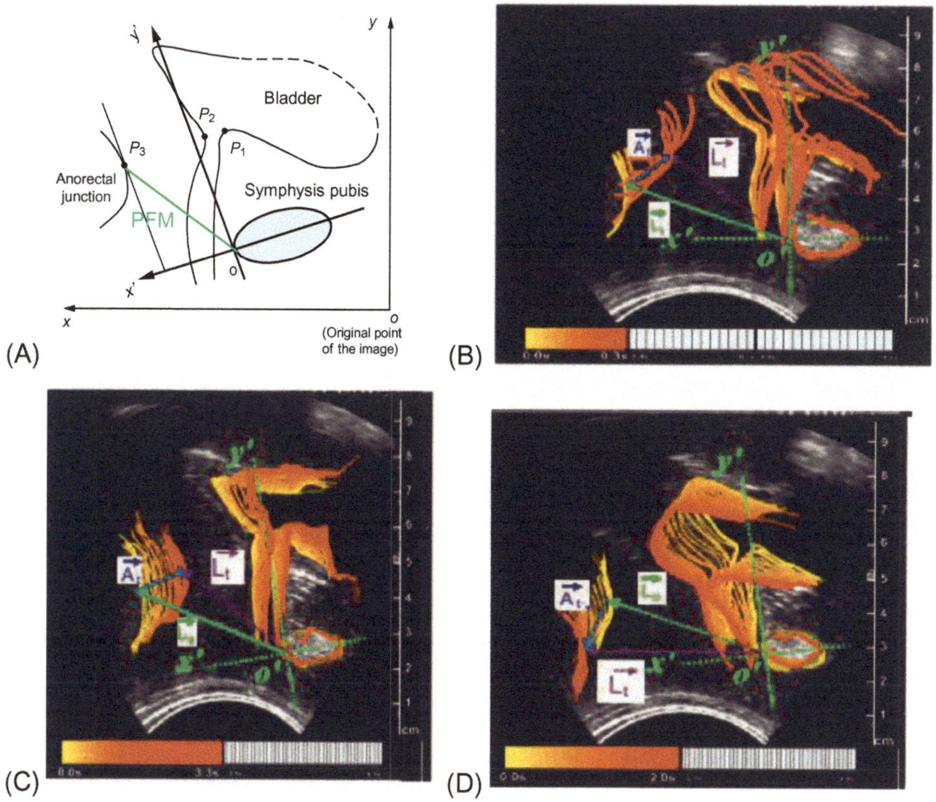

Figure 8 Diagrammatic presentation of measurements (A) to calculate strain indexed to ultrasound images identifying the outline of bladder/urethra and ARA trajectory [20]. (B) Cough response. (C) Voluntary Contraction. (D) Valsalva.

$$||\vec{L}_t|| = \sqrt{\left(\vec{l}_{t,x}\right)^2 + \left(\vec{l}_{t,y}\right)}. \tag{2}$$

$\dot{l}_{t,x}$ is the vector representing the dorsal-ventral component at time t and is measured by the equation

$$\dot{l}_{t,x} = \Delta\dot{x}_t + \dot{l}_{0,x}. \tag{3}$$

Likewise, $\dot{l}_{t,y}$ is the vector representing cardinal-caudal component at time t and is defined by the equation

$$\dot{l}_{t,y} = \Delta\dot{y}_t + \dot{l}_{0,y}, \tag{4}$$

where $\dot{l}_{0,x}$ and $\dot{l}_{0,y}$ are the dorsal-ventral and cardinal-caudal vector components at rest ($t=0$), respectively. $\Delta\dot{x}_t$ and $\Delta\dot{y}_t$ are the dorsal-ventral and cardinal-caudal displacements, respectively, at time t, which can be measured from the previous displacement measurements. $\dot{A}_t$ represents the vector $(\langle \Delta \dot{x}_t, \Delta \dot{y}_t \rangle)$.

To carry out the vector addition, it is necessary to retrieve the $\Delta \dot{x}_t$ and $\Delta \dot{y}_t$ values from the images and add the two vectors (dorsal-ventral and cardinal-caudal components) together; that is, the obtained displacement values of the two components need to be matched to the same time measurement. After the time measurements are matched, the mean of those values will be calculated, and an array made for the average time measurements. Displacement measurements in the dorsal-ventral and cardinal-caudal directions corresponding to those average time values will be added from the data files to that array. Vectors $\dot{l}_{0,x}$ and $\dot{l}_{0,y}$, shown in Fig. 5, will be added to the corresponding displacement measurements according to Eqs. (3), (4) and strain calculated according to Eq. (2) at each of the time values. A strain versus time graph will be plotted for each of the four maneuvers in the supine position and in the standing position for the case of cough in both asymptomatic and UI patients. Subsequently, the maximum strain of the PFMs for both groups is obtained and statistical comparisons between the two groups are carried out.

Strain rate (SR) measurements: Using these displacement measurements along with the velocity measurements, the *SR* of PFMs for asymptomatic and incontinent women during a cough was calculated.

SR can be measured using the following equation:

$$\frac{dL}{L} \approx \frac{v(r+\Delta r) - v(r)}{\Delta r} dt = SR\, dt, \tag{5}$$

where here r represents the position of the point on the ARA, thus $\Delta r = ||\dot{L}_t||$, which is defined by Eq. (2).

Also,

$$v(r+\Delta r) - v(r) = \Delta v = \sqrt{(\Delta v_x)^2 + (\Delta v_y)^2}. \tag{6}$$

SR is related to strain by the equation

$$\varepsilon = \exp\left(\int_{t_0}^{t} SR\, dt\right) - 1, \tag{7}$$

and that is because *SR* is defined as the time derivative of natural strain $\left(\ln \frac{L}{L_0}\right)$.

From Eq. (5) the *SR* will be calculated using the same algorithm and the matching algorithm as described earlier for strain measurement, with the addition that now the time measurement needs to be matched for both velocity and displacement in the dorsal-ventral and cardinal-caudal direction. Thus, the mean time value is calculated by averaging the four matched values, and the similar method described in the strain measurement using Microsoft Excel is used to apply Eq. (6) to all the collected data points. *SR* versus time is graphed for the two groups and the maximum *SR* is calculated and compared for significant differences.

Elastic Properties

The in vivo biomechanical analysis that is facilitated by the ultrasound imaging as described earlier, where the length and displacement are measured from the images, incorporates the elastic properties of the vaginal and urethral tissues. The response measured inevitably involves the complex interactions between mechanical forces, and neurological, muscular, and connective tissue factors [21]. Currently there is incomplete understanding of how tissue composition contributes to its function or its biomechanical properties. Due to methodological limitations it is challenging to perform biomechanical and biochemical testing on the same tissue. Also, most biomechanical data are on tissue from affected individuals while little is known about asymptomatic, unaffected tissue. Safety considerations limit our ability to obtain large tissue samples from asymptomatic controls. Most evaluation of the biomechanical properties of vaginal tissue in patients with clinical problems is done using a variation of the uniaxial loading technique to test tissue biomechanics. This technique is commonly used to characterize the elastic, viscous, and plastic properties of vaginal tissue. This methodology is limited, in that it only provides 1-D force displacement characteristics requiring large pieces to obtain a homogeneous stress state.

A novel evaluation of tissue properties was done where an evaluation of human vaginal wall stiffness was correlated with measurements to collagen expression in women with and without pelvic organ prolapse (POP) [21]. The results of these studies are demonstrated in Fig. 11, which shows the difference in the 2-D distribution of stiffness between prolapse patients and controls.

A biomechanical comparison of the elasticity of vaginal wall tissues in terms of the tangent modulus of elasticity can then be thus compared using vaginal wall tissue mapping (Fig. 9A). Vertical axis and color maps show stiffness in kilopascal. Red areas represent greater stiffness. Blue areas represent less stiff material. A stiffness distribution histogram shows that tissue from a patient with prolapse was on average stiffer than control tissue.

In this study, we evaluated vaginal tissue biomechanics in women with and without POP and confirmed that collagen type I and III composition is associated with measureable tissue biomechanics. We observed that the vaginal wall was stiffer in cases than in controls. Results are consistent with data from a study using a proof of concept prototype of the vaginal tactile imager ofEgorov et al. [8] and with data obtained by traditional uniaxial tension tests. Overall the traditional uniaxial tension studies indicate that prolapse confers less elasticity and greater

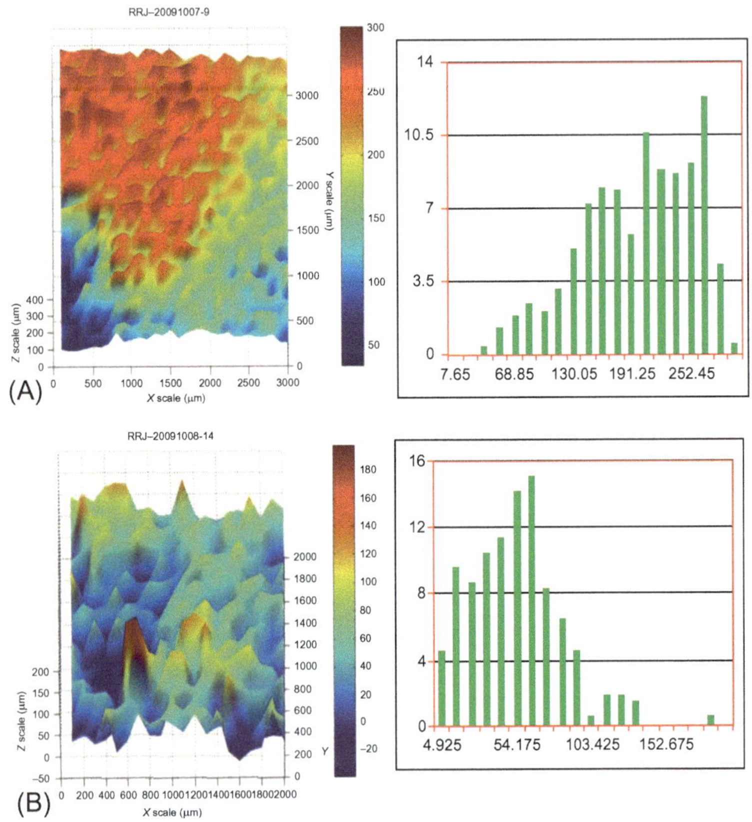

Figure 9 (A) Distribution of stiffness in a 78-year-old woman with prolapse. (B) Distribution of stiffness of an 80-year-old continent woman.

stiffness with low forces at failure. Our data are consistent with this and indicate that the use of a piezoelectric scanning probe is a reliable way to measure tissue biomechanics without the need for large tissue samples. From the biomechanics point of view, collagen type I is nonelastic and confers great resistance to tensile forces while collagen III has elastic properties and prevails in more flexible tissue [21,22].

Vaginal Probe

While aspects of PF function can be evaluated using the imaging techniques afforded by ultrasound above with the possibility of identifying mechanism of function and anatomical defects, in current clinical practice, digital vaginal examinations are necessarily performed. Such testing provides qualitative information about pathology but invariably do not generate quantitative measures to characterize the ability of the subject to function, as such conservative approaches to strengthen PF function remain subjective and operator dependant [10–14]. There is therefore a need to document the magnitude and direction of the biomechanical forces produced around the length of the vagina as well as active forces during voluntary contraction or the reflexes during coughing. Because the accuracy of the digital vaginal palpation is highly dependent on the experiences of the examiner and the ability to document it, it has become essential to design and build devices that can accomplish a more biomechanically accurate measure. A number of such devices have been evaluated and the results reported in the literature [6–9], each of which has specific characteristics and usefulness to its creator. Commercial devices are beginning to enter the marketplace. In this chapter, we will present the principles of our sensor system, shown in Fig. 10, designed to measure the mechanical characteristics of the vaginal wall and the amplitude of the pressure produced by the PFMs. The system is made of a probe with four force and displacement sensors and a six-degrees-of-freedom measuring device that can track the position (x, y, z) and orientation (azimuth, elevation,

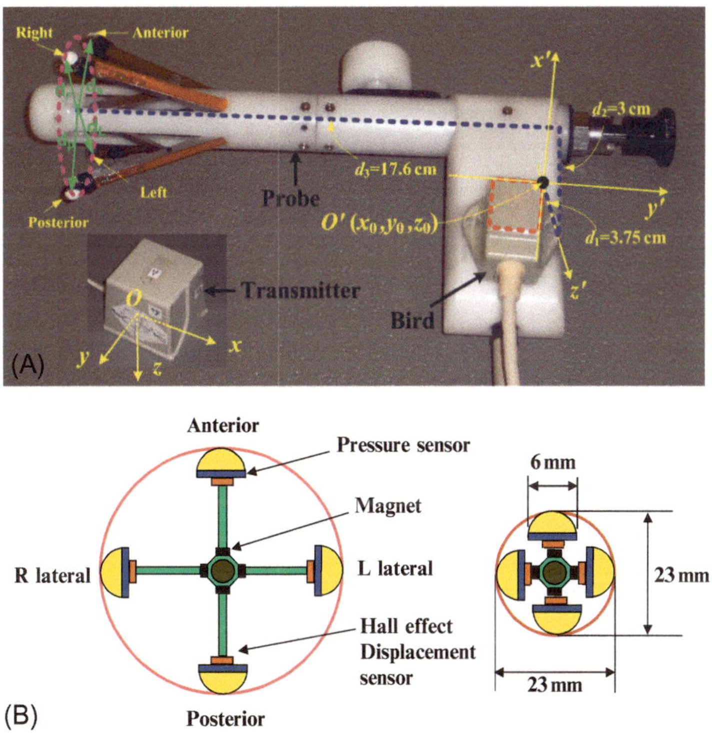

Figure 10 (A) The picture of the sensor system and (B) the cross section of the probe.

and roll angles) of the probe. The probe is made of an applicator with retractable low inertia cantilevered force and displacement transducers in four directions (anterior, posterior, left, and right). For patient protection, the probe is inserted in the vagina using a female condom placed over the sensors. Once the probe is in place, a diameter adjustor is gradually released to allow opening from its retracted position so that the transducers make contact with the vaginal wall. The vaginal pressure of 25 normal subjects and 10 patients with SUI are measured in supine by using this sensor system.

The results show that the normal subjects have significantly higher mean contact pressure of the PFMs side of the vagina both at rest (healthy: 1.67 ± 0.13 N/cm^2, SUI: 1.15 ± 0.18 N/cm^2, $p=0.048$) and with 50% PFMs contraction (healthy: 2.05 ± 0.13 N/cm^2, SUI: 1.35 ± 0.20 N/cm^2, $p=0.009$). Using this probe, a profile of contact pressures of the normal subjects and the SUI patients is illustrated by Fig. 11, showing the distribution of pressures is higher in the anterior aspect of the vagina compared to the posterior. Activation of PF contraction at 50% maximum shows the relative increase.

In recent times, a number of vaginal probes using different construction have been developed, some of which evaluate the tissue elastic properties instead of contact pressure [8]. It is important to indicate that probing the vaginal wall asymmetric values of closure are obtained. Fig. 12 graphically shows that maximal closure is measured in the anterior aspect of the vaginal wall.

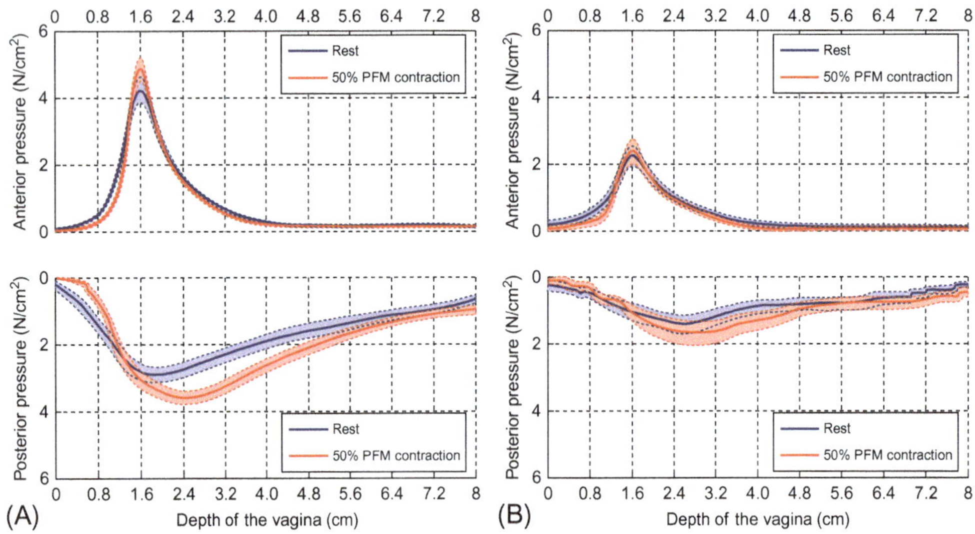

Figure 11 Vaginal contact pressure profile of (A) controls and (B) SUI subjects. Horizontal axis is depth into the vagina and vertical axis is the contact pressure. Mechanically, the probe shown in Fig. 1 contains four sensors and interpolation of the contact forces measured can be used to demonstrate the full 360 degrees distribution of closure.

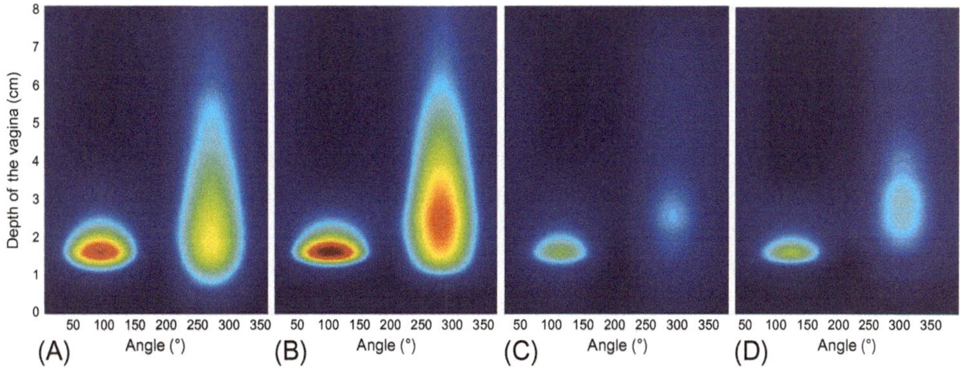

Figure 12 Distribution of vaginal closure over 360 degrees. (A) Shows closure at rest and (B) following PF contraction. Similarly (C) and (D) but in subjects with SUI.

As shown in Fig. 12, maximal closure is located 1.5 cm from the introitus in its anterior subject for all subjects and lower at the posterior aspect. Incontinent women possess numerically lower closure with a region of lower stiffness as measured by Egorov in a different chapter of this book.

Clearly with this and similar technology such as the instrumented speculum, incontinent women demonstrated lower values in passive force, endurance, and speed of contraction than continent women [6,7]. Given that PFMs strengthening exercises diminish the symptoms of SUI [10,11,23], it is therefore important to resolve whether the mechanisms involved in therapeutic changes to help identify the specific critical muscle components involved. It is unknown whether the PF exercises mimics the normal physiological behavior of the PFMs or is a compensation strategy. Nonetheless a program of awareness is indeed the most efficient method of implementing rehabilitation. To do so it is appropriate to best understand the many facets of function and to impart this information to the patient.

Summary of Ultrasound Imaging Parameters

Perineal ultrasound imaging provides a convenient and noninvasive way to evaluate PF function documenting the biomechanical metrics involved. It was shown that timing of reflex PF contractions is critical in maintaining urinary continence. Pelvic organ displacement from a resting state as a result of Valsalva indicates the state of the ligaments and the stiffness of the supporting structures as the body of the uterus, the bladder, and the rectal portion of the PFMs evidenced the highest displacement during Valsalva. The magnitude of the displacement of the bladder during voluntary PF contraction can be considered an indicator of muscle strength. It was shown that the tissue elasticity of the vaginal wall plays a significant role in the overall mechanisms involved in the maintenance of continence. Biomechanically functional parameters were identified using a number of analytical approaches.

In conclusion we have demonstrated that in focusing on the biomechanics of PF function using noninvasive ultrasound, the following metrics for each testing protocol have been identified: displacement, velocity, acceleration, trajectory, timing, and strain. However, these values depend on other parameters such as the elastic properties of the tissues and the strength of the contraction. Invasive methods such as a vaginal probe are necessary to document the strength of PF contractility. Such probing devices can be used to assess physiotherapy interventions or other treatment regiments.

ACKNOWLEDGMENTS

The support of NIH/NIBIB grants 1 R21 EB001654 and 1 R01 EB006170 is gratefully acknowledged.

REFERENCES

[1] J.O. DeLancey, The hidden epidemic of pelvic floor dysfunction: achievable goals for improved prevention and treatment, Am. J. Obstet. Gynecol. 192 (5) (2005) 1488–1495.

[2] S.H. Yang, W.C. Huang, S.Y. Yang, E. Yang, J.M. Yang, Validation of new ultrasound parameters for quantifying pelvic floor muscle contraction, Ultrasound Obstet. Gynecol. 144 (Suppl. 1) (2009) S159–S165. Review.

[3] C.E. Constantinou, Dynamics of female pelvic floor function using urodynamics, ultrasound and magnetic resonance imaging (MRI). Eur. J. Obstet. Gynecol. Reprod. Biol. 144 (Suppl. 1) (2009) S159–S165, http://dx.doi.org/10.1016/j.ejogrb.2009.02.021. Review.

[4] R.C. Lovegrove Jones, Q. Peng, M. Stokes, V.F. Humphrey, C. Payne, C.E. Constantinou, Mechanisms of pelvic floor muscle function and the effect on the urethra during a cough, Eur. Urol. 57 (6) (2010) 1101–1110, http://dx.doi.org/10.1016/j.eururo.2009.06.011. PMID:19560261.

[5] H.P. Dietz, M. Erdmann, K.L. Shek, Reflex contraction of the levator ani in women symptomatic for pelvic floor disorders, Ultrasound Obstet. Gynecol. 40 (2) (2012) 215–218, http://dx.doi.org/10.1002/uog.11087. PMID:22223551.

[6] J.M. Miller, J.A. Ashton-Miller, D. Perruchini, J.O. DeLancey, Test-retest reliability of an instrumented speculum for measuring vaginal closure force, Neurourol. Urodyn. 26 (6) (2007) 858–863.

[7] M. Morin, D. Gravel, D. Bourbonnais, C. Dumoulin, S. Ouellet, Reliability of dynamometric passive properties of the pelvic floor muscles in postmenopausal women with stress urinary incontinence, Neurourol. Urodyn. 27 (8) (2008) 819–825.

[8] V. Egorov, H. van Raalte, A.P. Sarvazyan, Vaginal tactile imaging, IEEE Trans. Biomed. Eng. 57 (7) (2010) 1736–1744, http://dx.doi.org/10.1109/TBME.2010.2045757. PMID:20483695.

[9] C.E. Constantinou, S. Omata, Direction sensitive sensor probe for the evaluation of voluntary and reflex pelvic floor contractions, Neurourol. Urodyn. 26 (3) (2007) 386–391. PMID:17301962.

[10] K. Bø, G. Hilde, J. Stær-Jensen, F. Siafarikas, M.K. Tennfjord, M.E. Engh, Postpartum pelvic floor muscle training and pelvic organ prolapse — a randomized trial of primiparous women, Am. J. Obstet. Gynecol. 212 (1) (2015) 38.e1–38.e7, http://dx.doi.org/10.1016/j.ajog.2014.06.049. PMID:24983687.

[11] K. Bø, G. Hilde, M.K. Tennfjord, J. Stær-Jensen, F. Siafarikas, M.E. Engh, Pelvic floor muscle variables and levator hiatus dimensions: a 3/4D transperineal ultrasound cross-sectional study on 300 nulliparous pregnant women, Int. Urogynecol. J. 25 (10) (2014) 1357–1361, http://dx.doi.org/10.1007/s00192-014-2408-8. PMID:24828605.

[12] F. Siafarikas, J. Stær-Jensen, G. Hilde, K. Bø, M. Ellström Engh, Levator hiatus dimensions in late pregnancy and the process of labor: a 3- and 4-dimensional transperineal ultrasound study, Am. J. Obstet. Gynecol. 210 (5) (2014) 484.e1–484.e7, http://dx.doi.org/10.1016/j.ajog.2014.02.021. PMID:24569040.

[13] J.A. Thompson, P.B. O'Sullivan, N.K. Briffa, P. Neumann, Differences in muscle activation patterns during pelvic floor muscle contraction and Valsalva maneuver, Neurourol. Urodyn. 25 (2) (2006) 148–155. PMID:16302270.

[14] I. Volløyhaug, S. Mørkved, Ø. Salvesen, K.Å. Salvesen, Assessment of pelvic floor muscle contraction with palpation, perineometry and transperineal ultrasound: a cross-sectional study. Ultrasound Obstet. Gynecol. (2015), http://dx.doi.org/10.1002/uog.15731.

[15] J.M. Miller, J.A. Ashton-Miller, J.O. DeLancey, A pelvic muscle precontraction can reduce cough-related urine loss in selected women with mild SUI, J. Am. Geriatr. Soc. 46 (7) (1998) 870–874. PMID:9670874.

[16] D. Howard, J.M. Miller, J.O. Delancey, J.A. Ashton-Miller, Differential effects of cough, valsalva, and continence status on vesical neck movement, Obstet. Gynecol. 95 (4) (2000) 535–540. PMID:10725485.

[17] C.E. Constantinou, D.E. Govan, Spatial distribution and timing of transmitted and reflexly generated urethral pressures in healthy women, J. Urol. 127 (5) (1982) 964–969. PMID:7201031.

[18] G.C. Nesbitt, S. Mankad, Strain and strain rate imaging in cardiomyopathy, Echocardiography 26 (2009) 337–344.

[19] S. Rahmanian, R. Jones, Q. Peng, C.E. Constantinou, Visualization of biomechanical properties of female pelvic floor function using video motion tracking of ultrasound imaging, Stud. Health Technol. Inform. 132 (2008) 390–395. PMID:18391328.

[20] L. Zhou, J.H. Lee, Y. Wen, C. Constantinou, M. Yoshinobu, S. Omata, B. Chen, Biomechanical properties and associated collagen composition in vaginal tissue of women with pelvic organ prolapse, J. Urol. 188 (3) (2012) 875–880, http://dx.doi.org/10.1016/j.juro.2012.05.017. PMID:22819408.

[21] P. Chantereau, M. Brieu, M. Kammal, J. Farthmann, B. Gabriel, M. Cosson, Mechanical properties of pelvic soft tissue of young women and impact of aging, Int. Urogynecol. J. 25 (11) (2014) 1547–1553, http://dx.doi.org/10.1007/s00192-014-2439-1. PMID:25007897.

[22] P.A. Norton, Pelvic floor disorders: the role of fascia and ligaments, Clin. Obstet. Gynecol. 36 (4) (1993) 926–938. PMID:8293593. Review.

CHAPTER THIRTEEN

Magnetic Resonance Imaging of the Pelvic Floor

J. Fielding*, H.C. Ming†,‡
*UTSouthwestern, Dallas, TX, United States of America
†Singapore General Hospital, Singapore
‡Duke-National University of Singapore, Singapore

INTRODUCTION

Advancement in magnetic resonance imaging (MRI) has resulted in higher signal to noise ratio, faster image acquisition, and improved capability for cinematic studies. These advancements have led to an increased acceptance of MRI in elucidating the pelvic anatomy of women with symptoms of pelvic floor dysfunction, especially those involving defects in multiple compartments.

MRI allows concurrent assessment of structural abnormalities of the pelvic floor and determines the degree of abnormal pelvic floor descent, which can be correlated to the severity of symptoms involving respective compartments [1].

The dynamic pelvic floor protocol and interpretation primarily relies on use of T2-weighted midsagittal MRI and identification of the positions of the pelvic viscera at rest and in maximal strain. Biometric markers such as the pubococcygeal line (PCL), M line, H line, and the levator plate, quantify this anatomic and functional information.

This chapter seeks to improve understanding of the complex anatomy of the pelvic floor in the interpretation and reporting of such studies. The normal and abnormal appearances of pelvic floor descent are also discussed.

MRI OF THE PELVIC FLOOR

Historically, pelvic floor function can was assessed by physical examination and fluoroscopy, the former limited by sensitivity, especially for enterocele [2]. The latter provides real-time imaging of the urinary bladder or the anorectal canal, at rest and strain, outlined by radio-opaque material, and further reviewed on video recordings. Disadvantages include radiation exposure and the requirement of a radiolucent commode in the fluoroscopy room. Ultrasound evaluation, which requires specialized training and is less widely prevalent, is discussed in an earlier chapter.

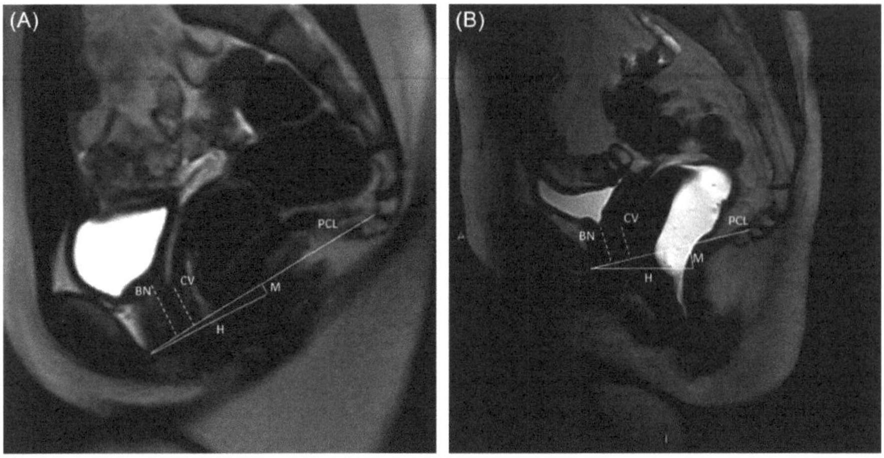

Figure 1 Normal sagittal appearance of the female pelvis at rest and upon strain. (A) The pubococcygeal line (PCL) is drawn from the inferior margin of the public symphysis to the last coccygeal joint. H line represents the anteroposterior length of the levator hiatus from and the inferior margin of the pubic symphysis to the posterior wall of the anorectal junction. The perpendicular distance between the PCL and H line constitutes the M line. (B) The rectum descends upon evacuation and the anal canal opens to allow passage of stool. The bladder neck (BN) and cervical lip (CV) are remains above the PCL line.

Dynamic MRI imaging uses the same diagnostic principles as fluoroscopy, with the cinematic mode consisting of numerous images that can be viewed as a video, and has the advantages of assessing multiple compartments simultaneously. MRI is best suited in clinical situation for preoperative assessment of multicompartmental or complex pathology, and for evaluation of recurrent symptoms in postoperative cases [3].

Advantages over other tests also include multiplanar capability and superior soft tissue contrast in T2-weighted sequences. Fluid (urine and intraluminal gel) and fat are of bright signal, which outlines the intermediate-to-low signal of the wall of the pelvic viscera, and separates them from the bony cortices and muscles that are lower in signal intensity (Fig. 1).

MRI is a noninvasive, safe modality, with no ionizing radiation and does not require intravenous contrast. Major contraindications to MRI are MRI-incompatible implants and claustrophobia.

TECHNIQUE

No patient preparation is required. The patient is in a supine or left lateral position during a conventional MRI study. The less prevalent open-magnet system allows the more physiological sitting position and increases detection rate of rectal prolapse [2], small rectoceles, and rectal intussusception [4], but is not superior for assessing clinically relevant rectoceles and abnormal bladder descent as the straining forces are far greater than gravity [5].

Phased-array coils are applied covering the perineum and proximal thighs. Endovaginal coils are optional, but their rigid nature raises the concern for reduced vaginal compliance, although they improve assessment of stress incontinence [6]. A wedge placed beneath the patient's knees [7] aids in strain maneuver. Typically, 60 cm^3 of ultrasound gel is placed in the rectum using a straight tip flexible catheter. This can be mixed with gadolinum chelate to increase the visibility of the rectal wall. Vaginal opacification is usually not necessary, but can be performed using barium paste if further delineation is needed.

Sagittal T2-weighted images 10 mm in thickness are acquired using a 30 cm field of view (FOV); dynamic sequences vary by vendor and include SSFSE (single shot fast spin echo, GE Healthcare scanners) or HASTE (Half fourier Acquisition Single shot Turbo spin Echo, Siemens Medical Solution scanners) [8]. Scanning is first performed at rest and midline is defined as an image containing the symphysis, vagina, and sacrum. T2-weighted images are then obtained during a Valsalva maneuver and again during bowel evacuation (Fig. 2). Coaching the patient in performing the Valsalva maneuver prior to scanning improves cooperation and image quality. Real-time truFISP (true fast imaging with steady state precession, Siemens Medical Solution scanners), and other vendor-specific techniques are routinely performed, and these are used to assess dynamic movement of the pelvic floor [9].

Axial images are performed aligned with the plane of the puborectalis, using a small FOV of 20–24 cm across both hip joints (Fig. 2). Standard 5 mm T2-weighted FSE (fast spin echo, GE Healthcare scanners) or TSE (turbo spin echo, Siemens Solution scanners) images are usually adequate [8]. Three-dimensional T2-sequences allow for thinner slices, but the decreased signal contrast reduces the ability to differentiate the soft tissues. High resolution T2-weighted imaging depicts the pelvic muscles and the fascial

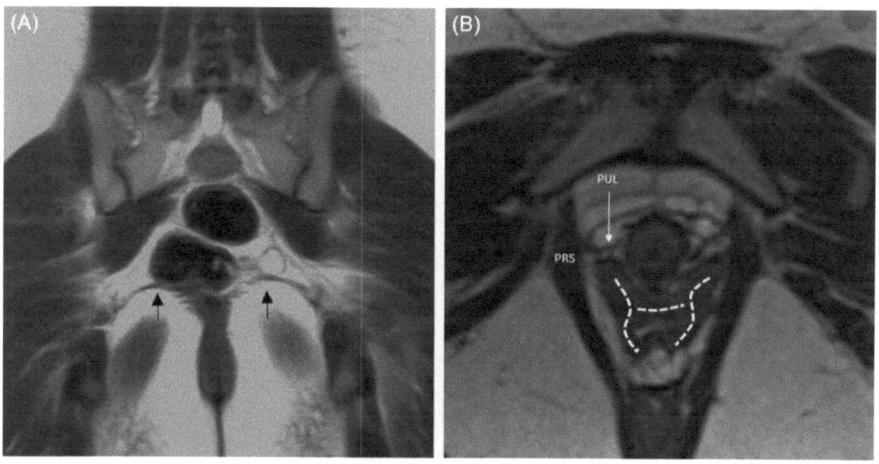

Figure 2 Normal coronal and axial appearance of the female pelvis. (A) The levator ani muscles usually show an upward-bulging curve *(black arrows)*. (B) The normal H-shaped or butterfly configuration *(curved line)* on the axial T2W SPC image, aligned along the plane of the puborectalis sling (PRS), is maintained in another patient. The periurethral ligaments (PUL) are also well visualized in this high resolution image.

Table 1 Suggested protocol for dynamic MRI of pelvic floor dysfunction

Sequence	Plane	TR	TE	Slice thickness/gap	FOV	Matrix	Remarks
Localizer	Sag, Ax, Cor	10	5	10 mm/0	350–400	128 × 256	
Localizer — pelvic floor	Sag, Ax, Cor	10	5	10 mm/0	250–300	128 × 256	Center at pubic symphysis, 60% phase oversampling
HASTE	Sag	2000	96	5 mm/0	280	205 × 256	100% phase oversampling breath hold/Valsalva/2 concatenations/IPAT
T2 Spc 3D	Ax	2000	123	1 mm	256	256 × 256	75% phase oversampling/IPAT
T2 TSE	Cor	6000	114	4 mm/0.8 mm	250	288 × 320	100% phase oversampling/IPAT
FISP	Sag	4.45	1.98	4 mm/0	210	256 × 256	100% phase oversampling breath hold/Valsalva/IPAT

condensations especially around the urethra (Fig. 2). Imaging in the coronal plane is optional but sometimes useful in assessing the normally upward-convex curve of the levator muscles. Overall, these sequences require 20 min to complete, and a suggested MRI protocol is summarized in Table 1.

INTERPRETATION

The sagittal plane in the midline provides the basis of assessment of most of the parameters required in assessing the positions of the pelvic viscera. This is readily appreciated by depiction of the fixed structures such as the urethra, cervix, vagina, anorectal junction, levator plate, and the coccyx, which can all be visualized in a single image (Fig. 1).

The puborectalis muscles are paramedian structures well seen in axial images. Though not directly visualized in the midline sagittal image, its position can be inferred from its attachment to the posterior wall of the pubic symphysis. The puborectalis wraps around the posterior aspect of the anorectal junction, and displaces it in a direction toward the symphysis pubis (Fig. 3).

Once the midline is identified, a reference line can be annotated on the images. The most widely accepted and highly reproducible marker is the PCL [10]. The PCL (Fig. 1) is drawn from the inferior margin of the symphysis pubis to the last visualized coccygeal joint, both of which are easily identified with very low signal on a T2-weighted image.

In addition to the PCL, the H (hiatus) line is drawn from the pubic symphysis to the anorectal junction. This is usually less than 4.5 cm, and if more than 5 cm [11], implies

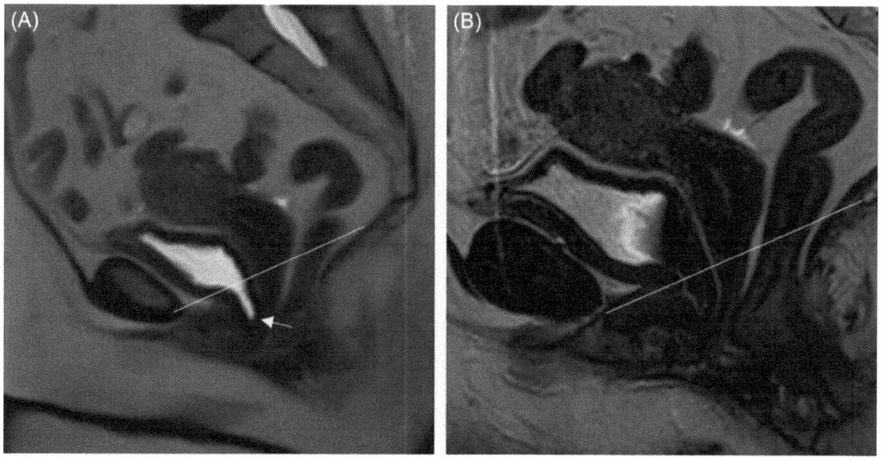

Figure 3 Small cystocele. (A) A 61-year-old female presented with constipation and urinary incontinence. A small cystocele *(arrow)*, measuring 1.8 cm below the PCL *(white line)*, is easily identified in the HASTE image with large field of view (FOV), due to the inherent high T2-weighted signal of the urine. (B) A high resolution T2W TSE (turbo spin echo) image with smaller FOV better demonstrates the anterior cervical lip which just abuts the PCL.

anteroposterior elongation of the levator hiatus and laxity of the puborectalis muscle band, possibly also tearing of the fascial supports. The hiatus can be widened and elongated in its transverse and anteroposterior dimension associated with weakening of the puborectalis sling, which can be seen on the axial images (Fig. 4).

The perpendicular distance from the PCL to the H line is the M line, which measures the degree of pelvic floor descent. The puborectalis hiatus normally does not descend more than 2 cm below the PCL at rest [12]. Severity of the prolapse can be arbitrarily graded into mild, moderate, and severe by a descent of less than 3 cm, less than 6 cm, and 6 cm or more [13].

An alternative to the PCL is the midpubic line (MPL), which corresponds to the level of the vaginal hymen and is similar to the clinical examination [14]. The choice for PCL or MPL may depend on the preference and familiarity of the referring physician and the reporting radiologist.

Reference points of the degree of pelvic organ prolapse in maximal strain or evacuation can be made from the perpendicular distance from the reference lines discussed earlier. These are namely the posteroinferior aspect of the bladder base, the anterior cervical lip, or posterosuperior vaginal apex if a hysterectomy was performed, and the anterior aspect of the anorectal junction.

Abnormal position and contour of the viscera can also be observed in the form of outpouching of the rectum, sigmoid colon, small bowel, and urinary bladder as rectocele, sigmoidocele, enterocele, and cystocele (Figs. 5–8).

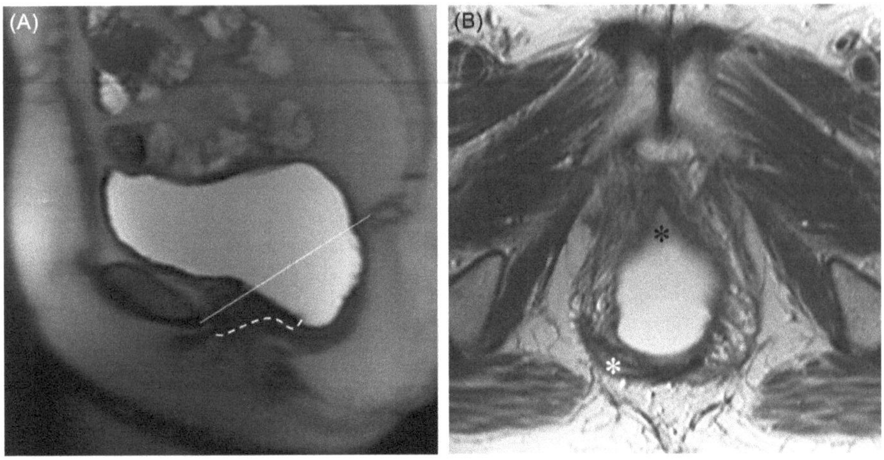

Figure 4 Large cystocele. (A) A 62-year-old patient with anal carcinoma underwent abdominopelvic resection and total hysterectomy. She subsequently presented with difficulty in voiding and increased frequency over several months. A follow up MRI shows a new cystocele far below the PCL *(solid line)*, with increased horizontal orientation of the urethra *(dotted line)*. (B) The beak of the urinary bladder is sometimes better seen in the axial view *(black asterisk)*. The vaginal vault *(white asterisk)* is displaced posteriorly due to the large cystocele.

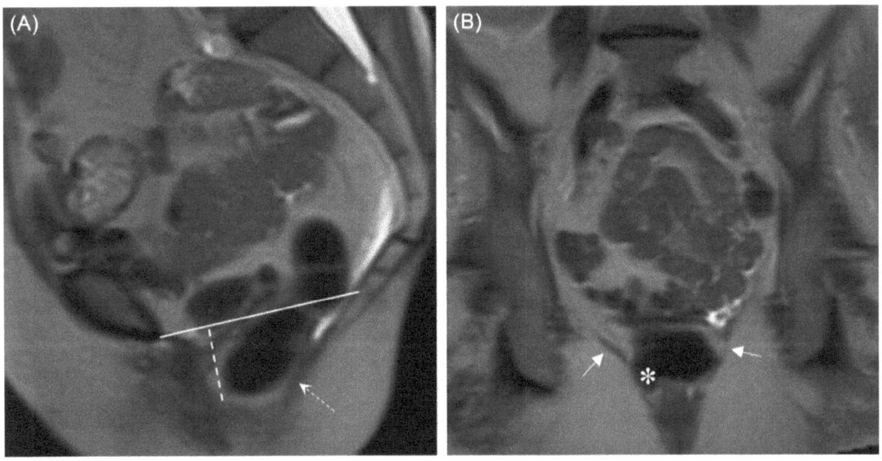

Figure 5 Rectal prolapse. (A) This 77-year-old patient underwent cystectomy and TAHBSO for carcinoma of the urinary bladder. There is abnormal position *(dotted line)* of the anorectal junction, 3.5 cm inferior to the PCL *(solid line)* as well as increased caudal inclination of the levator plate *(dotted line with arrow)*. (B) A small rectocele is present on the right lateral aspect *(asterisk)* on the coronal T2W HASTE image. There is also loss of the upward-convex curve of the iliococcygeus muscle *(arrows)*, indicative of poor pelvic floor support in the posterior compartment.

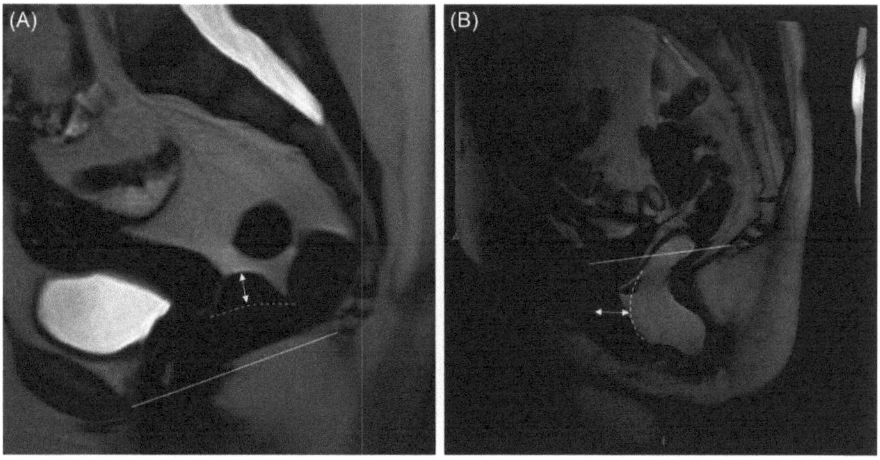

Figure 6 Anterior rectocele. Two patients respectively showing anterior rectoceles (A) above and (B) below the PCL *(solid line)*, with bulges *(line with arrowheads)* projecting out of the expected contour *(dotted lines)* of the rectum. This is often better visualized when the rectum is sufficiently distended with air or with intraluminal gel, especially in the sitting position.

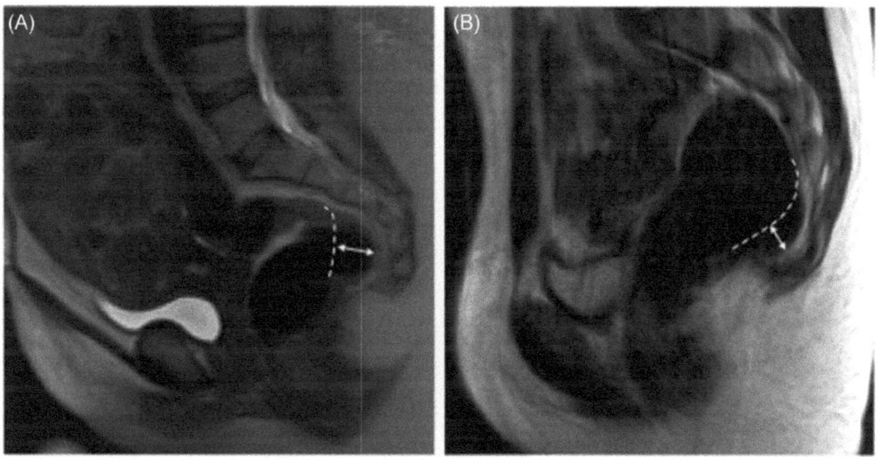

Figure 7 Posterior rectocele. A 52-year-old patient with numerous uterine leiomyomata presents with persistent constipation. (A) A moderate posterior rectocele measuring 2.4 cm (*line with arrowheads*) beyond the expected contour (*dotted line*) is evident, consistent with her symptoms. (B) A small bulge is seen in an asymptomatic patient, illustrating the need to correlate the imaging findings with clinical presentation.

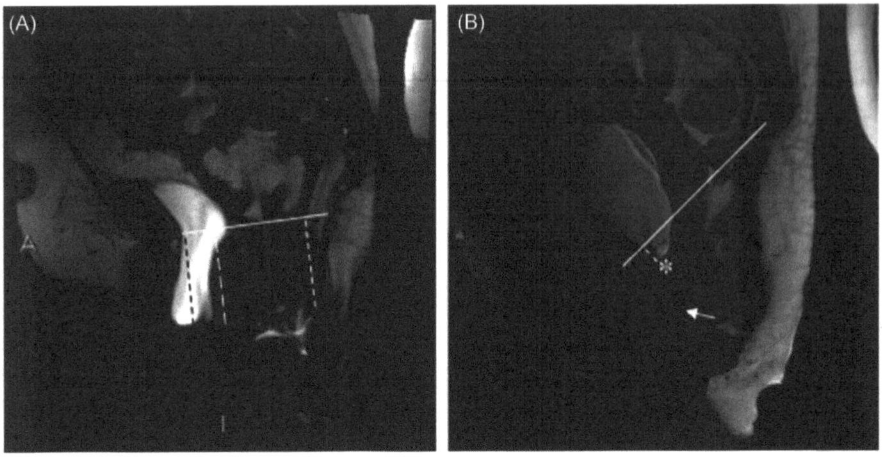

Figure 8 Abnormal descent in multiple organs. (A) This patient shows complete prolapse of all three compartments, with a large cystocele (*black dotted line*), complete uterine procidentia (*white dashed line*), and an abnormally low anorectal junction (*white dotted line*), displaced and compressed by the retroverted uterus. (B) A full thickness intussusception (*arrow*) is demonstrated in another patient, also showing an abnormally low position of the bladder neck (*asterisk*) below the PCL (*white solid line*).

STRESS INCONTINENCE

The junction between the urinary bladder and the urethra is usually located at the upper third or half of the pubic tubercle. While a narrow beaking of the bladder neck and urethral funneling are common, a cystocele is usually obvious as a wide protrusion and can be easily identified as the urine shows high T2-weighted signal (Fig. 4).

At rest, the urethrovesical junction rests at least 2 cm above the PCL. The urethra is usually perpendicular or angles slightly toward the sacrum. With normal void, the urethrovesical junction does extend below the PCL. The urethra remains vertical or minimally displaced toward the symphysis. Rotation toward the symphysis at the level of the urethrovesical junction indicates a hypermobile urethra.

Abnormal descent of the pelvic floor can cause the urethra to be rotated into a more horizontal orientation (Fig. 6) when seen on a sagittal MRI image. This kinking effect has the potential to block urination even during straining. The periurethral, paraurethral, and pubourethral ligaments can also be distorted and are best seen on the axial images (Fig. 4).

BLADDER AND UTERINE PROLAPSE

The apical, mid, and lower thirds of the vagina are respectively supported by the parametrium/paracolpium, arcus tendineus, and perineal body, the latter of which can be identified in the midline. The anterior vaginal wall is also closely apposed with the urethra, hence concurrent pathology is common (Fig. 3).

The vagina typically shows a butterfly or H-shaped configuration (Fig. 2), which can be distorted or lost in an abnormal pelvic floor descent, implying loss of urethral support [15–17].

At strain, the normal uterovaginal junction is usually more than 1 cm above the PCL. Approximation toward the PCL or descent beyond the PCL, along with posterior displacement of vaginal fornices, suggests poor pelvic floor support.

The normally acute anorectal angle indicates the integrity of the puborectalis muscle body and pubic attachments. This angle is widened associated with an elongated H line and M line in pelvic floor dysfunction. In addition, the rectovaginal space widen in pelvic floor dysfunction, along with widening of the levator hiatus and lateral bulging of the puborectalis. On axial images, the right puborectalis muscle may appear thinner than the left, attributed to effects of variable chemical shift artifact [18].

RECTAL PROLAPSE AND CONSTIPATION

Abnormal bulges of the bowel contour, often manifesting as anterior rectocele (Fig. 6), can be observed and can be associated with chronic constipation. Enteroceles arising from the small bowel extend into the pouch of Douglas. Rectoceles are classified radiologically according to the excess anteroposterior distance from the rectal wall with respect to the expected boundary. Mild, moderate, or severe rectoceles are indicated by measurements of up to 2 cm, less than 4 cm, and 4 cm or more. Residual material within the rectocele following evacuation is often an indication for a surgical repair in symptomatic women (Figs. 5–8) [19].

Increased caudal inclination of levator plate [20] (parallel to the PCL in normal individuals) of more than 10 degrees on evacuation [16], is usually indicative of deficiency of the support in the posterior compartment (Figs. 7 and 8), and accompanies weakening of the iliococcygeus and puborectalis muscle.

Excellent soft tissue contrast of MRI, upon adequate evacuation phase, and with adequate strain, allows identification of intussusception involving only the mucosa versus that involving the entire wall (Fig. 8) [21], which can aid in the decision making for mucosal excision versus rectopexy [22].

Evacuation of contrast can be limited by the supine position and may require episodes of Valsalva. True pelvic dyssnergia with paradoxical contraction of the puborectalis muscle can sometimes be differentiated [23] from difficulty defecating in the supine position, by identifying the lack of obtuse opening (usually by more than 15 degrees between rest and defecation) [13] of the anorectal angle during evacuation attempts.

Other than the functional assessment of movement of pelvic viscera, MRI offers the opportunity to directly assess the muscles of the pelvic floor. On coronal imaging, the levator ani muscles normally show an upward-convex bulge (Fig. 2), which indicates the integrity of support provided by the iliococcygeus muscle. Axial imaging, on the other hand, best demonstrates the integrity of the puborectalis muscle (Fig. 2).

Three-dimensional MRI studies of the morphologic features of the levator ani in different grades of prolapse allow evolving understanding of additional parameters [24]. For example, the levator-symphysis gap, measuring the anteroposterior distance of the gap between the pubic symphysis and the anterior-most aspect of the levator ani sling, is assessed on the axial view. This gap is a marker for the attachment of the puborectalis muscle to the pubis, which helps to maintain the function of the urethra, vagina, and external anal sphincter, and when widened, is associated with increasing severity of prolapse [25].

In addition, muscle tears can be evaluated by assessing for morphological integrity such as fraying and reduction in thickness, features of edema with increased T2-weighted signal inhomogeneity, and resultant loss of function with asymmetry (Fig. 4).

CONCLUSION

Static and dynamic MRI is a unique diagnostic tool allowing identification of pelvic muscle, viscera, and fascial condensations. Movement of the floor with strain demonstrates the relationships among the structures. In addition to surgical planning, this geometric assessment may be used to drive computational analysis of extreme biomechanical events, such as childbirth injuries, which cannot ethically be studied experimentally in living women.

REFERENCES

[1] M. Li, T. Jiang, P. Peng, X.Q. Yang, W.C. Wang, Association of compartment defects in anorectal and pelvic floor dysfunction with female outlet obstruction constipation (OOC) by dynamic MR defecography, Eur. Rev. Med. Pharmacol. Sci. 19 (8) (2015) 1407–1415.

[2] G.T. Bitti, G.M. Argiolas, N. Ballicu, E. Caddeo, M. Cecconi, G. Demurtas, G. Matta, M.T. Peltz, S. Secci, P. Siotto, Pelvic floor failure: MR imaging evaluation of anatomic and functional abnormalities, Radiographics 34 (2) (2014) 429–448, http://dx.doi.org/10.1148/rg.342125050.

[3] H.K. Pannu, M.C. Javitt, P. Glanc, P.R. Bhosale, M.G. Harisinghani, N.J. Khati, D.G. Mitchell, D. A. Nyberg, P.V. Pandharipande, T.D. Shipp, C.L. Siegel, L. Simpson, D.J. Wall, J.J. Wong-You-Cheong, ACR appropriateness criteria pelvic floor dysfunction, J. Am. Coll. Radiol. 12 (2) (2015) 134–142, http://dx.doi.org/10.1016/j.jacr.2014.10.021 (Epub 2014 Oct 31).

[4] L.S. Dvorkin, F. Hetzer, S.M. Scott, N.S. Williams, W. Gedroye, P.J. Lunniss, Open-magnet MR defecography compared with evacuation proctography in the diagnosis and management of patients with rectal intussusception, Colorectal Dis. 6 (2004) 45–53.

[5] R. Seynaeve, I. Billiet, P. Vossaert, P. Verleyen, A. Steegmans, MR imaging of the pelvic floor, JBR-BTR 89 (2006) 182–189.

[6] N. Tasali, R. Cubuk, O. Sinanoğlu, K. Sahin, B. Saydam, MRI in stress urinary incontinence: endovaginal MRI with an intracavitary coil and dynamic pelvic MRI, Urol. J. 9 (1) (2012) 397–404.

[7] C.A. Woodfield, S. Krishnamoorthy, B.S. Hampton, J.M. Brody, Imaging pelvic floor disorders: trend toward comprehensive MRI, AJR Am. J. Roentgenol. 194 (6) (2010) 1640–1649.

[8] Y.M. Law, J.R. Fielding, MRI of pelvic floor dysfunction: review, AJR Am. J. Roentgenol. 191 (Suppl. 6) (2008) S45–S53.
[9] E.M. Hecht, V.S. Lee, T.P. Tanpitukpongse, et al., MRI of pelvic floor dysfunction: dynamic true fast imaging with steady-state precession versus HASTE, AJR Am. J. Roentgenol. 191 (2008) 352–358.
[10] A. Yang, J.L. Mostwin, N.B. Rosenheim, E.A. Zerhouni, Pelvic floor descent in women: dynamic evaluation with fast MR imaging and cinematic display, Radiology 179 (1991) 25–33.
[11] C.V. Comiter, S.P. Vasavada, Z.L. Barbaric, A.E. Gousse, S. Raz, Grading pelvic floor prolapse and pelvic floor relaxation using dynamic magnetic resonance imaging, Urology 54 (1999) 454–457.
[12] R.B. Thapar, R.V. Patankar, R.D. Kamat, R.R. Thapar, V. Chemburkar, MR defecography for obstructed defecation syndrome, Indian J. Radiol. Imaging 25 (1) (2015) 25–30, http://dx.doi.org/10.4103/0971-3026.
[13] M.C. Colaiacomo, G. Masselli, E. Polettini, S. Lanciotti, E. Casciani, L. Bertini, G. Gualdi, Dynamic MR imaging of the pelvic floor: a pictorial review, Radiographics 29 (3) (2009) e35.
[14] C.A. Woodfield, B.S. Hampton, V. Sung, J.M. Brody, Magnetic resonance imaging of pelvic organ prolapse: comparing pubococcygeal and midpubic lines with clinical staging, Int. Urogynecol. J. Pelvic Floor Dysfunct. 20 (2009) 695–701.
[15] A.A. Tillack, B.N. Joe, B.M. Yeh, S.L. Jun, J. Kornak, S. Zhao, D. Deng, Vaginal shape at resting pelvic MRI: predictor of pelvic floor weakness? Clin. Imaging 39 (2) (2015) 285–288, http://dx.doi.org/10.1016/j.clinimag.2014.10.007 (Epub 2014 Oct 25).
[16] J.R. Fielding, MR imaging of pelvic floor relaxation, Radiol. Clin. N. Am. 41 (2003) 747–756.
[17] J. Stoker, S. Halligan, C.I. Bartram, Pelvic floor imaging, Radiology 218 (2001) 621–641.
[18] L. Boyadzhyan, S.S. Raman, S. Raz, Role of static and dynamic MR imaging in surgical pelvic floor dysfunction, Radiographics 28 (4) (2008) 949–967, http://dx.doi.org/10.1148/rg.284075139.
[19] C.W. Hicks, M. Weinstein, M. Wakamatsu, L. Savitt, S. Pulliam, L. Bordeianou, In patients with rectoceles and obstructed defecation syndrome, surgery should be the option of last resort, Surgery 155 (4) (2014) 659–667, http://dx.doi.org/10.1016/j.surg.2013.11.013 (Epub 2013 Dec 5).
[20] L. Hoyte, L. Schierlitz, K. Zou, G. Flesh, J.R. Fielding, Two- and 3-dimensional MRI comparison of levator ani structure, volume, and integrity in women with stress incontinence and prolapse, Am. J. Obstet. Gynecol. 185 (1) (2001) 11–19.
[21] M. Flusberg, V.A. Sahni, S.M. Erturk, K.J. Mortele, Dynamic MR defecography: assessment of the usefulness of the defecation phase, AJR Am. J. Roentgenol. 196 (4) (2011) W394–W399.
[22] J. Tsiaoussis, E. Chrysos, M. Glynos, J.S. Vassilakis, E. Xynos, Pathophysiology and treatment of anterior rectal mucosal prolapse syndrome, Br. J. Surg. 85 (1998) 1699–1702.
[23] K.J. Mortele, J. Fairhurst, Dynamic MR defecography of the posterior compartment: indications, techniques and MRI features, Eur. J. Radiol. 61 (2007) 462–472.
[24] L. Ping, C. Ruolan, C. Chunlin, H. Lu, G. Chuanjia, C. Lan, P. Cheng, W. Jun, L. Kedan, L. Xuan, W. Jianping, R. Daokun, T. Huanqing, T. Lei, Z. Shizhen, Anatomical characteristics of the pelvic floor muscles in young nulliparous women based on three-dimensional MRI, Zhonghua Fu Chan Ke Za Zhi 49 (5) (2014) 336–340.
[25] K. Singh, M. Jakab, W.M. Reid, L.A. Berger, L. Hoyte, Three-dimensional magnetic resonance imaging assessment of levator ani morphologic features in different grades of prolapse, Am. J. Obstet. Gynecol. 188 (4) (2003) 910–915.

CHAPTER FOURTEEN

Geometric Representation of the Pelvic Organs

L. Hoyte*, C. Lisle†,‡

*Director, Urogynecology and Female Pelvic Reconstructive Surgery, USF College of Medicine, Tampa, FL, United States of America; Medical Director, Urogynecology and Robotic Surgery, Tampa General Hospital, Tampa, FL, United States of America
†KnowledgeVis, LLC, Maitland, FL, United States of America
‡University of Central Florida, Orlando, FL, United States of America

MR IMAGE ACQUISITION

To be available for segmentation, a magnetic resonance imaging (MRI) source image data set must be acquired. Modern MR image acquisition is usually accomplished by placing the patient in the MRI scanner, where the interesting body part is surrounded by a high field magnet, with a receiving "coil" placed either around the body (eg, torso phased array coil) or inside a body cavity (eg, endovaginal coil).

In the case of MRI, each "slice" in the image stack is a matrix of 512 × 512 pixels each able to represent at least 256 shades of gray. Usually, proton density (or T2 weighted) MRI sequences are used as the source images for segmentation because of the excellent tissue definition and relatively high resolution of the images. Magnetic resonance images are normally acquired in one or more of three planes — axial, sagittal, or coronal. Depending on the details of the MRI sequence used, pixel dimensions usually range from 0.2 to 1.0 mm in the plane of each two-dimensional (2-D) "slice," and the slice "thickness" will range from 1 to 5 mm. The smaller the slice thickness, the more data is captured from the subject being scanned, increasing the resolution of the scanned image. However, this increased resolution comes at the expense of increased scanning time, and/or increased cost of the MRI machine, due to the need for a stronger magnet. In the case of MRI, the scan dimensions (field of view) are usually adequate to span the width (transverse dimension) and height (anterio-posterior dimension) of the vast majority of female pelvises.

The "in-plane" scan dimensions are defined by the pixel resolution (512 × 512 pixels), and the "height" of the image stack is determined by the slice thickness and the number of slices obtained. As an example, for a typical pelvic axial scan area covering 15 cm, at a slice thickness of 5 mm, there will be 30 slices of images in the MRI data set. An organ or tissue of interest will need to be identified and outlined on each slice in which it appears, to completely represent it by segmentation. The desired result is to establish the 3-D boundary enclosing the interesting portions of each tissue or organ.

An example of a T2-weighted MRI protocol in use at our center is given in Table 1. This sequence is suitable for segmentation and reconstruction of the pelvic organs.

Table 1 GE 3 T MR scanner model HDXT — scanning sequence parameters — torso phased array coil

Sequence	Plane	TR	TE	Slice thickness/ gap (mm)	Field of view (cm)	Matrix	Remarks
Localizer	Sagittal, axial, coronal	Min	100	8	40	256 × 128	
FRSE-XL	Sagittal	2000	85	3/0	26	352 × 192	100% phase oversampling
FRSE-XL	Axial	4900	85	3/0	26	352 × 192	100% phase oversampling
FRSE-XL	Coronal	4900	85	3/0	26	352 × 192	100% phase oversampling

IDENTIFYING THE PELVIC ORGANS

Segmentation is the process of outlining interesting organs and tissues on each slice of the MRI data set. This requires a thorough understanding of the MRI anatomy such that the correct tissues can be identified on each image slice.

The attached figures show the grayscale MR images (Figs. 1 and 2), without and with selected segmented organs, respectively. The axial, sagittal, and coronal plane images are shown along the right and in 3-D perspective along the left of the user interface.

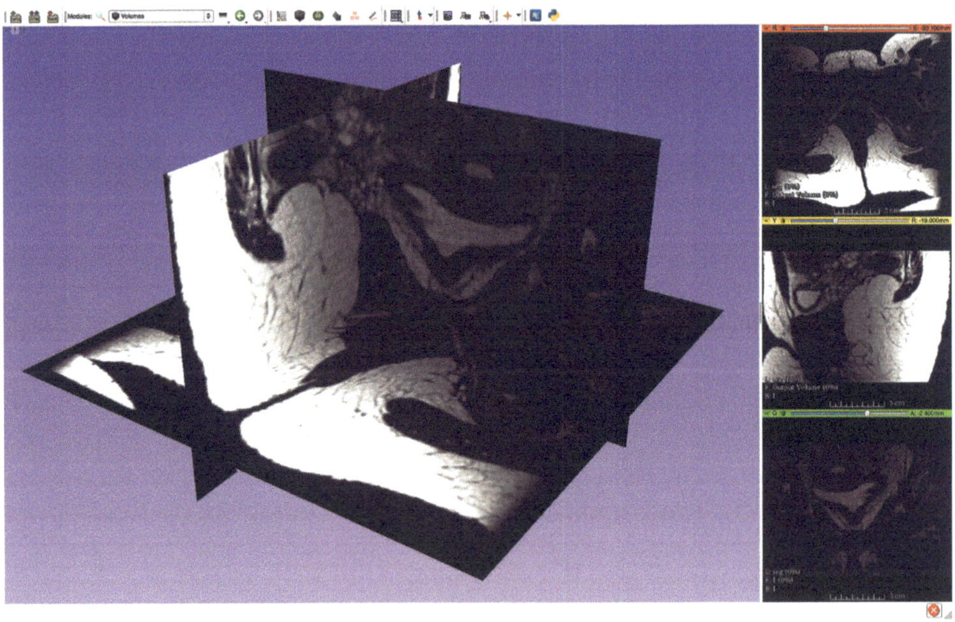

Figure 1 Planar views of the female pelvic floor anatomy. Left image: the three-plane view with intersecting axial, coronal, and sagittal images. Right images (top to bottom): axial, sagittal, and coronal plane views.

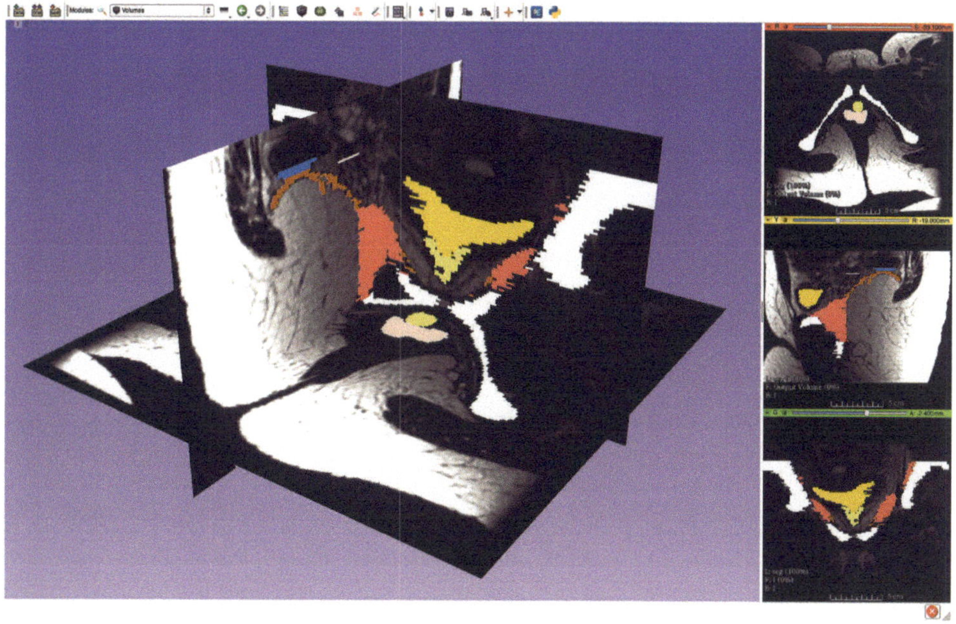

Figure 2 Planar views of the female pelvic anatomy with segmented tissues shaded.

Many software applications exist for the purpose of performing segmentation on gray-scale images. At our center, we use the 3-D Slicer (www.slicer.org), a free public domain application, for viewing, segmenting, and 3-D rendering of radiologic image stacks [1]. The featured segmentations in this chapter were performed using the 3-D Slicer.

SEGMENTING THE PELVIC ORGANS

Segmentation can be achieved by manual methods, in which an individual segmenter manually outlines each organ of interest, on each relevant slice in an image data set. This requires that the segmenter have an intimate knowledge of the anatomy of the body region being studied. If the segmenter is working on one slice at a time, it is also challenging to maintain slice-by-slice consistency as the tissue boundaries move around normally from one slice to another. Fig. 3 shows an axial slice from a scan showing the hand-drawn segmentation regions for the bladder, levator, obturator, rectum, vagina, and pelvic bone.

TYPES OF IMAGE SEGMENTATION

Segmentation is the process of partitioning the MR scan into multiple segments, corresponding to tissue and organ structures. Manual segmentation, where an individual segmenter manually reviews each slide and outlines the boundaries of important tissues, is

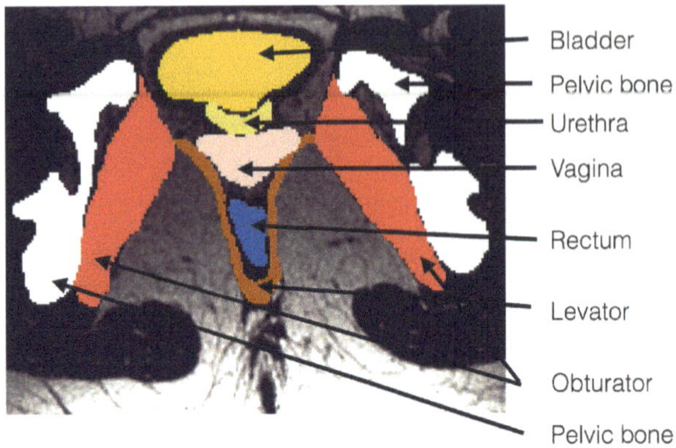

Figure 3 Organs obtained by manual segmentation.

the most commonly used type of segmentation, at the time of this writing. However, computer-assisted and automatic segmentation processes are also being increasingly used as these technologies mature. The different segmentation types are further described in the following paragraphs.

Manual Segmentation

Manual segmentation is traditionally performed one slide at a time, so the organ tracing operations must be repeated many times for a single pelvic data set. The exact number of repetitions depends on how many slices are in a particular MR acquisition and how many of these slices contain the image of the target organ. Initial interfaces for manual segmentation required the segmenter to paint within the region of the desired tissue to segment it. With this technique, it is required to visit each pixel inside a tissue of interest by dragging a "paint brush" over the desired areas in each slice of the volume. An example painting operation during segmentation of the bladder is shown partially completed in Fig. 4.

Because of the time-consuming nature of manual segmentation, various tools have been developed by different segmentation programs to allow the segmenter to more quickly mark pixels of interest. For example, a smart *wand* tool in 3-D Slicer extends out from its center location and automatically selects surrounding pixels with similar intensity [2]. The left panel in Fig. 5 shows the effect of a single click of the wand tool selecting an entire group of pixels, yet stopping within a tissue boundary because the pixel intensity values changed noticeably at the boundary. Alternatively, a *level-tracing* tool, shown in the center panel of Fig. 5, demonstrates the segmenter positioning the cursor and reviewing the outline contour suggested automatically by 3-D Slicer. Finally, the right panel shows the selection result after a single click of the *level-tracing* tool.

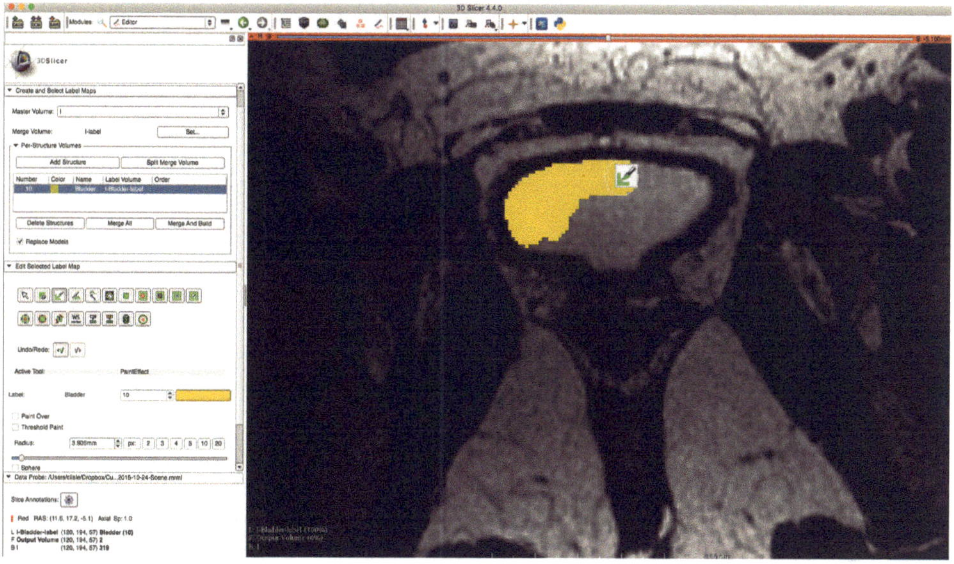

Figure 4 Manual segmentation by painting a label.

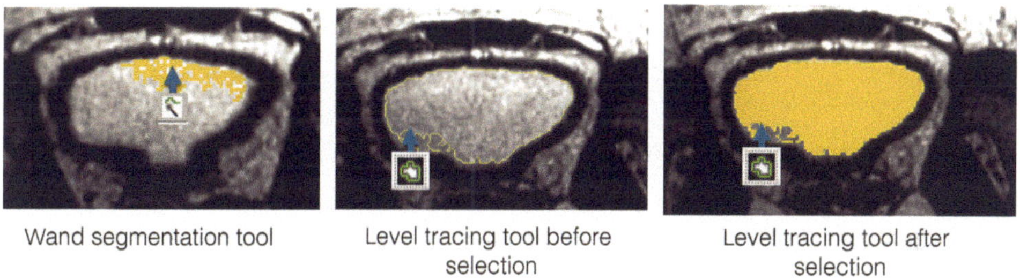

Figure 5 Smart manual selection tools in action.

The Labelmap Volume

When all desired tissues have been identified through marking contained pixels on each slice, the resulting collection of geometric data is called a *labelmap*. A labelmap is a volume of the same dimensions as the original MR acquisition, but its volume elements, referred to as *voxels*, encode the locations of the organs. Each voxel in the labelmap volume is assigned either a background value or an index corresponding to a segmented region (Fig. 6). The color tissues in the rendered labelmap correspond to different segmentation labels. In practice, the labelmap begins as an empty volume, with all voxels assigned to the background value, and then voxel values are assigned as segmentation is performed through either manual or automatic means.

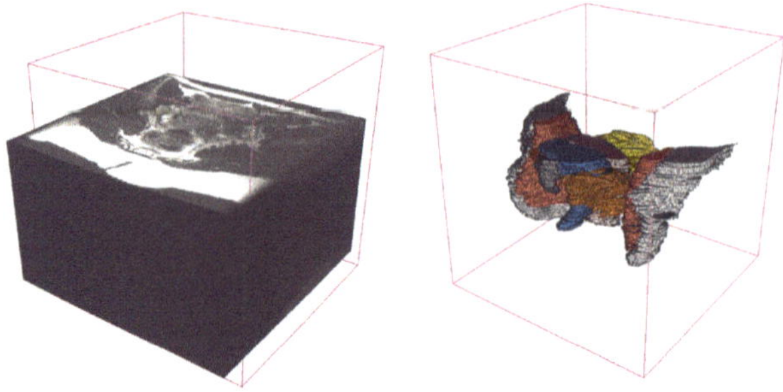

Figure 6 MR grayscale stack (volume) and labelmap stack — the labelmap stack is formed by "stacking" the labelmap slices together in order. Areas not explicitly segmented are given the "background" value, corresponding to clear spaces in the labelmap.

Limitations of Manual Segmentation

Manual segmentation allows an expert to fully identify and label tissues of interest. However, this method can suffer from lower interobserver and intraobserver repeatability than may be considered acceptable in certain applications [3]. Lack of operator skill can be a limiting factor to optimal results. Furthermore, manual segmentation can be tedious, often requiring many hours to obtain a labelmap of the female pelvic organs, thereby limiting the number of data sets that can be comparatively analyzed in any given time period.

Semiautomatic Segmentation

Semiautomatic segmentation is the process where the locations of features of interest within a source volume are determined through interaction between a computer algorithm analyzing the voxel intensities and an expert segmenter guiding the algorithm. In semiautomatic segmentation, the segmenter uses either 2-D or 3-D image analysis algorithms to identify regions of interest. There are many different image analysis algorithms used for automatic and semiautomatic segmentation, and as this is a challenging area in computer science; new algorithms are constantly being developed. For our purposes, we will group the majority of automated segmentation algorithms into three classes: intensity-based segmentation, region-growing segmentation, and template-driven segmentation.

Intensity-based segmentation algorithms directly examine voxels and select or discard them based on value (or shade of gray — recall each pixel is assigned one of 256 or more shades of gray by the MRI scanner during acquisition), such as selecting all voxels with value over a given shade threshold or between a given lower and upper bound. These algorithms do not generally perform adequately for uses in pelvic floor segmentation so are not further discussed.

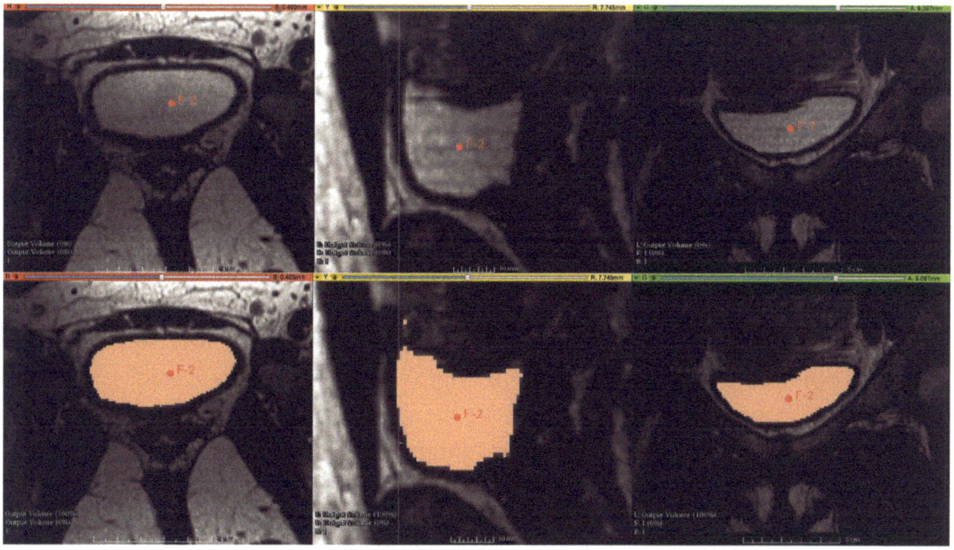

Figure 7 Bladder segmentation by 3-D region growing. Upper panel (left to right) grayscale axial, sagittal, and coronal images. The urine in the bladder is lightly colored in the upper images, and segmented peach colored in the lower images. The "seed" points F-1 and F-2 were supplied by the segmenter. The dark boundary around the bladder is the bladder wall.

One of the simplest region-growing examples is seeded region-growing, where seed point locations inside a tissue are hand-selected, and then the algorithm automatically expands from the seed points, trying to fill regions of similar intensity, but stop at any boundaries (where values are changing rapidly between voxels). This approach can fill relatively homogeneous regions (eg, a bladder filled with urine or a contrast-filled rectum) well, but is also subject to "leaking" into adjacent tissues if the voxel intensities do not change sharply enough at the tissue boundaries. In Fig. 7, a bladder is segmented using a 3-D region-growing algorithm. In the top set of axial views, the segmenter has placed seed points (labeled as F-1 and F-2) inside the tissue to be segmented. The bottom row demonstrates how the algorithm has grown out from the seed points, yet stayed within the tissue bounds. Recent progress in semiautomatic segmentation algorithms, such as the *GrowCut* algorithm [4], has addressed limitations in traditional region-growing methods and show promise to somewhat reduce the effort required for semiautomatic segmentation in the future.

Semiautomatic segmentation uses the light intensity of a given tissue as a guide for segmentation algorithms. This is a good option for tissues that have excellent contrast with the surrounding tissues; for example, the fluid-distended bladder or the rectum filled with contrast. However, as many pelvic tissues have overlapping light intensities, this method is not optimally reliable for segmenting most other pelvic floor organs and tissues. An exception to this is the vagina, which can be opacified with a contrast gel or paste in order to better define the vaginal anatomy. In fact, semiautomatic segmentation often requires intensive human intervention to guide the process, and this guidance needs to be repeated for every data set to be segmented, limiting the numbers of data sets that can be comparatively analyzed in a reasonable period of time.

The final class of segmentation algorithms, template-driven segmentation, is an evolving alternative to earlier segmentation approaches. In this technique, the algorithm learns how to segment tissues in a body region from comparing multiple matched pairs of grayscale and segmented images [5]. When presented with an unknown data set from a known body region, the automatic segmentation algorithm uses its learning from the template/grayscale pairs to estimate a segmentation of the new data set. An expert segmenter can then guide the algorithm to improve the segmentation results as needed. This "guiding" process will not only improve the segmentation results, but will also add a new template to the algorithm's library for future use. The advantage of this approach is the potential for segmenting large numbers of data sets relatively quickly, thus opening the door for comparative analysis of large populations of images. Automatic segmentation algorithms are discussed in greater detail in a later chapter.

MODEL GEOMETRY

After the segmentation step is completed, a labelmap volume containing all identified pelvic floor tissues is the resulting structure. Some geometric analysis operations can work directly off the labelmap, others require a polygonal surface to be generated that encloses the voxels. For 3-D rendering or visualization of the organs, a surface is generally needed. The surfaces generated from segmentations are referred to as *models*.

Polygonal surface models are created by a voxel to surface conversion algorithm. One of the earliest examples of this algorithm is the marching cubes algorithm [6]. Marching cubes and similar algorithms produce *isosurfaces*. These are connected 3-D surfaces that trace the boundary between regions [7]. Depending on the type of geometric surface needed, there are multiple techniques to produce the isosurface. Generally a surface is first created by an isosurface algorithm, and then smoothed to reduce the "stair-step" patterns that come from exactly contouring voxel boundaries. Both smoothed and unsmoothed surfaces produced from the region-growing bladder segmentation (from Fig. 7) are shown in Fig. 8. Note how the "stair-step" patterns are visible on the unsmoothed model.

Close inspection of the surfaces show they are composed entirely of connected triangles. Higher-order polygons (polygons with more than three sides) are generally not used in these surfaces because arbitrary boundaries can always be fit using triangles, whereas it is not always possible to fit any arbitrary shape using higher-order polygons. The most common triangular surface representations are either regular grids of connected triangles or Delaunay triangulation, where larger triangles are used to represent areas with

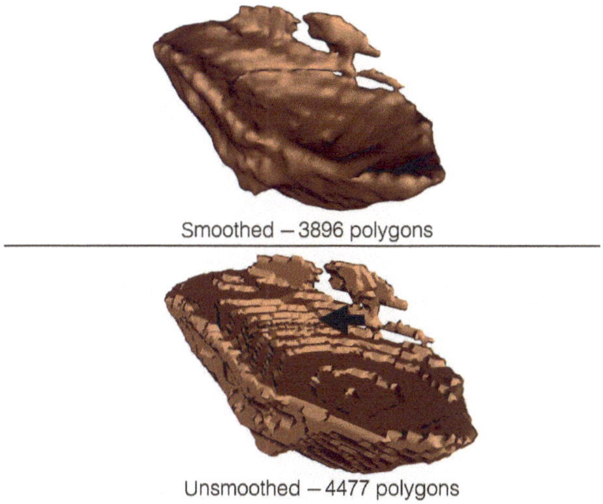

Figure 8 Tissue surface models of the bladder from Fig. 7. Upper panel: smoothed 3-D surface. Lower panel: unsmoothed 3-D surface. Note the sharp surface discontinuities, known as "stair stepping" *(black arrow)*.

less curvature [8]. Delaunay triangulations are popular because they require fewer triangles, even though there is some positional error induced during the relaxation process to create larger triangles. The smoothed model from Fig. 8 uses a Delaunay-style triangulation — with larger triangles used in the smoother areas of the model.

As is expected, relaxing the triangulated mesh can result in some localized shape changes. However, in the case of pelvic floor tissues, smoothed models better represent the true shape of the anatomy because smoothing operations remove error induced by the limited image resolution of the MR scanner. The execution behavior of the smoothing algorithm and the isosurface algorithm can be modified through parameters, which must be adjusted to produce models optimal for different analysis tasks. For example, rendering for visual review can be accomplished successfully using lower polygon counts than what would be required for precise biomechanical contact analysis.

The computer graphics community has repeatedly improved polygon reduction and polygon smoothing algorithms over the past 30 years. Early polygon reduction work [9] has been adapted for the needs of today's medical and scientific analysis communities [10]. Exactly what geometric representation is required for different mechanical computations is now relatively well understood by engineering and biomechanics specialists. For example, in the case where the surface models are used for contact forces between organs, further processing will be needed to create internal structural information to support computational analytic simulation methods [11,12].

To complete a geometric model of the pelvic floor organs, a separate surface model is created for each organ, and these models are then combined into a multiorgan model data

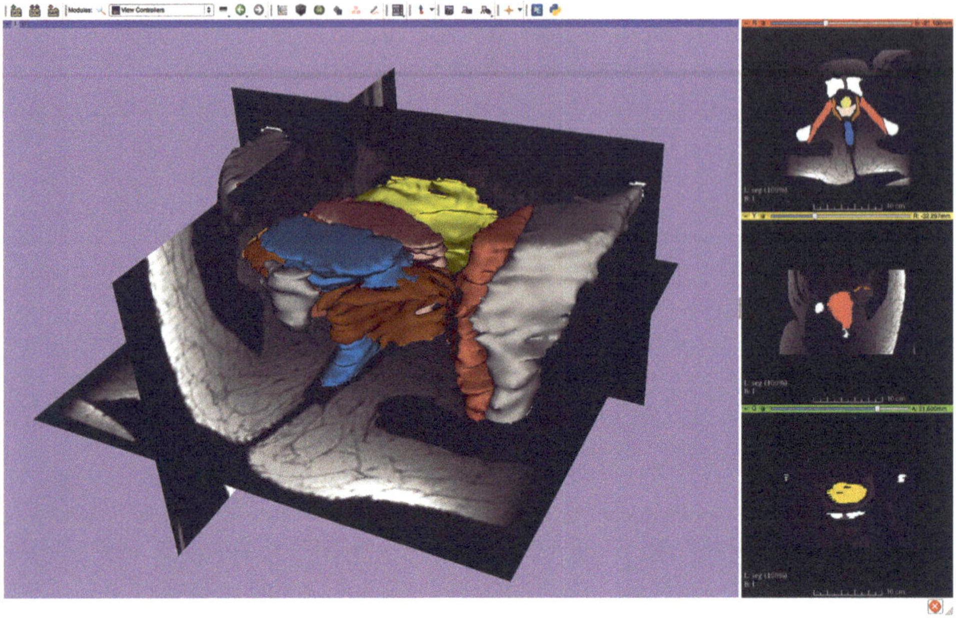

Figure 9 A complete pelvic floor geometric representation.

set. The final set of 3-D pelvic floor models are shown together, using the 3-D Slicer user interface along with the original MR imagery in Fig. 9.

BARRIERS TO OPTIMAL GEOMETRIC REPRESENTATION

There can be a number of reasons that a particular pelvic tissue geometric model could diverge from an accurate representation of the true anatomy. First, the MR imaging protocol can affect the amount of intensity contrast available for segmentation to be performed. This would make it difficult to differentiate the tissues from the images. In such a case, it would be difficult for either manual segmentation or semiautomatic segmentation to be conducted correctly when the source data set has insufficient contrast. Additionally, the gantry system is in motion during the acquisition of an MR series. Positional error, due to mechanical shift, or patient movement, can occur between the slices of an image stack. If this occurs, one or more of the slices will be shifted or misregistered, with respect to the others. If the images are not adequately registered with each other across the entire image sequence, the quality of the segmentation result will be degraded. In this section, we will present several examples of how errors can enter into the segmentation process.

With most MR protocols, the interslice distance along the axis orthogonal to the scan plane can be substantially greater than the in-plane pixel dimensions. It is not unusual for

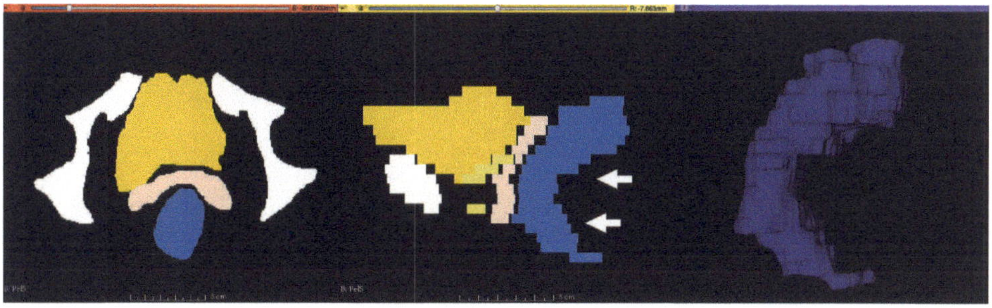

Figure 10 The effect of low orthogonal resolution due to high interslice distances. Left to right: axial, sagittal views, and 3-D rendering of the rectum. *White*, bones; *Yellow*, bladder; *Pink*, vagina; *Blue*, rectum. "Stair stepping" is shown by the *white arrowheads*.

the axial distance to be five times greater than the in-plane pixel dimensions. This compromises the accuracy of the segmentation as there is a pronounced stair-step effect. This is demonstrated in Fig. 10, where the in-plane resolution in the axial plane (the left panel) offers fine detail at 0.67 mm/pixel, yet the interslice distance is 5.0 mm/pixel. Note the stair-step effect in the sagittal view (the center panel — white arrowhead) and in the resulting 3-D rectum model (right panel, white arrowhead). For biomechanics analysis applications, modifying the MR acquisition protocol toward uniform voxel size in all three planes will result in more optimal geometric results.

Reliability/Repeatability of Manual Segmentations

As the reference standard for correct segmentation of tissues today is still manual segmentation, expert segmenters with deep knowledge of pelvic anatomy are needed to produce good quality segmentations. Knowledge of the pelvic floor anatomy is necessary to correctly delineate tissue boundaries when structures are not obvious in the source imagery.

Even with expert segmenters, reproducibility remains challenging. Interrater and intrarater reproducibility was examined for 20 pelvic MRI data sets, using three expert segmenters, each of whom segmented out the pelvic organs on three separate occasions, in random order [3].

Low sensitivity was noted for all organs with the exception of the bladder, but the positive predictive value (PPV) was high. The high PPV meant that the segmenters agreed on the location of the organs, and the low sensitivity indicated that they disagreed on the location of their boundaries. This finding is exemplified in Fig. 11 for structures with high and low agreement, respectively.

Computing Volumes of Segmented Tissues

Since a data set's voxel sizes are established during the MR scan acquisition, it is straightforward to count voxels assigned each separate label value in the labelmap (remember, the

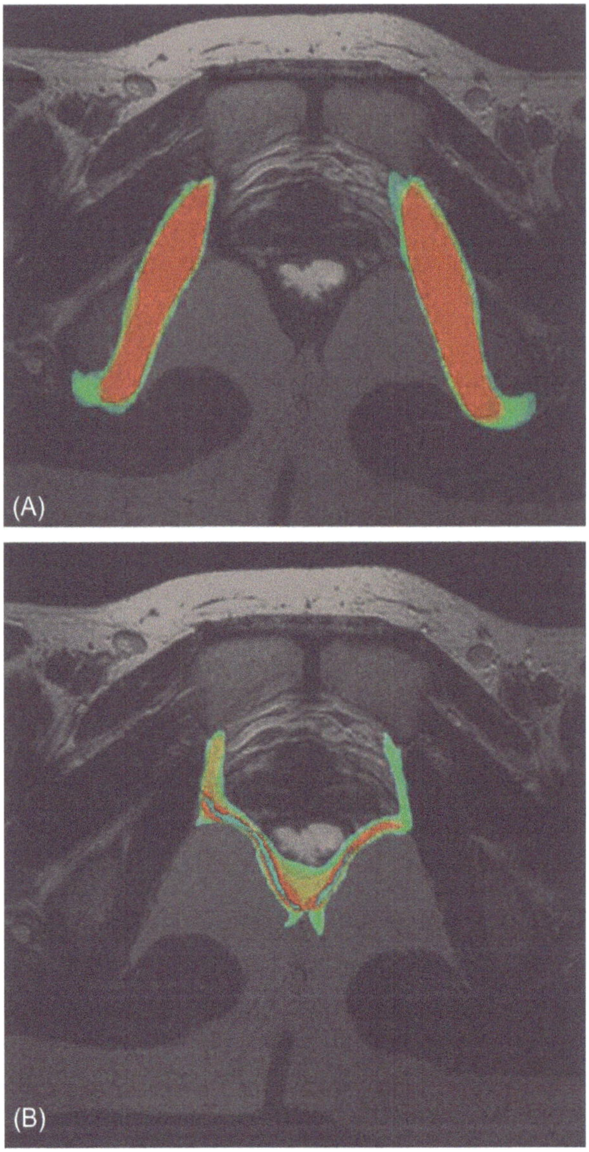

Figure 11 Segmentation reproducibility across experts. The outcome for an organ with high agreement (obturator internus) is shown in the upper image (A), and an organ with low agreement (levator ani) is shown in the lower image (B). For the slice shown, areas in *red* indicate complete agreement among all nine segmentations, and *orange* indicates a region of less agreement. Decreasing agreement areas are colored in the order *yellow*, *green*, and *blue*, with *blue* being the lowest agreement.

Table 2 Organ volumes from a manually segmented MRI data set

Tissue name	Voxel count	Volume (cm^3)	Min voxel shade value	Max voxel shade value	Mean voxel shade value	Standard deviation of voxel shade values
Pelvic bone	214,577	130.97	4	344	94.97	33.51
Vagina	55,075	33.62	13	284	82.81	30.94
Obturator	155,912	95.16	0	373	59.62	29.42
Urethra	7967	4.86	14	210	66.52	22.10
Bladder	52,748	32.19	13	391	252.79	57.71
Rectum	43,860	26.77	12	344	89.16	31.32
Levator	34,478	21.04	10	394	105.59	55.74
Coccyx	4467	2.73	13	537	213.72	125.11
Symphysis	3668	2.24	13	167	55.52	22.05

The range of intensity (shade of gray) values for each tissue type shows that there is considerable overlap in shade intensity across the different tissue types.

labelmap has the same voxel dimensions as the source image) and calculate the volume sizes of each segmented organ. The resulting organ volume table for the data set used in the previous figures, using manual segmentation by an expert segmenter, is given in Table 2. Understanding the anatomy is important for proper segmentation, because it can be seen from the table that segmentation is not only about segmenting regions with similar voxel intensities into separate classes (several tissues here contain voxels of varying intensities), but about following appropriate tissue boundaries.

The statistics columns (voxel min, max, mean, and standard deviation) help us to further understand the voxels mapping to each tissue. The standard deviation gives us a measure of the type of intensity variation across each tissue type. Specifically, tissues with a lower standard deviation will have more uniform voxel intensities throughout the volume. A region with a lot of uniform intensity and a few *outliers* of different intensity will still exhibit a low standard deviation. Let's consider the segmentation of the symphysis in this data set as a further illustration (Fig. 11). The symphysis has low standard deviation, so the voxels are relatively uniform, with the mean value (55.52) noticeably lower than the average of the minimum and maximum value found (13 and 167, respectively). This tells us the tissue is largely composed of low-value, dark voxels, but there are a few high-intensity pixels present in the label as well. This is confirmed visually by examining one of the axial slices, shown in Fig. 12. In the slice shown, an area inside the symphysis label appears to be occupied by a region from an adjacent bone. This overlap is a source of the high-intensity pixels (some measured at value = 151, in a dark region where most values range from 30 to 60). The effect of the overlap can be observed, even aggregated, at the level of the tissue volume table. Therefore, even small segmentation errors can have a noticeable effect in volume calculations.

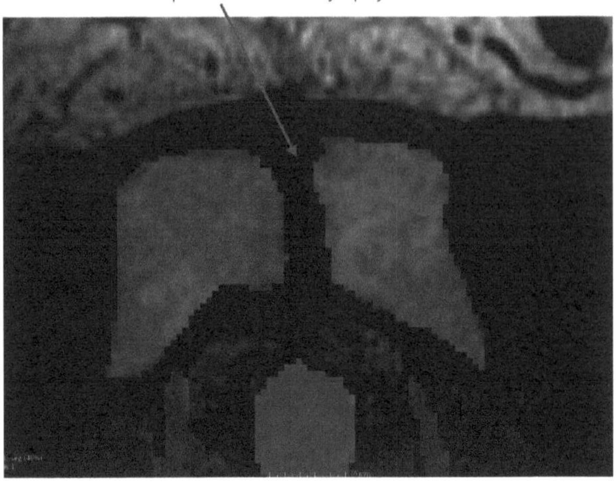

Figure 12 Symphysis label and bone overlap.

Additionally, to achieve optimal results, the visual interfaces for browsing and labeling anatomical structures must be accurate to ground truth, or it can also induce error. We illustrate this with a situation where a region-growing segmentation was applied to the same MR stack previously segmented by an expert segmenter. This algorithm failed to segment some organs, but was able to run successfully on the bladder. The algorithm's resulting volume was approximately 10% smaller when compared with manual segmentation. The quantitative comparison details are shown in Table 3.

To what do we attribute this difference? Close inspection of the manual and semiautomatic labelmaps (Fig. 13) shows the manual segmentation, which was performed using an older version of the segmentation tool, and seems to be drawn at a lower resolution. The manually selected regions are more "blocky" and have extra pixels outside the algorithmic boundary selected. This hypothesis is supported by the minimum voxel value results, where the manual segmentation included low-intensity pixels (note the intensity = 13). These were likely from the region outside the ideal tissue boundary.

Table 3 Manual versus automatic segmentation of the bladder — variations in segmented volumes

Tissue name	Voxel count	Organ volume (cm^3)	Min voxel value	Max voxel value	Mean voxel value	Standard deviation of values
(A) Expert segmenter's bladder	52,748	32.19	13	391	252.79	57.71
(B) Region-growing algorithm bladder	47,445	28.96	107	391	266.66	40.88
(A – B) Volume difference	5303	3.24				
Volume difference in percentage	10.05					

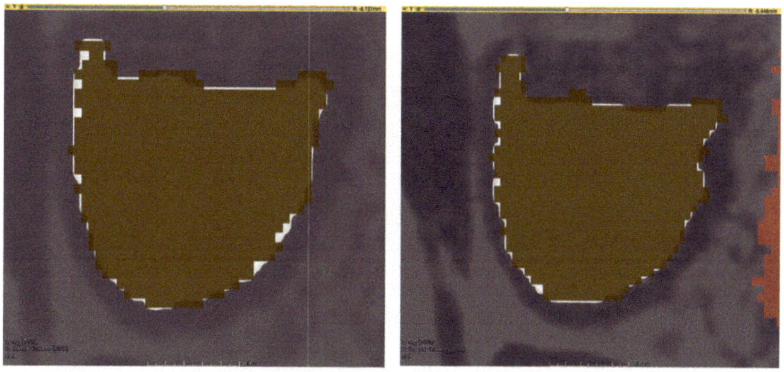

Manual segmentation — (foreground) green color
Automatic segmentation—(background) peach color

Figure 13 Accuracy of manual and semiautomatic segmentation.

It is our hypothesis that the segmenter was not presented with a full-resolution version of the volume at the time manual segmentation was performed. It so happens that the operation of the 2-D slice planes (the user interface element in 3-D Slicer where segmentation is performed) was changed between Slicer versions 3 and 4 and the resolution of the source imagery could sometimes be lost in Slicer version 3, whereas native resolution is always presented in Slicer version 4. This is based on operational memories of the tool history by the authors, but is substantiated by the measurement results presented. Clearly, a detailed understanding of the limitations of the user interface and presentation tools is required of the user, if high-fidelity segmentation results are to be obtained.

We present these illustrations to underscore the need for adequate algorithms, careful training of segmentation specialists, and appropriate tools for validation of segmentation fidelity, repeatability, and reliability of segmentation results.

Even with these factors in place, our community needs to be cognizant of the opportunity for errors to creep into the segmentation process and adversely affect the accuracy of geometric tissue models. This is an important point, because creation of the geometry is but a step in the process of creating models suitable for computational analysis. Errors in the geometric representation have the potential to be magnified during the model generation and analytic phases, thereby producing erroneous and unusable results. Future tools for validation will be critical to ensuring that the pathway from MRI image to computational analytic result is stable and produces appropriate results.

CONCLUSIONS

Labelmap generation is required for creating models suitable for visualization or computational analysis. Creating models strictly for visualization and learning is readily achievable using current methods. However, the generation of labelmaps of interesting tissues from source MRI data sets is a critical, and still evolving, part of creating computational analysis-ready geometric models suitable for pelvic floor biomechanical analysis.

Labelmap generation is currently complicated by three major issues: reproducibility, fidelity, and the time required to generate the labelmaps. A subsequent chapter will focus on the development of tools and algorithms for automating the process of segmentation, with pathways for continuously improving the reliability and fidelity of the results. By definition, these automated tools will dramatically decrease the requirement for manual interventions, thereby converting segmentation into a largely computational activity. In this scenario, the time to generate the segmentations will be governed by computational power, instead of the current manual bottleneck. This will increase the number of segmentations that can be generated in a given time, slowly paving the way for large-scale computational studies to be performed using pelvic floor imaging data.

REFERENCES

[1] R. Kikinis, S.D. Pieper, K.G. Vosburgh, 3D Slicer: a platform for subject-specific image analysis, visualization, and clinical support. in: F.A. Jolesz (Ed.), Intraoperative Imaging and Image-Guided Therapy, Springer, New York, 2013, pp. 277–289, http://dx.doi.org/10.1007/978-1-4614-7657-3_19.
[2] Slicer, 3D Slicer version 4.4 modules documentation tutorial, http://www.slicer.org/slicerWiki/index.php/Documentation/4.4/Modules/Editor#Tutorials, 2015 (accessed 23.10.15).
[3] L. Hoyte, et al., Segmentations of MRI images of the female pelvic floor, a study of inter- and intra-reader reliability. J. Magn. Reson. Imaging 33 (3) (2011) 684–691, http://dx.doi.org/10.1002/jmri.22478.
[4] A. Hamamci, N. Kucuk, K. Karaman, K. Engin, G. Unal, Tumor-cut: segmentation of brain tumors on contrast enhanced MR images for radiosurgery applications, IEEE Trans. Med. Imaging 31 (3) (2012) 790–804.
[5] O. Commowick, A. Akhondi-Asi, S.K. Warfield, Estimating a reference standard segmentation with spatially varying performance parameters: local MAP STAPLE. IEEE Trans. Med. Imaging 31 (8) (2012) 1593–1606, http://dx.doi.org/10.1109/TMI.2012.2197406 (Epub 2012 May 2).
[6] W.E. Lorensen, H.E. Cline, Marching cubes: a high resolution 3D surface construction algorithm. in: Proceedings of the 14th Annual Conference on Computer Graphics and Interactive Techniques — SIGGRAPH '87, 1987, http://dx.doi.org/10.1145/37401.37422.
[7] W. Schroeder, K.W. Martin, B. Lorensen, The Visualization Toolkit: An Object-Oriented Approach to 3D Graphics, fourth ed., Kitware, Inc., 2006. ISBN-13: 978-1930934191, ISBN-10: 193093419X.
[8] H. Edelsbrunner, E.P. Mucke, Three-dimensional alpha shapes, ACM Trans. Graph. 13 (1994) 43–72.
[9] W.J. Schroeder, J.A. Zarge, W.E. Lorensen, Decimation of triangle meshes. in: Proceedings of the 19th Annual Conference on Computer Graphics and Interactive Techniques — SIGGRAPH '92, 1992, http://dx.doi.org/10.1145/133994.134010.
[10] P.P. Pébay, D. Thompson, J. Shepherd, P. Knupp, C. Lisle, V.A. Magnotta, N.M. Grosland, New applications of the verdict library for standardized mesh verification pre, post, and end-to-end processing, in: Proceedings of the 16th International Meshing Roundtable, Springer, Berlin, Heidelberg, ISBN 978-3-540-75102-1, 2008, pp. 535–552.
[11] D.C. Simkins, A. Kumar, N. Collier, L.B. Whitenack, Geometry representation, modification and iterative design using RKEM, Comput. Methods Appl. Mech. Eng. 196 (41–44) (2007) 4304–4320.
[12] N.M. Grosland, K.H. Shivanna, V.A. Magnotta, N.A. Kallemeyn, N.A. Devries, S.C. Tadepalli, C.R. Lisle, IA-FEMesh: an open-source, interactive, multiblock approach to anatomic finite element model development, Comput. Methods Prog. Biomed. 94 (1) (2009) 96–107.

CHAPTER FIFTEEN

Segmentation of Images of the Pelvic Floor

S.K. Warfield, A. Akhondi-Asl
Department of Radiology, Boston, MA, United States of America

INTRODUCTION

Magnetic resonance imaging (MRI)-based three-dimensional (3D) reconstruction has been used successfully to evaluate the female pelvic floor muscles and tissues in women with and without pelvic floor dysfunction [1,2]. Furthermore, MRI-based 3D reconstructed models have been used to generate finite element and element-free computational models suitable for simulating vaginal childbirth [3,4], offering insight into risk factors for childbirth-related pelvic floor injury. The finite element method has also been used to evaluate anterior vaginal wall support and the mechanisms underlying cystocele formation [5]. Precise segmentation of pelvic MRIs is an important step in building the 3D reconstructed structural models suitable for computational analysis. Segmentation consists of outlining each organ or tissue of interest from the underlying grayscale images on each slice in the data set.

The 3D reconstructed models are currently generated from manually segmented label-maps, which currently require multiple hours of tedious manual segmentation to produce each reconstructed 3D model. This manual segmentation bottleneck limits the number of computational models that can be reliably produced in a timely manner, thereby limiting the number of study subjects available for the kind of statistical comparisons that can potentially lead to clinically meaningful insight. In addition, the structures of the female pelvic floor are very complex, which makes manual segmentation of the structures challenging, even for the expert raters. Moreover, manual segmentation suffers from inter- and intrarater variability [6]. This combination of low interrater repeatability and the long time required to segment the data sets limit the ability to perform meaningful computational analysis across large numbers of data sets. To address these limitations of manual segmentation, much attention has been focused on automatic segmentation techniques [7–10].

Reliable automatic segmentation has the potential to remove the manual segmentation bottleneck, which would dramatically increase the number of study simulation models available for analysis and comparison, allowing us to draw meaningful conclusions from observations involving large groups of study subjects. As an example, automatic segmentation can permit comparison of differences in the quantity and distribution on levator ani stretch during childbirth [3].

Algorithms that achieve segmentation by the alignment and fusion of multiple templates offer promise for automatic segmentation of pelvic MRI. In this class of segmentation techniques, a collection of templates is aligned to a target, and the segmentation of the target is inferred by a fusion algorithm that exploits similarities between the templates and the target. However, these algorithms have been limited by imperfections in the alignment of each template, consisting of an image and its segmentation, to the target, and by template segmentation errors. A number of algorithms sought to improve segmentation performance by combining image intensities and template labels as two independent sources of information, carrying out decision fusion through local intensity weighted voting schemes. This class of approach is a form of linear opinion pooling, but achieves unsatisfactory performance for this application [11]. We hypothesized that better decision fusion could be achieved by assessing the contribution of each template in comparison to a reference standard segmentation of the target image and developed a novel segmentation algorithm to enable automatic segmentation of MRI of the female pelvic floor. The algorithm achieves high performance by estimating and compensating for both imperfect registration of the templates to the target image and template segmentation inaccuracies. The algorithm is a generalization of the simultaneous truth and performance level estimation (STAPLE) algorithm [12] in which a reference segmentation is estimated and used to infer an optimal weighting for fusion of templates. A local image similarity measure is used to infer a local reliability weight, which contributes to the fusion through a novel logarithmic opinion pooling.

IMAGE ACQUISITION

Image segmentation begins with an MR grayscale data set, sometimes called source images. Ideally, the grayscale images need to be of sufficient resolution in plane and between slices to allow for discrimination of the different tissue types. The scanning protocol that allows for the best resolution tissue discrimination is the T2/proton density weighted imaging protocol, with in-plane resolution in the range of 0.5–1 mm, with slice thicknesses on the order of 2–3 mm. A representative axial plane image acquisition protocol for a 1.5 T magnet machine with a phased array surface coil is as follows: T2 Turbo SE axial images with TR/TE = 5000/132 ms, FOV = 200 cm, slice thickness = 3 mm/interleaved, no gap, flip angle = 180 degree. This, or a similar protocol, produces a stack of MR images suitable for segmentation.

MANUAL SEGMENTATION/TEMPLATE POOL GENERATION

Interactive manual segmentation is well known to demonstrate intraexpert and interexpert variability. Even disciplined experts and established segmentation protocols can generate inconsistent segmentations, especially in areas in which the image contrast is reduced, such as at the boundaries between structures of interest. Moreover, the repeated segmentations of the experts have regions that show local intrarater and interrater variability [12]. Therefore, repeated manual segmentations from multiple raters should be obtained for each data set. Then the manual segmentations of each scan should be fused to make a reference segmentation of the corresponding image data set. Local MAP STAPLE, a local implementation of maximum a posteriori formulation of STAPLE where the prior probabilities for the performance parameters are modeled by a beta distribution, is an excellent way to estimate a reference standard segmentation for each scan [13]. Local MAP STAPLE has the lowest coefficient of variation of Dice overlap [14,15], indicating it is more similar to each of the input segmentations. An example of this procedure is shown in Fig. 1, where nine structures are manually segmented three times, using three different raters, and utilized for the fusion using local MAP STAPLE.

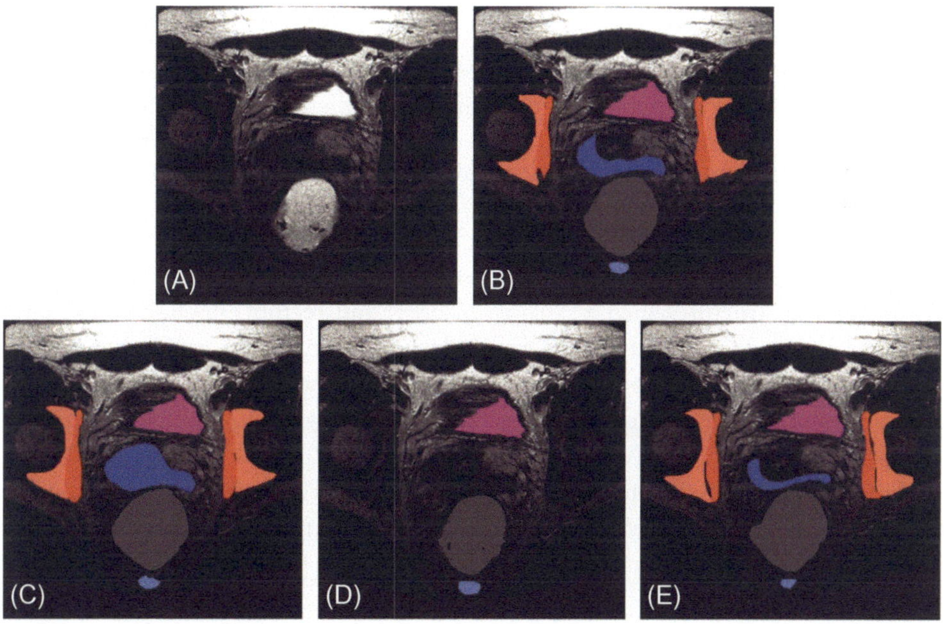

Figure 1 (A) MRI sample slice of female pelvis. (B) Estimated segmentation using local MAP STAPLE. (C-E) illustrate manual segmentations from each one of the experts. Legend: pink, pelvic bones; dark blue, vagina; red, obturator internus; violet, bladder; blue, coccyx; gray, rectum.

Image Registration

Registration can be defined as the maximization of the alignment of the corresponding features of the two images in the space of the topology-preserving maps. We have developed a registration method based on the block matching algorithm (BMA), which iteratively increases the weights assigned to the regions with more reliable correspondences [16]. In contrast to the regular BMA, which seeks the optimal correspondence in each neighborhood, we look for multiple correspondences and iteratively adjust the assigned weight to each correspondence in an expectation-maximization (EM) process. This feature is of utmost importance, especially when some regions are without any correspondence (eg, a portion of colon and intestine), which might appear in the grayscale image, but not in the segmented templates. In this situation, the most similar block correspondence is not necessarily the correct one. Further, our algorithm is suitable for intersubject image registration in the presence of some structures with only partial correspondence, which might be problematic in other registration algorithms.

There are two major challenges specific to the intersubject pelvic floor MR image registration: the first arises when there are significant anatomical differences between two subjects. The second challenge arises in the presence of regions without correspondence, depending on the field of view.

Based on the BMA, our registration method calculates the similarity between each block in one image to all the blocks in a neighborhood in the second image. The measured similarity illustrates an initial correspondence between each pair of blocks. Next, our method iteratively assigns more weights to the blocks with more reliable correspondences using an EM strategy in a multiresolution framework, which lowers the incorrect effect of regions with no correspondence.

Intensity and Label Fusion

Pelvic floor segmentation algorithms have been largely semiautomatic and restricted to a few anatomical structures such as vagina, bladder, and rectum, because of the interindividual anatomic variability of the pelvic floor and complexity of structures such as obturator internus [7–10]. To overcome these limitations, we have developed a powerful multiple-template-based fusion algorithm to segment pelvic floor structures with high accuracy [17].

The fusion of multiple templates with a target is a very popular family of techniques for image segmentation. In this approach, a set of template images is registered to a target image and a weighted combination of the templates is used to generate a segmentation of the target image. Majority voting is the simplest fusion algorithm that can be used for this purpose [18]. However, it considers the same contribution for each template, and this is only accurate if each is an equally faithful reproduction of the anatomy of the target. However, both the segmentation of template and the registration of the template to the target contain errors. Therefore, the fusion algorithm should detect the errors and decrease their effects upon the segmentation of the target image.

Here, we describe the sources of these errors. Registration can be defined as the maximization of the alignment of the corresponding features of the two images in the space of the topology-preserving maps. Diffeomorphic registration algorithms assure topology-preserving maps by making smooth and invertible transformation between the two images. The motivation is that there is one true anatomy for every tissue due to normal development. Therefore, every normal tissue can be diffeomorphically mapped to any other equivalent normal tissue. However, in practice, none of these registration algorithms is able to do a precise alignment of the anatomy of the two images. Consequently, it is impossible to completely align a template to the target image in these types of algorithms. There are two reasons for this major problem of the topology-preserving registration algorithms and in particular diffeomorphic image registration methods.

First, the relation between the image intensity and the underlying anatomy is very complex. Medical imaging devices can be utilized to illustrate spatial distribution of some physical quantity. However, there is an imperfect relation between the measured physical quantity and the anatomy of the tissue. This is caused by the lack of specificity in MRI due to the similarity of MRI properties of different tissues and also by problems such as noise, intensity inhomogeneities, artifacts, and partial voluming. Therefore, while the anatomies of two similar tissues can be aligned with some diffeomorphic maps, their corresponding intensity images may not be diffeomorphically mapped to each other. This means that the target image and the templates have uncaptureable interindividual differences in some regions that cannot be captured by the smooth and invertible transformations.

Second, nonrigid registration algorithms have parameters, which need to be set. The optimum values of these parameters are unknown and differ for each pair of the images. Also, finding the global optimal value of these parameters might not be practically achievable. Therefore, it is not possible to optimize the parameters perfectly in a practical registration problem. In addition, even with the optimum parameters there would be fundamental errors caused by inadequate model selection. This causes unwanted intrinsic registration errors due to the suboptimal registration parameters.

Unsatisfactory segmentation of the templates is the other source of error, which is independent of the registration accuracy. Manual or automatic segmentation algorithms can be used for the segmentation of a template. The utilized segmentation may have both random and systematic errors [12,19,20]. These errors cannot be detected by measuring the local intensity similarity of the target image and the template. In fact, with a perfect registration of the templates to the target image, all of the templates will have identical intensity similarity to the target image; however, due to the errors in the segmentation of the templates, their corresponding segmentations would not be identical. Therefore, taking these unavoidable errors into account in the fusion process is very important for a successful multiple-template-based segmentation algorithm.

Warfield et al. developed the STAPLE algorithm where the optimal weighting of the aligned templates is estimated by comparison with an estimate of the desired target segmentation. This fusion method is an EM algorithm that iteratively estimates the performance of the templates and the target segmentation [12]. In each iteration, templates' performance parameters are estimated based on comparison to the estimated reference standard segmentation, and they are then utilized to update the reference standard segmentation. To improve the final segmentation quality, a number of generalizations of this fusion approach have been developed. In particular, spatially adaptive performance estimation and effective estimation of priors have been found to be important [13,21,22]. Many alternative formulations of the fusion problem have been developed [13,20,21,23–27].

Locally weighted voting methods use local intensity similarity of the templates and the target image to estimate the reliability weight of the templates [11,28–33]. The motivation for this approach is that visual inspection of the image registration errors implies that there are local intensity dissimilarities between the target image and the templates in poorly aligned anatomical regions. In addition, incompatibility of the labels in the corresponding segmentation image of the aligned template and the target image is due to anatomical differences. Therefore, it is reasonable to use the local intensity similarity of the target image and the templates to estimate a local reliability weight of the templates.

Locally weighted voting fusion algorithms utilize the intensity similarities to compute the weight of each template to compensate for the misalignments. However, because intensity and label information are combined as two independent sources of information, their performance regarding template errors is constrained by the performance of the majority-voting algorithm. In fact, if the alignment is perfect, these fusion methods will be identical to majority voting. On the other hand, STAPLE-based fusion algorithms are able to identify and compensate for both template and registration errors as the contribution of each template is evaluated in comparison to a reference standard segmentation of the target image.

In [17], we directly employed intensity information of the templates and the target image in the STAPLE framework to improve the accuracy of the estimation of the performance of templates and segmentation of the target image. Intensity information assists the fusion algorithm to better detect and compensate for the registration errors. We have used intensity information as a new source of information to achieve a more accurate and robust STAPLE-based fusion algorithm.

To form the weighted votes, two well-known models can be used: linear and logarithmic opinion pools [34,35]. All of the locally weighted voting fusion methods utilize the linear opinion pool to integrate reliability weights in the fusion process where the linear weighted average of the rater probabilities is used as the aggregation method [35]. One of the problems associated with the linear opinion pool is that it is typically multimodal [35]. Also, it is relatively insensitive to the selection of weights and it is not externally Bayesian [35]. On the other hand, in the logarithmic opinion pool the linear weighted average of the logarithm of rater probabilities is used as the aggregating method. Contrary to the linear opinion pool, the logarithmic opinion pool is typically unimodal and can be externally Bayesian [35].

Therefore, we have formulated a decision fusion algorithm that uses a logarithmic opinion pool to fuse the segmentations with associated local intensity similarity-based reliability weights. We have demonstrated that a logarithmic opinion pool allows a tractable formulation of this problem, and that it can be solved exactly with a fixed-point solution, and it can be solved rapidly using a closed form approximation. Unlike linear opinion pooling fusion algorithms, in our fusion algorithm the contribution of vote and intensity is optimally weighted by estimation of the similarity to the progressing reference standard segmentation. Our fusion algorithm compensates for both template and registration errors in the fusion procedure, which significantly improves the accuracy of the target segmentation. Fig. 2 shows an example of automatic segmentation of nine pelvic floor structures using our method and manual segmentation.

Shape and Volume Analysis

Subtle changes and variations in the anatomy of healthy and diseased patient populations are often difficult to discover and quantify through subjective clinical evaluation.

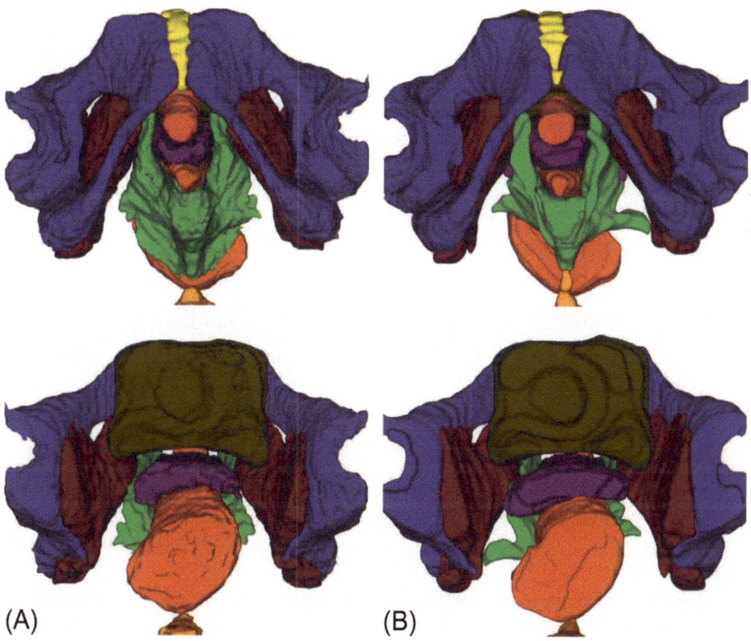

Figure 2 3D visualization of segmentations of nine pelvic floor structures. (A) LOP-STAPLE. (B) Manual segmentation. Legend: blue violet, pelvic bones; indigo, vagina; brown, obturator internus; red, urethra; olive, bladder; dark orange, rectum; green, levator ani; orange, coccyx; yellow, symphysis.

Therefore, quantitative evaluation based on shape and volume metrics has been increasingly used for the assessment of anatomic variations in groups of subjects. Variability in a class of shapes can be learned from the distribution of linearly uncorrelated shape vectors obtained from a principal component analysis (PCA) method. Assuming that the distribution of shape vectors among populations can be modeled by a multivariate Gaussian, linear PCA may be used; however, the aforementioned assumption does not usually hold. In fact, anatomical shape variations often follow complex and nonlinear deformations that cannot be accurately modeled through linear PCA. The use of a nonlinear PCA approach is suitable for anatomical shape analysis. For overcoming this problem nonlinear extension of the PCA can be used. One of the most important and interesting choice is the kernel PCA (KPCA) [36]. Using KPCA allows us to build shape priors capturing nonlinear variations.

CONCLUSIONS

Precise, reliable segmentation of pelvic floor tissues is crucial for enabling computational analytic tools to noninvasively study variations in childbirth parameters, which might contribute to childbirth-related injury to the tissues. However, manual segmentation is time-consuming, and minimally reliable. Automatic pelvic floor tissue segmentation methods offer an opportunity to reliably segment large numbers of MRI data sets, suitable for computational analysis, which can help improve our understanding of the mechanisms that may be the root of childbirth-related injury.

We presented an automatic segmentation algorithm, which simultaneously assesses both segmentation labels and intensity similarity. The algorithm estimates both the target image segmentation and the performance of each template. To estimate local reliability weights of the templates, the local intensity similarity of the target image and the templates is used. We have shown that the fusion segmentation algorithm is statistically superior to other methods in the literature [17]. This automated segmentation algorithm is expected to enable evaluation of large numbers of the female pelvic MRI data sets in the future.

REFERENCES

[1] A.A. Rodrigues Jr., R. Bassaly, M. McCullough, H.L. Terwilliger, S. Hart, K. Downes, L. Hoyte, Levator ani subtended volume: a novel parameter to evaluate levator ani muscle laxity in pelvic organ prolapse, Am. J. Obstet. Gynecol. 206 (3) (2012) 244-e1.

[2] L. Hoyte, L. Schierlitz, K. Zou, G. Flesh, J.R. Fielding, Two- and 3-dimensional MRI comparison of levator ani structure, volume, and integrity in women with stress incontinence and prolapse, Am. J. Obstet. Gynecol. 185 (1) (2001) 11–19.

[3] L. Hoyte, M.S. Damaser, S.K. Warfield, G. Chukkapalli, A. Majumdar, D.J. Choi, A. Trivedi, P. Krysl, Quantity and distribution of levator ani stretch during simulated vaginal childbirth, Am. J. Obstet. Gynecol. 199 (2) (2008) 198-e1.

[4] K.-C. Lien, B. Mooney, J.O. DeLancey, J.A. Ashton-Miller, Levator ani muscle stretch induced by simulated vaginal birth, Obstet. Gynecol. 103 (1) (2004) 31.
[5] L. Chen, J.A. Ashton-Miller, J.O. DeLancey, A 3D finite element model of anterior vaginal wall support to evaluate mechanisms underlying cystocele formation, J. Biomech. 42 (10) (2009) 1371–1377.
[6] L. Hoyte, W. Ye, L. Brubaker, J.R. Fielding, M.E. Lockhart, M.E. Heilbrun, M.B. Brown, S.K. Warfield, Segmentations of MRI images of the female pelvic floor: a study of inter- and intra-reader reliability, J. Magn. Reson. Imaging 33 (3) (2011) 684–691.
[7] J.M. Venuti, C. Imielinska, P. Molholt, New views of male pelvic anatomy: role of computer-generated 3D images, Clin. Anat. 17 (3) (2004) 261–271.
[8] X. Li, Semi-automatic segmentation of normal female pelvic floor structures from magnetic resonance images, Ph.D. Dissertation, Cleveland State University, 2009.
[9] Z. Ma, R.N.M. Jorge, et al., A shape guided CV model to segment the levator ani muscle in axial magnetic resonance images, Med. Eng. Phys. 32 (7) (2010) 766–774.
[10] Z. Ma, J.M.R.S. Tavares, R.N. Jorge, T. Mascarenhas, A review of algorithms for medical image segmentation and their applications to the female pelvic cavity, Comput. Methods Biomech. Biomed. Eng. 13 (2) (2010) 235–246.
[11] X. Artaechevarria, A. Munoz-Barrutia, Combination strategies in multi-atlas image segmentation: application to brain MR data, IEEE Trans. Med. Imaging 28 (8) (2009) 1266–1277.
[12] S.K. Warfield, K.H. Zou, W.M. Wells, Simultaneous truth and performance level estimation (STAPLE): an algorithm for the validation of image segmentation, IEEE Trans. Med. Imaging 23 (7) (2004) 903–921.
[13] O. Commowick, A. Akhondi-Asl, S.K. Warfield, Estimating a reference standard segmentation with spatially varying performance parameters: local MAP STAPLE, IEEE Trans. Med. Imaging 31 (8) (2012) 1593–1606.
[14] L.R. Dice, Measures of the amount of ecologic association between species, Ecology 26 (3) (1945) 297–302.
[15] A.G. Bedeian, K. Mossholder, On the use of the coefficient of variation as a measure of diversity, Org. Res. Methods 3 (3) (2000) 285–297.
[16] S. Ourselin, A. Roche, S. Prima, N. Ayache, Block matching: a general framework to improve robustness of rigid registration of medical images, in: Medical Image Computing and Computer-Assisted Intervention — MICCAI, 2000, pp. 557–566.
[17] A. Akhondi-Asl, L. Hoyte, M.E. Lockhart, S.K. Warfield, A logarithmic opinion pool based STAPLE algorithm for the fusion of segmentations with associated reliability weights, IEEE Trans. Med. Imaging 33 (10) (2014) 1997–2009.
[18] T. Rohlfing, D.B. Russakoff, C.R. Maurer Jr., Expectation maximization strategies for multi-atlas multi-label segmentation, in: Information Processing in Medical Imaging, vol. 2732, 2003, pp. 210–221.
[19] S.K. Warfield, K.H. Zou, W.M. Wells, Validation of image segmentation by estimating rater bias and variance, Philos. Trans. R. Soc. A Math. Phys. Eng. Sci. 366 (1874) (2008) 2361–2375.
[20] O. Commowick, S.K. Warfield, A continuous STAPLE for scalar, vector and tensor images: an application to DTI analysis, IEEE Trans. Med. Imaging 28 (6) (2009) 838–846.
[21] A.J. Asman, B.A. Landman, Robust statistical label fusion through consensus level, labeler accuracy, and truth estimation (COLLATE), IEEE Trans. Med. Imaging 30 (10) (2011) 1779–1794.
[22] B.A. Landman, A.J. Asman, A.G. Scoggins, J.A. Bogovic, F. Xing, J.L. Prince, Robust statistical fusion of image labels, IEEE Trans. Med. Imaging 31 (2) (2011) 512–522.
[23] O. Commowick, S.K. Warfield, Incorporating priors on expert performance parameters for segmentation validation and label fusion: a maximum a posteriori STAPLE, in: Medical Image Computing and Computer-Assisted Intervention (MICCAI'10), ser. LNCS, vol. 6363, 2010, pp. 25–32.
[24] A. Akhondi-Asl, S.K. Warfield, Estimation of the prior distribution of ground truth in the STAPLE algorithm: an empirical Bayesian approach, in: Medical Image Computing and Computer-Assisted Intervention — MICCAI, 2012, pp. 593–600.
[25] T.R. Langerak, U.A. van der Heide, A.N.T.J. Kotte, M.A. Viergever, M. van Vulpen, J.P. Pluim, Label fusion in atlas-based segmentation using a selective and iterative method for performance level estimation (SIMPLE), IEEE Trans. Med. Imaging 29 (12) (2010) 2000–2008.

[26] A.J. Asman, B.A. Landman, Formulating spatially varying performance in the statistical fusion framework, IEEE Trans. Med. Imaging 31 (6) (2012) 1326–1336.
[27] A. Akhondi-Asl, S.K. Warfield, Simultaneous truth and performance level estimation through fusion of probabilistic segmentations, IEEE Trans. Med. Imaging 32 (10) (2013) 1840–1852.
[28] M. Sabuncu, B. Yeo, K. Van Leemput, B. Fischl, P. Golland, A generative model for image segmentation based on label fusion, IEEE Trans. Med. Imaging 99 (2010) 1.
[29] E.M. van Rikxoort, I. Isgum, Y. Arzhaeva, M. Staring, S. Klein, M.A. Viergever, J.P. Pluim, B. van Ginneken, Adaptive local multi-atlas segmentation: application to the heart and the caudate nucleus, Med. Image Anal. 14 (1) (2010) 39–49.
[30] J.M.P. Lötjonen, R. Wolz, J.R. Koikkalainen, L. Thurfjell, G. Waldemar, H. Soininen, D. Rueckert, Fast and robust multi-atlas segmentation of brain magnetic resonance images, Neuroimage 49 (3) (2010) 2352–2365.
[31] P.A. Yushkevich, H. Wang, J. Pluta, S.R. Das, C. Craige, B.B. Avants, M. Weiner, S. Mueller, Nearly automatic segmentation of hippocampal subfields in in vivo focal T2-weighted MRI, Neuroimage 53 (4) (2010) 1208–1224.
[32] A.R. Khan, N. Cherbuin, W. Wen, K.J. Anstey, P. Sachdev, M.F. Beg, Optimal weights for local multi-atlas fusion using supervised learning and dynamic information (SuperDyn): validation on hippocampus segmentation, Neuroimage 56 (1) (2011) 126–139.
[33] H. Wang, J.W. Suh, S.R. Das, J.B. Pluta, C. Craige, P.A. Yushkevich, Multi-atlas segmentation with joint label fusion. IEEE Trans. Pattern Anal. Mach. Intell. 35 (3) (2013) 611–623. http://dx.doi.org/10.1109/TPAMI.2012.143.
[34] C. Genest, J.V. Zidek, Combining probability distributions: a critique and an annotated bibliography, Stat. Sci. 1 (1) (1986) 114–135.
[35] M. Rufo, J. Martín, C. Pérez, Log-linear pool to combine prior distributions: a suggestion for a calibration-based approach, Bayesian Anal. 7 (2) (2012) 411–438.
[36] L.J. Cao, K.S. Chua, W.K. Chong, H.P. Lee, Q.M. Gu, A comparison of PCA, KPCA and ICA for dimensionality reduction in support vector machine, Neurocomputing 55 (1) (2003) 321–336.

CHAPTER SIXTEEN

Biomechanical Characterization of the Pelvic Floor Using Tactile Imaging

V. Egorov*, H. van Raalte†, V. Lucente‡, A. Sarvazyan*
*Artann Laboratories, Trenton, NJ, United States of America
†Princeton Urogynecology, Princeton, NJ, United States of America
‡The Institute for Female Pelvic Medicine & Reconstructive Surgery, Allentown, PA, United States of America

ABBREVIATIONS

E Young's modulus
FTI functional tactile imaging
MRI magnetic resonance imaging
P pressure
POP pelvic organ prolapse
SUI stress urinary incontinence
TI tactile imaging
VTI vaginal tactile imager

TACTILE IMAGING

Introduction

Over the centuries, soft tissue palpation has been the most prevalent and successful medical diagnostic technique [1]. Human sense of touch can distinguish structural mechanical changes in soft tissues that correlate with different disease conditions [2]. However, one of the problems with manual palpation is that it lacks quantification, objectivity, and reproducibility across diverse examiners.

By the end of the last century, a new technology named elastography, or elasticity imaging, for measuring and visualizing soft tissue viscoelastic characteristics emerged and the ancient art of palpation gained new life [3]. A review of the literature suggests the birth of a new chapter of Biomechanics evidenced by the substantial growth of elastography-related publications from 1 article in 1992 to 1019 in 2014 (PubMed — search criteria "Elastography").

V. Egorov: CEO and cofounder of Advanced Tactile Imaging Inc.; H. van Raalte: cofounder of Advanced Tactile Imaging Inc.; V. Lucente has nothing to disclose; A. Sarvazyan: cofounder of Advanced Tactile Imaging Inc.

Elasticity imaging involves the application of a stress to soft tissue, and the measurement of the resulting tissue mechanical response. There are multiple methods for applying the stress to the tissue and measuring the response. The stress can be generated in at least four different ways: from an external source such as a compression probe, an external vibrator, acoustic radiation force, and physiological sources of motion (eg, cardiac motion, fluid flow) [3]. The most common types of measurement methods of tissue response include ultrasound, magnetic resonance imaging (MRI), and tactile imaging (TI). Table 1 shows the different methods, their basic characteristics, equipment cost assessment, and clinical applications.

The female pelvic organs are accessible to manual palpation, and may have varying viscoelastic characteristics across physiological states [5,6], healthy and diseased conditions [7,8]. This makes the pelvic floor tissues especially amenable to evaluation by elasticity imaging techniques [9]. It can potentially be applied to map the elasticity of pelvic floor tissues, and this opens up new possibilities for quantitative and reproducible biomechanical assessment and monitoring of pelvic floor conditions.

Table 1 Elasticity imaging methods and their basic characteristics

Method	Stress generation	Measurements	Imaging and resolution	Imaging depth	Equipment cost	Current clinical applications
MRI elastography	Application of stress to the tissue surface, tissue vibration	3-D displacement patterns of induced waves	3-D 0.5–5 mm	200 mm	$$$	Breast, liver, brain, heart, lung, muscles, prostate, thyroid [3]
Ultrasound elastography	Application of stress to the tissue surface, physiological motion	2-D/3-D tissue displacement/strain	2-D/3-D 1–3 mm	60 mm	$$	Breast, thyroid, lymph nodes, arterial wall, heart liver, prostate [3]
Share wave ultrasound elastography	Tissue vibration, radiation force	1-D/2-D share wave speed	1-D/2-D 1–2 mm	60 mm	$$	Breast, liver, muscle, prostate, thyroid [3]
Tactile imaging	Application of stress to the tissue surface	2-D pressure pattern on the tissue surface, sensor location	2-D/3-D 1–2 mm	40 mm	$	Breast, prostate, muscle, vagina [4]

Definition

TI, also called "Mechanical Imaging," is a medical imaging modality that translates the sense of touch into a digital image. The tactile image is a function of $P(x,y,z)$, where P is the pressure on soft tissue surface under applied deformation and x, y, and z are coordinates where pressure P was measured [9–13]. The tactile image is a pressure map on which the direction of tissue deformation must be specified [13]. TI closely mimics manual palpation, because the probe of the device with a pressure sensor array mounted on its face acts similarly to human fingers during clinical examination, deforming soft tissue by the probe and detecting resulting changes in the pressure pattern. The sensor head is moved over the surface of the tissue to be studied, and the pressure response is evaluated at multiple locations along the tissue under study. The results are used to generate 2-D or 3-D images showing the pressure distribution over the area of tissue under study.

Examples of TI can be seen at:

http://link.springer.com/content/esm/art:10.1007/s00192-014-2549-9/file/MediaObjects/192_2014_2549_MOESM1_ESM.mp4.

Functional Tactile Imaging (FTI) is a variation of TI that translates muscle activity into a dynamic pressure pattern $P(x,y,t)$ for an area of interest, where t is time and x, y are coordinates where pressure P was measured. Muscle activity to be studied may include a voluntary contraction (eg, a pelvic floor squeeze), an involuntary reflex contraction (eg, due to a cough), an involuntary relaxation, or a Valsalva (veering down) maneuver.

Interpretation

To fully characterize tissue as a mechanical system, a great number of parameters are needed. These include the shear and Young's moduli bulk compressional modulus, nonlinearity, Poisson's ratio, viscosity, poroelastic parameters, anisotropy and heterogeneity indices [14–16]. However, in most practical cases, even just one elasticity parameter, such as Young's modulus (E), may be sufficient to address diagnostic tasks [17–20]. Detection of a mechanical heterogeneity by manual palpation is based exclusively on sensing the variations of the Young's modulus of tissue (or shear elasticity modulus, μ, which is approximately equal to $E/3$ for soft tissues) [3,21–24]. Soft tissues are called "incompressible" because their bulk compressional modulus (K) is generally several orders of magnitude larger than the shear modulus. As a result, a short external stress applied to soft tissues causes mainly a change in the shape of the stressed tissue, while the volume remains constant with a high degree of precision. If a soft tissue is deformed, the relationship between the stress and strain patterns is completely defined by the Young's modulus only, regardless of the K value for the tissue. Bulk compressibility and shear elasticity are dependent on different features of tissue. Bulk compressibility modulus depends on short-range molecular interactions and is defined mainly by tissue molecular composition, while shear elasticity is defined by structural peculiarities of tissue, its cellular and higher level of architecture [23]. The bulk modulus for all the soft tissues is close to that of water and varies within only about 10% [23]. In contrast to that, the range of variability of structural features of tissues, such as geometrical parameters of cells in different tissues and the degrees of heterogeneity and anisotropy, is much greater. The shear elasticity for different soft tissues varies over four orders of magnitude and, even within one tissue, may change by hundreds of percent during such processes as development of a tumor or an ordinary muscle contraction [25].

Generally, an inverse problem solution for the 3-D tactile image $P(x,y,z)$ would allow reconstruction of the tissue elasticity distribution as function of the coordinates x, y, and z. Unfortunately, the inverse problem solution is hardly possible for most real objects because it is nonlinear and ill posed problem. TI reveals tissue or organ anatomy and elasticity distribution because it keeps the stress-strain relationship for deformed tissue [26,27]. It appears that the 3-D tactile image can be transformed into an elasticity image with the use a linear transformation for a region of interest. That means, in general, the spatial gradients $\partial P(x,y,z)/\partial x$, $\partial P(x,y,z)/\partial y$, and $\partial P(x,y,z)/\partial z$ can be used in practice for quantitative assessment of tissue elasticity because they have a validated background [9–13,26–31] allowing quantitative comparison and analysis for different tissues with anatomical variations.

History

Review of the literature reveals that the first description of a technical implementation related to TI was given in 1979 by Frei et al. [32,33], who proposed an instrument for breast examination. That device used a plurality of spaced piezoelectric force sensors. The sensors were pressed against the breast by a pressure member, to apply a periodic or steady stress to the soft tissue. A different principle for evaluating the pattern of pressure distribution over a compressed breast was proposed by Gentle [34]. The pressure distribution was monitored optically by using the principle of frustrated total internal reflection to generate a brightness distribution. Using this technique, simulated lumps in breast prostheses were detected down to a diameter of 6 mm. But the author was unable to obtain any quantitative data on lumps in a real breast. The failure was explained by the insufficient sensitivity of the registration system, and that "the load that the volunteers could comfortably tolerate, was less than that used in the simulation." Then, Dario et al. (1988) and Sabatini et al. (1990) designed a robotic system capable of carrying out complex sensory-motor sequences intended for collecting information on some body functions through palpation with an articulated finger incorporating joint force and position sensors as well as piezoelectric polymer skinlike sensors aimed at detecting hardened regions embedded in soft biological tissues

[35,36]. Another attempt to use a robotic system to detect the presence of a lump in the breast was described by Koganezawa et al. [37].

Sarvazyan in 1992 proposed a method, mechanical imaging, that can be used for both screening purposes and documentation purposes. That method utilizes distributed pressure measurements at the surface of the tissue to estimate the properties of breast lesions and prostate nodules [7]. Mechanical imaging, as a modality of medical diagnostics and the imaging device with mechanical sensors, was introduced in 1997 [10,38]. The PhD theses "Tactile Imaging" by Wellman (1999) and "Mapping Tactile Imaging Information" by Galea (2004) described a system for documentation of the properties of palpable breast lumps to improve the objectivity of clinical breast examination [39,40]. Despite the first FDA approval for a tactile breast imaging device based on the Tekscan pressure array in 2003 [41], that device proved unreliable due to low sensitivity and poor reproducibility of electrical resistive sensors for that specific application. Many physical principles have been explored for the realization of tactile sensors: resistive, inductive, capacitive, optical optoelectric, magnetic, piezoelectric, and electroacoustic principles, in a variety of configurations [42,43]. Oie et al. [44] designed a tactile mapping system (scanning haptic microscope) with a spatial resolution of 5 μm (the tactile tip had a 5 μm diameter and used a lead zirconate titanate-based sensor). Its applicability for elasticity assessment was demonstrated in a porcine artery model.

During the last decade, several devices based on TI technology have been developed and clinically validated for soft tissue imaging and elasticity assessment for prostate [26], breast [27], myofascial trigger points [30], and vagina [9]. The capacitive pressure sensor arrays were used in all these applications. Each pressure sensor has sensing area of about 2 mm by 2 mm, the sensitivity of 20–70 Pa in the operational range up to 60 kPa. Probe design, number of sensors in the array (up to 256 sensors), and data processing algorithms were adapted to specific needs.

A directional vaginal probe with four force sensors was developed by Constantinou and Omata [45] to evaluate the force and displacement produced during isotonic pelvic floor contractions. The ManoScan 360 probe with 256 tactile-sensitive microtransducers developed for anorectal application was tested for vaginal imaging [46].

Applications

Breast

The lifetime probability of developing breast cancer is evaluated as 13% [47]. A critical key to a continued reduction in mortality is early detection and accurate diagnosis made in a cost-effective manner. The tactile breast probe has a 2-D pressure sensor array (12×16 sensors) to acquire TI responses from the breast, with the aim of detecting worrisome lesions [27]. The probe head is lubricated and slid over the breast surface. No motion tracking or orientation system was used because tissue heterogeneity itself can be used as reference coordinates. When a suspicious site is detected, the manipulation of the probe is switched to a local scan accomplished by two procedure variations: probe pressings against the breast over the detected abnormality and circular motion of the probe. The examiner observes in real time accumulated cross-sectional images of a mass/lesion in orthogonal projections. The 3-D image composition and feature calculations for the breast tissue are accomplished in real time [27]. A multisite clinical study on 179 subjects demonstrated that the breast mechanical imager provides a reliable image formation of breast tissue abnormalities and calculation of lesion features. Malignant breast lesions (histologically confirmed) demonstrated increased hardness and strain hardening as well as decreased mobility and longer boundary length in comparison with benign lesions. Statistical analysis of differentiation capability for 147 benign and 32 malignant lesions revealed sensitivity of 91.4% and specificity of 86.8% [28].

Prostate

We designed the transrectal probe with 2-D pressure sensor array (8 × 16 sensors) to acquire pressure patterns from the prostate. It has additional 2-D pressure sensor array (3 × 16 sensors) to measure probe position relative to sphincter and 3-D orientation sensor composed of accelerometers, magnetometers, and gyroscopes. The probe provides a real-time 3-D tactile image of the prostate by capturing its geometrical and elastic characteristics, and reveals the tissue abnormalities within the gland [26]. The correlation between the device and manual palpation (digital rectal examination [DRE]) in detection of palpable nodularity was 81%, as indicated by the area under the receiver operating characteristic curve [29].

Myofascial Trigger Points

Presence of exquisitely tender nodules in a taut band of skeletal muscle and an associated symptom cluster may cause myofascial pain syndrome, which is prevalent and clinically significant [48]. We designed the muscle TI probe adapted for tender muscles with 192 durable tactile sensors. The clinical study on 50 subjects (27 healthy controls and 23 with symptomatic chronic neck pain) demonstrated that symptomatic muscle tissue in subjects with neck pain is mechanically more heterogeneous and stiffer compared to normal muscle in control subjects ($p < 0.05$) [30]. At the same time, difficulties ware noted in 3-D tactile image composition and reliable trigger point discrimination by TI.

Female Pelvic Floor

Pelvic floor disorders affect the majority of women in their lifetime [49–51]. There is an 11% lifetime risk of undergoing surgery for prolapse or incontinence [52,53]. Market data collected by FDA from manufacturers indicate that in 2010 approximately 300,000 women underwent surgical procedures in the US to repair pelvic organ prolapse (POP) and approximately 260,000 underwent surgical procedures to repair stress urinary incontinence (SUI) [54]. The risk of reoperation is high [55,56]. Pelvic floor biomechanical changes occur in the course of applied nonsurgical therapy, in pregnancy and after giving birth, with age, and under a heavy load in sporting and military activities. Sections below describe the TI approach to assess biomechanical properties and results of clinical research in this field.

Limitations

The first limitation of the TI method is the dependence of acquired tactile image or function $P(x,y,z)$ on the probe size, tissue or organ anatomy, and boundary conditions. Using lubrication, probes with contact area of 15 μm^2 [44] and 20 cm^2 [27] demonstrate different absolute values of $P(x,y,z)$ acquired for the same tissue. But, comparison of the two image data sets reveals a lot of similarity and common features. Both probes show close relative distribution within $P(x,y,z)$ and enable tissue characterization.

In general, an examination with a TI probe is highly operator dependent, similar to colonoscopy [57]. Operator training is required to quantify and standardize operator skills [58]. Minimization of the operator dependence is also possible by intentional probe design, data processing algorithms, and real-time feedback to the operator. For example, for the breast TI probe, the probe curvature was optimized to reduce the influence of applied tissue deformation on TI performance. In addition, probe skewness was calculated and subtracted as an image background at the software level [27]. The prostate probe was optimized for reproducible contact conditions with the prostate. To define and visualize probe location relative to the prostate in real time, we placed an additional tactile array for imaging of the urethral sphincter as a reference structure [26]. The vaginal probe, shown in Fig. 1 (probe 3), was intentionally designed to eliminate dependence on operator skills as much as possible.

TI reproducibility is an important consideration for imaging of soft tissue, which may exhibit nonlinear and viscoelastic properties with possible hysteresis. The technical

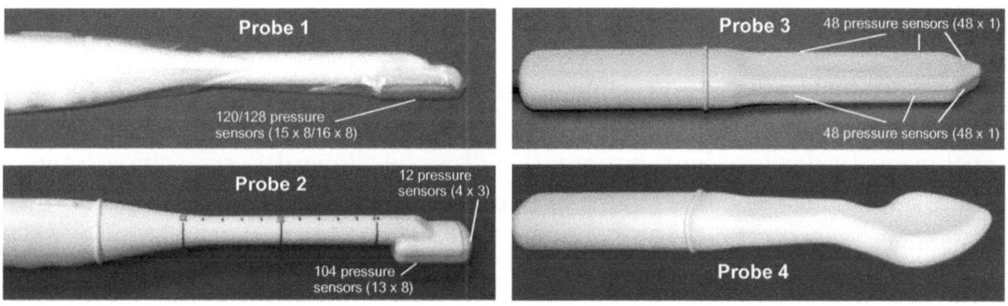

Figure 1 Vaginal tactile imaging probes. Probe 1 has a 2-D pressure sensor array (sensor size 2.5×2.0 mm); Probe 2 has two 2-D pressure sensor arrays (sensor size 2.5×2.0 mm); Probe 3 has two linear pressure sensor arrays (sensor size 2.5×4.0 mm); Probe 4 is designed for perineum imaging.

reproducibility of pressure sensors, incorporated within the TI probe, is easy to establish in repeat calibrations using an air-pumped chamber with a high-precision reference air pressure sensor. Reimaging of silicone models allows reproducibility assessment for elastic media. Clinical reproducibility evaluation requires completion of specific clinical protocols in controlled conditions. Establishing new diagnostic thresholds or imaging markers, sensitive to specific diseased conditions, does not require knowledge about clinical reproducibility, as long as the marker demonstrates statistically significant sensitivity/specificity or diagnostic accuracy or dependence on the diseased conditions.

TI depth, detection capability, and spatial resolution may contribute added limitations. These parameters depend on the probe/sensor geometry and targeted application. For the breast TI probe, we observed superior sensitivity of the TI versus a human finger; the depth of nodules detected by TI has reached 40 mm, which is more than double the depth achieved with manual palpation [27]. A human finger can distinguish stiffness in pairs of elastic samples in the range from 0.13 to 0.46 [59–61] of the Weber fraction (defined as the relative increase above a baseline value that can be reliably detected). That means the human finger cannot reliably discriminate the difference in tissue stiffness up to 46% of the time, using manual palpation cues [62]. These numbers do not take into account that the human mind remembers the perceived information for less than 2 sec after a haptic exploration [63]. The stiffness discrimination of the pressure sensors, used in the vaginal TI probe [12] ranges from 2 to 10% for the entire elasticity range of tissues within the female pelvic floor. TI software can support resolutions of 1.0 mm, but current pressure sensors are 2.0 mm or larger. It has been demonstrated that, using dynamic and spatial motion components over the tissue in mechanical imaging significant increases (10 times and more) in spatial resolution relative to the sensor size may be achieved [10].

Vaginal TI can clearly report $P(x,y,z)$ as a function of the deformation applied to vaginal tissue. However, the probe heads presented cannot "see" all the layers of anatomic structures or the full context underneath the tissues under study. Therefore, the probe is essentially reading the response of the tissue composite underneath the probe head (in the case of the distal posterior wall — vaginal mucosa, muscularis, serosa, rectovaginal septum, rectal serosa, muscularis, mucosa, lumen, stool if present, mucosa, muscularis, serosa, levator plate, anococcygeal raphe, depending on how hard you press). However, vaginal TI allows quantitative observations of all composite structures detectable by intravaginal manual palpation. Furthermore, the TI probe allows data acquisition from two opposite sides along the entire vagina in one visual frame with higher resolution than a human finger can provide (see Figs. 14–16). The vaginal walls look "transparent" for the tactile probe; they are easily deformable allowing visualization of the pelvic floor support structures [13]. As demonstrated earlier, application of a 5–10 mm deformation to tissue models allows the tactile probe to "see" 20–35 mm in depth [4,27].

BENCH TESTING

Probe Design

Over 10 different TI probes have been designed and clinically tested over the past decade for prostate [26], breast [27], muscle [30], vaginal, and pelvic floor assessments [9]. Four vaginal probes are shown in Fig. 1. Probes 1 and 2, intended for 3-D imaging, were designed and tested subsequently. Probe 2 has an additional sensor array designed for contact with the cervix. Probe 3 was designed to reduce operator dependence and to image opposite sides of vagina along the entire length, including muscle function. Probe 4 is designed for 3-D perineal imaging under large deformations before delivery (prototype). As an example, probe 3 has capacitive pressure sensors with sensitivity, defined as an average noise level, of about 0.1 mmHg, accuracy $\pm$ 3 mmHg, spatial resolution 1–2 mm, dynamic response 40 ms, with a measuring range up to 500 mmHg.

TI probes need to be used with a lubricating gel to provide reproducible boundary/contact conditions with the deformed tissue; these conditions are classified as slip boundary conditions. The tactile probe measures an applied pressure, but not force. Force is a vector and by definition has amplitude and direction. The pressure sensors designed for tactile probes are not sensitive to the tangential component of a force they may arise during probe motion. The sensors measure pressure, defined as the orthogonal component of force divided by area. The probe may be used not only for tissue compression in the orthogonal direction to the tissue surface, but it can be used for sliding over the tissue and/or probe elevation. These probe maneuvers allow accumulation of multiple pressure patterns from the tissue surface to compose an integrated tactile image for the investigated area or organ [11,26,27].

Accuracy and Reproducibility

The vaginal models, shown in Fig. 2B, were built from two-component silicones [SEMICOSIL by Wacker Chemical Corporation (Adrian, MI) and RTV6166, 6186 by GE Silicones (Albany, New York)] [64]. The range of Young's moduli of the parts manufactured from these components was varied from 2 to 100 kPa and controlled by the tissue elastometer [65]. Soft tissue specimens were compressed between the object plate of an electronic balance and a linearly actuated indenter with a small rounded tip. The elastometer was designed such that a deformation model for semiinfinite media is applicable for calculating the Young's modulus of test specimens from their collected force-displacement data (see page 367 in [14]). The used silicones can be considered as elastic, homogeneous, and isotropic materials. The tested models were designed to approximate the pressure patterns collected under clinical conditions for normal and diseased pelvic floor conditions by the same TI probe. A nonlinear model for transformation

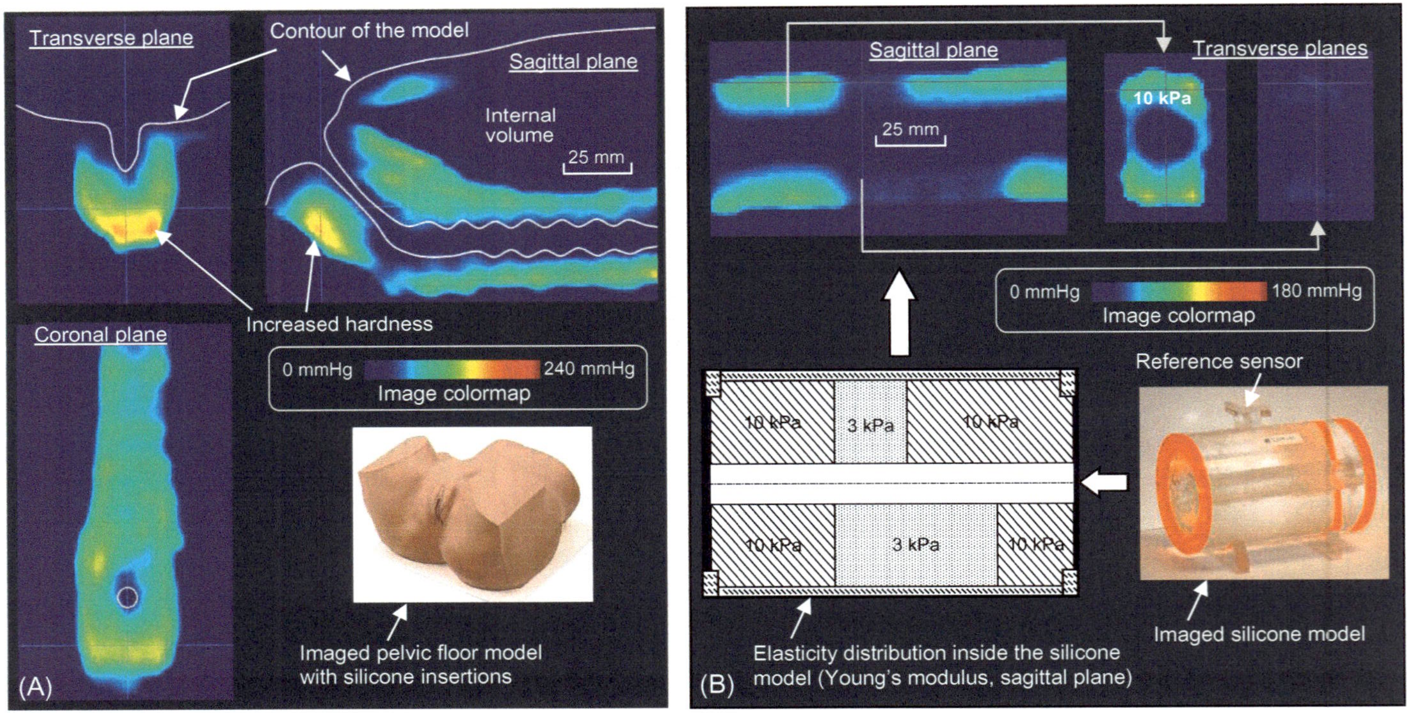

Figure 2 Tactile imaging of designed pelvic floor model (A) and one of the vaginal modes (B) used in VTI bench testing and validation.

of spatial pressure gradients with the tactile images into Young's modulus was developed and validated on multiple silicone models.

Algorithms for 2-D image filtering were applied, along with 3-D coordinate translation to transform motion tracking data into the 3-D coordinate of each pressure sensor, to create the 3-D tactile image with a resolution of 1.0 mm [27]. Image rendering on the display is presented as in MRI by three orthogonal cross-sections.

Fig. 2 demonstrates the imaging capability of probe 1 (Fig. 1) with a pelvic floor model (Fig. 2A), which is modified from a commercially available setup in which the internal rubber parts are replaced with silicone ones with realistic mechanical characteristics [11]. This model was used for imaging verification of anatomical features. A set of vaginal models, shown in Fig. 2B, was used for imaging accuracy and reproducibility assessment.

The following numerical procedures were applied for processing the 2-D/3-D tactile images:

(a) best fit as a mutual position of two images [27],

(b) average image from multiple images acquired from the same tissue model or subject [26], and

(c) image deviation for the analyzed image (I_{an}) relative to a reference image (I_{ref}).

The tactile image deviation (*TID*) is calculated as

$$TID(x, y, z) = \frac{\sum_{x,y,z} (I_{an}(x, y, z) - I_{ref}(x, y, z))^2}{\sum_{x,y,z} I_{ref}(x, y, z)^2} * 100\% \qquad (1)$$

The best fit procedure (a, above) was used before averaging of two images according to procedure (b, defined above). Averaging of multiple images is accompanied by recurrent application of procedures (a) and (b). Procedure (c) is used for imaging reproducibility assessment relative to an average image (I_{ref}).

Accuracy

The examination procedure included multiple compressions of the internal walls of a vaginal model, similar in design to the model shown in Fig. 2B, until a circumferential 3-D tactile image was acquired. The operator observed in real time the probe location within the vaginal model and three cross-sections from the 3-D tactile image on display. Imaging accuracy was evaluated using a visual subjective score ranging from 1 to 5 (No image, Poor, Fair, Good, Excellent) to characterize relative comparison of 3-D tactile image with an actual map of the vaginal model. The average imaging score 3.9 was calculated for 35 examinations with TI probe 1 (Fig. 2).

Calculation of *TID* for I_{an} (see Eq. (1)) relative to the real elasticity map (I_{em}) for each vaginal model with two correction factors (constant and linear members), demonstrated deviations in the range of 30–40%, which can be considered as a good result taking into account that we compare the 3-D pressure map (tactile image) with the 3-D elasticity distribution. Looking at the sagittal planes of the elasticity map and the tactile image in Fig. 2B, one may see a detectable correspondence. However, it is more appropriate to compare images of elasticity distribution, created as the result of the inverse mechanical problem solution for tactile image I_{an}, with the I_{ref}, or to compare a tactile image, created in the result of a direct mechanical problem solution for I_{an}, with an acquired tactile image I_{an}. Based on experiments with vaginal tissue models of known composition we found that the vaginal tactile imager (VTI) allows differentiation of tissue elasticity within 2–10% and measurement accuracy of 10–15%, which is sufficient for quantitative elasticity assessment given that elasticity deviations related to diseased conditions are known to vary by up to 10,000% [4,31,65].

By incorporating softer parts into vaginal tissue models to mimic prolapse conditions, we estimated TI accuracy in size measurements to be ± 2–5 mm; for stiffer parts, we found VTI accuracy in size measurement as ±2–3 mm.

Reproducibility

Imaging reproducibility was calculated as a standard deviation (SD) for *TID* defined in Eq. (1) where an average image composed from multiple examinations of the same vaginal models are used as the reference. We found that the SD ranged from 4.2 to 5.8%, depending on the model design and elasticity of included components. The reproducibility in measuring anatomical dimensions is within 1–3 mm; reproducibility in pressure measurements is 2 mm Hg; and reproducibility in Young's modulus measurement for the range from 2 to 20 kPa is 0.3–1.8 kPa.

CLINICAL APPLICATION

Following clinical evaluation of the feasibility of the VTI method, efforts focused on exploring findings in different stages of pelvic support, evaluating the possibility of identifying conditions that might predict the development of POP, and studying response of pelvic floor tissues across ranges of age and parity. TI markers were identified for evaluating POP, and the response of the pelvic floor tissues to functional maneuvers was investigated.

Feasibility Clinical Study

Clinical feasibility of the VTI technique was evaluated in a clinical study with 13 subjects using probe 1 (Fig. 1) equipped with a 2-D pressure sensor array and a tilt sensor. Examinations were performed with an empty bladder and rectum. The probe was held against the anterior or posterior vaginal wall (Fig. 3), and the applied force versus the probe elevation angle was obtained and plotted. The elevation angle is proportional to the displacement of the probe head, which creates the applied stress. The results for two

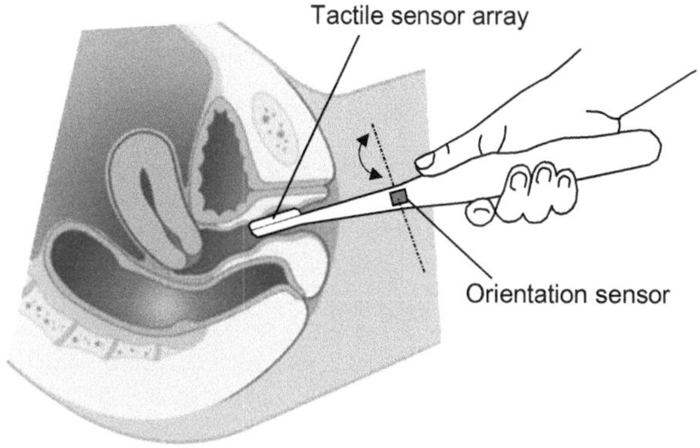

Figure 3 Two-dimensional pressure sensor array mounted on the probe allows acquisition of pressure (stress) patterns on the vaginal wall under manually applied pressure as a function of the angular position. *Adapted from V. Egorov, H. van Raalte, A. Sarvazyan, Vaginal tactile imaging, IEEE Trans. Biomed. Eng. 57 (7) (2010) 1736–1744.*

cases with normal and prolapse conditions are shown in Figs. 4 and 5. A tissue elasticity index (E_s) was defined as the slope of the scanhead applied force versus elevation angle to characterize the tissue elasticity as stress to strain ratio [15]. These results show that the E_s for the anterior vaginal compartment in normal conditions was 7.2 (1.8 N/deg vs. 0.25 N/deg) higher than in Stage 3 prolapse conditions; the value of E_s for the posterior vaginal compartment in normal conditions is 3.1 (1.1 N/deg vs 0.35 N/deg) higher than in State 2 prolapse conditions. Such significant differences in tissue elasticity (720 and 310%) *exceed the estimated measurement error of used approach* [9].

The abovementioned TI probe was used to observe the tissue elasticity contrast between the native tissue and implanted mesh sections of a vaginal wall repair [9]. One such case (Fig. 6) shows pressure patterns (tactile images) of the anterior and posterior compartments for a 59-year-old woman with a history of a mesh augmented anterior repair, and a native tissue posterior colporrhaphy using a linear vertical posterior wall incision. The color pattern represents a contact pressure distribution under the applied load to the tissue. The spatial *X*, *Y*-coordinates for the color images show the transverse (*X*) and longitudinal (*Y*) direction for a vaginal wall over which the elasticity is being characterized. These patterns allow visualization and quantitative evaluation of increased rigidity at the mesh graft. TI also shows a small rigidity increase (yellow-colored zone) at the posterior vaginal wall after the posterior colporrhaphy, correlating with the incisional "scar" tissue detected by manual palpation. Quantitative evaluation of the most rigid tissue is presented in a graph where the horizontal axis is proportional to the total applied force to the scanhead, while the vertical axis is the peak pressure related to the maximum measured value inside the observed zone (see right panels** in Fig. 6). The slope of these

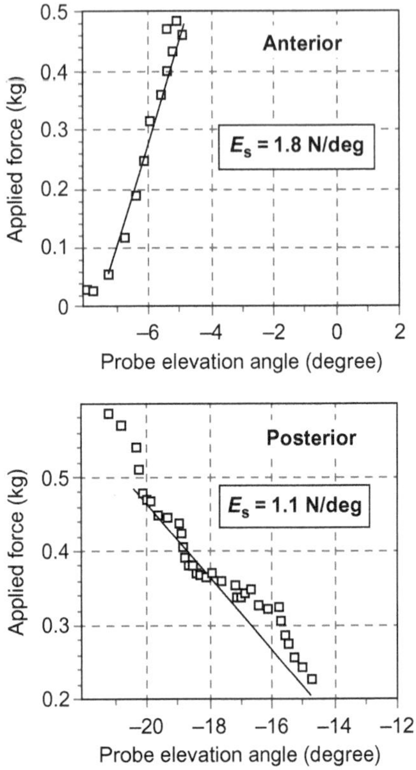

Figure 4 Loading curves obtained for 39-year-old women with normal pelvic floor conditions. The slope (E_s) of the measured force versus elevation angle of the probe characterizes an average elasticity of the vaginal tissue in the mid part of the vagina. *Adapted from V. Egorov, H. van Raalte, A. Sarvazyan, Vaginal tactile imaging, IEEE Trans. Biomed. Eng. 57 (7) (2010) 1736–1744.*

load curves characterizes the elasticity deviation of an embedded structure inside the soft tissue [26,27].

This study shows that vaginal TI (a) has the potential to contribute to quantitative biomechanical characterization of the tissues, (b) has the potential for differentiating normal tissue from tissues involved in prolapse, and (c) allows the characterization of implanted structures in vivo following reconstructive surgery.

Quantifying Prolapse Conditions

The basic purpose of the next study was to assess the clinical suitability of 3-D TI of the vagina and tissue elasticity quantification under normal and prolapse conditions. An improved TI design was used in this clinical study of 31 subjects [11].

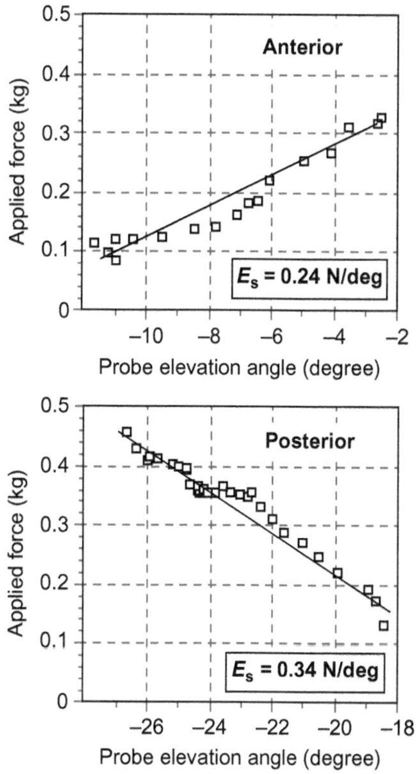

Figure 5 Loading curves obtained for 69-year-old women with Stage III anterior prolapse and Stage II posterior prolapse. The slope (E_S) of the measured force versus elevation angle of the probe characterizes an average elasticity of the vaginal tissue in the mid part of the vagina. *Adapted from V. Egorov, H. van Raalte, A. Sarvazyan, Vaginal tactile imaging, IEEE Trans. Biomed. Eng. 57 (7) (2010) 1736–1744.*

The updated system includes a transvaginal probe, a motion tracking system, a data acquisition electronic unit, and a computer with touch screen monitor. The 2-D pressure sensor array was installed on the probe head surface contacting with the vaginal wall during the examination procedure. The probe head measures 45 mm in length and 20 mm in diameter. The pressure sensor array is comprised of 128 capacitive pressure sensors (probe 1, Fig. 2). On average, each pressure sensor has a sensitivity of 0.15 mm Hg; reproducibility is about 2.2 mm Hg and an operational range of 225 mm Hg. The six-degree-of-freedom motion tracking system provided probe positioning accuracy better than 1 mm and angular accuracy of about 0.25 degree. The electronic unit provided data acquisition from the pressure sensors, synchronized data from motion tracking sensors, and communicated with the computer through a USB port. The data acquisition rate was about 25 pressure patterns per second.

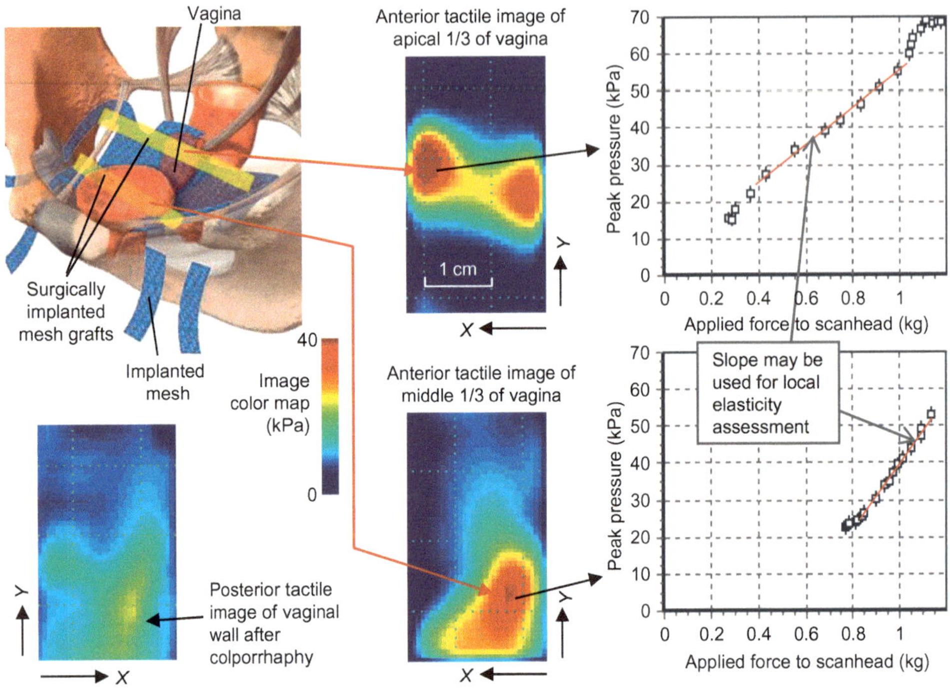

Figure 6 Tactile imaging of anterior and posterior vaginal walls allows visualization and quantitative elasticity evaluation of increased rigidity at both mesh grafts. The pressure values are shown in kPa. *Adapted from V. Egorov, H. van Raalte, A. Sarvazyan, Vaginal tactile imaging, IEEE Trans. Biomed. Eng. 57 (7) (2010) 1736–1744.*

The VTI examination was performed on patients in a standard position for a routine gynecologic exam. Examinations were performed with an empty bladder. During examination, the VTI probe was covered by a disposable plastic sheath with a lubricant. The full VTI examination required 3–5 min for completion. Three orthogonal projections of the 3-D vaginal pressure map with VTI probe location were observed by the operator in real time. The VTI clinical operators were trained on pelvic floor models prior to VTI clinical application to standardize imaging techniques. The examination procedure included multiple compressions of the vaginal walls and allowed composing a circumferential 3-D tactile image or pressure map of the vagina and storing the acquired data in digital format.

Thirty-one women were enrolled in this study. Of these, 23 were considered normal, with an average age of 58 years. Of the eight women with prolapse, two each had stage I and stage II support, and four had stage III support. The average age of the eight women was 70 years. Differences in the anatomy and tissue elasticity were found between the normal patients and those with prolapse. Fig. 7 presents examples of examination results

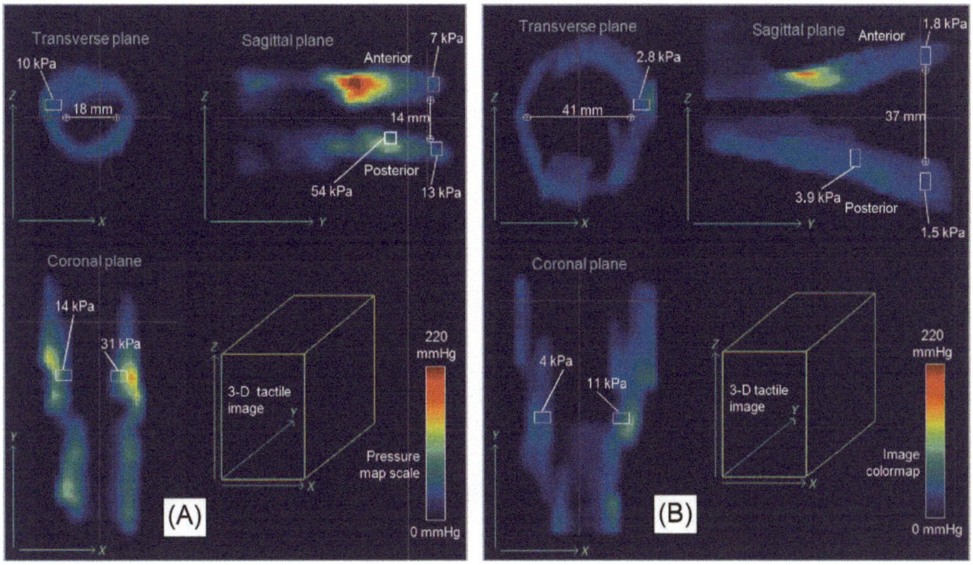

Figure 7 Cross-sections of 3-D vaginal tactile image received with VTI for a patient with normal pelvic floor conditions (A) and Stage III prolapse (B). Young's modulus (kPa) was calculated for areas specified by a rectangle. See text.

for normal and prolapse conditions. Specifically, Fig. 7A shows transverse and sagittal cross-sections of the 3-D vaginal tactile image for a 63-year-old patient with normal pelvic floor anatomy on manual palpation during physical examination. Young's modulus (E) was calculated for areas specified by the rectangles in the figure, and was found to be 7 kPa at the apical anterior location and 13 kPa at the apical posterior location. On the right apical vagina, E was 10 kPa (transverse plane in Fig. 7A). The anterior/posterior spacing at the apical vagina was measured as 14 mm (sagittal plane in Fig. 7A). Fig. 7B shows transverse and sagittal planes of 3-D vaginal tactile image received with VTI for a 77-year-old patient with Stage III prolapse in the anterior and upper half of the posterior compartment that recurred less than a year from a vaginal hysterectomy and anterior repair. For this case, that E was 1.8 kPa at the apical anterior wall and 1.5 kPa at the apical posterior wall. The left apical vaginal wall had E of 2.8 kPa (transverse plane in Fig. 7B). The anterior/posterior spacing at the apical vagina was measured as 37 mm (sagittal plane in Fig. 7B).

A notched box plot shows the confidence interval for the median value (central horizontal line), 25 and 75% quartiles (Figs. 8 and 9). The spacings between the different parts of the box help to compare a variance. The boxplot also identifies skewness (asymmetry) and outliers (small circles). The intersection or divergence of confidence intervals for two patient samples is a visual analog of the paired t-test. Fig. 8 presents the vaginal tissue elasticity distribution for the apical and midanterior aspects of the vagina across varying POP

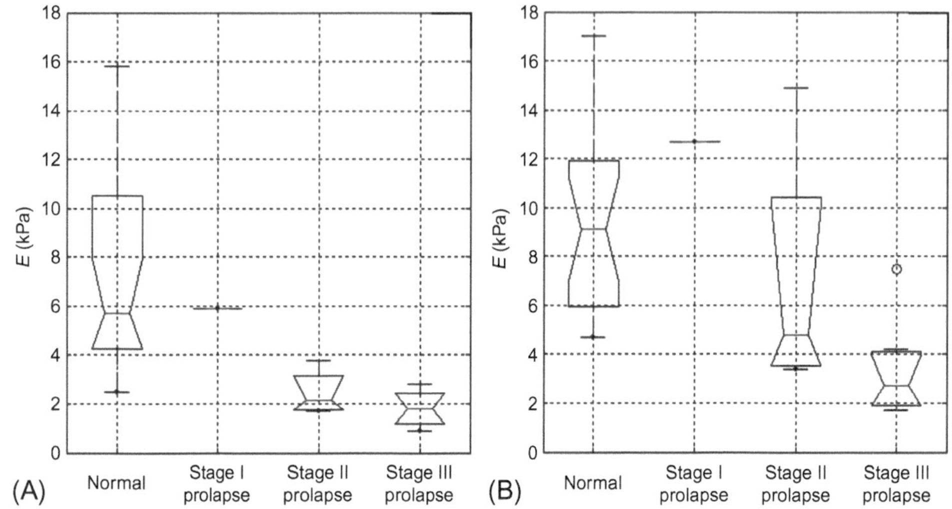

Figure 8 Anterior tissue elasticity, Young's modulus (*E*), versus anterior conditions defined by physical examination. (A) Apical anterior, (B) Midanterior. *Adapted from V. Egorov, H. van Raalte, V. Lucente, Quantifying vaginal tissue elasticity under normal and prolapse conditions by tactile imaging, Int. Urogynecol. J. 23(4) (2012) 459–466.*

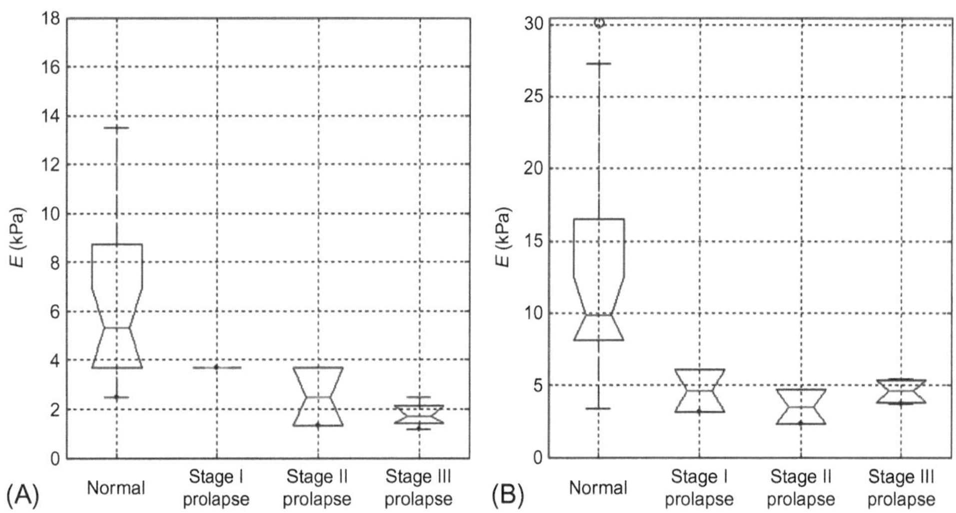

Figure 9 Posterior tissue elasticity, Young's modulus (*E*), versus anterior conditions defined by physical examination. (A) Apical posterior, (B) Midposterior. *Adapted from V. Egorov, H. van Raalte, V. Lucente, Quantifying vaginal tissue elasticity under normal and prolapse conditions by tactile imaging, Int. Urogynecol. J. 23 (4) (2012) 459–466.*

conditions; among these were 19 normal (average age 55 years old), 1 Stage I, 4 Stage II, and 7 Stage III (average age 70 years old) cases. The data demonstrate the respective elasticity modulus decrease of 3.2 times for the apical anterior walls and up to 3.4 times for the midanterior walls with Stage III prolapse relative to normal. From the apical to the midanterior vaginal walls, the average elasticity modulus is increased by 80%. Fig. 9 presents the vaginal tissue elasticity distribution for the apical and midposterior vaginal walls across varying POP conditions; among them (23 normal (average age 58 years old), 2 Stage I, 2 Stage II, and 4 Stage III (average age 70 years old) cases). The data demonstrate the respective elasticity modulus decrease of 3.1 times for the apical posterior and up to 2.2 times for the midposterior with Stage III prolapse relative to the normal conditions. From the apical to the midposterior vaginal walls, the average elasticity modulus is increased by 95%. The average values for tissue elasticity for the anterior and posterior compartments for normal conditions were 7.4 ± 4.3 kPa and 6.2 ± 3.1 kPa, respectively. For Stage III prolapse, the average values for tissue elasticity for the anterior and posterior compartments were 1.8 ± 0.7 kPa and 1.8 ± 0.5 kPa, respectively.

Figs. 8 and 9 demonstrate significant differences in anterior and posterior vaginal tissue elasticity with POP stage. This difference is statistically significant as demonstrated by both the visual comparison of the confidence intervals for the presented sample median values and ANOVA testing ($p < 0.0001$). The most affected locations are the mid and apical aspects of the anterior vaginal walls, where elasticity is decreasing up to 3.4 times from normal to POP Stage III. The lesser affected is the midposterior part, where elasticity is decreased up to 2.2 times; the apical side walls of the vagina, where elasticity is decreased by 50%; and the mid sides of the vagina, where no statistically significant tissue elasticity decrease was detected [11]. That means the horizontal (anterior, posterior) support structures weaken the most under POP conditions.

These studies suggest that VTI may serve as a means for 3-D imaging of the vagina and a quantitative assessment of vaginal tissue elasticity, providing important information for furthering our understanding of POP and surgical treatment.

Detection of Preprolapse Conditions

The objective of the next study was to estimate the ranges of normality for tissue elasticity of the vagina and pelvic floor support structures, and to explore the possibilities of 3-D TI in early prolapse detection [66].

One hundred and thirty-six women were enrolled and examined using VTI in a case-control study to evaluate the method. Study subjects included 36 women with normal pelvic support, 11 women with Stage I, 43 women with Stage II, and 46 women with Stage III prolapse. The average age was 56 ± 22 years, with a range of 21–90 years. The VTI used the transvaginal probe shown in Fig. 2 (probe 2) and a six-degree-of-freedom motion tracking system. The transvaginal probe is comprised of two pressure sensor arrays, a temperature sensor, a microheater, and a motion tracking sensor. The first array contains 104 pressure sensors and is installed on the probe surface to contact the vaginal wall. The second array contains 12 pressure sensors and is mounted in the probe tip to contact the uterus during the examination procedure. The VTI examination was performed on patients in the dorsal lithotomy position with an empty bladder and rectum. The full VTI examination required 3–5 min to complete. Three orthogonal projections of the 3-D vaginal pressure map with the VTI probe location were observed in real time by the operator. The VTI clinical operators were trained on pelvic floor models prior to the VTI clinical application to standardize imaging techniques. The examination procedure includes multiple compressions of the vaginal walls and allows a circumferential 3-D tactile image or pressure map of the vagina to be composed. The tissue elasticity, Young's modulus (E), was calculated from the spatial gradients in the resulting 3-D tactile images. A nonlinear tissue deformation model used after having been validated with silicone samples with a known elasticity distribution (see section "Accuracy and reproducibility").

All 136 enrolled women were successfully examined with the VTI. Three-dimensional images of the vagina and the surrounding structures were recorded and stored. The elasticity distribution and its variation for normal pelvic floor conditions were established. Substantial differences were found in the anatomy and tissue elasticity between normal and prolapse conditions. Specifically, changes from 150 to 300% in the vaginal tissue elasticity were observed for Stage II and III prolapse compared to normal conditions. The range of normality (normal pelvic support) for the tissue elasticity (Young's modulus; kPa) of the apical anterior is [5.3; 27]; the apical posterior is [4.4; 23]; the midanterior is [6.7; 32]; the midposterior is [8.0; 67]; and the mid lateral vaginal walls is [7.5; 31]. The results show that Stage I prolapse cases have about 50% overlap with the ranges of the tissue elasticity for the normal conditions (Fig. 10). The average value of the tissue elasticity for the anterior and posterior walls under prolapse Stages II and III was estimated to be 2.9 ± 1.8 kPa; these cases have only about 5% overlap with the normal conditions [66].

The VTI enables the quantification of the vaginal tissue elasticity and the strong differentiation between normal and Stage II and III prolapse conditions. The overlap in tissue elasticity between normal and Stage I prolapse conditions means that (a) in some cases under Stage I prolapse the tissue elasticity is the same as in normal conditions; or (b) the normal case, as defined by the POPQ system, with lower values of Young's modulus, is already in the range for Stage I prolapse. In other words, the lower level of the normality range with the decreased tissue elasticity in Fig. 10 may enable the detection of preprolapse conditions that require attention to delay the development of anatomically not observable prolapse conditions.

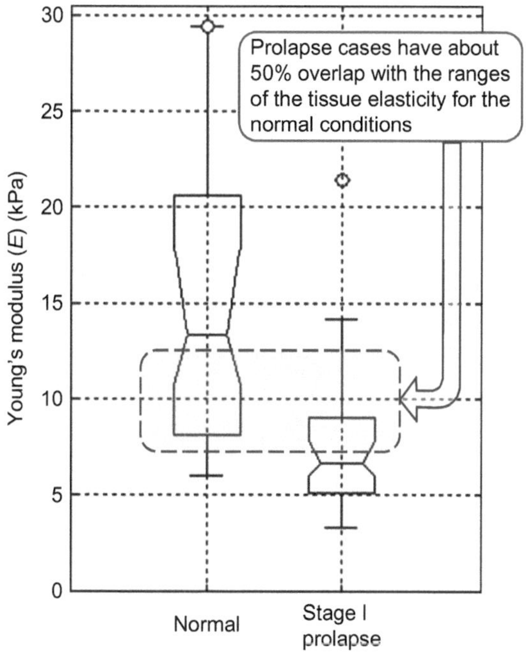

Figure 10 Tissue elasticity of normal versus Stage I prolapse conditions for midposterior and anterior combined.

The pressure pattern on the surface of the vaginal wall under applied load reveals not only the elasticity conditions of the vaginal wall itself, but basically the elasticity distribution of the underlying structures. The greater the applied load from the probe, the deeper the structures surrounding the vagina may be visualized. The softer the tissues, the deeper the structures may be visualized by TI because these softer tissues may be deformed more. The elasticity values calculated at the mid lateral vaginal walls may be related to the conditions of the underlying puborectalis, levator ani, and obturator internus muscles; the midposterior elasticity to the rectovaginal fascia and perineal body; the apical posterior elasticity to the uterosacral ligaments; and the anterior elasticity to vesical and pubocervical fascia. The pressure patterns on the surface of the vaginal walls together with the tissue displacement under the applied deformations can be considered as a documentation of the current elasticity state of the vagina and the pelvic floor support structures that are elements of the biomechanical system providing a critically important set of physiological processes.

This study concluded that the normality ranges for the tissue elasticity of the vagina and the pelvic floor support structures evaluated by the VTI may be used as markers for characterizing pelvic floor conditions. It seems possible, at least in 50% of the cases, to use VTI for the detection of early preprolapse conditions not observed by the pelvic organ prolapse quantification (POPQ) approach.

Changes with Age and Parity

The clinical data set presented in section Quantifying Prolapse Conditions was analyzed to assess vaginal tissue elasticity measured with VTI in women across age and parity. Fig. 11 presents the vaginal tissue elasticity distribution for the midposterior region of the vagina across a range of age and parity. The average values of Young's modulus are reduced with age from 13.1 ± 7.8 to 6.14 ± 3.0 kPa and with parity from 15.3 ± 9.7 to 7.0 ± 4.4 kPa.

In the apical anterior and posterior compartments, larger tissue elasticity changes were found. The average value of Young's modulus for the apical anterior vaginal compartment was 13.2 ± 3.1 kPa for a group aged 28–35 years old, and 3.0 ± 1.5 kPa for a group aged 76–90 years. The average value of tissue elasticity for the apical posterior vaginal compartment for the same age groups in comparison was 8.8 ± 2.6 kPa and 2.6 ± 1.0 kPa, respectively. The apical anterior and posterior vaginal compartments of nulliparous women showed values of Young's modulus of 10.4 ± 4.2 kPa and 7.0 ± 3.4 kPa, respectively. For women having 5–6 pregnancies, the average values for tissue elasticity for the apical anterior and posterior compartments were 3.6 ± 2.5 kPa and 2.8 ± 2.2 kPa, respectively. It is important to note that the analyzed data set has poor correlation ($r = 0.21$) between age and parity.

This study suggests that, in vaginal tissues, Young's modulus may be affected by 77% due to age and by 66% due to childbirth.

TI Markers

In another clinical study, VTI was performed on 22 subjects with probe 3 (Fig. 1) [12,13]. The examination consisted of four tests: (1) probe insertion, (2) elevation, (3) rotation, and (4) muscle contraction. These tests yielded the following information:

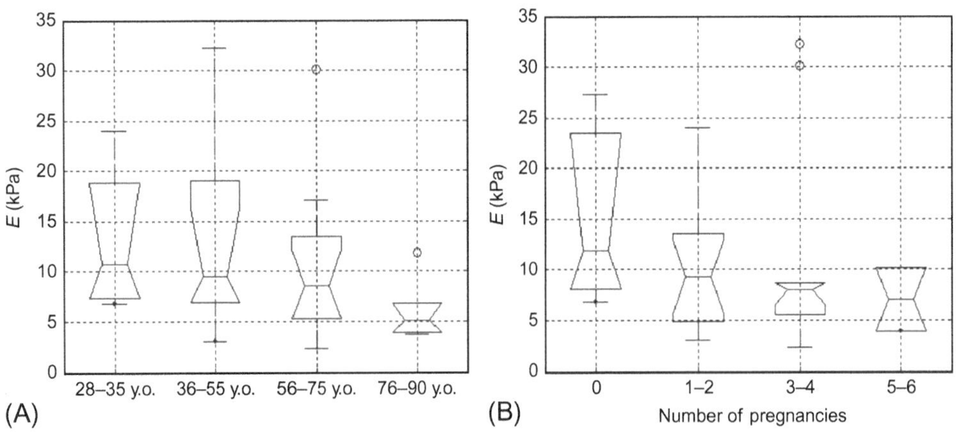

Figure 11 Midposterior tissue elasticity, Young's modulus (E), versus patient age (A) and parity (B).

- Test 1: Tactile image for the vaginal anterior and posterior compartments along the entire vagina; pressure gradients (elasticity) and anatomical measures.
- Test 2: Tactile image for apical anterior and posterior compartments related to pelvic floor support structures; pressure gradients and anatomical measurements.
- Test 3: Tactile images for left and right sides of vagina (circumferential tactile image from vaginal walls); anatomical measurements.
- Test 4: Dynamic pressure response from pelvic floor muscle contractions recorded from the opposite sides along the entire vagina; static and dynamic components.

Review of the tactile images revealed several anatomic areas with consistently observed pressure peaks across the VTI scans. These areas were selected as marker sites for further analysis. There were two anterior marker sites (A1, A2), two posterior sites (P1, P2), and one lateral site (L1). Specifically, the following locations along the pelvic floor were used for marker analysis: A1 — anterior in the vicinity of the hymen with maximum pressure feedback; A2 — anterior and proximal to A1; P1 — posterior in the vicinity of hymen; P2 — posterior and proximal to P1; L1 — vaginal sides with maximum pressure peaks in the vicinity of hymen. Fig. 12 illustrates the listed locations. Unlike the static locations of POPQ points (eg, Aa and Ba), the pressure peaks used for A1, A2, B1, B2, and L1 are not in fixed locations but varied among the patients.

Figs. 13–16 summarize the study results in boxplot panels and provide examples of acquired images for four tests [13].

The data showed that, in patients with prolapse, pressure gradient measurements decreased twofold to fourfold (200 to 400%) at specific locations (see Fig. 13 panels B and D), which can be interpreted as being twofold to fourfold softer compared to patients

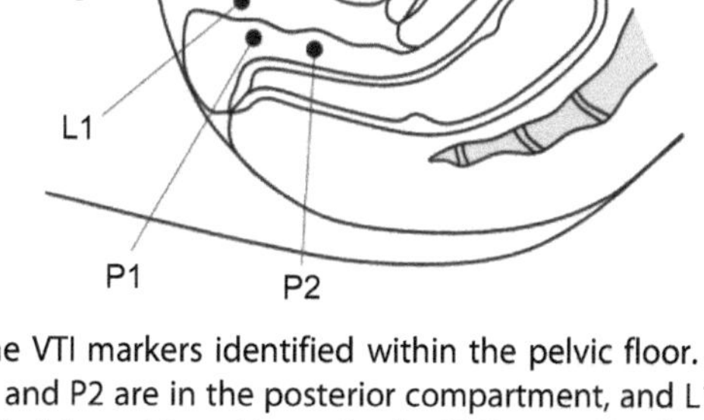

Figure 12 Locations of the VTI markers identified within the pelvic floor. A1 and A2 are within the anterior compartment, P1 and P2 are in the posterior compartment, and L1 is in the lateral compartments (left and right sides). *Adapted from H. van Raalte, V. Egorov, Tactile imaging markers to characterize female pelvic floor conditions, Open J. Obstet. Gynecol. 5 (2015) 505–515.*

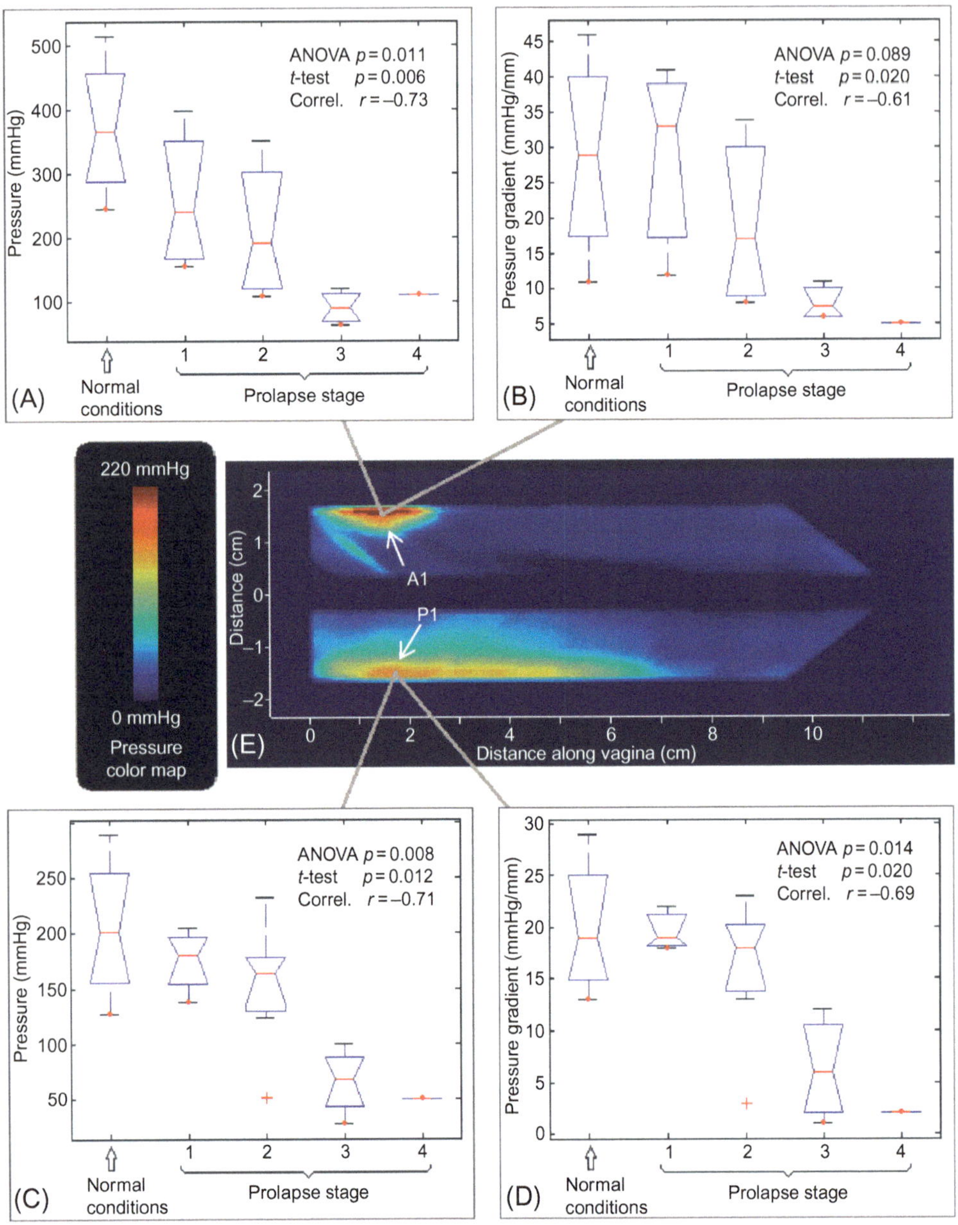

Figure 13 VTI Test 1 — Probe insertion results. Tactile imaging markers at distal anterior (panels A and B) and distal posterior (panels C and D), which are sensitive to varying degrees of prolapse. Panel E shows a typical pressure response map (tactile image) for this test. *Adapted from H. van Raalte, V. Egorov, Tactile imaging markers to characterize female pelvic floor conditions, Open J. Obstet. Gynecol. 5 (2015) 505–515.*

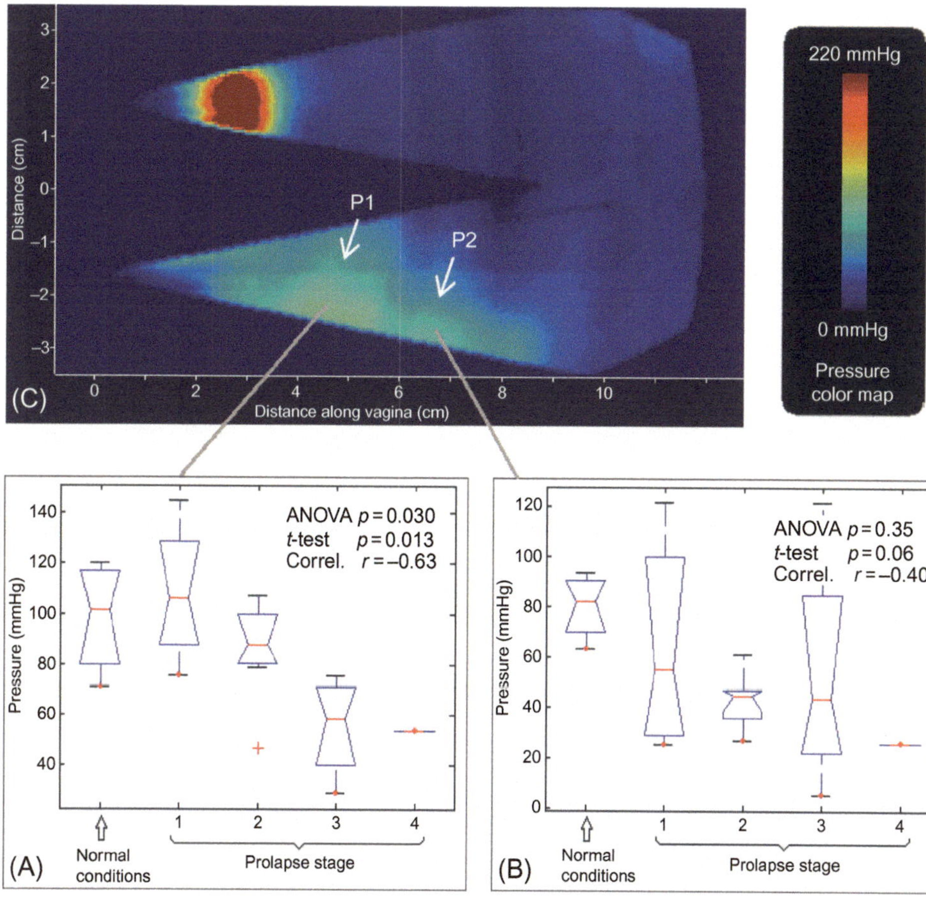

Figure 14 Test 2 — Probe elevation results. Tactile imaging markers at posterior (panels A and B), which are sensitive to varying degrees of prolapse. Panel C shows a typical pressure response map (tactile image) for this test. *Adapted from H. van Raalte, V. Egorov, Tactile imaging markers to characterize female pelvic floor conditions, Open J. Obstet. Gynecol. 5 (2015) 505–515.*

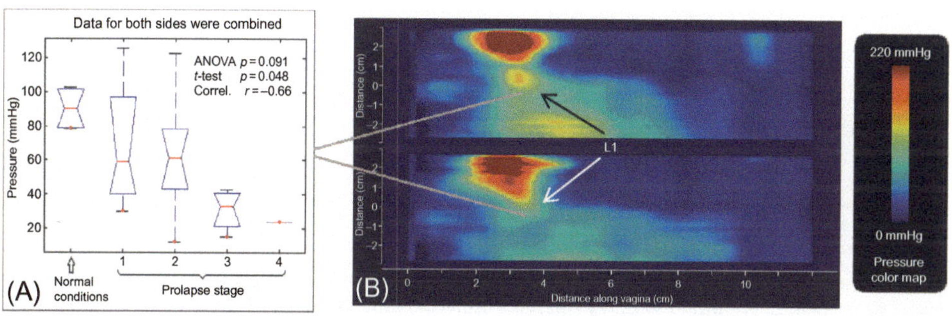

Figure 15 Test 3 — Probe rotation results. Tactile imaging marker at distal lateral location (panel B), which is sensitive to varying degrees of prolapse. Panel A shows a typical pressure response map (tactile image) for this test. *Adapted from H. van Raalte, V. Egorov, Tactile imaging markers to characterize female pelvic floor conditions, Open J. Obstet. Gynecol. 5 (2015) 505–515.*

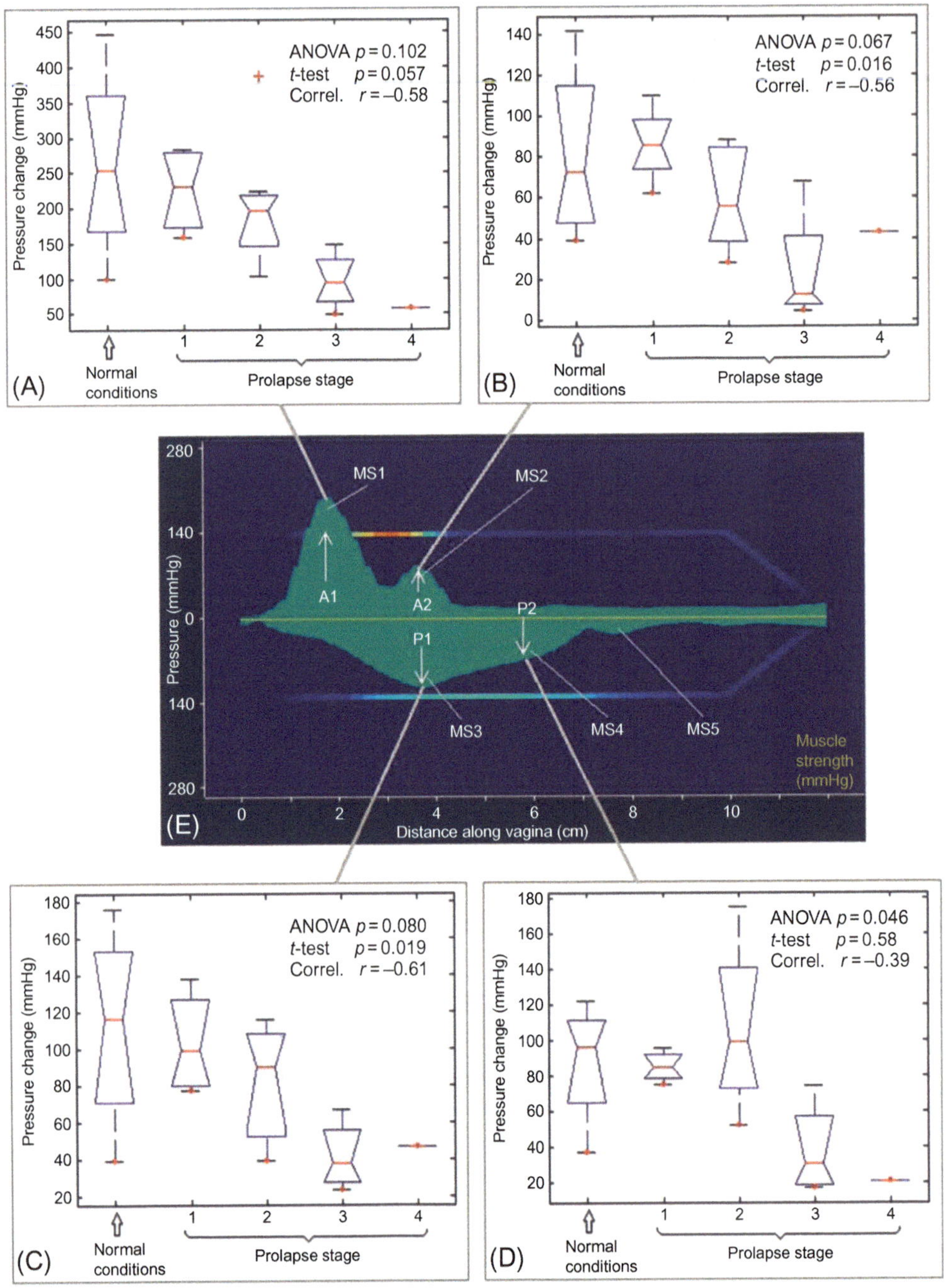

Figure 16 Test 4 — Voluntary PFM contractions. Muscle contraction capabilities at anterior (panels A and B) and posterior (panels C and D), which are sensitive to varying degrees of prolapse. Panel E shows distribution of maximum pressure response for anterior and posterior compartments for this test. *Adapted from H. van Raalte, V. Egorov, Tactile imaging markers to characterize female pelvic floor conditions, Open J. Obstet. Gynecol. 5 (2015) 505–515.*

with normal support. Pelvic floor muscle contractive capabilities (muscle strengths) decrease up to five times (500%) in prolapse subjects compared to normal subjects (see Fig. 16). These results demonstrate that women with prolapse have significant mechanical differences within the vaginal and surrounding pelvic floor supportive systems. In addition to recording tactile feedback during the tissue deformation, the VTI obtains measurement of muscle strength and allows evaluation of the relative functional impact of muscle contraction on measured biomechanical properties. Of the five pressure peaks observed during pelvic floor muscle squeezing (see Fig. 16 panel E), site A1 (see Fig. 16 panel E) is potentially exaggerated or an artifact because of the pubic bone, but this peak does have a lateral component that contradicts sole resistance from the static structure. These peaks have a complex, dynamic pattern and require further investigation.

The results of this study demonstrate that the vaginal TIs can be acquired and coupled with functional pelvic muscle assessment in one VTI examination [13].

Functional Imaging

Vaginal TI was used to investigate the behavior of the pelvic floor muscles during the Valsalva maneuver, voluntary muscle contraction, involuntary relaxation, and involuntary contraction, in a group of 77 women, using probe 3 (Fig. 1). The subjects were classified as being normal, having POP, and having SUI [68,69]. The data analysis reveals the following findings.

1. During Valsalva, nonuniform pressure patterns are observed along the anterior and posterior compartments that are substantially different than those observed during pelvic floor muscle contraction. The high-pressure zones correspond to softer tissues with low support capability; these high-pressure zones (30–50 mm Hg) appear to have been created by penetration of surrounding load through an internal strain (easily flexed). That means that this test allows detection of biomechanically weak structures and their locations.
2. Significant amplitude difference (ie, the ratio in voluntary muscle contractions) for anterior versus posterior and left versus right side, which may allow recognizing of muscle avulsion and further characterization of their functional conditions. It seems possible to recreate their dynamics in 3-D.
3. During the involuntary muscle relaxation, the patient was asked to maintain a sustained pelvic muscle strain. By quantifying the pressure decline versus time (angle) at a region of interest or for specific muscle, it seems possible to characterize functional conditions of multiple pelvic dynamic structures.
4. The pressure patterns during involuntary muscle contraction (cough) are substantially different from voluntary contractions in amplitudes as well as in peak locations.
5. Patients with normal pelvic floor conditions demonstrate higher pressure applied amplitudes in both voluntary and involuntary muscle contractions than the patients with SUI.

6. The pressure patterns during involuntary muscle contraction (cough) with SUI conditions have distinctive structure from the patterns without SUI.
7. It seems possible to characterize functional conditions of multiple dynamic pelvic structures.

SUMMARY

There is a complex relationship between the supportive ligaments, fascias, and pelvic floor muscles to provide the resulting support and contribute to diseased conditions. TI allows allocation and quantitative assessment of biomechanical defects such as changes in tissue elasticity at specified locations, possible muscle avulsion, and weakness in muscle function within the pelvic floor.

VTI offers an opportunity to assess the vaginal support along the entire length of the anterior, posterior, and lateral vaginal walls at rest, with manually applied deflection pressures and with voluntary and involuntary muscle contraction, involuntary relaxation, and Valsalva maneuvers. This allows a large body of measurements to evaluate individual variations in support defects as well as identify specific potential markers to measure tissue properties as they correlate to pelvic floor support in patients with POP and/or SUI. This technology gives the ability to measure pelvic floor muscle strength at specific locations along the vaginal wall and help correlate the relative contributions to measured tissue properties. These measurements provide insight into the pelvic floor function and conditions of the support structures, which in turn will allow quantitative diagnostics of the affected or defective biomechanical components. There may be very important distinctions in muscle function between a patient with a large distention defect of the vaginal wall and a primary apical defect, symptomatic and asymptomatic prolapse, or among patients with associated SUI or rectal complaints. For future studies, it would be important to evaluate symptom severity for pelvic floor disorders to determine whether there is a correlation between pelvic floor muscles function, resting tone, and associated elasticity measurements of the underlying tissue. This may help further differentiate the types of pelvic floor conditions, their underlying severity, and understand how to tailor treatments for the individual patient in the most effective manner.

To date, it has been challenging to combine functional anatomy with in vivo tissue properties. It would be extremely useful to correlate mechanical properties measured by the VTI with dynamic MRI and ultrasound measurements. The combination of these modalities, image fusions, anatomic and functional valuation will provide the best assessment of pelvic floor conditions before making any decision about application of a cure.

CONCLUSIONS

- TI VTI may serve as a means for 3-D imaging of the vagina and a quantitative assessment of vaginal tissue.
- TI provides biomechanical insight into POP development by evaluating pelvic floor conditions and functions.
- TI may allow assessment of pelvic muscle conditions/defects that contribute to SUI by quantitative, anatomically sensitive, and specific manner.

ACKNOWLEDGMENTS

The authors would like to thank Robin Haff, RN, BSN, and Alison Cooke, RN, for study coordination; Lioudmila Lipetskaia, MD, Miles Murphy, MD, Cristina Saiz, MD, Vladimir Kalis, MD, Zdenek Rusavy, MD, and Andrea Barker, PA, for participation in clinical studies; Randee Wood, MS, RDMS, for clinical research documentation and data management; Milind Patel, MS, for technical assistance with the device; Victor Kurtenok, BS, for electronics and firmware/software development; and Noune Sarvazyan, PhD, MBA, for text editing assistance and project management.

The work was supported by the National Institute on Aging, USA, grant AG034714 "Vaginal Tactile Imager for Pelvic Floor Biomechanical Assessment."

REFERENCES

[1] W.L. Johnston, The role of static and motion palpation in structural diagnosis, J. Am. Osteopath. Assoc. 75 (4) (1975) 421–424.

[2] A. Sarvazyan, V. Egorov, N. Sarvazyan, Tactile sensing and tactile imaging in detection of cancer, in: K. E. Harold, A. Rasooly (Eds.), Biosensors and Molecular Technologies for Cancer Diagnostics, Taylor & Francis, Boca Raton, FL, 2012, pp. 337–352.

[3] A. Sarvazyan, T.J. Hall, M.W. Urban, M. Fatemi, S.R. Aglyamov, B.S. Garra, An overview of elastography — an emerging branch of medical imaging, Curr. Med. Imag. Rev. 7 (4) (2011) 255–282.

[4] A. Sarvazyan, V. Egorov, Mechanical imaging — a technology for 3-D visualization and characterization of soft tissue abnormalities: a review, Curr. Med. Imag. Rev. 8 (1) (2012) 64–73.

[5] D.D. Rahn, M.D. Ruff, S.A. Brown, H.F. Tibbals, R.A. Word, Biomechanical properties of the vaginal wall: effect of pregnancy, elastic fiber deficiency, and pelvic organ prolapse, Am. J. Obstet. Gynecol. 198 (5) (2008). 590.e1-6.

[6] X. Jiang, P. Asbach, K.J. Streitberger, A. Thomas, B. Hamm, J. Braun, I. Sack, J. Guo, In vivo high-resolution magnetic resonance elastography of the uterine corpus and cervix, Eur. Radiol. 24 (12) (2014) 3025–3033.

[7] L. de Landsheere, S. Blacher, C. Munaut, B. Nusgens, C. Rubod, A. Noel, J.M. Foidart, M. Cosson, M. Nisolle, Changes in elastin density in different locations of the vaginal wall in women with pelvic organ prolapse, Int. Urogynecol. J. 25 (12) (2014) 1673–1681.

[8] J. Luo, T.M. Smith, J.A. Ashton-Miller, J.O. DeLancey, In vivo properties of uterine suspensory tissue in pelvic organ prolapse, J. Biomech. Eng. 136 (2) (2014) 021016.

[9] V. Egorov, H. van Raalte, A. Sarvazyan, Vaginal tactile imaging, IEEE Trans. Biomed. Eng. 57 (7) (2010) 1736–1744.

[10] A. Sarvazyan, Mechanical imaging: a new technology for medical diagnostics, Int. J. Med. Inform. 49 (2) (1998) 195–216.

[11] V. Egorov, H. van Raalte, V. Lucente, Quantifying vaginal tissue elasticity under normal and prolapse conditions by tactile imaging, Int. Urogynecol. J. 23 (4) (2012) 459–466.

[12] H. van Raalte, V. Egorov, Characterizing female pelvic floor conditions by tactile imaging, Int. Urogynecol. J. 26 (4) (2015) 607–609. with Video Supplement.

[13] H. van Raalte, V. Egorov, Tactile imaging markers to characterize female pelvic floor conditions, Open J. Obstet. Gynecol. 5 (2015) 505–515.

[14] S. Timoshenko, J.N. Goodier, Theory of Elasticity, McGraw-Hill Book Company, New York, 1951. 1–508.

[15] L.D. Landau, E.M. Lipshitz, Theory of Elasticity, third ed., (1970). 1-172.

[16] A.T. Taber, Nonlinear Theory of Elasticity. Applications in Biomechanics, World Scientific Pub. Co. Inc., Toh Tuck Link, Singapore, 2004, pp. 1–416

[17] A.R. Skovoroda, A.N. Klishko, D.A. Gusakian, E.I. Maevskii, V.D. Ermilova, G.A. Oranskaia, A.P. Sarvazyan, Quantitative analysis of the mechanical characteristics of pathologically changed biological tissues, Biophysics 40 (6) (1995) 1359–1364.

[18] T.A. Krouskop, T.M. Wheeler, F. Kaller, B.S. Garra, T. Hall, Elastic moduli of breast and prostate tissues under compression, Ultrason. Imaging 20 (4) (1998) 260–274.

[19] R.Q. Erkamp, P. Wiggins, A.R. Skovoroda, S.Y. Emelianov, M. O'Donnell, Measuring the elastic modulus of small tissue samples, Ultrason. Imaging 20 (1) (1998) 17–28.

[20] P.S. Wellman, R.D. Howe, E. Dalton, K.A. Kern, Breast tissue stiffness in compression is correlated to histological diagnosis. Technical Report. Harvard BioRobotics Laboratory, 1999; pp. 1–15.

[21] F.A. Duck, Physical properties of tissue: a comprehensive reference book, in: Mechanical Properties of Tissue, Academic Press, London, San Diego, 1990, pp. 137–165 (Chapter 5).

[22] Y.C. Fung, Biomechanics: Mechanical Properties of Living Tissues, second ed., Springer-Verlag, New York, 1993, pp. 1–568.

[23] A. Sarvazyan, M. Levy, H.E. Bass, R.R. Stern (Eds.), Elastic Properties of Soft Tissues Handbook of Elastic Properties of Solids, Liquids, and Gases, vol. III, Academic Press, New York, 2001, pp. 107–127 (Chapter 5).

[24] S.F. Leung, Y.P. Zheng, C.Y.K. Choi, S.S.S. Mak, S.K.W. Chiu, B. Zee, et al., Quantitative measurement of post-irradiation neck fibrosis based on Young's modulus: descriptive of a new method and clinical results, Cancer 95 (2002) 656–662.

[25] T.A. Krouskop, D.R. Dougherty, F.S. Vinson, A pulsed Doppler ultrasonic system for making non-invasive measurements of the mechanical properties of soft tissue, J. Rehabil. Res. Dev. 24 (1987) 1–8.

[26] V. Egorov, S. Ayrapetyan, A.P. Sarvazyan, Prostate Mechanical Imaging: 3-D image composition and feature calculations, IEEE Trans. Med. Imaging 25 (10) (2006) 1329–1340.

[27] V. Egorov, A.P. Sarvazyan, Mechanical imaging of the breast, IEEE Trans. Med. Imaging 27 (9) (2008) 1275–1287.

[28] V. Egorov, T. Kearney, S.B. Pollak, C. Rohatgi, N. Sarvazyan, S. Airapetian, S. Browning, A. Sarvazyan, Differentiation of benign and malignant breast lesions by mechanical imaging, Breast Cancer Res. Treat. 118 (1) (2009) 67–80.

[29] R.E. Weiss, V. Egorov, S. Ayrapetyan, N. Sarvazyan, A. Sarvazyan, Prostate mechanical imaging: a new method for prostate assessment, Urology 71 (3) (2008) 425–429.

[30] D. Turo, P. Otto, V. Egorov, A. Sarvazyan, L.H. Gerber, S. Sikdar, Elastography and tactile imaging for mechanical characterization of superficial muscles, J. Acoust. Soc. Am. 132 (3) (2012) 1983.

[31] V. Egorov, M. Patel, A.P. Sarvazyan, Vaginal tactile imager for direct modulus estimation, in: Proceedings of the 10th International Tissue Elasticity Conference, Arlington, Texas, Oct 12-15, 2011, p. 59.

[32] E.H. Frei, B.D. Sollish, S. Yerushalmi, Inventors; Yeda Research and Development Co. Ltd., assignee. Instrument for viscoelastic measurement. U.S. Patent 4,144,877; March 1979.

[33] E.H. Frei, B.D. Sollish, S. Yerushalmi, Inventors; Yeda Research and Development Co. Ltd., assignee. Instrument for viscoelastic measurement. U.S. Patent 4,250,894; Feb 1981.

[34] C.R. Gentle, Mammobarography: a possible method of mass breast screening, J. Biomed. Eng. 10 (2) (1988) 124–126.

[35] P. Dario, M. Bergamasco, A. Sabatini, Sensing body structures by an advanced robot system. Robotic and Automation, Proc. IEEE Int. Conf. 3 (1988) 1758–1763.

[36] A.M. Sabatini, P. Dario, M. Bergamasco, Interpretation of mechanical properties of soft tissues from tactile measurements, First Int. Symp. Exp. Robotics 139 (1990) 452–462.
[37] K. Koganezawa, A. Takanishi, S. Sugano, Development of Waseda Robot: The Study of Biomechanisms and Kato Laboratory, third ed., Waseda University, 1991.
[38] A.P. Sarvazyan, A.R. Skovoroda, Method and apparatus for elasticity imaging. U.S. Patent 5,524,636; filing date: Dec 21, 1992; publication date: June 11, 1996.
[39] P.S. Wellman, Tactile imaging. PhD Thesis. Harvard University's Division of Engineering and Applied Sciences Cambridge, MA, 1999, pp. 1–158.
[40] A.M. Galea, Mapping tactile imaging information: parameter estimation and deformable registration. PhD Thesis. Harvard University's Division of Engineering and Applied Sciences Cambridge, MA, 2004, pp. 1–235.
[41] P.S. Wellman, E.P. Dalton, D. Krag, K.A. Kern, R.D. Howe, Tactile imaging of breast masses: first clinical report, Arch. Surg. 136 (2) (2001) 204–208.
[42] P.P.L. Regtien, Tactile imaging, Sens. Actuators A Phys. 31 (1–3) (1992) 83–89.
[43] J. Tegin, J. Wikander, Tactile sensing in intelligent robotic manipulation — a review, Ind. Rob. 32 (1) (2005) 64–70.
[44] T. Oie, H. Suzuki, T. Fukuda, Y. Murayama, S. Omata, K. Kanda, Y. Nakayama, Tactile mapping system: a novel imaging technology for surface topography and elasticity of tissues or organs, Innovations (Phila) 4 (6) (2009) 345–350.
[45] C.E. Constantinou, S. Omata, Direction sensitive sensor probe for the evaluation of voluntary and reflex pelvic floor contractions, Neurourol.Urodyn. 26 (2007) 386–391.
[46] V. Raizada, V. Bhargava, S.A. Jung, A. Karstens, D. Pretorius, P. Krysl, R.K. Mittal, Dynamic assessment of the vaginal high-pressure zone using high-definition manometery, 3-dimensional ultrasound, and magnetic resonance imaging of the pelvic floor muscles, Am. J. Obstet. Gynecol. 203 (2) (2010). 172.e1-8.
[47] American Cancer Society, Breast Cancer Facts & Figures 2009 2010, American Cancer Society, Inc., Atlanta, GA, 2010, pp. 1–40.
[48] S.C. Han, P. Harrison, Myofascial pain syndrome and trigger-point management, Reg. Anesth. 22 (1) (1997) 89–101.
[49] S.E. Swift, The distribution of pelvic organ support in a population of female subjects seen for routine gynecologic health care, Am. J. Obstet. Gynecol. 183 (2000) 277–285.
[50] J.E. Jelovsek, C. Maher, M.D. Barber, Pelvic organ prolapse, Lancet 369 (9566) (2007) 1027–1038.
[51] I. Nygaard, M.D. Barber, K.L. Burgio, K. Kenton, S. Meikle, J. Schaffer, C. Spino, W.E. Whitehead, J. Wu, D.J. Brody, Pelvic Floor Disorders Network. Prevalence of Symptomatic Pelvic Floor Disorders in US Women, JAMA 300 (11) (2008) 1311–1316.
[52] F.J. Smith, C.D. Holman, R.E. Moorin, N. Tsokos, Lifetime risk of undergoing surgery for pelvic organ prolapse, Obstet. Gynecol. 116 (5) (2010) 1096–1100.
[53] A.L. Olsen, V.J. Smith, J.O. Bergstrom, et al., Epidemiology of surgically managed pelvic organ prolapse and urinary incontinence, Obstet. Gynecol. 89 (1997) 501–506.
[54] Urogynecologic Surgical Mesh: Update on the Safety and Effectiveness of Transvaginal Placement for Pelvic Organ Prolapse, *FDA*, *CDRH*, July 2011. Accessed on-line on August 14, 2015: http://www.fda.gov/downloads/MedicalDevices/Safety/AlertsandNotices/UCM262760.pdf.
[55] S.H. Boyles, A.M. Weber, L. Meyn, Procedures for pelvic organ prolapse and urinary incontinence in the United States, 1979–1997, Am. J. Obstet. Gynecol. 188 (2003) 108–115.
[56] I. Nygaard, L. Brubaker, M. Halina, et al., Long-term outcomes following abdominal sacrocolpopexy for pelvic organ prolapse, JAMA 309 (19) (2013) 2016–2024.
[57] L.Y. Korman, V. Egorov, S. Tsuryupa, B. Corbin, M. Anderson, N. Sarvazyan, A. Sarvazyan, Characterization of forces applied by endoscopists during colonoscopy by using a wireless colonoscopy force monitor, Gastrointest. Endosc. 71 (2) (2010) 327–334.
[58] V. Egorov, A. Sarvazyan, Operator training and performance descriptor for prostate mechanical imaging, in: Proceedings of the 7th International Conference on the Ultrasonic Measurement and Imaging of Tissue Elasticity, Austin, Texas, Oct 27-30, 2008, p. 27.

[59] M.A. Srinivasan, R.H. LaMotte, Tactual discrimination of softness, J. Neurophysiol. 73 (1995) 88–101.
[60] F.K.B. Freyberger, B. Färber, Compliance discrimination of deformable objects by squeezing with one and two fingers, in: Proceedings of EuroHaptics, 2006, pp. 271–276.
[61] W.M. Bergmann Tiest, A.M.L. Kappers, Cues for haptic perception of compliance, IEEE Trans. Haptics 2 (2009) 189–199.
[62] E.H. Weber, E.R. Ross, D.J. Murray, E.H. Weber on Sense of Touch, Taylor & Francis, East Sussex, UK, 1996, pp. 1–260.
[63] R. Shih, A. Dubrowski, H. Carnahan, Evidence for haptic memory. In: EuroHaptics Conference, 2009 and Symposium on Haptic Interfaces for Virtual Environment and Teleoperator Systems. World Haptics 2009. Third Joint. 18-20 March 2009. IEEE, Salt Lake City, UT, pp. 145–149
[64] A.P. Sarvazyan, V. Egorov, Human tissue phantoms and methods for manufacturing thereof. U.S. Patent No. 7,419,376; Sept 2, 2008.
[65] V. Egorov, S. Tsyuryupa, S. Kanilo, M. Kogit, A. Sarvazyan, Soft tissue elastometer, Med. Eng. Phys. 30 (2) (2008) 206–212.
[66] H. van Raalte, V. Egorov, V. Lucente, M. Murphy, C. Saiz, 3D tactile imaging in early prolapse detection, in: International Continence Society 43rd Annual Meeting. Barcelona, Spain, August 26-30, 2013.
[67] A.P. Sarvazyan, Knowledge-Based Mechanical Imaging, in: Proceedings of the 10th IEEE Symposium on Computer-Based Medical Systems, 1997, pp. 120–125.
[68] H. van Raalte, V. Lucente, V. Egorov, High definition pressure mapping of the pelvic floor muscles during Valsalva manever, voluntary muscle contraction and involuntary relaxation, in: American Urogynecologic Society 36th Annual Meeting, Seattle, WA, October 13-17, 2015.
[69] H. van Raalte, V. Lucente, V. Egorov, Pressure mapping of voluntary and involuntary muscle contraction for assessment of SUI conditions, in: International Continence Society 45th Annual Meeting. Montreal, Canada, October 6-9, 2015.

SECTION 4

Biomechanical Modeling and Simulation of the Pelvic Floor

J.A. Ashton-Miller
University of Michigan, Ann Arbor, MI, United States of America

CHAPTER SEVENTEEN

The Role of Computational Tools in Biomechanics

D.C. Simkins Jr.*,†, J.B. Alford*
*University of South Florida, Tampa, FL, United States of America
†Formerics, LLC, Tampa, FL, United States of America

THE NEED FOR BETTER SOFTWARE TOOLS

Computational tools for biomechanics can be classified into two broad groups, models for imagery and visualization and analytical models capable of also representing the behavior of the object. Biological objects are first described by medical images that do not possess a mathematical description of their geometry. Hence, one must be developed. Typically, an imaging system, for example, MRI or CT, is used to obtain an image stack with a set of discrete points in space representing each object. From the discrete point set, a mathematical description is developed. In this way, a set of discrete points (eg, a segmented MRI of the pelvis) can be converted into either a visualization model for display purposes, or an analytic model, suitable for computational modeling analysis. Visualization models are most often developed as surface meshes from the label map data. The development of an analytical model, even starting from a visualization model, can be very time-consuming.

The main purpose for a visual model, as the name suggests, is to allow one to see, or visualize, a three-dimensional rendering of the object. Large collections of points are difficult to interpret, and a visual model can significantly contribute to understanding the shape and features of the object. For such a model it "is good" if it "looks good."

Analytic models, on the other hand, serve to provide a basis for simulating the response of a biological structure to some physical process, such as chemical diffusion or electromagnetic radiation. In the pelvic floor, this often means applying a deformation to tissue and determining the induced stresses and strains from that deformation.

Generally, analytic models can also be used as visual models, but they are much more than that. They allow an investigator to study and predict behavior that would be impossible or ethically impermissible, to study experimentally. With such models, many variations in treatment or surgical options can be compared, as well as tradeoffs in prosthetic types and designs. Greater understanding and discovery of mechanisms may be obtained by performing many virtual tests in an epidemiological study, provided that a sufficient number of image datasets can be converted into analytic models suitable for simulation.

Further, analytic models, by nature, contain a fully detailed mechanical history of deformation and the derived quantities. With an appropriately created analytic model, one can slice through portions of an organ to see what happens inside, or peel back layers of complexity that obscure clearer understanding.

A significant impediment to development of analytic models is the construction of the mathematical geometric description from the discrete point set. The details will be covered more fully later in this chapter, but this difficulty has necessitated practitioners to take several alternative approaches [1].

Ideally, every biomechanical analysis would be performed on a detailed analytical model of the precise organs or objects of interest. Such a model would supplement the detailed geometric information with specification of the constitutive parameters for each tissue and connection element. Often, this is too costly or time-consuming for routine use. One can make simplifications of the geometry to allow easier model development, but this leads to an idealized, generic model. For purposes of education and training, or studies of a general nature, this may be sufficient. Much can be learned from such first-order models [2].

To bridge the gap between these two approaches, some make use of a template [1,3–6]. In this method, a detailed analytical model is built for a single organ. Though it is time-consuming, it is done only once. Then, for another model, one takes the discrete point set and compares it to the original template point set. From that comparison, one then deforms or morphs the template-detailed analytical model into the shape of the object of interest. This is faster than developing a detailed model from scratch, but closer to the actual object than just the generic model. There are, however, still draw backs. First, the template may not be able to be deformed into the new object's shape and still retain the important analytical qualities of the representation. Further, a feature that does not exist in the template model, but does in the object of interest (or vice versa), will be lost. For example, if the template model has a hole in a certain spot, but the actual model does not, the morphed model will contain a hole where there should not be one. Clearly, there are instances where this could be quite unsatisfactory.

What is needed, therefore, is the ability to quickly and accurately develop analytical models from discrete point sets. Computational tools for studying the biomechanics of the pelvic floor must possess several characteristics. They must be:

1. Fast, not only in computer time, but human time and effort as well
2. Accurate in faithfully reproducing the geometric features of each specific, individual model
3. Easy to use for researchers and clinicians
4. Readily available to the user community

The first two items are solved by new approaches to be discussed in Meshfree Analysis section and demonstrated in Applications section. The last two are best addressed by commercial offerings of products based on the technology solving items 1 and 2.

METHODS AND CAPABILITIES

Process Overview

The sequence of steps to develop a model from medical images is depicted in Fig. 1. One begins with a stack of raw slices from an imaging machine, for example, MRI or CT (not shown in the figure). These slices must be segmented into *label maps* (first figure on the left) that represent pixel-level outlines of each tissue or organ of interest overlaid on the source MRI stack. The final set of label maps are processed into a point cloud or discrete point set by knowing the physical size represented by each pixel and the slice spacing, shown in the center figure. The discrete point set is the representation of the object as determined by the imaging machine and the segmented label map. The final

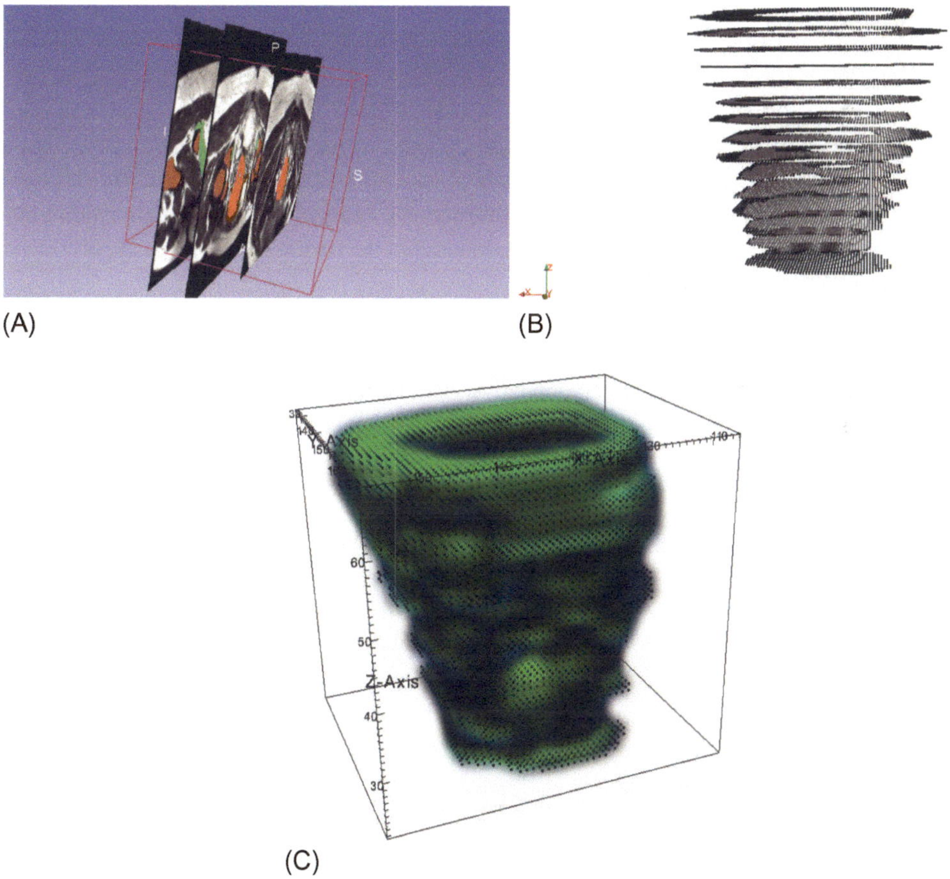

Figure 1 Overview of processing medical images to analysis-suitable geometry. (A) Label maps, (B) point set, and (C) 3D model.

step is to develop a continuous mathematical description of the geometry that best represents the object. Once the mathematical description is available, one can use it as a visual model or, if suitable, as an analytic model.

Visual Model

The most common visual representation is a surface mesh. A surface mesh consists of a set of polygonal faces, often triangles, that, taken together, form a surface covering of the object. The advantage of a surface mesh is that it is fast to generate and works very well for visualization. In fact, modern graphics cards have specialized hardware specifically for fast processing of triangular surface meshes. However, such a representation has several deficiencies:

1. Only a surface is represented, not the interior of the structure
2. Automated processing often leads to significant topological and geometric imperfections, including
   - **(a)** gaps between adjacent triangles
   - **(b)** duplicate vertices and/or triangles
   - **(c)** overlapping triangles
   - **(d)** selfintersecting surfaces

For primarily visual graphics applications, these problems do not matter and the speed and simplicity of a surface mesh outweighs its detractions. However, for applications in which the structural response to deformations needs to be determined computationally, negative aspects of the surface meshing technique can become significant.

Definition of Analysis-Suitable Geometry

To obtain the benefits of engineering analysis in understanding pelvic floor biomechanics, evaluating potential childbirth injury, and vetting different treatment options, an analysis-suitable geometric (ASG) model must be developed. Specifically, the term ASG is used to distinguish among various geometric representations of discrete point sets. For this document, the term *analysis* is used to mean the solution of the equations governing the physical process of interest, for example, mechanics, electromagnetics, etc., and is usually formulated as a partial differential equation. To be analysis suitable, the geometry must be a full solid three-dimensional representation, not merely a surface. One must be able to construct a set of mathematical functions on the geometric domain that are differentiable. This latter requirement requires that geometric gaps, jumps, and other irregularities be resolved prior to analysis.

Finite Element Analysis

Perhaps the most popular ASG representation is that of a finite element (FE) mesh [7–12]. A FE mesh extends the idea of a surface mesh to include a full polyhedral three-dimensional representation. For many complex geometries, especially those common to biologic systems, the automated mesh generation software used to create the finite element model (FEM) mesh suffers from many of the same issues as the surface mesh. A valid FEM mesh must meet the same requirements as any other ASG, so one is faced with a tedious manual intervention to guide the meshing software in generating and fixing the mesh until it meets the requirements of ASG. This human-intensive activity can take much more time than the computer time required to solve the governing equations of the physical problem and postprocess the analysis. For example, in the author's own research on shark teeth [13], the micro-CT scan took approximately 1 h to perform, and the resulting FE analyses consumed about 2 h of CPU time. The meshing portion required about 40 h of manual editing to convert the segmented micro-CT scan into an ASG for FE processing. That is about a 20:1 ratio of human to CPU time for processing the discrete point set from the micro-CT. In a pelvic floor example [2], the birthing simulation project took about 18 months to complete, 15 months of which was spent in manual development of ASG models for the relevant organs. This manual human intervention time is the single largest impediment to applying engineering analysis in pelvic floor applications, where the geometries can be complex, irregular, and contain tissue gaps. The accuracy of solution of an FEM analysis is dependent on the aspect ratio of the polyhedra that make up the mesh. There are several ways to define aspect ratio, but essentially it is the ratio of the longest side to the shortest side of a polyhedron and is a measure of whether an element is "long and skinny" or "short and wide," with high aspect ratios indicating the former. Thus, one desires a mesh filled with small aspect ratio elements, and such a mesh is called a *quality mesh*. The human time spent in meshing is mostly spent guiding the meshing software in creating a quality mesh. Poor quality meshes suffer not only from degraded accuracy, but they also perform poorly under large deformation, such as seen in pelvic floor muscles during childbirth. On the other hand, FEM is a very well-established technique backed by years of research and a robust mathematical foundation. As the dominate engineering analysis technique, there are many commercial product offerings and a number of open-source options, as well. Further, properly coded FEM software is computationally efficient.

Meshfree Analysis

A number of alternatives to FEM have been explored over the years to overcome some of the difficulties of FEM, in particular the restrictions imposed by the requirement that the FEM domain be represented by a quality mesh. For applications to geometries initially described by a discrete point set, the class of methods known as *meshfree* methods seems well-suited to the task. A number of meshfree methods exist, including the original method, smoothed particle hydrodynamics (SPH) [14,15], element free Galerkin [16], reproducing kernel particle method (RKPM) [17], diffuse particle method [18], hp-cloud methods [19], material point method [20], and many more. This chapter focuses on the RKPM as the underlying numerical method.

All of the expense involved in generating FEM solutions to problems in the pelvic floor comes from the model development stage. Developing a FEM generally involves tedious manual work when dealing with the complex geometry of the pelvic floor. Thus, often times, patient-specific models of the pelvic floor are highly idealized to simplify meshing and hide irregularities in the data that may lead to gaps, discontinuities, or other topological problems. In order to cut down on the cost associated with developing a highly detailed FE mesh, it is standard practice to either simplify the geometry down to a shell model [7–9,11,12,21,22], or to generate a volume mesh that does not necessarily incorporate all of the patient-specific features [10]. Meshfree methods, based on point sets, can avoid the trouble associated with accurately representing complex geometry with a mesh. A meshfree method can replace both the expensive FE mesh, as well as provide a computational advantage in large deformation problems over the FE analysis capabilities. Meshfree methods such as the RKPM use a point (particle) set discretization of an object in place of the FE mesh. Since no mesh is developed, meshfree models can be generated automatically with minutes of computer time. To re-emphasize, the adjective "automatic" means that no human intervention or guidance is required to develop the ASG from the point set. The reduction in model development time gained by switching to a meshfree method brings about the possibility of analyzing a large number of data sets.

Meshfree Correction Implicit Geometry

Meshfree models are composed of a set of discrete points, called meshfree particles, and an example of a collection of meshfree particles is shown in Fig. 2. Each meshfree particle is associated with a length parameter, ρ, and a modified kernel function, $\mathcal{K}(x)$, as defined in Refs. [17,23]. The kernel function is termed "modified" because the original kernel function from the SPH method does not satisfy consistency conditions, thus the originators of RKPM used a function to modify to the kernel to restore consistency in the meshfree approximation. A function f defined on the domain is approximated by an expression

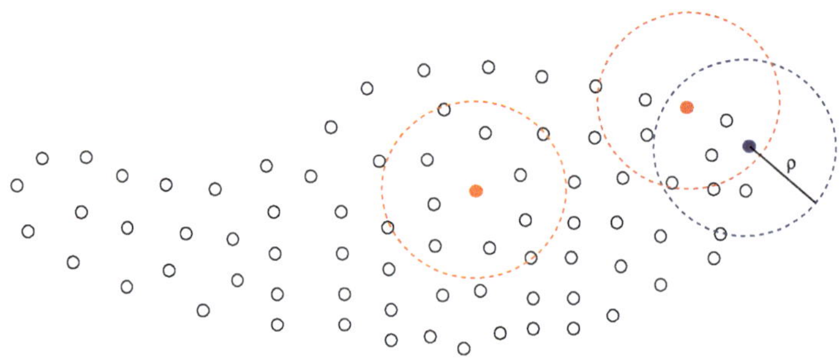

Figure 2 Collection of meshfree particles.

$$\mathcal{I}f(x) = \int_{\Omega} \mathcal{K}_{\rho}(x-y,x)f(y)\,dy \tag{1}$$

where $\mathcal{K}_{\rho}$ is the kernel function defined as

$$\mathcal{K}_{\rho}(x-y,x) = C_{\rho}(x-y,x)\boldsymbol{\Phi}_{\rho}(x-y) \tag{2}$$

where C_{ρ} is the kernel correction function defined as

$$C_{\rho}(x-y,x) = \mathbf{P}^T\left(\frac{x-y}{\rho}\right)\mathbf{b}(x) \tag{3}$$

The vector **P** contains the monomial terms of a polynomial of some order. In the present work, a trilinear basis is used, so

$$\mathbf{P}^T(x) = [1 \;\; x \;\; y \;\; z \;\; xy \;\; yz \;\; zx \;\; xyz] \tag{4}$$

Finally, the vector **b**, which we term the normalizer, is computed at each evaluation point as a correction function to enforce consistency conditions on the approximation. The consistency conditions lead to a set of equations

$$\left.\begin{aligned} \sum_I \mathcal{K}_{\rho}(x-x_I,x)\Delta x_I &= 1 \\ \sum_I \left(\frac{x-x_I}{\rho}\right)\mathcal{K}_{\rho}(x-x_I,x)\Delta x_I &= 0 \\ \sum_I \left(\frac{y-y_I}{\rho}\right)\mathcal{K}_{\rho}(x-x_I,x)\Delta x_I &= 0 \\ \sum_I \left(\frac{z-z_I}{\rho}\right)\mathcal{K}_{\rho}(x-x_I,x)\Delta x_I &= 0 \\ &\vdots \\ \sum_I \left(\frac{x-x_I}{\rho}\right)^{\alpha}\mathcal{K}_{\rho}(x-x_I,x)\Delta x_I &= 0 \end{aligned}\right\} \tag{5}$$

where α is a multiindex with $||\alpha|| = n$. Substituting Eqs. (2) and (3) into Eq. (5) leads to the following set of linear equations

$$\begin{pmatrix} m_0(x) & m_1(x) & \cdots & m_n(x) \\ m_1(x) & m_2(x) & \cdots & m_{n+1}(x) \\ \vdots & \vdots & \vdots & \vdots \\ m_n(x) & m_{n+1}(x) & \cdots & m_{2n}(x) \end{pmatrix}\begin{pmatrix} b_0(x) \\ b_1(x) \\ \vdots \\ b_n(x) \end{pmatrix} = \begin{pmatrix} 1 \\ 0 \\ \vdots \\ 0 \end{pmatrix} \tag{6}$$

or $\mathbf{M}(x)\mathbf{b}(x) = \mathbf{P}(0)$. The individual entries $m_i(x)$ are computed from

$$m_i(x) = \sum_I \left(\frac{x-x_I}{\rho}\right)^i \boldsymbol{\Phi}\left(\frac{x-x_I}{\rho}\right)\Delta V_i$$

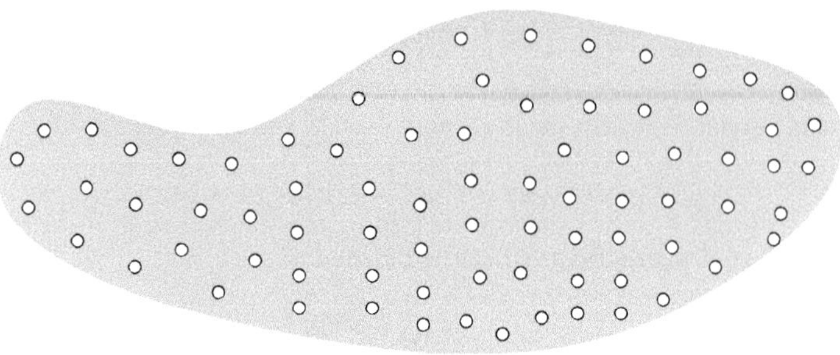

Figure 3 Approximation domain of meshfree particles.

The system in Eq. (6) is called the moment equation whose solution yields the values for the normalizer $\mathbf{b}(x)$. Once the normalizer is determined, completeness is restored to the meshfree approximation and the chosen polynomial field can be reproduced exactly. The meshfree correction implicit geometry (MCIG) is based upon the observation that the approximation domain, shown in Fig. 3, is defined where the moment matrix is non-singular, and thus a meshfree approximation can be made. Conversely, where the moment matrix is singular, there is no body. Though the object is represented as a discrete point set, it is shown that the approximation domain is smooth and continuous. The approximation domain is defined in MCIG by using the *condition number* of the moment matrix. The condition number is a measure of the error generated in matrix inversion and is defined in Ref. [24] as

$$\kappa(\mathbf{A}) := ||\mathbf{A}||\,||\mathbf{A}^{-1}||, \quad \kappa \in (1, +\infty) \tag{7}$$

The best conditioned matrix has $\kappa = 1$, a singular matrix has $\kappa \rightarrow \infty$. We define the MCIG function

$$\mathcal{F}(x) = \kappa(\mathbf{M}(x))^{-1} \tag{8}$$

where $\mathbf{M}(x)$ is the moment matrix at a point and κ is a function returning the condition. Notice that in Eq. (8), the reciprocal condition number is used and the range of the function is (0, 1], with $\kappa^{-1} = 0$ indicating a singular matrix. A particular surface can be defined for the geometry by selecting a level-set of the MCIG function, which defines a continuous surface in three dimensions or a line in two dimensions as shown in Fig. 4. The following is a demonstration of the MCIG for a simple geometric object — a cube. The cube is represented with 27 particles as shown in Fig. 5A. A particle MCIG surface of the meshfree cube is shown in Fig. 5B and C. Notice how the corners and edges of this cube are rounded and smooth which is a fundamental feature of the MCIG model and is very beneficial for recreating the smooth features in the pelvic floor from binary image data. The MCIG is directly computed from imaging data, yet it does not suffer from aliasing artifacts such as those found in voxel models. Further refinement of the number

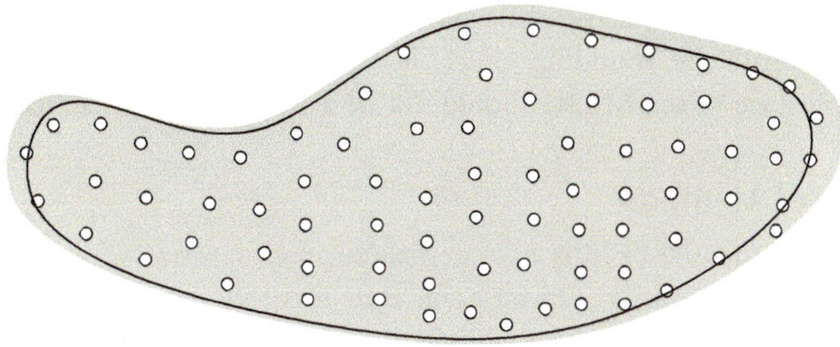

Figure 4 Approximation domain of meshfree particles with level-set.

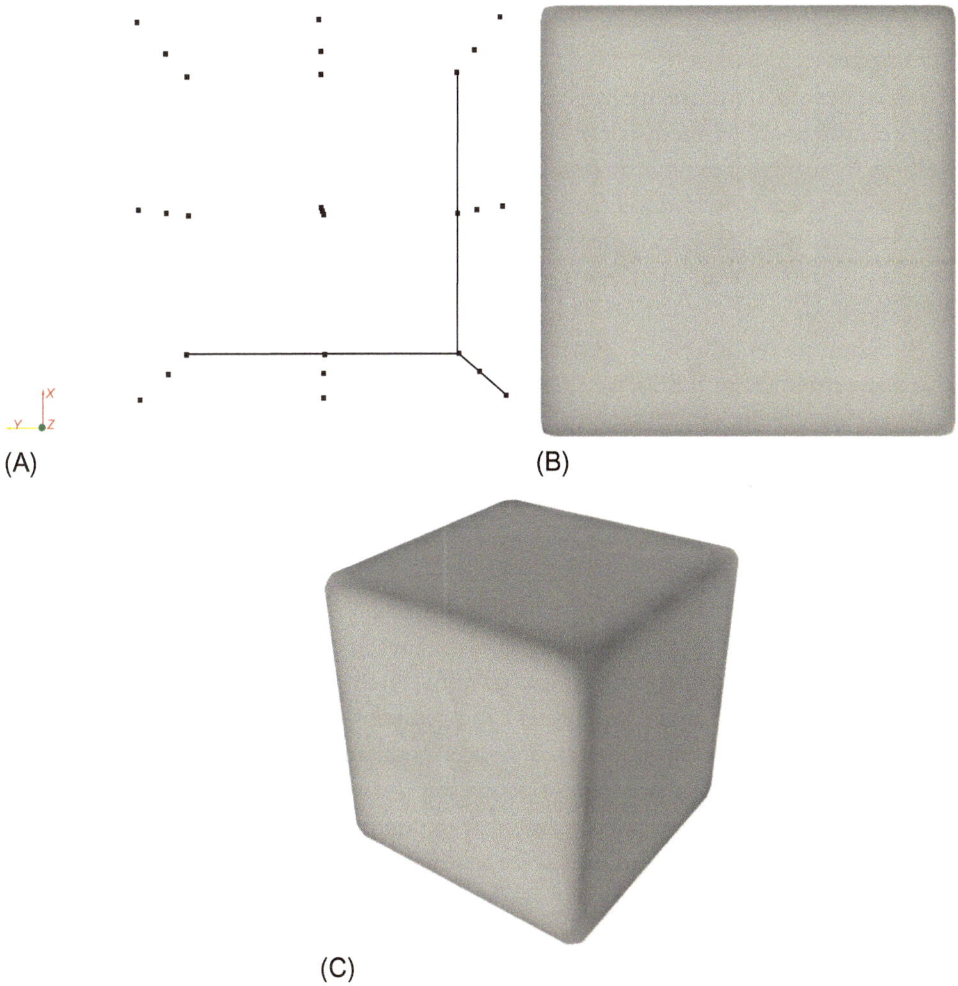

Figure 5 Meshfree correction implicit geometry representation of meshfree cube. (A) Reproducing kernel particle method particles, (B) front, and (C) perspective.

of particles used to represent the cube will sharpen the edges and corners, but never completely eliminate the rounding effect. Few biological objects have abrupt, sharp edges, so this feature of the MCIG should not be a concern in biomechanics.

Comparison of Methods

It should be noted that the FEM is an extremely successful, robust tool, and in problems where it is most suitable it should be used. But, it is still a numerical approximation method, and as such, it must have some limitations. The following sections will highlight the relative advantages and disadvantages of using a meshfree method for biomechanical analysis over the FEM.

Advantages

The main benefit of using a meshfree method is automation of the ASG. The tedious step of generating an FEM mesh is obviated, and replaced by simple point set generation. Meshing the irregular geometric biologic data tends to cover up any anomalous features in the data, which may lead to errors in the final simulation. MCIG, on the other hand, often shows a more faithful representation of what the data looks like. In the author's experience, the error in meshfree geometry is typically due to poor segmentation of the images.

Meshfree methods are also superior when modeling large deformation situations, which are common in the pelvic floor (eg, in childbirth, where the tissues can stretch by factors of 3 or more [25,26]). FE simulations can fail under large deformation due to element distortion. Since meshfree methods do not rely on a mesh, they are much less sensitive to large deformation [23].

Disadvantages

One difficulty that comes with using a meshfree method is exactly the thing that makes it efficient to use, the lack of a mesh to represent the object. The lack of a mesh makes visualization in three dimensions difficult. Virtually all computer graphics applications are based upon using a mesh to represent the geometry. Without a mesh, it is difficult to visualize either the initial geometry, deformed configuration, or plot the values of solution quantities, for example, stress or strain, on the object. In an analytical context, meshfree shape functions are more expensive to compute per evaluation than FE shape functions. For example, the author's code that implements FE and RKPM in the same structure, the RKPM solution is two to three times slower than the pure FE solution. However, noting the relative human to machine times in Table 1, tripling the CPU time in analysis is negligible compared to the time that would be saved in ASG development. Further, human time is expensive compared to computer time, so trading human effort for machine effort is always a win.

Table 1 Machine and human time to perform shark tooth analysis

Task	Segmenting	Meshing	Analysis
Machine time (min)	<1	<5	120
Human time (min)	60	2400	0

A mathematical issue arises with meshfree methods when applying Dirichlet, or essential boundary conditions [23]. The issue is that evaluation of shape functions along a boundary between particles do not linearly interpolate between the particles. The net effect is that if boundary particles are specified to have, say, zero displacement along the boundary, between the particles the displacement will not be exactly zero. This is termed weak enforcement, as opposed to strong enforcement of essential boundary conditions. In mechanics, St. Venant's principle states that local disturbances in the solution only yield local effects. Thus, a small imperfection along the boundary will only have a small effect near the boundary. Thus, in most biomechanical analyses, this is unlikely to be an issue. If it is an issue, though, there are ways to combine meshfree and FE methods so that FEM can be used only along the boundary to strongly enforce essential boundary conditions, and meshfree elsewhere. The requirement of a mesh only along certain portions of a boundary should be a much easier meshing problem and allows one to get the best of both worlds [27,28].

APPLICATIONS

The efficiency with which patient-specific models of the pelvic floor can be developed leads to the possibility of many new applications. These applications can be predictive, both in the sense of how a perturbation may affect an individual and in the sense of how a population may be affected by a specific perturbation. Patient-specific computational modeling could be used to determine the optimal surgical intervention for that patient, or it could be used to determine risk factors for a particular outcome (eg, childbirth injury) for a population of patients. The application could also be used to help design implanted prostheses, specifically predicting the behavior of different prosthesis designs in the body.

Types of Applications

The applications can be divided into different categories depending on what the investigator is seeking to understand. For example, to understand basic system behavior, a simplified, generic representative model is sufficient. Such an application might be the teaching of basic interactions among organs and muscles in the pelvic floor. Generic anatomical models can be simplified FE based, or meshfree based, and may help quickly understand behavior that would be unknown without some invasive procedure.

Understandably, the scope of uses of these generic models is limited by their simplicity. The pelvic floor is a very complex structure, and oversimplification of the model may lead to drastically incorrect results if the results from the generic model are applied to a specific patient with significant differences in her anatomy compared to the generic model.

The next type of application is modeling the actual geometry of an individual, that is, patient-specific modeling. This is a highly detailed model of an individual's anatomy and necessarily requires capture of the unique details of the anatomy. Since minimal simplification of the geometry is done for the patient-specific model, it can be assumed that any error associated with simplification is avoided. This type of analysis is demonstrated in modeling childbirth. Several stages of deformation of the levator ani during the passage of a 9 cm sphere through the birth canal is shown in Fig. 6. Other work in this problem area is published in Refs. [2,29].

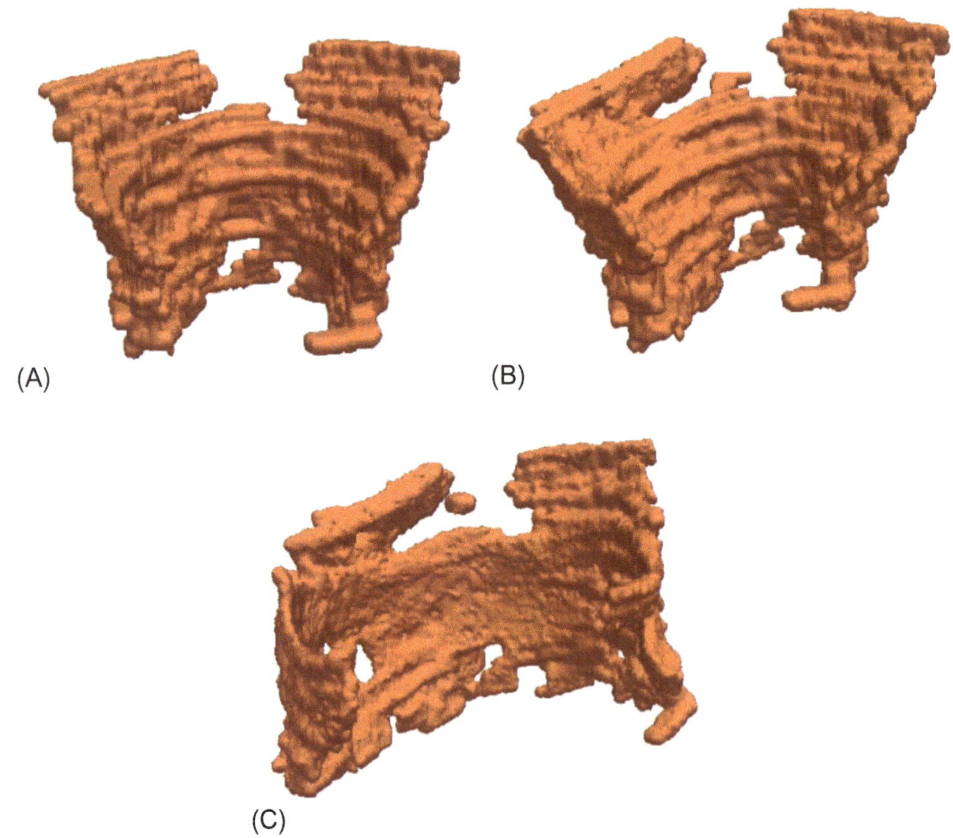

Figure 6 Various stages of childbirth simulation. (A) Stage 1, (B) stage 2, and (C) stage 3.

A third type of application is to study aggregate, or statistical information. In this case, the investigator seeks to run many individual models and extract a distribution of responses of the same process or stimuli and compare the differences in responses due solely to the differences in individual anatomies. This requires a large number of patient-specific models to be generated and analyzed to obtain the response distributions. An example of this use might be in an epidemiological study looking for correlations between patient trait and response.

Multipatient example studies of childbirth and vaginal contraction are shown in Fig. 7. The study included 12 patient-specific models and the maximum distortional strain in the vagina due to simulated radiation therapy of cervical cancer is plotted in Fig. 7A, and maximum distortional strain over all stages of delivery in the levator ani during the childbirth example above is plotted in Fig. 7B. These specific results are simply for proof of concept, and should not be viewed as accurate, but the figures do a good job of illustrating the benefit of this type of application. In Fig. 7A, there is clearly a patient that stands out from the rest of the group. This could be an indicator that this specific patient should be analyzed in further detail to discover the reason for the higher strain compared to the remainder of the population studied. On the other hand, in Fig. 7B, no clear pattern emerges and this may be an indication that the quantity of interest is either not clearly understood, or not a clear indicator in the problem studied.

CONCLUSIONS

In the world of traditional engineering design of products and processes, computational modeling has provided design engineers the ability to perform virtual tests that are either too expensive or physically impossible to perform. Further, the computational models provide a wealth of detailed information about nearly any aspect of the behavior or performance of the product under a wide variety of circumstances. All this has led to an unprecedented increase in quality, safety, and reliability of these products. The promise of computational modeling in biomechanics offers the same possible strides forward in understanding the pelvic floor, its components, and their mechanical characteristics. Biomechanical problems share many similarities with traditional engineering design problems, but they differ in several important ways that lead to new challenges. These differences mean that it is difficult to use the traditional tools of engineering analysis directly in biomechanical problems, and new tools are needed. The first fundamental step in both traditional engineering design and biomechanics is the ability to obtain realistic ASG models. In product design, this comes from the computer-aided design system used to create the geometry. In biomechanics, it begins with a stack of gray-scale images, generally obtained from a diagnostic imaging system, such as MRI or CT. The process for converting the images into geometric models involves creation of label maps, which in and of itself is a human-intensive effort. To truly automate the geometry creation for

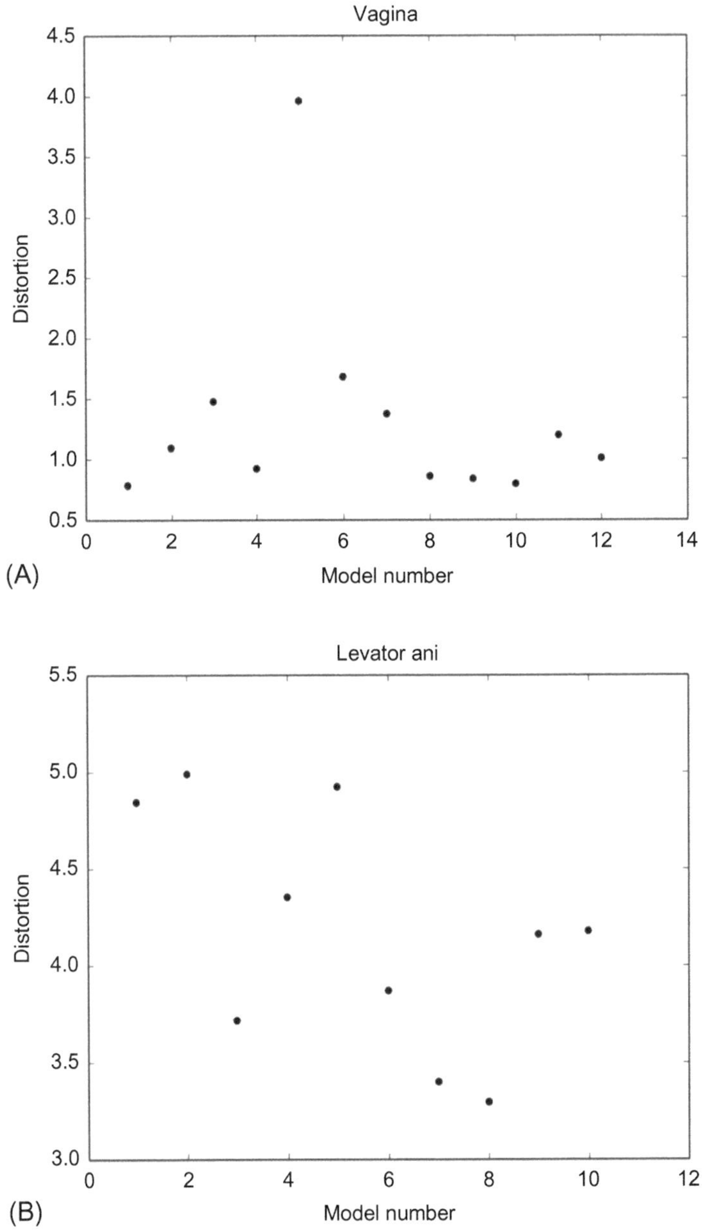

Figure 7 Epidemiological studies. (A) Vaginal contraction study involving 12 patients and (B) childbirth study involving 12 patients.

biomechanics, one also needs automated segmentation. Another chapter in this book describes the state of the art in automated segmentation.

To fully realize the vision of engineering design for biomechanics, one also needs suitable material response models and macroscopic process models to account for the many complications that living tissue and biological systems have that inanimate objects do not exhibit. Nonetheless, the community can now begin to explore the various geometric aspects of the pelvic floor system with the simple material and process models that do exist today. Large-scale studies can be undertaken to hunt through big data for correlations and related factors for various conditions. Development of normative ranges of response of the pelvic floor to such things as childbirth can begin to be developed to look for risk factors for a patient to develop various conditions, prolapse, for example.

ACKNOWLEDGMENT

The lead author is a cofounder and principal of Formerics LLC. The software and tools described in this chapter are owned and sold by Formerics LLC. Some of the material in this chapter is covered under US patent number 8,854,366 and additional pending patent.

REFERENCES

[1] L. Baghdadi, D.A. Steinman, H.M. Ladak, Template-based finite-element mesh generation from medical images, Comput. Methods Prog. Biomed. 77 (1) (2005) 11–21.

[2] L. Hoyte, M.S. Damaser, S.K. Warfield, G. Chukkapalli, A. Majumdar, D.J. Choi, A. Trivedi, P. Krysl, Quantity and distribution of levator ani stretch during simulated vaginal childbirth, Am. J. Obstet. Gynecol. 199 (2) (2008) 198.e1–198.e5.

[3] M. Bucki, C. Lobos, Y. Payan, A fast and robust patient specific finite element mesh registration technique: application to 60 clinical cases, Med. Image Anal. 14 (3) (2010) 303–317.

[4] B. Couteau, Y. Payan, S. Lavalle, The mesh-matching algorithm: an automatic 3D mesh generator for finite element structures, J. Biomech. 33 (8) (2000) 1005–1009.

[5] S. Ji, J.C. Ford, R.M. Greenwald, J.G. Beckwith, K.D. Paulsen, L.A. Flashman, T.W. McAllister, Automated subject-specific, hexahedral mesh generation via image registration, Finite Elem. Anal. Des. 47 (10) (2011) 1178–1185.

[6] Z. Salo, M. Beek, C.M. Whyne, Evaluation of mesh morphing and mapping techniques in patient specific modeling of the human pelvis, Int. J. Numer. Methods Biomed. Eng. 29 (1) (2013) 104–113.

[7] D. d'Aulignac, J.A.C. Martins, E.B. Pires, T. Mascarenhas, R.M. Natal Jorge, A shell finite element model of the pelvic floor muscles, Comput. Methods Biomech. Biomed. Eng. 8 (5) (2005) 339–347. PMID: 16298856.

[8] D. Jing, J.A. Ashton-Miller, J.O. DeLancey, A subject-specific anisotropic visco-hyperelastic finite element model of female pelvic floor stress and strain during the second stage of labor, J. Biomech. 45 (3) (2012) 455–460.

[9] S.-L. Lee, A. Darzi, G.-Z. Yang, Subject specific finite element modelling of the levator ani, in: J.S. Duncan, G. Gerig (Eds.), Medical Image Computing and Computer-Assisted Intervention MICCAI 2005, vol. 3749, Lecture Notes in Computer Science, Springer, Berlin/Heidelberg, 2005, pp. 360–367.

[10] X. Li, J.A. Kruger, M.P. Nash, P.M.F. Nielsen, Effects of fetal head motion on pelvic floor mechanics, in: K. Miller, P.M.F. Nielsen (Eds.), Computational Biomechanics for Medicine, Springer, New York, NY, 2010, , pp. 129–137.

[11] J.A.C. Martins, M.P.M. Pato, E.B. Pires, R.M. Natal Jorge, M. Parente, T. Mascarenhas, Finite element studies of the deformation of the pelvic floor, Ann. N. Y. Acad. Sci. 1101 (1) (2007) 316–334.

[12] M.P.L. Parente, R.M. Natal Jorge, T. Mascarenhas, A.A. Fernandes, J.A.C. Martins, Deformation of the pelvic floor muscles during a vaginal delivery, Int. Urogynecol. J. 19 (2008) 65–71.
[13] L.B. Whitenack, D.C. Simkins, P.J. Motta, Biology meets engineering: the structural mechanics of fossil and extant shark teeth, J. Morphol. 272 (2) (2011) 169–179.
[14] L.B. Lucy, A numerical approach to the testing of the fission hypothesis, Astron. J. 82 (1977) 1013–1024.
[15] R.A. Gingold, J.J. Monaghan, Smoothed particle hydrodynamics: theory and application to non-spherical stars, Mon. Not. R. Astron. Soc. 181 (3) (1977) 375–389.
[16] T. Belytschko, Y.Y. Liu, L. Gu, Element-free galerkin methods, Int. J. Numer. Methods Eng. 37 (1994) 229–256.
[17] W.K. Liu, S. Jun, Y.F. Zhang, Reproducing kernel particle methods, Int. J. Numer. Methods Fluids 20 (1995) 1081–1106.
[18] G. Touzot, P. Villon, B. Nayroles, Generalizing the finite element method: diffuse approximation and diffuse elements, Comput. Mech. 10 (5) (1992) 307.
[19] C. Armando Duarte, J. Tinsley Oden, An h-p adaptive method using clouds, Comput. Methods Appl. Mech. Eng. 139 (14) (1996) 237–262.
[20] D. Sulsky, Z. Chen, H.L. Schreyer, A particle method for history-dependent materials, Comput. Methods Appl. Mech. Eng. 118 (1) (1994) 179–196.
[21] P.Y. Noritomi, J.V. Lopes. da Silva, R.C. Ardoz Dellai, A. Fiorentino, L. Giorleo, E. Ceretti, Virtual modeling of a female pelvic floor and hypothesis for simulating biomechanical behavior during natural delivery, Procedia CIRP 5 (2013) 300–304. First CIRP Conference on BioManufacturing.
[22] G. Venugopala Rao, C. Rubod, M. Brieu, N. Bhatnagar, M. Cosson, Experiments and finite element modelling for the study of prolapse in the pelvic floor system, Comput. Methods Biomech. Biomed. Eng. 13 (3) (2010) 349–357. PMID: 20099169.
[23] S. Li, W.K. Liu, Meshfree and particle methods and their applications, Appl. Mech. Rev. 55 (1) (2002) 1–34.
[24] Y. Saad, Iterative Methods for Sparse Linear Systems, second ed., Society for Industrial and Applied Mathematics, Philadelphia, PA, 2003.
[25] K.-C. Lien, B. Mooney, J.O.L. DeLancey, J.A. Ashton-Miller, Levator ani muscle stretch induced by simulated vaginal birth, Obstet. Gynecol. 103 (1) (2004) 31–40.
[26] J.A. Ashton-Miller, J.O.L. DeLancey, Functional anatomy of the female pelvic floor, Ann. N. Y. Acad. Sci. 1101 (1) (2007) 266–296.
[27] A. Huerta, S. Fernández-Méndez, Enrichment and coupling of the finite element and meshless methods, Int. J. Numer. Methods Eng. 48 (11) (2000) 1615–1636.
[28] S. Fernández-Méndez, A. Huerta, Imposing essential boundary conditions in mesh-free methods, Comput. Methods Appl. Mech. Eng. 193 (12) (2004) 1257–1275.
[29] L. Hoyte, P. Krysl, G. Chukkapalli, A. Majumdar, D.J. Choi, A. Trivedi, S.K. Warfield, M.S. Damaser, Computational model of levator ani muscle stretch during vaginal delivery, J. Biomech. 39 (2006) S348.

CHAPTER EIGHTEEN

Applications of Pelvic Floor Modeling and Simulation

J.A. Kruger*, X. Yan*, X. Li†, P.M.F. Nielsen*, M.P. Nash*
*University of Auckland, Auckland, New Zealand
†University of Sheffield, Sheffield, United Kingdom

INTRODUCTION

Computational biomechanical models of the pelvic floor offer novel and quantitative ways to reliably assess pelvic floor mechanics during childbirth, and have the potential to elucidate mechanisms for the development of pelvic floor disorders and offer structural analysis of different surgical repair options in cases of pathology, for example, for pelvic organ prolapse (POP) repair. Computational modeling involves a complex integration of the anatomical, mechanical, and physiological properties of the pelvic floor muscles (PFMs). Defining the geometry of the muscles is relatively straightforward, and has been carried out using magnetic resonance images (MRIs) of the region of interest [1–5]. Most MRI scanners now provide high-quality images whereby the muscles, organs, and bones can be readily identified and segmented. Anatomically accurate models are now possible to produce. However, incorporating the functionality of the muscles and other structures remains challenging.

This chapter will introduce (a) development of 3-D computational models of pelvic floor anatomy and illustrate how they can be used to quantify normal and abnormal anatomy; (b) integration of anatomical structural information and mathematical inputs needed to develop the functional computational models of the pelvic floor; (c) the application of these models to investigate the mechanics of childbirth; and (d) potential new applications (eg, patient-specific evaluation of different surgical options in pathology such as prolapse). Other clinical applications of pelvic floor modeling and simulation are dealt with in the following chapters under pelvic floor dysfunction, which includes POP and incontinence.

DEVELOPING ANATOMICAL 3-D MODELS OF THE PELVIC FLOOR

The ability to segment anatomical structures from MRI or computed tomography scans, and create 3-D models, has enabled more quantitative approaches to describing the spatial relations between anatomical structures. This capability has become readily available with the burgeoning of commercial segmentation software such as 3-D Slicer (http://www.slicer.org/), ITK-snap (http://www.itksnap.org), Amira (http://www.fei.com/software/amira-3d-for-life-sciences/), ScanIP (http://www.simpleware.com/software), and others.

To obtain 3-D geometries of the pelvic organs and muscles, researchers usually start from a set of MR images of the pelvic floor region. The structures of interest can then be segmented either manually or semiautomatically [6–9]. The segmented geometries are then imported into a meshing (or mesh free) software to obtain descriptions of the structure of interest suitable for simulation. Initially, most simulation models of the pelvic floor were created using MRI of normal or asymptomatic patients, representing the normal anatomy. This was later expanded to include pathological conditions of the pelvic floor as the technology became better established and more data sets became available for abnormal conditions.

3-D Models of the Normal and Abnormal Anatomy of the Pelvic Floor

Methods to develop geometrically accurate models are reasonably well established. The anatomical geometry of the levator ani (LA) muscle has been comprehensively described [10–14], as have the other structures that complete the pelvic floor.

Early work by Fielding et al. [6] and Hoyte et al. [15] described the normal anatomy of the pelvic floor, using a combination of axial MRIs and 3-D models, obtained from asymptomatic nulliparous women. They found that the volume and integrity of the LA muscle were demonstrated to be negatively correlated with pelvic floor dysfunction (Fig. 1). This established a range of normative values for the volume of the LA muscle, hiatal dimensions, and other relevant anatomical structures within the pelvic floor. These models have provided not only educational tools, but a reference to compare to abnormal anatomy, with respect to pelvic floor dysfunction.

These early models highlighted the anatomical shape variability in normal subjects, and the importance of establishing a range of values for the structures of interest. Statistical shape analysis, using a population-based approach, can add to our knowledge of the anatomical differences across a population and help to explain some of the variability.

Lee et al. [16–18] conducted extensive statistical modeling of in vivo LA muscle shapes using MRIs obtained from normal subjects, and from patients with obstructed defecation. The urogenital hiatus was found to be wider in the group of patients with obstructed defecation, compared to the asymptomatic subjects. Using this statistical shape analysis approach, the authors demonstrated that it was possible to differentiate between patients with obstructed defecation and asymptomatic patients, based on their shape models [18]. In addition, simulation of LA contraction and straining observed stresses to mostly occur in the pubococcygeus and puborectalis, corresponding to the damage observed clinically using ultrasound and MRI [19,20].

Figure 1 Three-dimensional reconstructed images. Showing a lithotomy view of a reconstructed 3-D model (note the subject was scanned in dorsal supine position). Color legend: white, pelvic bones; gray, symphysis and coccyx; yellow, bladder/urethra; pink-tan, vagina; red-brown, levator ani; purple-brown, obturator internus; blue, recto-sigmoid. *Modified from L. Hoyte, L. Schierlitz, K. Zou, G. Flesh, J.R. Fielding, Two-and 3-dimensional MRI comparison of levator ani structure, volume, and integrity in women with stress incontinence and prolapse, Am. J. Obstet. Gynecol. 185 (1) (2001) 11–19.*

Hoyte et al. [15,21] created volume-rendered surface meshes of the LA muscles from segmentation data, and performed statistical analysis on manually quantified parameters that indicated LA muscle morphology. A three-way comparison was conducted on those parameters among asymptomatic subjects, patients with stress incontinence, and patients with prolapse. The "bulkiest" and most intact LA muscles were found in the asymptomatic subjects, and the least intact LA muscles were found in the patients with prolapse [21]. The group then investigated the LA muscle geometry in patients with different grades of prolapse, and concluded that the level of asymmetry and abnormality was positively related to the severity of the prolapse [22]. In a more recent study, color maps of the LA muscle thickness were constructed to quantitatively describe the LA muscle bulkiness for both the asymptomatic subjects and patients with pelvic floor dysfunction (stress incontinence or POP). Loss of insertion points and LA muscle atrophy were found to be more prevalent in women with pelvic floor dysfunction [23].

Zhuang et al. [24] developed a geometric 3-D model describing the muscle detachment (avulsion injury) using MRIs of women who had experienced a difficult vaginal delivery. These authors used ultrasound, MRI, and a 3-D model representation to illustrate the change in geometry of the LA muscle in those women with an avulsion injury. These models provide a powerful visual tool for both clinician and trainee (Fig. 2) [24].

Three-dimensional models of the pelvic floor have provided quantitative and useful representations of the variation in geometry of the pelvic floor in both normal and abnormal subjects. This area is continually growing as imaging quality improves, and as the accessibility to segmentation software increases with enhanced functionality.

An accurate representation of the 3-D anatomy of pelvic floor structures enables simulation models to be created for continuum biomechanics applications. However, to investigate the biomechanics of the muscles, and in particular their contractile function, it is essential to include the muscular structures that generate the force within the muscle. Work has recently started to incorporate quantitative descriptions of muscle fiber structure into the computational models [35].

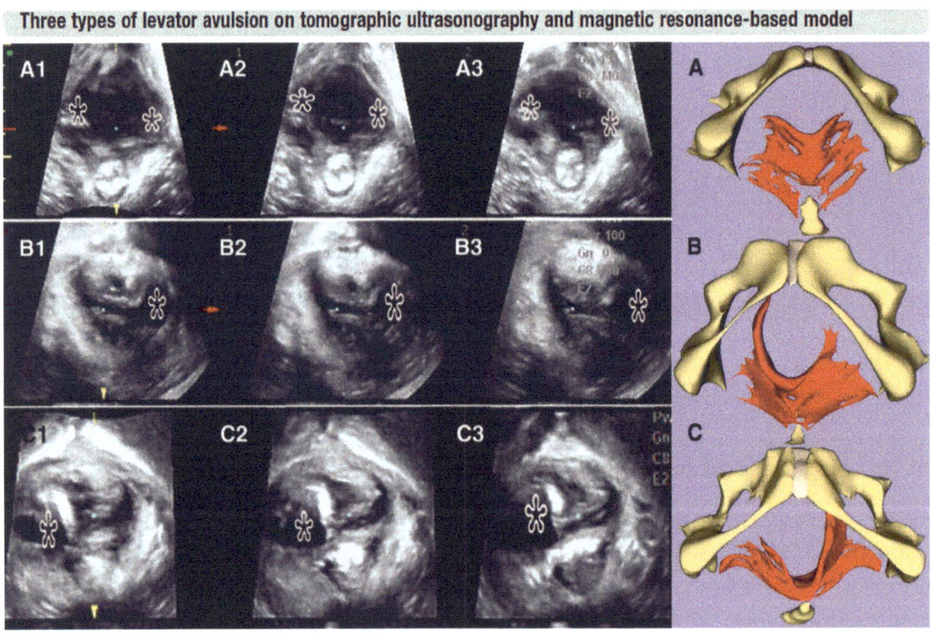

Figure 2 (A1–3 and A) Represent levator avulsion on both sides, (B1–3 and B) represent levator avulsion on the left side, (C1–3 and C) represent levator avulsion on the right side. The star denotes the levator defect. *Modified from R.R. Zhuang, Y.F. Song, Z.Q. Chen, M. Ma, H.J. Huang, J.H. Chen, Y.M. Li, Levator avulsion using a tomographic ultrasound and magnetic resonance-based model, Am. J. Obstet. Gynecol. 205 (3) (2011) 232.e231–232.e238.*

Muscle Fiber Organization

The majority of computer modeling studies have assumed that muscle fiber bundles were either aligned with the directions of origin-insertion pairs established in the literature [33], or that the muscles' mechanical properties were considered to be passive and isotropic [26].

Early attempts to construct models of muscle fiber organization relied on cadaver dissections. Janda et al. [12] were among the first to qualitatively describe the arrangement of the major LA muscle bundles measured with a palpator and strings from an embalmed cadaver [12]. Although useful for the initial stages of describing the muscle fiber architecture, the embalming process is known to affect muscle structure. Furthermore there is little 3-D quantitative information included in the published descriptions [27]. Recent research has used data from either high-resolution anatomical images with clearly identifiable muscle bundles, or diffusion tensor imaging (DTI) [28,29]. DTI has been used to study the pubovisceral muscle, perineal body, anal and urethral sphincters, and the obturator internus muscle [30]. However, the use of DTI to determine PFM fiber orientations has not yet been validated. In particular, the application of DTI to in vivo imaging remains speculative [30]. Although DTI may adequately represent the global anatomy of several PFMs, the corresponding information cannot be clearly identified. Furthermore, the procedure is currently limited by the manual seeding process, which is both time-consuming and subjective [29,30]. The complex muscle structures in the LA muscle, and its close proximity to the surrounding soft tissues, are likely to introduce errors in fiber tracking [30]. These factors suggest that DTI data on the PFMs should be treated with caution [29,30]. In addition, more research is required in this area to speed up the procedure and to investigate potential errors in fiber tracking.

Several studies have used the Visible Human Project® data set (https://www.nlm.nih.gov/research/visible/visible_human.html) to extract information on skeletal muscle fiber orientations of limbs using a range of image processing techniques [31–33]. Dong et al. [31] constructed computer models of limb muscles based on segmentation data from the Visible Man and demonstrated the use of Hough transform to obtain the muscle fiber orientations. However, the information on the fiber organization was not used in simulating muscle deformation and was simply used for visual realism. Sachse et al. [32] used an image gradient-based Sobel operator to determine the fiber orientation at individual pixels and a Hotelling transform to statistically determine the direction of principal axes for a local group of pixels. This technique still relies on a fundamentally sound knowledge of the muscle anatomy and a manual detection of the orientation. More recently, anatomical images of the pelvic floor region from the Visible Woman data set (https://www.nlm.nih.gov/research/visible/visible_human.html) were used to generate a quantitative description of the PFM fiber architecture using a structure tensor approach [34,35]. However, imaging techniques for quantifying fiber orientations from the Visible

Human data set are limited; by the image resolution, slicing of the specimen, and the subsequent distortion of the muscle fibers. This, and extrapolations from gradients of the images, is likely to result in an unrealistic representation of the in vivo fiber organization. It is clear that further work is required in this area to provide more accurate and personalized information about muscle fiber structure.

BIOMECHANICAL MODELS OF THE PELVIC FLOOR

Biomechanical modeling enables quantitative predictions of the mechanical function of the pelvic floor, and has the potential to provide indices for the prediction of LA muscle injury and development of pathologies such as POP. However, to fully realize this potential the geometric anatomical information, including connectivity and constitutive models of the tissues, has to be readily available. Geometric information is available from MRIs but constitutive models are not. Thus, before these models can be useful in a clinical setting, these parameters (ie, the constitutive tissue models) need to be readily available as inputs. Outputs from model simulations also need to be generated within a reasonable clinical response time. A comprehensive framework for modeling the pelvic floor, incorporating detailed essential information, is necessary before we can begin to realize a patient-specific paradigm for prediction, diagnosis, and treatment. Some of these requirements are discussed in this section.

Constitutive Modeling of the Pelvic Floor Muscles

The mechanical behavior of a material is described by its constitutive relation, which is the relationship between strain/deformation and stress/force. Large deformations are observed in the soft tissues of the female pelvic floor, particularly during the second stage of labor, but also during other physiological events that raise abdominal pressure such as laughing, sneezing, or coughing. Hyperelastic constitutive relations, with either isotropic or anisotropic material properties, are often used to model soft tissues undergoing large deformations. Commonly used hyperelastic constitutive relations for modeling the PFMs include neo-Hookean, Mooney-Rivlin, and exponential constitutive models. The effect of nonlinearity on the mechanics of childbirth was investigated by Li et al. [36], who used the neo-Hookean and the exponential constitutive relations to compare the response of the PFMs to a simulated vaginal delivery [36]. The authors demonstrated that increased nonlinearity in the elastic response of the tissues led to considerably higher estimated forces required for delivery, accompanied by a more homogeneous spatial distribution of strains throughout the LA muscles. They demonstrated that the way the constitutive behaviors are extrapolated beyond the presently available experimental data markedly affects the predicted function of the LA muscle during a vaginal delivery, due to the large strains that occur.

Transversely isotropic constitutive models, including those proposed by Holzapfel, Humphrey, and Yin, have also been used for modeling the PFMs [8,25,37,38]. The strain energy functions in these models are composed of two parts: (a) an isotropic part that describes the relatively linear mechanical response of the ground substance matrix; and (b) an anisotropic part that quantifies the additional nonlinear stiffness provided by aligned collagen and muscle fibers. To investigate the effect of anisotropy on the mechanics of childbirth, Li et al. [38] varied the relative stiffness between the fiber and the matrix components, but kept the overall stress-strain response in the fiber direction constant to match the available experimental data [38]. The authors demonstrated that, as the fiber anisotropy was increased, the forces required for delivery increased. This highlights an urgent need to obtain experimental data on muscle mechanical properties to better approximate the mechanical function of the pelvic floor for quantitative analysis.

There are obvious difficulties in obtaining suitable experimental data necessary for mechanics modeling. Despite these difficulties, Jing et al. [25] conducted a series of biaxial tests on the LA muscle from a human female cadaver. The derived material properties were incorporated into an anisotropic visco-hyperelastic constitutive model to estimate the stress during the second stage of labor [25]. Although the model takes into account the time-dependent viscoelastic properties of the PFMs, the material parameters were estimated from nonpregnant cadaver tissue, which is unlikely to accurately represent the muscle properties during late stage pregnancy. However, to try to account for the effect of pregnancy, the values determined from the biaxial tests were then scaled based on results from a pregnant squirrel monkey [39].

There has been preliminary work investigating the in vivo passive stretch of vaginal tissue and underlying LA muscle during the third trimester of pregnancy [40]. Tissue stiffness was measured using a novel elastometer, which records the passive force exerted by the tissues against the speculum tips of the elastometer during controlled displacements of the vaginal and underlying tissue. Muscle stiffness was estimated by regressing the force/displacement data (Fig. 3). The elastometer has found significant differences in passive

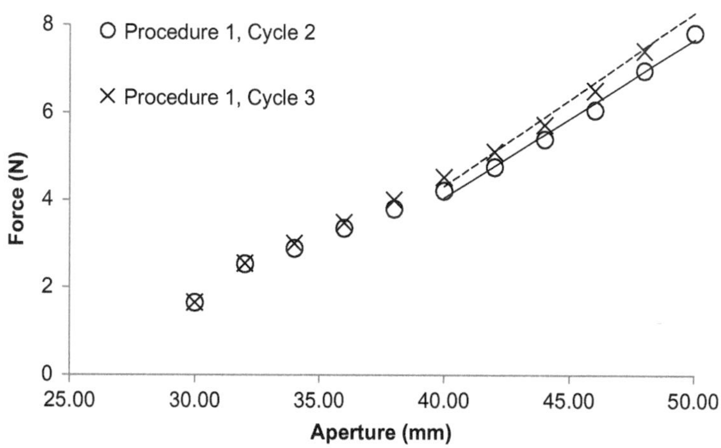

Figure 3 Representative plot of force/displacement measurements using the elastometer. Linear trend lines are fitted to the measurements spanning 40–50 mm.

stiffness between pregnant and nonpregnant women, which may prove to be a useful clinical prediction tool, but the results have yet to be validated for use in computer simulations [41]. Although an advancement over using nonpregnant cadaver tissue, it is unlikely that the higher strains needed to characterize the constitutive behavior of the muscles during vaginal delivery can be achieved during the displacement of the elastometer. Nevertheless, this is the first attempt to quantify in vivo tissue stiffness during pregnancy in a large cohort of women. The results could be used in future biomechanical models to study the effect of muscle stiffness on the mechanics of childbirth.

Kinematic Constraints

Application of kinematic constraints is important for all computational models. The specification of these constraints is often based on anatomical or clinical knowledge. In most cases, this entails fixing a position on the model so that it does not move during a simulation, often at a muscle or tendon attachment point to a bone. Defining where to fix the constraints for the pelvic floor models is difficult, primarily due to the complex muscle and connective tissue structures that comprise the pelvic floor. In addition, inclusion of other pelvic organs such as the bladder and rectum may alter the mechanical response, and ideally should be included in pelvic floor modeling. These structures would provide more realistic boundary constraints but also require advanced numerical solutions for the contact mechanics.

Constraints on the Pelvic Floor Muscles

Laterally, the LA muscles are connected to the arcus tendinous by the fascia covering the obturator internus muscle. To account for this, some studies have included part of the obturator internus in the model to move the fixed boundary at the superior aspect of the pelvic floor away from the pubococcygeus and puborectalis where most birth-induced injuries occur. This also allowed the iliococcygeus to deform at the attachment along the arcus tendinous [9,26,42]. Recent work by Kim et al. [43] has demonstrated the origin of the LA muscle changes from the medial to lateral regions of attachment. The medial portion appears to be thicker and more closely attached to the periosteum of the symphysis, while the more lateral fibers attach to the levator arch where they then transition into the iliococcygeal muscle [43].

When modeling childbirth, few models have included the entire pelvis as it is thought that usually only the pelvic outlet is of clinical importance. For some models, the pubic bones have been included either partially [7], or in full [44], to provide anterior constraint to the movement of the fetal head (Fig. 4). Inclusion of the full pubis significantly increased the predicted force required for delivery and initiated a different motion for the fetal head, compared to the model with only a partial pubis [44]. Berardi et al. [37] spatially fixed the attachment of the LA muscles to the pubic bones and the coccyx, providing both anterior and posterior constraints for the passage of the fetal head through

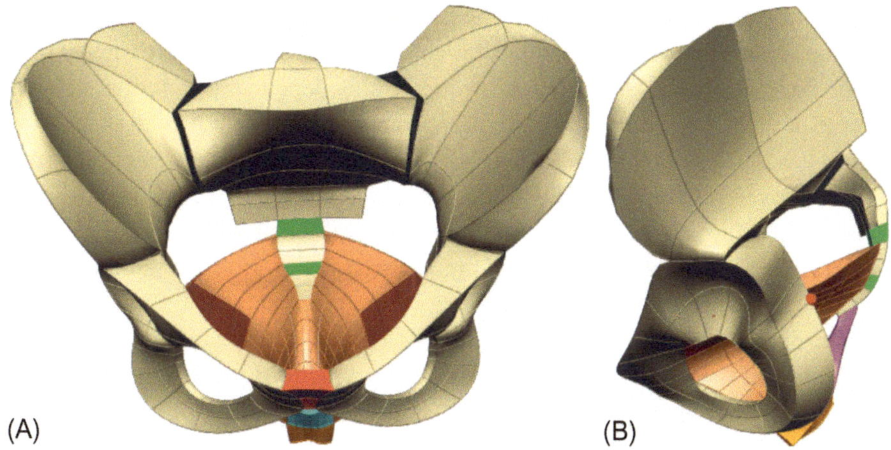

Figure 4 Finite element model of a female pelvis in anterior view (A) and lateral view (B). Illustrated are the pelvic bones *(cream)*, levator ani muscles *(beige)*, part of obturator internus muscles *(maroon)*, external sphincter muscles *(gold)*, joints of sacrum *(green)*, perineal body *(aqua)*, sacroiliac joints *(black)*, anococcygeal ligament *(purple)*, and position of the ischial spines *(red)*.

the birth canal [37]. However, in this model, the passage of the fetal head was still prescribed. Accurate and complete geometric models of the pelvis and pelvic floor are fundamentally important for reliable simulations of function.

Controlling Fetal Head Descent

The mechanism of second stage labor was first described by several obstetricians early in the 18th century, and much of our understanding has been based on these early pioneers. William Smellie (1697–1763) documented that the primary considerations for a vaginal delivery were the size and shape of the pelvic bones and the fetal head. He published a set of anatomical tables in 1754, describing the relationship between the bony pelvic outlet diameters and the fetal head diameters [45]. Smellie's measurements were confirmed by a radiological study in the 1960s in which the second stage of labor was progressively X-rayed and the bony dimensions of the fetal head and pelvic dimensions were measured [46]. Today, it is still the bony pelvis and the fetal head that provide the means for describing the "passage" (the birth canal) and the "passenger" (the fetus) during the second stage of labor, with little consideration of the role of the PFMs [47,48]. Most computational models of childbirth prescribe the pathway of the fetal head as it is presented in a clinical context, via a series of rotations, flexion, and extension [3,9,37]. This pathway is typically described as the "curve of Carus" which was first described by the German Anatomist Karl G. Carus, 1789 [49]. The curve of Carus is an imaginary line created by connecting the center points of the planes of pelvic inlet, the pelvic cavity, and the outlet and has provided a clinical tool by which the progress of labor and descent of fetus through the birth canal can be monitored [48]. The minimal energy state of the childbirth model can only be achieved when the fetal head is able to determine its own pathway through the pelvic floor. The passage of the fetal head is highly dependent on individual-specific geometries of the pelvic bones; the maternal pelvic floor; and the position, shape, and size of the fetal head. Therefore, models that completely prescribe the passage of the fetal head will typically overestimate the resistive forces and stretches that occur during childbirth.

Li et al. [26,50] constructed the first model of childbirth where the fetal head was able to translate and rotate with minimal prescription of descent through the birth canal [50]. The model prescribed only one degree of freedom in the vertical direction (to achieve descent), and the head was allowed to rotate and translate in all other directions. The results showed substantial fetal head movements during descent through the pelvic floor, which appeared to differ depending on pelvic floor geometries. This suggested that there is unlikely to be a uniform fetal head descent path across all individuals.

APPLICATIONS OF PELVIC FLOOR MODELS

Analysis of Childbirth Mechanics

It was not until recently that the role of the LA muscle during childbirth has been investigated using computational modeling. This has been driven partly by a better understanding of the function of the LA muscle, and by an increasing awareness of the association between vaginal delivery, LA muscle damage, incontinence, and POP.

The application of computational modeling to increase our understanding of the mechanics of second stage labor has contributed to advances in this field. Some of the essential factors have already been investigated and incorporated into models of childbirth, such as more incorporation of different attitudes and position of the fetal head, and more accurate geometry of pelvic floor structures [15,21,51–53], while others have yet to be quantitatively described. The limited availability of robust and reliable experimental data (such as the in vivo mechanical muscle properties during pregnancy and muscle fiber structure) to inform the pelvic floor models is an important limitation of this approach. Similar limitations apply to modeling the fetus. Although simplifications are likely to have an impact on the accuracy of the simulations, several groups have already produced biomechanical models of childbirth where predictions of muscle deformation, fetal head molding, and the effect of different fetal positions have been simulated [8,9,26,37,52]. These model predictions have given rise to a number of indices of muscle stretch/strain and force, but these have yet to be validated. Thus, the caveats already mentioned in this chapter must be applied to all these models, and their application in the clinical setting is limited at present, but highly likely to increase as research advances in this area.

Effects of Physiological Variability on Childbirth Mechanics

One of the challenges in developing computational models is accounting for the variability that occurs with the anatomical structures, but also with the variability of other factors that will impact on outcomes. Several of these factors have been investigated using different approaches, and will be discussed in this section.

Implications of Uterine Contraction

Lien et al. [54] used a simplified viscoelastic biomechanical model of the PFMs, and an incompressible sphere to represent the fetal head, to investigate the effectiveness of different patterns of maternal effort during the second stage of labor. Forces from the uterine contractions, and maternal effort, were summed and applied to the fetal head to advance it through a prescribed pathway. The effect of different combinations of pushes on the duration of second stage labor were then assessed. The triple push pattern, which consisted of three voluntary pushes during a uterine contraction, resulted in the shortest delivery time, although this was the least efficient in terms of energy expenditure. A voluntary push at peak uterine contraction had the highest efficiency, as it combined maternal effort with the uterine contractions to yield higher push forces and only required one push for each uterine contraction [54].

Implications of Fetal Head Presentation

Parente et al. [52] and Jing et al. [39] studied the effect of occipito-posterior (OP) malposition on the mechanics of childbirth. The former study concluded that, compared to childbirth with occipito-anterior (OA) position, the simulation of OP presentation gave higher maximum stretch ratios of the PFMs, indicating a higher risk of muscle injury [52]. The later study also found that the OP presentation led to a 8.6% increase in stretch ratio at the enthesis (connective tissue attachment of) of the pubovisceral muscle, compared to an OA presentation.

Implications of Fetal Head Molding

Fetal head molding during the first stage of labor has been investigated in two studies. Lapeer and Prager [55] constructed a finite element model of the fetal skull based on a glass replica [55]. The model consisted of stiff bony plates, modeled using an orthotropic linear elastic material and deformable fontanelles and sutures, modeled using a Mooney-Rivlin material model. Peak labor force, including intrauterine and head-to-cervix pressure, was applied to the model and the changes in maxillo-vertical diameter, biparietal diameter, suboccipito-bregmatic diameter, and orbito-vertical diameter were measured. The simulation results were in close agreement with clinical experiments. Pu et al. [56] adopted a similar modeling approach and studied the effect of different labor forces on the fetal head molding. A modified molding index was used to describe the degree of molding and this index was found to increase as the labor force increased [56].

Risk Prediction Tool for Clinical Decision Making

Determining the risk of damage to the PFMs during a vaginal delivery, and providing better clinical advice to reduce such risk, is an important goal for childbirth modeling. This could be approached using a subject-specific computational model, or a population-based approach. Several large epidemiological studies have identified risk factors for pelvic floor damage, and eventual pelvic floor dysfunction [57–59]. This approach relies on statistical correlations between risk factors and disease development, and a probability score for any particular risk factor. There is no clear understanding of the underlying mechanisms involved in PFM damage, or the development of pelvic floor dysfunction.

A statistical model that is able to rapidly predict important indices of childbirth mechanics is required for real-time evaluation of intrapartum obstetric risks. The combination of a computational approach with a statistical prediction model will generate information much more quickly, and can be used on a patient-specific basis.

Predictive Statistical Model

Subject-specific computational models of childbirth have the ability to predict the desired mechanical indices for vaginal delivery, such as total force required, maximum principal stretch ratios, and maximum principal stresses. This requires not only individualized construction of anatomically accurate geometric models of the pelvic floor and fetal skull, but also the computational modeling of their interactions. This process is often time-consuming, computationally expensive, and in general not feasible to use in a clinical setting. Using a statistical modeling approach, in combination with results from the biomechanical simulations, may be one way of overcoming this limitation.

Yan et al. [34,35] conducted a statistical shape analysis on the morphology of the PFMs, using a principal component analysis approach, and performed childbirth simulations from the derived models. The statistical model captured the primary shape variation of the PFMs in a group of 12 nulliparous women, and quantified the effect of these shape variations on the mechanics of childbirth [34,35]. This initial approach was extended by constructing partial least squares regression (PLSR) models to predict the important mechanical indices of childbirth based on the shape and size of the fetal skull. A series of childbirth biomechanical simulations were conducted based on 26 subject-specific models of fetal skulls. Statistical PLSR models were constructed using the geometries of the fetal skulls, together with the mechanical indices obtained from the simulation results. One of the PLSR models used routinely measured antenatal biometrical parameters, including biparietal diameters and fetal head circumferences, as inputs and predicted total force required for delivery, maximum principal stretch ratios, and maximum principal stresses [42]. The use of such statistical models to rapidly predict obstetric outcomes, based on precomputed databases of biomechanical models, has great potential. Validation is still required, and it remains to be seen if this approach would be acceptable in the clinical setting.

SUMMARY

Computational modeling of the pelvic floor and simulation of childbirth and pelvic floor disorders has made considerable progress over a short period of time [1,37,60]. Accurate representation of the complex anatomical structures of both the pelvic floor musculature and the organ systems is now possible [25,60]. The use of statistical shape analysis for the PFMs [18,34,35], the bony pelvis [44], and fetal head shape [42] has facilitated a population-based approach to model generation, for simulation of childbirth. This approach will enable rapid generation of patient-specific geometries, based on these generic models. Potential applications of these models are to gain an understanding of the mechanisms involved in various pathologies such as POP, urinary incontinence, and pelvic floor injury during childbirth. This knowledge will help us understand the mechanical environment of tissue mechanics and, together with clinical knowledge, this presents an exciting new paradigm for treatment and prevention strategies. There remain many challenges to overcome before pelvic floor modeling and simulation can be applied in clinical settings, but many of these are being addressed. With advancing technology, and the inevitable improved ability to collect quantitative experimental data, these models are likely to be well informed and functional within the next decade. This is a new field with immense potential. The application of computational modeling in clinical scenarios is being extensively researched in many other organ systems: in cardiovascular disease, musculo-skeletal medicine, pulmonary disease, and others [61–63].

Many of the same caveats apply when using these models as risk predictors for disease states. Great strides have been made in recent years, and as more modeling tools become available and the computational power increases, these models may soon be an important component of the physicians' tool box.

REFERENCES

[1] S. Brandão, M. Parente, T. Mascarenhas, A.R.G. da Silva, I. Ramos, R.N. Jorge, Biomechanical study on the bladder neck and urethral positions: simulation of impairment of the pelvic ligaments, J. Biomech. 48 (2) (2015) 217–223.
[2] L. Hoyte, M.S. Damaser, S.K. Warfield, G. Chukkapalli, A. Majumdar, D.J. Choi, A. Trivedi, P. Krysl, Quantity and distribution of levator ani stretch during simulated vaginal childbirth, Am. J. Obstet. Gynecol. 99 (2) (2008) 198.e191–198.e195.
[3] K. Lien, B. Mooney, J.O.L. DeLancey, J. Ashton-Miller, Levator ani muscle stretch induced by simulated vaginal birth, Obstet. Gynecol. 103 (1) (2004) 31–40.
[4] G. Venugopala Rao, C. Rubod, M. Brieu, N. Bhatnagar, M. Cosson, Experiments and finite element modelling for the study of prolapse in the pelvic floor system, Comput. Methods Biomech. Biomed. Engin. 13 (3) (2010) 349–357.
[5] Y. Zhang, S. Kim, A.G. Erdman, K.P. Roberts, G.W. Timm, Feasibility of using a computer modeling approach to study SUI induced by landing a jump, Ann. Biomed. Eng. 37 (7) (2009) 1425–1433.

[6] J.R. Fielding, H. Dumanli, A.G. Schreyer, S. Okuda, D.T. Gering, K.H. Zou, R. Kikinis, F.A. Jolesz, MR-based three-dimensional modeling of the normal pelvic floor in women: quantification of muscle mass, Am. J. Roentgenol. 174 (3) (2000) 657–660.

[7] X. Li, J. Kruger, J. Chung, M. Nash, P. Nielsen, X. Li, J. Kruger, J. Chung, M. Nash, P. Nielsen, Modelling the pelvic floor for investigating difficulties in childbirth, in: Proc. SPIE (Medical Imaging) 6916, 69160V, 2008.

[8] J.A. Martins, M.P. Pato, E.B. Pires, R.M. Jorge, M. Parente, T. Mascarenhas, Finite element studies of the deformation of the pelvic floor, Ann. N. Y. Acad. Sci. 1101 (2007) 316–334.

[9] M. Parente, R. Jorge, T. Mascarenhas, A. Fernandes, J. Martins, Deformation of the pelvic floor muscles during a vaginal delivery, Int. Urogynecol. J. 19 (1) (2008) 65–71.

[10] C. Betschart, J. Kim, J.M. Miller, J.A. Ashton-Miller, J.O. DeLancey, Comparison of muscle fiber directions between different levator ani muscle subdivisions: in vivo MRI measurements in women, Int. Urogynecol. J. 25 (9) (2014) 1263–1268.

[11] B. Fröhlich, H. Hötzinger, H. Fritsch, Tomographical anatomy of the pelvis, pelvic floor, and related structures, Clin. Anat. 10 (4) (1997) 223–230.

[12] Š. Janda, F.C. Van Der Helm, S.B. de Blok, Measuring morphological parameters of the pelvic floor for finite element modelling purposes, J. Biomech. 36 (6) (2003) 749–757.

[13] R. Kearney, R. Sawhney, J.O. DeLancey, Levator ani muscle anatomy evaluated by origin-insertion pairs, Obstet. Gynecol. 104 (1) (2004) 168.

[14] R.U. Margulies, Y. Hsu, R. Kearney, T. Stein, W.H. Umek, J.O. DeLancey, Appearance of the levator ani muscle subdivisions in magnetic resonance images, Obstet. Gynecol. 107 (5) (2006) 1064.

[15] L. Hoyte, J.R. Fielding, E. Versi, C. Mamisch, C. Kolvenbach, R. Kikinis, Variations in levator ani volume and geometry in women: the application of MR based 3-D reconstruction in evaluating pelvic floor dysfunction, Arch. Esp. Urol. 54 (6) (2001) 532–539.

[16] S. Lee, P. Horkaew, A. Darzi, G. Yang, Optimal scan planning with statistical shape modelling of the levator ani, in: Medical Image Computing and Computer-Assisted Intervention — MICCAI 2003, Springer, New York, 2003, pp. 714–721.

[17] S. Lee, P. Horkaew, A. Darzi, G. Yang, Statistical shape modelling of the levator ani with thickness variation, in: Medical Image Computing and Computer-Assisted Intervention — MICCAI 2004, Springer, New York, 2004, pp. 258–265.

[18] S. Lee, E. Tan, V. Khullar, W. Gedroyc, A. Darzi, G. Yang, Physical-based statistical shape modeling of the levator ani, IEEE Trans. Med. Imaging 28 (6) (2009) 926–936.

[19] H.P. Dietz, Pelvic floor ultrasound: a review, Am. J. Obstet. Gynecol. 202 (4) (2010) 321–334.

[20] R. Kearney, J.M. Miller, J.O. Delancey, Interrater reliability and physical examination of the pubovisceral portion of the levator ani muscle, validity comparisons using MR imaging. Neurourol. Urodyn. 25 (1) (2006) 50–54, http://dx.doi.org/10.1002/nau.20181.

[21] L. Hoyte, L. Schierlitz, K. Zou, G. Flesh, J.R. Fielding, Two- and 3-dimensional MRI comparison of levator ani structure, volume, and integrity in women with stress incontinence and prolapse, Am. J. Obstet. Gynecol. 185 (1) (2001) 11–19.

[22] K. Singh, M. Jakab, W.M. Reid, L.A. Berger, L. Hoyte, Three-dimensional magnetic resonance imaging assessment of levator ani morphologic features in different grades of prolapse, Am. J. Obstet. Gynecol. 188 (4) (2003) 910–915.

[23] L. Hoyte, M. Jakab, S.K. Warfield, S. Shott, G. Flesh, J.R. Fielding, Levator ani thickness variations in symptomatic and asymptomatic women using magnetic resonance-based 3-dimensional color mapping, Am. J. Obstet. Gynecol. 191 (3) (2004) 856–861.

[24] R.R. Zhuang, Y.F. Song, Z.Q. Chen, M. Ma, H.J. Huang, J.H. Chen, Y.M. Li, Levator avulsion using a tomographic ultrasound and magnetic resonance-based model. Am. J. Obstet. Gynecol. 205 (3) (2011) 232.e231–232.e238, http://dx.doi.org/10.1016/j.ajog.2011.03.052.

[25] D. Jing, J.A. Ashton-Miller, J.O. DeLancey, A subject-specific anisotropic visco-hyperelastic finite element model of female pelvic floor stress and strain during the second stage of labor, J. Biomech. 45 (3) (2012) 455–460.

[26] X. Li, J. Kruger, M. Nash, P. Nielsen, Effects of fetal head motion on pelvic floor mechanics, in: Computational Biomechanics for Medicine, IV, MICCAI Workshop, Springer, New York, 2010, pp. 129–137.

[27] D.C. Martin, M.K. Medri, R.S. Chow, V. Oxorn, R.N. Leekam, A.M. Agur, N.H. Mckee, Comparing human skeletal muscle architectural parameters of cadavers with in vivo ultrasonographic measurements. J. Anat. 199 (04) (2001) 429–434, http://dx.doi.org/10.1017/S0021878201008251.
[28] J.F. Budzik, V. Le Thuc, X. Demondion, M. Morel, D. Chechin, A. Cotten, In vivo MR tractography of thigh muscles using diffusion imaging: initial results. Eur. Radiol. 17 (12) (2007) 3079–3085, http://dx.doi.org/10.1007/s00330-007-0713-z.
[29] M. Froeling, J. Oudeman, S. van den Berg, K. Nicolay, M. Maas, G.J. Strijkers, M.R. Drost, A.J. Nederveen, Reproducibility of diffusion tensor imaging in human forearm muscles at 3.0 T in a clinical setting. Magn. Reson. Med. 64 (4) (2010) 1182–1190, http://dx.doi.org/10.1002/mrm.22477.
[30] F.M. Zijta, M. Froeling, M.P. van der Paardt, M.M.E. Lakeman, S. Bipat, A.D. Montauban van Swijndregt, A.J. Nederveen, J. Stoker, Feasibility of diffusion tensor imaging (DTI) with fibre tractography of the normal female pelvic floor, Eur. Radiol. 21 (6) (2011) 1243–1249, http://dx.doi.org/10.1007/s00330-010-2044-8.
[31] F. Dong, G.J. Clapworthy, M.A. Krokos, J. Yao, An anatomy-based approach to human muscle modeling and deformation, IEEE Trans. Vis. Comput. Graph. 8 (2) (2002) 154–170.
[32] F. Sachse, M. Wolf, C. Werner, K. Meyer-Waarden, Extension of anatomical models of the human body: three-dimensional interpolation of muscle fiber orientation based on restrictions, J. Comput. Inf. Technol. 6 (1) (1998) 95–101.
[33] C. Waldby, The Visible Human Project: Informatic Bodies and Posthuman Medicine, Psychology Press, Hove, 2000.
[34] X. Yan, J. Kruger, M. Nash, P. Nielsen, A 3D statistical shape analysis of pelvic floor morphology in nulliparous women, Neurourol. Urodyn. 30 (6) (2011) 1043–1045.
[35] X. Yan, J.A. Kruger, M.P. Nash, P.M. Nielsen, A Quantitative Description of Pelvic Floor Muscle Fibre Organisation Computational Biomechanics for Medicine, Springer, New York, 2011. pp. 119–130.
[36] X. Li, J.A. Kruger, M.P. Nash, P.M. Nielsen, Effects of nonlinear muscle elasticity on pelvic floor mechanics during vaginal childbirth, J. Biomech. Eng. 132 (11) (2010) 111010.
[37] M. Berardi, O. Martinez-Romero, A. Elías-Zúñiga, M. Rodríguez, E. Ceretti, A. Fiorentino, G. Donzella, A. Avanzini, Levator ani deformation during the second stage of labour, Proc. Inst. Mech. Eng. H J. Eng. Med. 228 (5) (2014) 501–508.
[38] X. Li, J.A. Kruger, M.P. Nash, P.M. Nielsen, Anisotropic effects of the levator ani muscle during childbirth, Biomech. Model. Mechanobiol. 10 (4) (2011) 485–494.
[39] D. Jing, Experimental and Theoretical Biomechanical Analyses of the Second Stage of Labor, University of Michigan, Ann Arbor, 2010.
[40] J.A. Kruger, P.M. Nielsen, S.C. Budgett, A.J. Taberner, An automated hand-held elastometer for quantifying the passive stiffness of the levator ani muscle in women. Neurourol. Urodyn. 34 (2) (2013) 133–138, http://dx.doi.org/10.1002/nau.22537.
[41] T. Tian, J. Kruger, V. Wong, P. Nielsen, L. Hayward, J. Smalldridge, M. Nash, S. Budgett, Functional changes of the pelvic floor muscles following vaginal delivery: the effect of ethnicity and avulsion injury, Neurourol. Urodyn. 34 (2015) S171–S172.
[42] X. Yan, J.A. Kruger, P.M. Nielsen, M.P. Nash, Effects of fetal head shape variation on the second stage of labour, J. Biomech. 48 (9) (2015) 1593–1599.
[43] J. Kim, C. Betschart, R. Ramanah, J.A. Ashton-Miller, J.O. DeLancey, Anatomy of the pubovisceral muscle origin: macroscopic and microscopic findings within the injury zone, Neurourol. Urodyn. (2014), http://dx.doi.org/10.1002/nau.22649.
[44] X. Yan, J. Kruger, X. Li, M. Nash, P. Nielsen, Modelling effect of bony pelvis on childbirth mechanics, Neurourol. Urodyn. 32 (2013) 531–532.
[45] W. Smellie, A Set of Anatomical Tables, With Explanations, and an Abridgment of the Practice of Midwifery, Isaiah Thomas, Worchester, MA, 1793.
[46] U. Borell, I. Fernstrom, The mechanism of labour, Radiol. Clin. N. Am. 5 (1) (1967) 73–85.

[47] R. Bennett, L. Brown, The first stage of labour: physiology and early care, in: V. Bennet, L. Brown (Eds.), Myles Textbook for Midwives, eleventh ed., Churchill Livingstone, Edinburgh, 1990, pp. 145–209.
[48] S. Clayton, D. Fraser, T. Lewis, Obstetrics by Ten Teachers, eleventh ed., Edward Arnold, London, 1970.
[49] Mosby's Medical Dictionary, Retrieved October 2015, from http://medical-dictionary.thefreedictionary.com/curve+of+Carus, 2009.
[50] X. Li, J. Kruger, M. Nash, P. Nielsen, Modelling fetal head motion and its mechanical interaction with the pelvic floor during childbirth, Neurourol. Urodyn. 28 (7) (2009) 672–673.
[51] K.F. Noakes, I.P. Bissett, A.J. Pullan, L.K. Cheng, Anatomically realistic three-dimensional meshes of the pelvic floor & anal canal for finite element analysis, Ann. Biomed. Eng. 36 (6) (2008) 1060–1071, http://dx.doi.org/10.1007/s10439-008-9471-6.
[52] M.P. Parente, R.M. Jorge, T. Mascarenhas, A.A. Fernandes, J.A. Martins, The influence of an occipito-posterior malposition on the biomechanical behavior of the pelvic floor, Eur. J. Obstet. Gynecol. Reprod. Biol. 144 (Suppl. 1) (2009) S166–S169.
[53] M.P. Parente, R.M. Natal Jorge, T. Mascarenhas, A.A. Fernandes, A.L. Silva-Filho, Computational modeling approach to study the effects of fetal head flexion during vaginal delivery, Am. J. Obstet. Gynecol. 203 (3) (2010) 217.e211–217.e216.
[54] K. Lien, J.O. DeLancey, J.A. Ashton-Miller, K.-C. Lien, J.O.L. DeLancey, J.A. Ashton-Miller, Biomechanical analyses of the efficacy of patterns of maternal effort on second-stage progress, Obstet. Gynecol. 113 (4) (2009) 873–880.
[55] R. Lapeer, R. Prager, Fetal head moulding: finite element analysis of a fetal skull subjected to uterine pressures during the first stage of labour, J. Biomech. 34 (9) (2001) 1125–1133.
[56] F. Pu, L. Xu, D. Li, S. Li, L. Sun, L. Wang, Y. Fan, Effect of different labor forces on fetal skull molding, Med. Eng. Phys. 33 (5) (2011) 620–625.
[57] C. Glazener, A. Elders, C. Macarthur, R.J. Lancashire, P. Herbison, S. Hagen, G. McPherson, D. Wilson, Childbirth and prolapse: long-term associations with the symptoms and objective measurement of pelvic organ prolapse, BJOG 120 (2) (2013) 161–168, http://dx.doi.org/10.1111/1471-0528.12075.
[58] J. Mant, R. Painter, M. Vessey, Epidemiology of genital prolapse: observations from the Oxford Family Planning Association Study, Br. J. Obstet. Gynaecol. 104 (5) (1997) 579–585.
[59] D.V. Valsky, M. Lipschuetz, A. Bord, I. Eldar, B. Messing, D. Hochner-Celnikier, S.M. Cohen, S. Yagel, Fetal head circumference and length of second stage of labor are risk factors for levator ani muscle injury, diagnosed by 3-dimensional transperineal ultrasound in primiparous women, Am. J. Obstet. Gynecol. 201 (1) (2009) 91.e91–91.e97.
[60] J. Luo, L. Chen, D.E. Fenner, J.A. Ashton-Miller, J.O. DeLancey, A multi-compartment 3-D finite element model of rectocele and its interaction with cystocele, J. Biomech. 48 (9) (2015) 1580–1586.
[61] K.S. Burrowes, A.R. Clark, M.L. Wilsher, D.G. Milne, M.H. Tawhai, Hypoxic pulmonary vasoconstriction as a contributor to response in acute pulmonary embolism, Ann. Biomed. Eng. (2014), http://dx.doi.org/10.1007/s10439-014-1011-y.
[62] J. Fernandez, P. Hunter, An anatomically based patient-specific finite element model of patella articulation: towards a diagnostic tool, Biomech. Model. Mechanobiol. 4 (1) (2005) 20–38.
[63] V.Y. Wang, H.I. Lam, D.B. Ennis, B.R. Cowan, A.A. Young, M.P. Nash, Modelling passive diastolic mechanics with quantitative MRI of cardiac structure and function, Med. Image Anal. 13 (5) (2009) 773–784.

CHAPTER NINETEEN

Building Pelvic Floor Biomechanical Simulations

J.B. Alford*, L. Hoyte†
*University of South Florida, Tampa, FL, United States of America
†Director, Urogynecology and Female Pelvic Reconstructive Surgery, USF College of Medicine, Tampa, FL, United States of America; Medical Director, Urogynecology and Robotic Surgery, Tampa General Hospital, Tampa, FL, United States of America

INTRODUCTION

The purpose of this chapter is to guide the practitioner in the development of patient-specific biomechanics simulations of the pelvic floor. Biomechanical structural information is derived from medical image data, usually computed tomography, or magnetic resonance imaging (MRI) data sets. Here, the visual medical data is acquired as a stack of radiologic images (slices), with each slice representing a cross section of a body region of interest. In the case of the pelvis, an axially acquired image stack would span the region from the tip of the pelvic bones (ischial tuberosities) to the pelvic brim. The slices are separated by a fixed distance, usually 5–10 mm. The tissues of interest on each slice are then identified and outlined to create a labelmap. This process is called segmentation. The labelmaps on each slice are aggregated across the image stack, and processed in software to generate the three-dimensional geometry for each tissue type (or label) in the image stack. The aggregation may introduce artifacts, such as unrealistic tissue discontinuities, caused when tissue contours change sharply across gaps between slices. These artifacts are a result of sampling error and some effort must be taken to repair them.

The finite element method (FEM) is a mathematical technique for solving the equations required to perform the simulation. The finite element (FE) method requires that the geometry be meshed into many smaller pieces, similar to the way that many small tiles may be aggregated to cover a large surface. The FE mesh defines the geometric component of the simulation. In addition to this geometry, processing of the FE simulation requires that the appropriate governing equations, material parameters, and boundary conditions be specified.

The initial three-dimensional geometry is a digitized representation that often contains artifacts related to image acquisition, including stair stepping (sharp tissue transitions) and tissue gaps (areas of apparent absence of tissue). These artifacts are unrealistic and deliver inaccurate results. For this reason, the geometry needs to be repaired and smoothed before generating the FE mesh. This repair can be done manually or using semiautomated tools.

Many simulations of the pelvic floor (eg, child birth) involve extremely large deformations as shown by the work in Ref. [1] where stretch ratios of greater than three were seen in the pelvic floor muscles during delivery of a simulated fetal head. In cases such as this, the FE mesh may become severely distorted causing the simulation to come to an end abruptly, without producing a useful result. In cases like these, meshfree methods such as the reproducing kernel particle method (RKPM) or element-free Galerkin (EFG) method may be applied with success. The pathway for generating meshfree models directly from the labeled image stack is discussed in the previous chapter and in this chapter we discuss converting an FEM mesh to a meshfree model. Meshfree methods use a point cloud to represent the tissue instead of the polyhedral mesh like FEM. This step involves converting the nodes of the FEM mesh to meshfree particles in a straightforward manner.

The process of developing biomechanics simulations from medical images is outlined in Fig. 1.

IMAGE PROCESSING

The data consists of a set of m two-dimensional grayscale images each consisting of an $n \times n$ array of pixels. An example grayscale image of the pelvic floor is shown in Fig. 2A. The images are segmented by tracing a line through the boundaries of any of the components of interest, such as in Fig. 2B. Then every pixel within the boundary of a given component is labeled as such, as shown in Fig. 2C. We call the images that have gone through this process labelmaps and the process of labeling segmentation. There are several methods for segmenting images, ranging from manual drawing of boundaries to sophisticated semiautomatic methods. Any of the methods will generally result in noise on the boundary of the components. In Ref. [2], a set of simple rules were established for statistical filtering of the noisy boundary.

OBTAINING THE GEOMETRY

Once the labelmaps are generated, the three-dimensional model can be constructed by orienting the labelmaps in a stack with the appropriate spacing between labelmaps. The three-dimensional counterpart to the pixel is called the voxel, so the stack of labelmaps is called the voxel model. A voxel model of the levator ani muscles is shown in Fig. 3A. At this point, the voxel model could be directly converted into an FEM mesh by generating one hexahedral brick element per voxel, leading to a large number of computational elements. Since the computational cost of the simulation is directly proportional to the number of elements in the model, great effort is taken to reduce the number of computational elements while preserving the geometric integrity of the model. In addition, voxel models tend to have unrealistically rough surfaces, leading

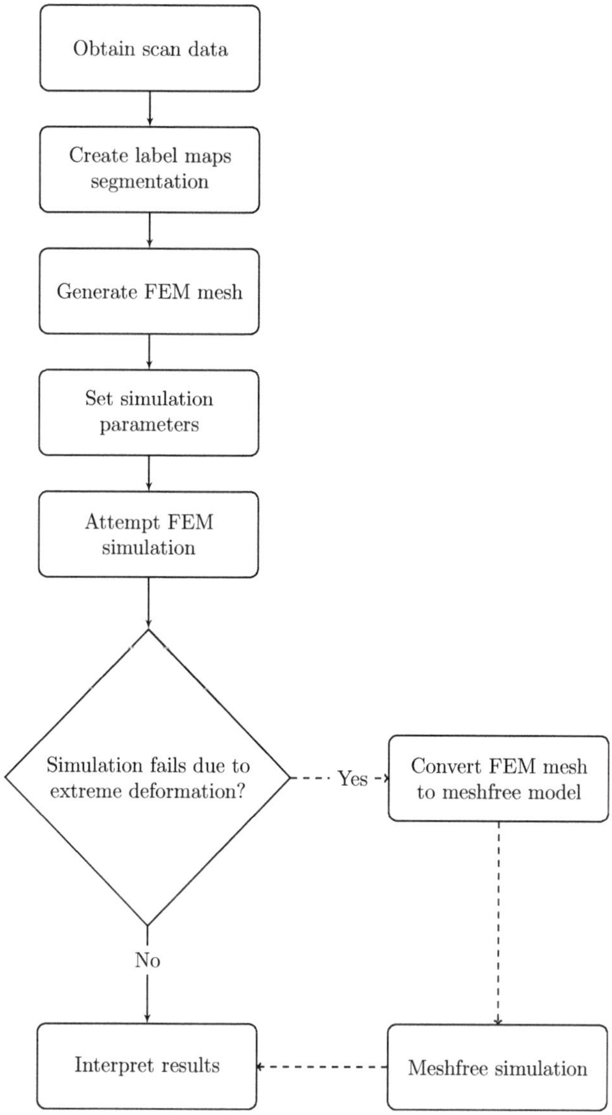

Figure 1 Flow of patient-specific biomechanics simulation.

to inaccurate results. Therefore, voxel models often need to be made smooth, in order to reduce inaccuracies.

Constructing the Smooth Geometry

In Fig. 3A, it can be seen that the surface of the voxel model has stair stepped features; this is a common feature of digitized models. This stair step is known as an aliasing artifact [3].

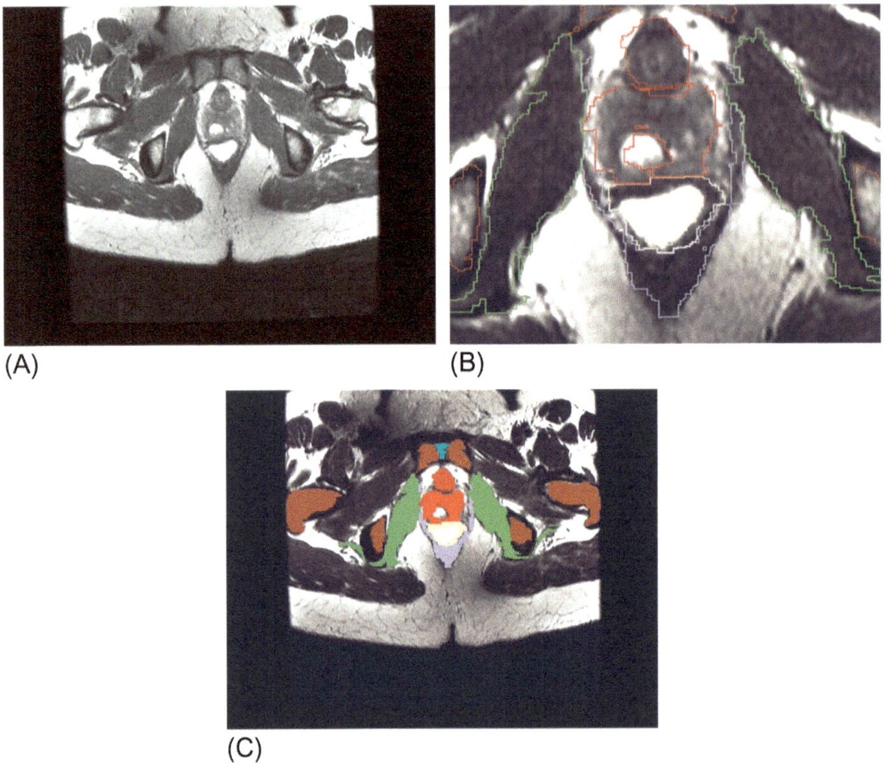

Figure 2 Axial MRI scan taken just distal to the bladder neck, from a young, asymptomatic parous woman. (A) Grayscale image, (B) semiautomatic segmented labels (legend: *brown* — bones, *red* — vagina, *gray* — levator ani, *green* — obturator internus, *white* — rectum, and *blue* — symphysis), and (C) labelmap — legend as in (B).

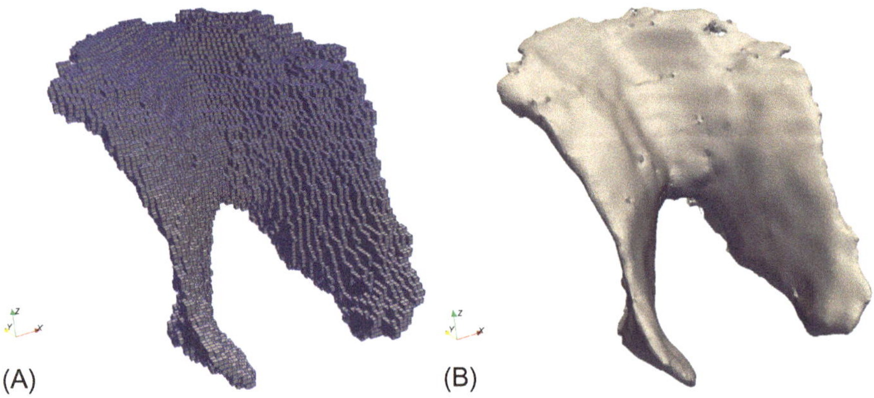

Figure 3 The voxel model and its smoothed counterpart. (A) Voxel model and (B) smoothed model.

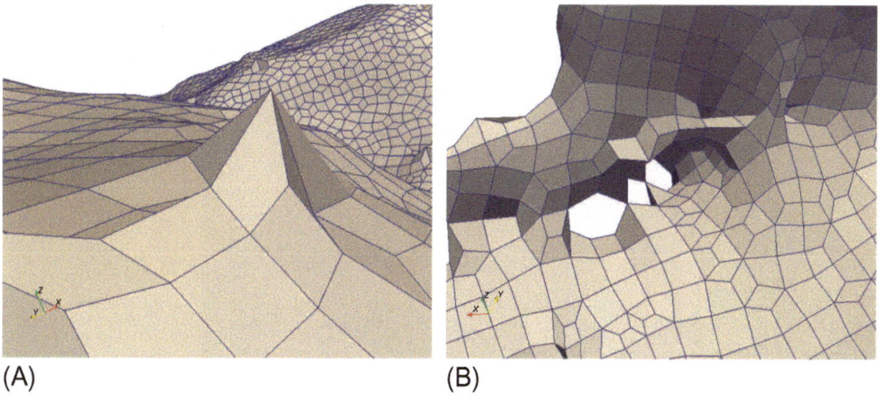

Figure 4 Undesirable geometric model features seen following smoothing. (A) Sharp anomaly and (B) holes.

Aliasing artifacts are undesirable in simulation models because the sharp features lead to unrealistic stress concentrations. If the voxel model is to be used directly to generate the FEM mesh, it must first be smoothed, as shown in Fig. 3B. The surface of the voxel model may be smoothed directly with an algorithm called Laplacian smoothing which is discussed in Refs. [2,4]. For a complete reference on Laplacian smoothing, see Ref. [5]. Alternatively, a smooth geometry may be produced with a contour type method. The contour methods work by computing some function on the voxel model, such as the distance from the assumed boundary points, then generating a mesh through the points where this function is zero via a marching cubes or similar algorithm. This results in a smooth geometric model as detailed in Refs. [6,7].

Once a smooth geometry is defined, the next step is cleaning up undesirable features. The smooth geometry may appear to be a good representation, but under scrutiny undesirable features may become evident. The types of undesirable features found are dependent on the algorithm employed to generate the smooth geometry. These may include artifacts from the digitized image (eg, sharp anomalies, seen in Fig. 4A), topological inconsistencies (eg, holes, seen in Fig. 4B), or other surface irregularities. These need to be repaired before attempting to generate a FE mesh. Such repairs are usually accomplished manually using mesh editing software.

Building the FE Mesh

Once the undesirable artifacts are repaired, the surface is tessellated (tiled) while retaining any anatomical surface features deemed important to the simulation. The most common element types (tiles) used for the FEM are the four-node linear tetrahedron and the eight-node trilinear hexahedral. These and other common element types are described in Ref. [8]. For a tetrahedral mesh, the surface would be tessellated with triangular elements, whereas for the hexahedral mesh the surface would be tessellated

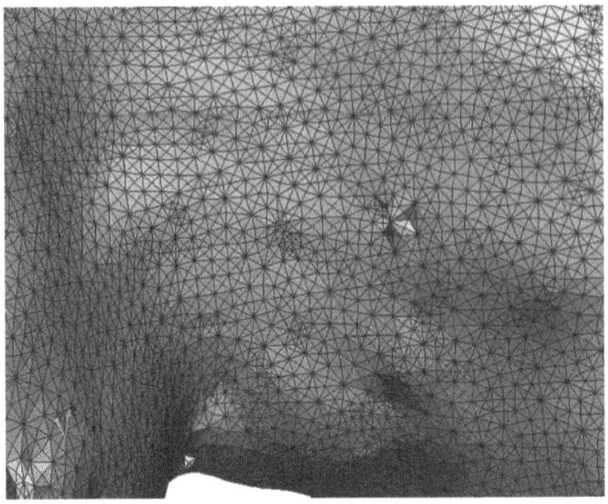

Figure 5 Surface tessellation of smooth voxel model.

with quadrilateral elements. Regardless of the approach, the elements must meet strict requirements on edge size ratios and minimum angles between edges. An example of a triangular surface tessellation is shown in Fig. 5. The interior of the model is meshed using an advancing front type algorithm such as in Ref. [2], and the interior volume elements must meet strict geometric requirements.

This process of first reconstructing the geometry from labelmaps, then constructing an FEM mesh, is tedious and generally requires manual editing. This makes the process of subject-specific FEM mesh generation quite time-consuming. There are several other methods that avoid generating a new mesh for each subject. These rely on mapping, or deforming, a preexisting atlas (or template) mesh to fit the geometry of a new subject (target) [9–12]. These methods are successful when the target anatomy has the same topology as the template anatomy, but tend to fail otherwise. When the template deformation method is applied to a new target, the resulting mesh may still have distorted elements that require editing, but this should require significantly less editing than building a mesh from a blank labelmap. For example, there are dramatic differences in levator and vaginal geometry between a young nullipara with normal pelvic anatomy compared to an elderly multipara with advanced prolapse and levator laxity. Manual editing would be required to deform a template derived from the anatomically normal nullipara into the geometry of a multipara with prolapse and levator laxity.

If the simulation results in extreme deformations, for example, like those seen in childbirth, the FEM mesh could become severely distorted. In this case, the elements may not meet the strict requirements on shape and size, causing the simulation to fail. In such a case, a solution would be to split the simulation into multiple smaller simulations, in which a new simulation is started at the point of failure of the previous simulation. This is costly in time and computational effort, because it involves remeshing the model before the start of each simulation. Alternatively, a meshfree model could be employed to accommodate the large deformations successfully in a single simulation. Fortunately, FEM meshes can be converted into meshfree models in a straightforward manner, described in the next sections.

A Brief Introduction to Meshfree Methods

Meshfree methods are a class of point-based techniques that were developed in the early to mid-1990s, such as RKPM, EFG, and h-p cloud method. An introduction to these methods can be found in Refs. [13–15]. A tutorial on programming meshfree methods can be found in Ref. [16]. In Ref. [17], it was shown that meshfree methods handle extremely large deformations with ease.

The domain is discretized with a set of points, called particles. Each particle is associated with a compact support of size ρ, called the radius of support. Fig. 6 depicts a collection of particles with a few of the compact supports shown.

The meshfree shape functions are defined as

$$\mathbf{N}_i(x) = \mathbf{P}\left(\frac{x - x_i}{\rho}\right)\mathbf{M}^{-1}(x)\mathbf{P}(0)\phi\left(\frac{x - x_i}{\rho}\right) \tag{1}$$

where the vector $\mathbf{P}$ includes a basis of monomial terms, ϕ is a compactly supported window function, and $\mathbf{M}$ is the moment matrix defined as:

$$\mathbf{M}(x) = \sum_j \mathbf{P}\left(\frac{x - x_j}{\rho}\right)\mathbf{P}^T\left(\frac{x - x_j}{\rho}\right)\phi\left(\frac{x - x_j}{\rho}\right)\Delta V_j \tag{2}$$

where ΔV is the representative volume of the particle. For a trilinear basis, the polynomial vector is $\mathbf{P} = \{1, x, y, z, xyz\}$.

Generating the Meshfree Model

Studies have found that meshfree methods are very well suited for use in biomechanics simulations of highly complex geometry with large deformations [18]. Meshfree methods have been used for biomechanics simulations in the heart [19,20], the brain [21], skeletal

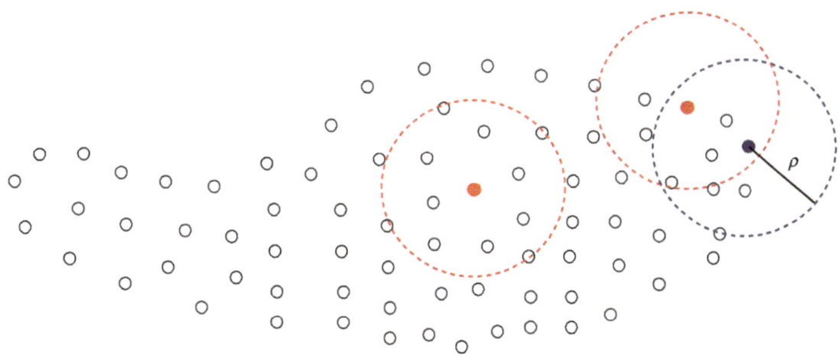

Figure 6 Particle supports.

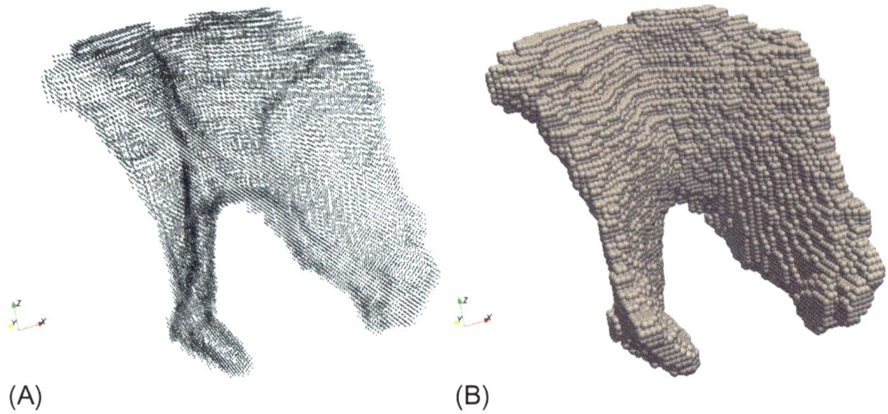

Figure 7 Meshfree model of levator ani. (A) Meshfree particles and (B) compact supports.

muscle [22], and of the pelvic floor [1,23]. In the event an FEM simulation fails, there is not much difficulty in converting the FEM mesh to a meshfree model. The mesh nodes are converted to particles by computing appropriate values for the radius of support and representative volume. The radius of support for a given particle is based on the size of the largest element connected to the particles' corresponding mesh node. The representative volume, for a given particle, is an accumulation of some portion of the element volume for each element connected to the particle's corresponding mesh node. Fig. 7 depicts the meshfree particles for the levator ani model beside a visualization of the compact support for each particle.

Generating the initial FEM mesh is usually an expensive process, involving multiple software tools and tedious manual editing. However, there is widespread community and software support for FEM analysis and this makes FEM tools more readily accessible. Furthermore, for equally dimensioned simulation models, FEM analysis is computationally more efficient than meshfree methods, meaning that FEM simulations can run faster than meshfree simulations.

For this reason, we take the approach of attempting to use an FEM model first, and if this fails, consider conversion to a meshfree model. Alternatively, if large deformations are anticipated, initial consideration is given to a meshfree approach.

SIMULATION

Executing the simulation involves solving a set of governing equations. These equations are independent of whether the model is of FEM or meshfree type. The simulated phenomena should be classified according to the types of expected motion. First, the effects of inertia on the solution should be evaluated; if these are thought to be minimal, a static load-stepping solution may be appropriate. Otherwise a dynamic solution may be required. Static solutions are much cheaper computationally than dynamic solutions, so it may be wise to first attempt a static solution if the effects of inertia are unknown, if this approach proves faulty, the simulation can be rerun as a dynamic simulation. For events that occur over longer time scales, for example a slow Valsalva, the inertial effect may be negligible; however for an impulse event (eg, a cough), inertial effects would be important.

Equations of Motion

For Galerkin type procedures, such as the FEM and some meshfree methods, a form of the conservation of momentum called the principle of virtual work is used to govern the motion. The equations of principle of virtual work can be expressed as

$$\int_{\Omega} \mathbf{P} : \frac{\partial(\delta u)}{\partial \mathbf{X}} d\Omega - \int_{\Omega} \mathbf{b} \cdot \delta \mathbf{u} \, d\Omega + \int_{\Gamma} \mathbf{t} \cdot \delta \mathbf{u} \, d\Gamma + \int_{\Omega} \rho \ddot{\mathbf{u}} \delta \mathbf{u} \, d\Omega = 0 \tag{3}$$

where $\mathbf{P}$ is the nominal stress tensor, $\delta\mathbf{u}$ is the virtual displacement, $\mathbf{X}$ is the original configuration coordinates, $\mathbf{b}$ is the body force per unit mass, and $\mathbf{t}$ is the surface traction vector. The nominal stress tensor is related to Cauchy stress through the relation

$$\boldsymbol{\sigma} = J^{-1} \mathbf{F} \cdot \mathbf{P} \tag{4}$$

where $\mathbf{F}$ is the deformation gradient tensor and $J = \det(\mathbf{F})$. The fourth term in Eq. (3) is the inertial term and is neglected in the case of static loading.

In general, there is no analytical solution to these equations, the solution is approximated by discretizing the domain Ω and the domain boundary Γ either through a FE mesh or meshfree particles. Both discretization techniques result in valid interpolation functions associated with each node, or particle, called shape functions. The approximated solution to the equations is then expressed as a combination of shape functions

$$\delta \mathbf{u} = \sum_{I} \delta u_I N_I(\mathbf{X}) \tag{5}$$

where $I \in \{1, 2, \ldots, n\}$, with n being the number of nodes, δu_I is the virtual displacement at node I, and $N_I(\mathbf{X})$ is the shape function for node I evaluated in the initial coordinates. This process leads to a set of linear equations that are then solved to get the displacement at each node. A complete description of nonlinear finite deformation analysis can be found in Ref. [24].

Simulation Parameters

A unique solution to the equations is found once an appropriate constitutive law (material response) is chosen and appropriate boundary conditions are applied. These two concepts are defined in the following paragraphs. The first term in Eq. (3), the work due to stress in the material, is dependent on the material properties of the tissue. The second and third terms in Eq. (3) are the work due to body forces and surface forces.

Constitutive laws relate the state of strain a body is in to the amount of stress in the material. These relationships are developed to model phenomena found in laboratory testing, thus each constitutive law comes with a set of parameters that tune the model to match real-world behavior. Modeling the material response of the pelvic floor muscles is a difficult undertaking. The tissue can be briefly described as incompressible, anisotropic, and viscoelastic. Muscles also have components that actively contract, further complicating the material model. Often times the material response is simplified, either for computational reasons or simply because much about the material is still unknown. A simple constitutive relation used to model the pelvic floor response is a hyperelastic model. Hyperelastic materials are characterized by a strain energy function relating the stress and strain:

$$\boldsymbol{\sigma} = \mathbf{F}\frac{\partial W}{\partial \mathbf{E}}\mathbf{F}^T \tag{6}$$

where $\mathbf{E}$ is the Lagrangian strain tensor and W is the strain energy function. Two forms of the strain energy function are commonly chosen for modeling the pelvic floor, Neo-Hookean and Mooney-Rivlin. For a comparison of the effects of the different strain energy functions, the reader should see Ref. [25]. The Neo-Hookean model is defined by the Lamé constants λ and ν. For simulations involving smaller time scales, such as impact-related phenomena (ie, coughing), the viscous properties of the tissue will become important to model. A viscoelastic model is more realistic, but much more complex and depends on the architectural relationships between the viscous and elastic components of the model [26]. Modeling active muscle contraction adds another layer of complexity and is considered in Ref. [27].

The boundary conditions come in the form of applied displacements or loads. Fixed displacement conditions should be applied so that the pelvic floor components work in conjunction to form a stable structure. An example of fixed displacement would be the attachments of the levator ani components to the bony pelvis and the levator arch. In Fig. 8, the red points are possible locations to apply fixed boundary conditions, but it would be more realistic to simulate connective tissue at these points and apply fixed conditions to the attachment points on the bony pelvis. Nonzero conditions should be applied that drive the simulated phenomena (ie, pressure load applied to simulate coughing event). In Fig. 8 the blue arrow glyphs represent possible load vectors for a step in a simulation, for example, fetal head making contact with the levator ani (childbirth).

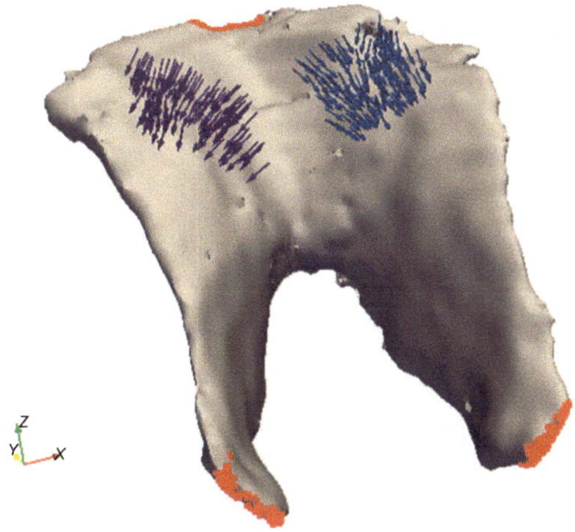

Figure 8 Possible loading and fixities on levator ani model.

Results

The solution strategy for the governing equations is dependent on whether a static or dynamic approach was taken. For the static case, a system of linear equations is formed and solved for the displacement at each node or particle. By updating the loading conditions, a series of static load solutions can be used to generate results over a pseudo time-dependent event. The dynamics problem requires explicit or implicit time integration. In time integration, the total time during which an event occurs is discretized into small steps. The displacement at every node or particle is computed for each time step. With explicit time integration, the computation of a single step is cheap but the step size needs to be very small for the solution to be stable. Implicit time integrators are fairly expensive to compute for each step, but are generally stable under larger time steps.

The material displacement in itself is not usually the quantity of interest. Instead the amount the material has strained is of interest because it portrays how much the material has stretched and distorted.

INTERPRETATION OF RESULTS

Subject-specific computational methods have proven effective in aiding understanding of the deformations that occur in the female pelvic floor muscles during vaginal childbirth. Using simulation models developed from manually segmented labelmaps, based on MRI grayscale data, Hoyte and Krysl [1] showed stretch factors of over 3, for the pubovisceral portion of levator ani during simulated passage of a fetal head (Figs. 9 and 10).

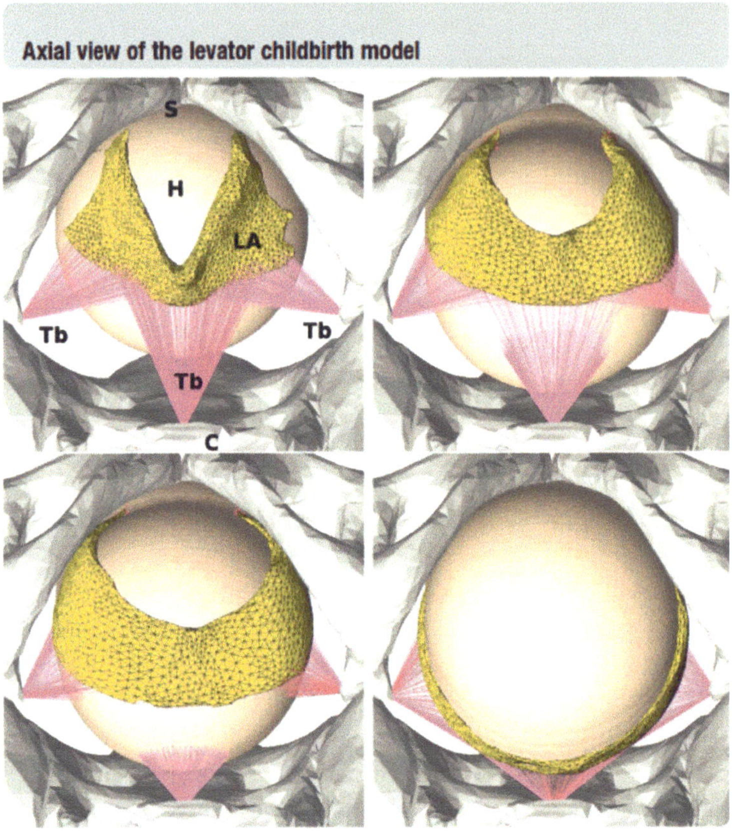

Figure 9 Computational analysis of passage of the simulated fetal head in labor. Upper panel: left — fetal head (H) above isocial spines, right — fetal head in contact with levator ani (LA), beginning of levator deformation. Lower panel: left — fetal head crowning, right — fetal head just past the levator hiatus (maximum stretch). Label legend: S — symphysis, H - fetal head, C — coccyx, Tb — muscle attachment ties.

Simulation results showed the areas of maximum stretch to be mostly in the distal and anterior portions of levator ani, corresponding to the pubovisceral portions that muscle complex 10. These findings correlate with the observed areas of levator disruption seen in 20% of primiparous women on MRI [28], shown in Fig. 11 as a loss of the levator muscle arm on the left side (A).

The impact of attachment stiffness on muscle stretch was also demonstrated using computational modeling, where the final equilibrium shape of the levator ani muscle was determined for two different stiffness values of the muscle attachments 12. The results show that levator stretch is increased as the stiffness of the attachments is increased (Petr Krysl — private communication). This finding is important because adaptations to pregnancy are known to alter the material characteristics of the pelvic

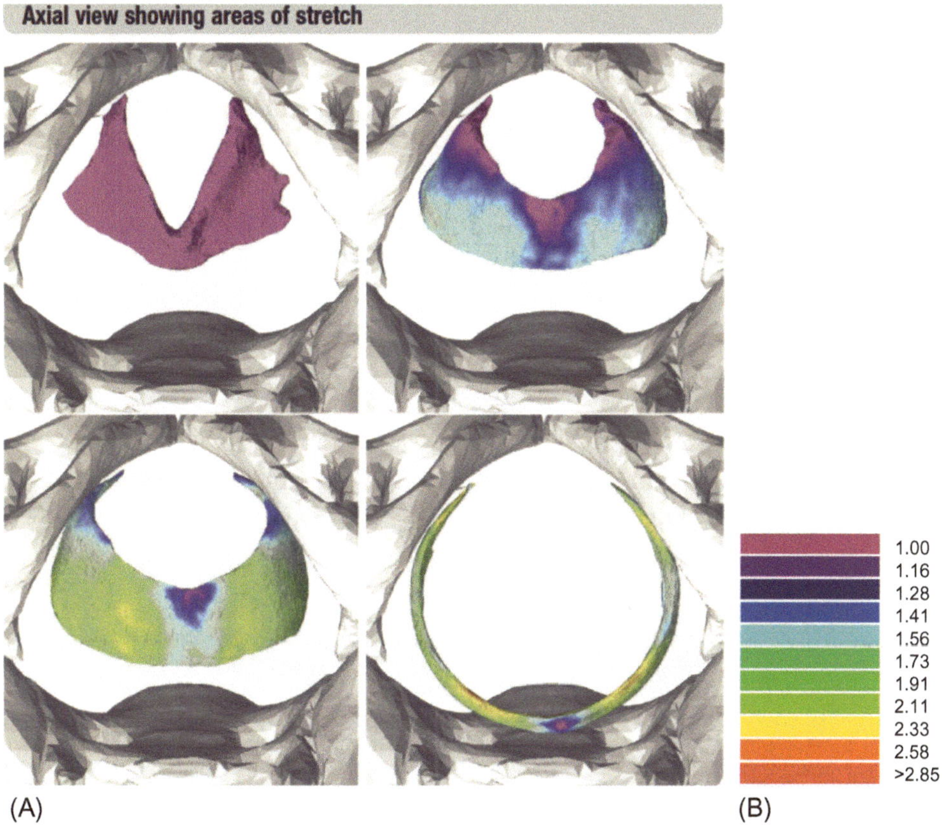

Figure 10 Computational analysis showing the color-coded distribution of levator ani stretch during vaginal delivery of a simulated fetal head. (A) Elements same as in Fig. 9 and (B) color legend of stretch ratio.

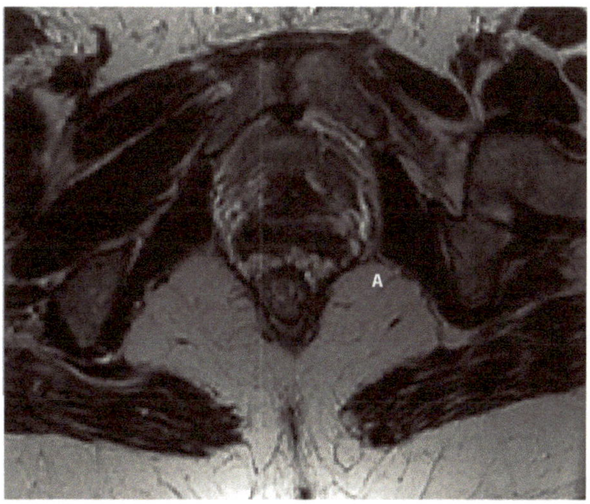

Figure 11 An axial slice from a vaginally parous woman with a left-sided levator ani disruption (A).

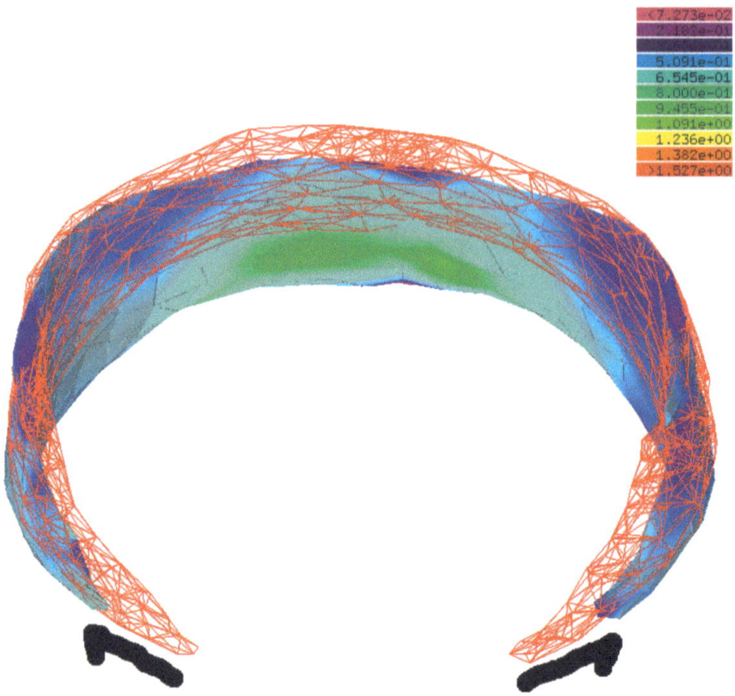

Figure 12 Final equilibrium state of the levator ani at maximum stretch. The wireframe rendering is the result for the stiffer attachments, whereas the colored rendering is for a fivefold reduction in attachment stiffness.

floor and other tissues, decreasing the stiffness of collagen elements, as discussed in detail in an earlier chapter of this text. The resulting reduction in levator stretch due to less stiff collagen attachments may reduce the risk for childbirth-related levator damage (Fig. 12).

Computational methods were also applied to study the importance of apical support in the development of cystocele. This ongoing work is described in detail by DeLancey in an earlier chapter of this text.

REFERENCES

[1] L. Hoyte, M.S. Damaser, S.K. Warfield, G. Chukkapalli, A. Majumdar, D.J. Choi, A. Trivedi, P. Krysl, Quantity and distribution of levator ani stretch during simulated vaginal childbirth, Am. J. Obstet. Gynecol. 199 (2) (2008) 198.e1–198.e5.

[2] J.R. Cebral, R. Löhner, From medical images to anatomically accurate finite element grids, Int. J. Numer. Methods Eng. 51 (8) (2001) 985–1008.

[3] M. Caon, Voxel-based computational models of real human anatomy: a review, Radiat. Environ. Biophys. 42 (4) (2004) 229–235.

[4] T. Bardyn, M. Reyes, X. Larrea, P. Bchler, Influence of smoothing on voxel-based mesh accuracy in micro-finite element, in: K. Miller, P.M.F. Nielsen (Eds.), Computational Biomechanics for Medicine, Springer, New York, 2010, pp. 85–93.
[5] G. Taubin, A signal processing approach to fair surface design, in: Proceedings of the 22nd Annual Conference on Computer Graphics and Interactive Techniques, SIGGRAPH '95, ACM, New York, 1995, pp. 351–358.
[6] V. d'Otreppe, R. Boman, J.-P. Ponthot, Generating smooth surface meshes from multi-region medical images, Int. J. Numer. Methods Biomed. Eng. 28 (6–7) (2012) 642–660.
[7] I. Braude, J. Marker, K. Museth, J. Nissanov, D. Breen, Contour-based surface reconstruction using {MPU} implicit models, Graph. Model. 69 (2) (2007) 139–157.
[8] T.J.R. Hughes, The Finite Element Method, Dover, Mineola, NY, 2000.
[9] S. Ji, J.C. Ford, R.M. Greenwald, J.G. Beckwith, K.D. Paulsen, L.A. Flashman, T.W. McAllister, Automated subject-specific, hexahedral mesh generation via image registration, Finite Elem. Anal. Des. 47 (10) (2011) 1178–1185.
[10] B. Couteau, Y. Payan, S. Lavalle, The mesh-matching algorithm: an automatic 3D mesh generator for finite element structures, J. Biomech. 33 (8) (2000) 1005–1009.
[11] L. Baghdadi, D.A. Steinman, H.M. Ladak, Template-based finite-element mesh generation from medical images, Comput. Methods Prog. Biomed. 77 (1) (2005) 11–21.
[12] M. Bucki, C. Lobos, Y. Payan, A fast and robust patient specific finite element mesh registration technique: application to 60 clinical cases, Med. Image Anal. 14 (3) (2010) 303–317.
[13] W.K. Liu, Y. Chen, S. Jun, J.S. Chen, T. Belytschko, C. Pan, R.A. Uras, C.T. Chang, Overview and applications of the reproducing kernel particle methods, Arch. Comput. Meth. Eng. 3 (1) (1996) 3–80.
[14] T. Belytschko, Y.Y. Lu, L. Gu, M. Tabbara, Element-free Galerkin methods for static and dynamic fracture, Int. J. Solids Struct. 32 (17–18) (1995) 2547–2570. A Symposium on the Dynamic Failure Mechanics of Modern Materials in Memory of Professor J. Duffy.
[15] T.J. Liszka, C.A.M. Duarte, W.W. Tworzydlo, hp-meshless cloud method, Comput. Methods Appl. Mech. Eng. 139 (14) (1996) 263–288.
[16] J. Dolbow, T. Belytschko, An introduction to programming the meshless element free Galerkin method, Arch. Comput. Meth. Eng. 5 (3) (1998) 207–241.
[17] J.-S. Chen, C. Pan, C.-T. Wu, Large deformation analysis of rubber based on a reproducing kernel particle method, Comput. Mech. 19 (3) (1997) 211–227.
[18] M. Doblaré, E. Cueto, B. Calvo, M.A. Martínez, J.M. Garcia, J. Cegoñino, On the employ of meshless methods in biomechanics, Comput. Methods Appl. Mech. Eng. 194 (68) (2005) 801–821.
[19] K.C. Wong, L. Wang, H. Zhang, H. Liu, P. Shi, Meshfree implementation of individualized active cardiac dynamics, Comput. Med. Imaging Graph. 34 (1) (2010) 91–103.
[20] H. Liu, P. Shi, Meshfree representation and computation: applications to cardiac motion analysis, in: C. Taylor, J. Alison Noble (Eds.), Information Processing in Medical Imaging, Lecture Notes in Computer Science, vol. 2732, Springer, Berlin, 2003, pp. 487–498.
[21] Z.Y. Johnny, G.R. Joldes, A. Wittek, K. Miller, Patient-specific computational biomechanics of the brain without segmentation and meshing, Int. J. Numer. Methods Biomed. Eng. 29 (2) (2013) 293–308.
[22] R.R. Basava, J.-S. Chen, Y. Zhang, S. Sinha, U. Sinha, J. Hodgson, R. Csapo, V. Malis, Pixel based meshfree modeling of skeletal muscles, in: Y.J. Zhang, J.M.R.S. Tavares (Eds.), Computational Modeling of Objects Presented in Images: Fundamentals, Methods, and Applications, Lecture Notes in Computer Science, vol. 8641, Springer, New York, 2014, pp. 316–327.
[23] J.B. Alford, D.C. Simkins, R.A. Rembert, L. Hoyte, Patient-specific modeling in urogynecology: a meshfree approach, Comput. Model. Eng. Sci. 98 (2014) 129–149.
[24] T. Belytschko, W.K. Liu, B. Moran, Nonlinear Finite Elements for Continua and Structures, Wiley, New York, 2000.
[25] X. Li, J.A. Kruger, M.P. Nash, P.M. Nielsen, Effects of nonlinear muscle elasticity on pelvic floor mechanics during vaginal childbirth, J. Biomech. Eng. 132 (11) (2010). Article number 111010.
[26] D. Jing, J.A. Ashton-Miller, J.O. DeLancey, A subject-specific anisotropic visco-hyperelastic finite element model of female pelvic floor stress and strain during the second stage of labor, J. Biomech. 45 (3) (2012) 455–460.

[27] P. Martins, E. Pea, B. Calvo, M. Doblar, T. Mascarenhas, R. Natal Jorge, A. Ferreira, Prediction of nonlinear elastic behaviour of vaginal tissue: experimental results and model formulation, Comput. Meth. Biomech. Biomed. Eng. 13 (3) (2010) 327–337. PMID: 20146131.
[28] J.O. DeLancey, R. Kearney, Q. Chou, S. Speights, S. Binno, The appearance of levator ani muscle abnormalities in magnetic resonance images after vaginal delivery, Obstet. Gynecol. 101 (1) (2003) 46–53.

CHAPTER TWENTY

Biomechanics of Pelvic Organ Prolapse

L. Chen, J. Luo, J.O.L. DeLancey, J.A. Ashton-Miller
University of Michigan, Ann Arbor, MI, United States of America

AN INTRODUCTION TO PELVIC ORGAN PROLAPSE

A pelvic organ prolapse is a type of hernia. Abdominal pressure, gravitational, and inertial forces combine to push one or more of the pelvic organs downward and out of the potential aperture represented by the levator hiatus. The hiatus is bounded by the pubic bone at the front and the most caudal levator ani muscles at the sides and toward the rear of the pelvis (see Chapter 2: Pelvic Floor Anatomy and Pathology). An important function of the levator ani muscles is to maintain closure of the levator hiatus during normal and strenuous activities, for it is only during micturition or defecation that these muscles relax to allow passage of body waste. One of the contributing factors to the development of prolapse appears to be an injury to a specific portion of the levator muscle, often when a mother gives vaginal birth for the first time (for review, see [1]). In this chapter we shall explore the biomechanical factors and mechanisms that may cause a prolapse to develop, because better insights can lead to improved treatments for this condition.

USING SYSTEMS ANALYSIS TO ASSESS BIOMECHANICAL FACTORS INVOLVED IN PELVIC ORGAN PROLAPSE

One can employ a systems analysis approach to consider the biomechanical interactions that may lead to a prolapse (Fig. 1). In this approach, one considers the inputs and outputs to a system, at first considering that system as a black box about which one knows very little. One can demonstrate the system using a simple rectangle to represent the black box, with an input to and an output from the box, completing what engineers call a block diagram. For example, if the input to the system is "x" and the output is "y," then one can define what is in the box by examining how y varies as a function of x. If one observed that y was simply proportional to x, then the ratio of y/x can be defined as "k," where $y = kx$. Thus by considering the relationship between the input and output of the black box, one can infer something about the function of what is in the black box. What is in

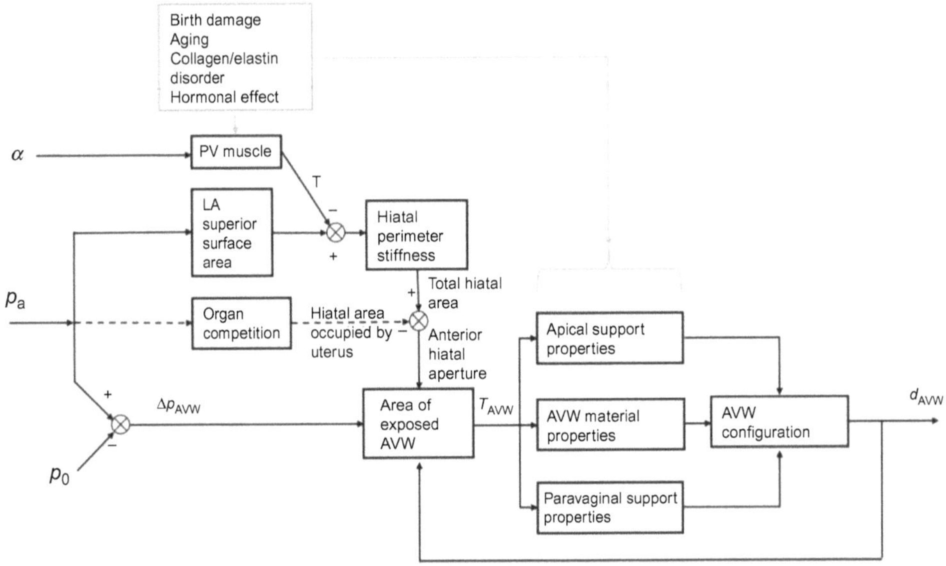

Figure 1 A conceptual model of the pathomechanics of cystocele. Inputs to the support system include the intensity (α) of the pubovisceral muscle (PVM) contraction (yielding tension, T, in the vaginal wall), intraabdominal pressure (p_a), and atmospheric pressure (p_o). 'p_a' acts on the surface of the levator ani muscle so as to increase T and help determine total hiatal area. It can also drive the uterus partially into the hiatus (dashed line), leaving the remaining hiatal aperture to be spanned by the distal anterior vaginal wall (AVW) and exposed to the pressure differential (Δp_{AVW}). The resulting tension (T_{AVW}) in the AVW then applies tension to its Apical and Paravaginal Supports helping to determine the size of the resulting cyctocele (d_{AVW}). Birth damage, aging, collagen/elastin disorders, and hormonal effects are all postulated to affect the properties of given structures. *Reproduced from L. Chen, J.A. Ashton-Miller, J.O. DeLancey, A 3D finite element model of anterior vaginal wall support to evaluate mechanisms underlying cystocele formation, J. Biomech. 42 (10) (2009) 1371–1377.*

the box may actually act nonlinearly, like $1/k^2$, or even exhibit a behavior that changes with time. Regardless, by breaking the overall system into logical subparts, each represented by a "'smaller" black box, one can consider the relationship between the input and output of each of these smaller black boxes. Finally, one can reassemble the subparts into a system block diagram that shows the interrelationships between each subpart of the system. When this is done in Fig. 1, one can clearly see the functional relationships between the intraabdominal pressure, atmospheric pressure, the levator muscle activation, hiatal size, tissue properties, and the resulting prolapse size.

Our working hypothesis has been that a prolapse will develop if the levator ani muscles are no longer able to maintain closure of the levator hiatus in the presence of intraabdominal pressure or increases in intraabdominal pressure. This loss of closure may be due to injury of certain levator muscles [2] that can include detachment at its origin on one or both sides, as for example, during vaginal birth, or levator ani muscle losing its strength due to aging. Once the hiatus no longer maintains its closure, the distal end of the vagina becomes exposed to the pressure differential between the intraabdominal pressure and the external atmospheric pressure. This is an example of a positive feedback loop. Hence, the tension that developed in the vaginal wall by the pressure differential pulls downward on the apical support provided by the cardinal and uterosacral "ligaments"[1] which suspend the uterus and upper vagina (see Chapter 2) thereby causing apical descent in which the uterus and upper vagina descend. This apical descent in turn causes a longer portion of the vaginal wall to be exposed to the pressure differential, which in turn causes larger tension in the vaginal wall and larger downward traction loads on the apical support until the system reaches a new equilibrium state. However, this new equilibrium is not necessarily a stable state if the vaginal wall and its supporting tissue are under an abnormal amount tension for extended period of time. This abnormal tension might trigger tissue adaptation and remodeling: these can lengthen the vaginal wall and apical supports, would further increase the portion of vagina being exposed the pressure differential, causing an ever larger apical descent and eventually a larger prolapse.

USING BIOMECHANICAL MODELING TO OBTAIN INSIGHTS INTO THE PATHOMECHANICS OF PROLAPSE

It has been written that "Just as the ability to devise simple but evocative models is the signature of the great scientist, so overelaboration and overparameterization is often the mark of mediocrity" [3]. While that was a statement that referred to statistical models, the sentiment also applies to biomechanical models used to capture the mechanics of the particular biology involved.

In broad outline, our approach to modeling the mechanics of a prolapse started by reconstructing the 3D geometry of a woman's pelvic anatomy who was known to be asymptomatic and having normal anatomy, on pelvic examination as captured in MRI scans. A series of simplifications and assumptions were then made, based on the current understanding of anatomy with respect to pelvic floor structures and their connections, in order to develop simplified representations of the pelvic floor support system. Thus tissue material properties were assigned to different support structures and the pelvic floor loading conditions were prescribed. Then the equations describing the biomechanical model were solved numerically in the usual manner. Next, the simulated tissue deformations under load were then compared with the deformation observed in patients with prolapse to verify that the assumptions made in the model result in anatomical deformations similar to those seen clinically. After verification, such a model can then be used to run virtual "what-if" experiments to test certain functional

[1] "Ligaments" in name only, these are actually mesenteries (see Chapter 2).

and structural hypotheses. When such a model is found to fall short, then more accurate models can be developed. So, given an absence of models of the mechanics of prolapse, we started with a simple 2D modeling approach, knowing that despite certain limitations, it might still provide useful insights.

2D Sagittal Plane Model of Cystocele

Motivated by clinical data demonstrating the strong associations between anterior vaginal wall prolapse and levator ani muscle impairment [2], our first biomechanical model aimed to examine the first-order effects of the presence of impairments in muscular support and connective tissue support on anterior vaginal wall prolapse [4]. An anterior vaginal wall prolapse, or cystocele, is the most common form of prolapse. In the model, we distilled the anterior vaginal wall support system to its essence by identifying four functional components and including them in a simple 2D, sagittally-symmetric, lumped parameter model (Fig. 2). The key components of the model included: a deformable and stretchable anterior vaginal wall, a rigid levator ani plate lifted and controlled by pubovisceral muscle

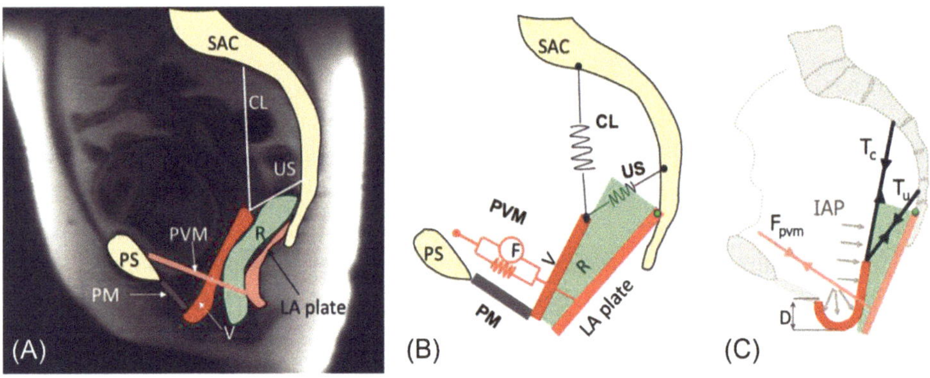

Figure 2 2D lumped parameter model. (A) Modeled element traced or projected on mid-sagittal MR image. (B) Lumped parameter biomechanical model. Pubovisceral muscle (PVM) is modeled as a spring in parallel with an active force generator. (C) Loaded pelvic floor with defective muscular support and part of vaginal wall exposed to intraabdominal pressure (light grey arrows). PS: pubic symphysis; SAC: sacrum; PM: perineal membrane; LA plate: levator plate; R: rectum; V: vagina; CL: cardinal ligament spring; US: uterosacral ligament spring; B: bladder; UT: uterus; IAP: intraabdominal pressure; F_{pvm}: tensile force generated by pubovisceral muscle between the projection of its origin on the pelvic side wall and insertion on the levator plate; Tc and Tu: tensile forces generated by the cardinal and uterosacral ligaments; D, the decent of the most dependent point of vaginal wall from end of perineal membrane (PM) which is used as the measurement of prolapse size in the simulation. Note descent of the vaginal apex as well as vaginal wall protrusion. *Reproduced from L. Chen, J.A. Ashton-Miller, Y. Hsu, J. O.L. DeLancey, Interaction among apical support, levatorani impairment, and anterior vaginal wall prolapse, Obstet. Gynecol. 108 (2006) 324–332.*

(also called the pubococcygeal muscle)[2] and two passive springs representing cardinal and uterosacral ligaments providing support at the top of vagina. Anterior vaginal wall, pubovisceral muscle, and the apical supports were assumed to have an exponential relationship between force and elongation, that is, $F = C_1 \times \left(e^{C_2 \times \left(\frac{l - l_0}{l_0}\right)} - 1 \right)$ based on the existing literature [5,6] where C_1 and C_2 are the material property parameters, 'l' is the current length and l_0 is the initial length. Impairment of the structure was modeled as a decrease in stiffness of the material (C_1).

To simulate a Valsalva maneuver, a distributed intraabdominal pressure was applied to the anterior vaginal wall. As shown in Fig. 2C, with increasing intraabdominal pressure, the levator plate is pushed posteriorly and downwards to a more vertical orientation, the anterior vaginal wall developed a bulge after being pushed out via the hiatus, and the top of vagina is pulled down. The model behavior matches well with that observed on MRI at maximum Valsalva, thereby helping to validate the model (Fig. 3).

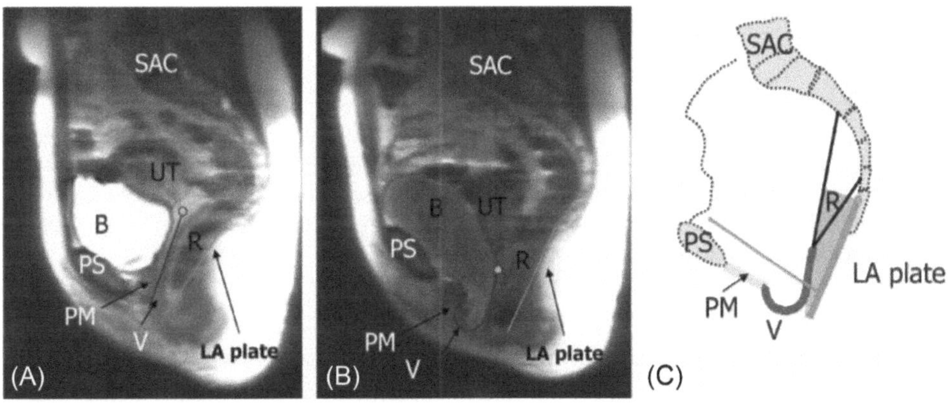

Figure 3 Validation of the model. (A) Mid-sagittal MR image of resting status; (B) mid-sagittal MR image at maximum Valsalva; (C) sample model simulation result having a similar configuration as in (B). In this figure PS denotes pubic symphysis; SAC: sacrum; PM: perineal membrane; LA plate: levator plate; R: rectum; V: vagina; B: bladder; and UT: uterus. *Copyright Biomechanics Research Lab, University of Michigan, 2006.*

[2] Terminologia Anatomica lists pubococcygeal muscle for this portion of the levator ani muscle; a term coined based on comparative anatomy, not on function. We have preferred the term pubovisceral muscle as it more accurately reflects the connections of the muscle between the pubic bones and the pelvis viscera. Since both of these terms have, at different times, been used in illustrations, we have not limited our use to only one term.

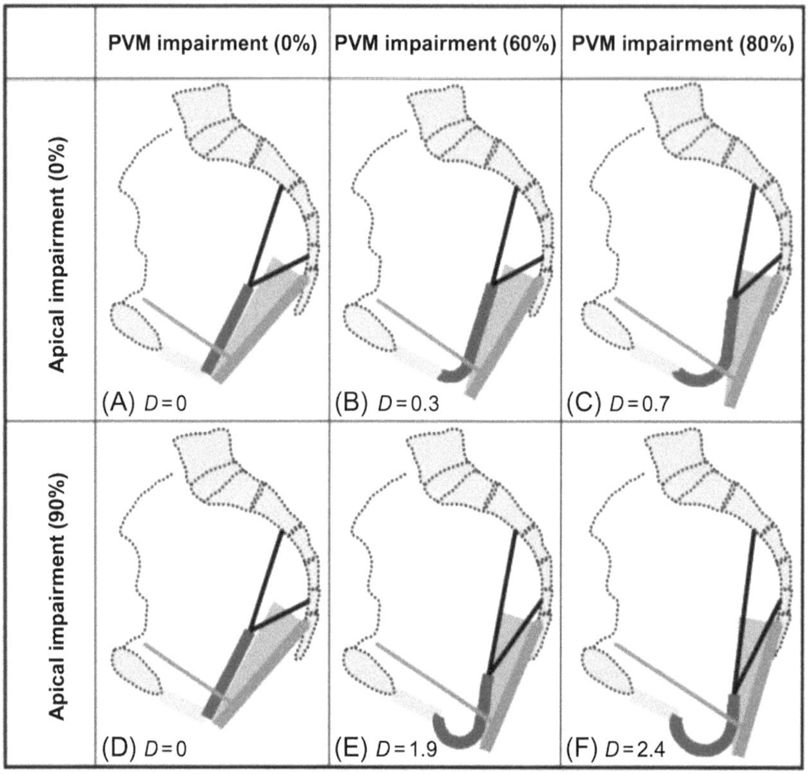

Figure 4 Simulated deformation of the model anterior vaginal wall, and its support system, under maximal Valsalva with various degrees of pubovisceral muscle (PVM) and cardinal and uterosacral ligament impairment (indicated in percent). The variable D presents the size of prolapse measured as the decent of the most dependent point of vaginal wall from end of perineal membrane. *Copyright Biomechanics Research Lab, University of Michigan, 2006.*

We then assigned various degrees of impairments in pubovisceral muscle and apical support to study the predicted deformation in the anterior vaginal wall for different combinations of impairment (Fig. 4). The model simulations suggest that, to some degree, normal muscle can compensate for impaired connective tissue function and vice versa. But it is when both systems have impairments that anterior vaginal wall prolapse is predicted.

This model helped to initiate useful investigations into muscle/connective interactions and it provided a systematic understanding linking the observation that the levator muscles are abnormal in many women with pelvic organ prolapse with observations of connective tissue abnormalities in this disease. Each of the functional components is modeled in their simplest form to keep the model focused on the most essential elements and their function. However, because it only represents midline function, we recognized that further work would be needed to add other important elements of the support system which are missing in the model. Among these are the lateral attachments of the vagina to the pelvic sidewall provided by paravaginal support.

3D Finite-Element Model of Cystocele

To address the main limitations of the 2D model, namely the lack of any paravaginal attachment to the pelvic side wall, the inability to study asymmetric levator injury, and the nondeformable levator plate and rectum, we developed a 3D finite-element model to test the hypothesis that factors such as muscular support, apical defect, and paravaginal defects, each affect cystocele formation [7].

First, we generated subject-specific 3D geometrical models representing anterior vaginal wall support system from a healthy nulliparous woman's MRI scans at rest. Manual segmentation in all three planes (axial, sagittal, and coronal) took into account the need to get more accurate representation of pelvic organs, levator ani muscle and connective fascia and ligaments. Then the volumetric 3D models were imported into Imageware (version 13.0; Siemens Product Lifecycle Management Software Inc., 2008) to generate simplified smoothed representations. This model was then imported into ABAQUS, finite-element meshes were generated, boundary conditions and connections between structures were defined, and hyperelastic tissue properties were assigned based on available data from the literature. Finally the model was solved numerically to predict the deformation under increasing intraabdominal pressure loading (Figs. 5 and 6).

It is worth mentioning that an iterative process was used to evaluate and refine the assumptions we made in the model. For example, since paravaginal support is difficult to visualize on MRI, the assumptions regarding the representation of the paravaginal support had to be made based on an expert's knowledge of functional anatomy. We put these assumptions into the model and tested its validity, then refined those assumptions until convergence was obtained between model behavior under increases of abdominal pressure and deformations observed on dynamic MRI (Fig. 7). A similar iterative process has also been used in later studies by, for example, Rau [8], Rubod [9], and Cosson [10] in their biomechanical models to evaluate the assumptions made regarding the connective tissue function anatomy.

To simulate cystocele formation, we set the 3D model to simulate impairments in the various supportive components (ie, cardinal and uterosacral ligaments, paravaginal supports, and pubovisceral muscle) in a similar way to that used in the 2D model. The magnitude of the cystocele was found to be sensitive to maximum abdominal pressure, the presence of impaired structures (levator ani muscle, apical connective tissue, paravaginal connective tissue, vaginal wall) and single versus multiple site impairments (Fig. 8). The finding that impairments in multiple domains (muscle impairment and mesenteric connective tissue impairment, including both apical and paravaginal) result in larger cystocele is similar to the result we found in a simpler 2D model [4].

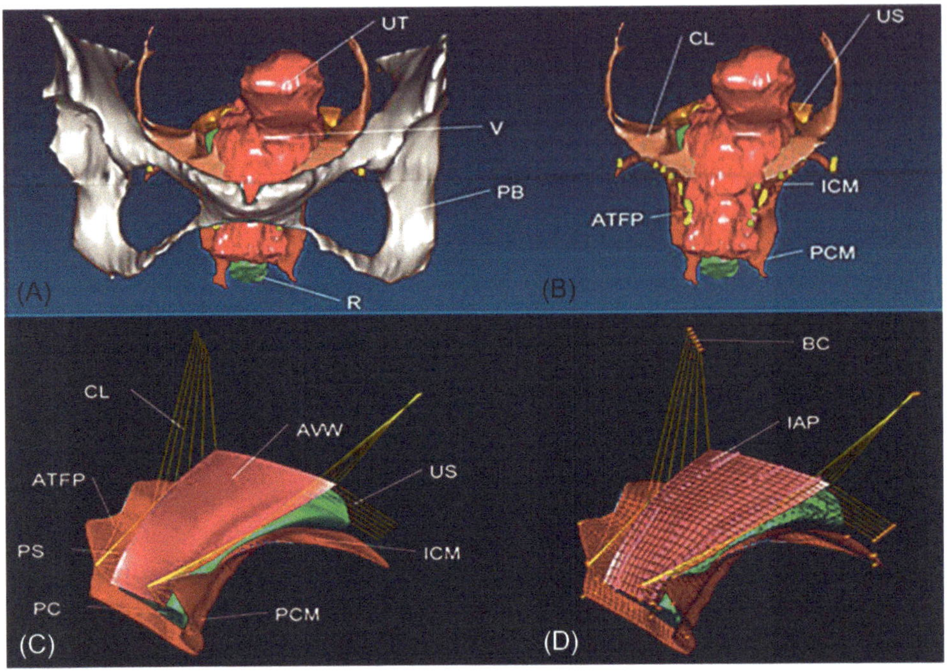

Figure 5 Model development. (A) 3D volume-rendered model of anterior support system including pubic bone; (B) 3D volume-rendered model without pubic bone; (C) geometrically simplified surface model; and (D) 3D finite-element model with mesh, boundary condition (orange pin representing ligaments and muscle origin is fixed to pubic bone and pelvic sidewall), and abdominal pressure loading. PB denotes pubic bone; UT: uterus (not included in surface model); V: vagina; R: rectum; CL: cardinal ligament; US: uterosacral ligament; ATFP: arcus tendineus fascia pelvis; ICM: iliococcygeus muscle; PCM: pubococcygeus muscle; AVW: anterior vaginal wall; PC: posterior compartment; PS: paravaginal support; and IAP: intraabdominal pressure. *Reproduced from L. Chen, J.A. Ashton-Miller, J.O. DeLancey, A 3D finite element model of anterior vaginal wall support to evaluate mechanisms underlying cystocele formation, J. Biomech. 42 (10) (2009) 1371–1377.*

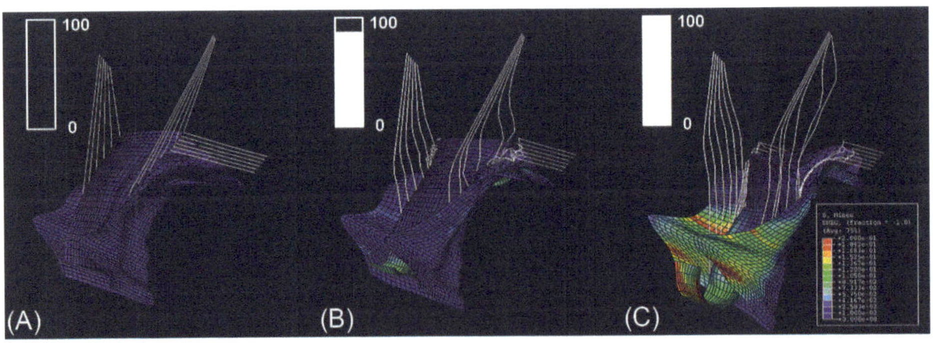

Figure 6 (A-C) A three-quarter, left, anterior view of these sequential development of a typical simulated cystocele during a gradually increasing load. In this simulation, the levator ani muscle had a 60% impairment, and apical and paravaginal support properties were set as having a 50% impairment. The color map shows the stress distributions in different regions, with blue indicating a low stress region and red indicating a high stress region. The box on the upper left corner shows the abdominal pressure (ordinate), increasing from 0 to 100 cm H_2O, over time corresponding to the three scenarios above. *Reproduced from L. Chen, J.A. Ashton-Miller, J.O. DeLancey, A 3D finite element model of anterior vaginal wall support to evaluate mechanisms underlying cystocele formation, J. Biomech. 42 (10) (2009) 1371–1377.*

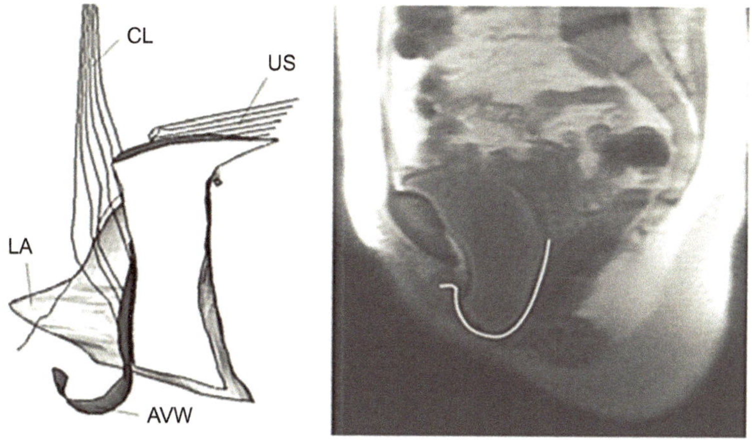

Figure 7 Model validation. On the left is a model-generated simulation result showing a cystocele formed in the manner of that seen clinically in one frame of a dynamic MRI on right side of the figure. *Reproduced from L. Chen, J.A. Ashton-Miller, J.O. DeLancey, A 3D finite element model of anterior vaginal wall support to evaluate mechanisms underlying cystocele formation, J. Biomech. 42 (10) (2009) 1371–1377.*

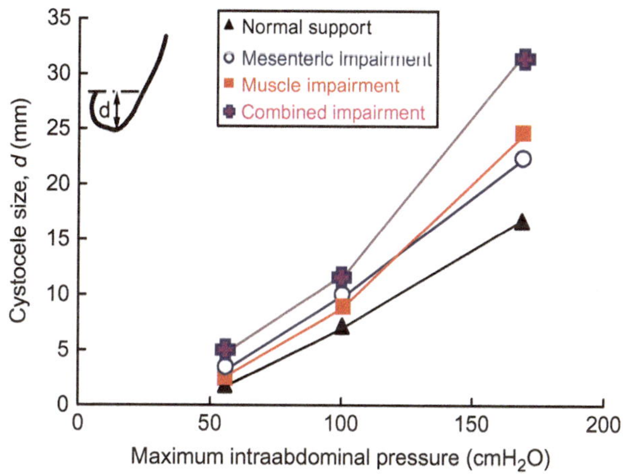

Figure 8 Simulated cystocele size for models with different impairment patterns at increasing values of maximum intraabdominal pressure loading. *Reproduced from L. Chen, J.A. Ashton-Miller, J.O. DeLancey, A 3D finite element model of anterior vaginal wall support to evaluate mechanisms underlying cystocele formation, J. Biomech. 42 (10) (2009) 1371–1377.*

Even though the model is still a simplified representation of realty, with simplified geometry, material properties, and loading conditions, this 3D finite-element model of the anterior vaginal wall support system had a more realistic anatomic geometry and reproduces realistic cystoceles similar to those seen on dynamical MRI in patients. It provided representation of more elements of the anterior vaginal wall support system so that it would be possible to analyze more potential failure mechanisms than was possible with the 2D model. Based on the simulation results, we have developed the conceptual model to help provide a framework to systematically investigate the interaction between muscle, connective tissue, and anterior vaginal wall, as shown earlier. Again, as in the 2D model, we represented each functional element as simply as possible to capture the essential behavior of the system being studied.

Multiple Compartment Pelvic Floor and Prolapse

The 3D finite-element model of the anterior vaginal support system provides valuable insights on the underlying pathomechanics of cystocele [7]. A knowledge gap remained, however, as to how and why a prolapse of the posterior compartment, a rectocele, forms, and whether there is interaction occurring between anterior and posterior prolapse. To address this knowledge gap, we developed a 3D finite-element multiple compartment model [11] which could develop either cystocele and/or rectocele, depending on the particular impairments of the anterior and posterior pelvic compartment structural support systems. Of particular interest were the factors leading to the development of rectocele and comparing the results with observations in women with this condition. Furthermore, some women in whom a cystocele is repaired, develop a new rectocele postoperatively, a generally acknowledged but poorly understood cause of surgical failure. We therefore sought to assess the model's ability to simulate rectocele interactions with the anterior vaginal wall under increasing intraabdominal pressures.

A more detailed 3D geometrical model of the pelvic floor was developed from our original model 3D model [12], which now included 23 structures. The vaginal walls, levator ani, and other pelvic floor structures were simplified and lofted to form smooth surfaces. The simplified models were then imported into ABAQUS to generate the structural mesh used to simulate the prolapse. In this model, we also updated the material properties based on our recent in vivo apical support pulling tests [13].

Detailed pelvic floor structures′ origins, insertions, and connections were evaluated and refined until the model could simulate the development of a rectocele with the characteristic of downward displacement, forward protrusion, and perineal descent as well as the "kneeling" profile matching those seen in 3D stress MRI (Fig. 9).

This multi-compartment model contains the structurally more complex posterior compartment, thereby extending previous models and allowing posterior compartment prolapse or rectocele to be simulated. Fig. 10 shows the biomechanical effects of changes in intraabdominal pressure and apical support on rectocele size. In addition, having this compartment represented now allows its interaction with cystocele to be quantified in a systematic manner so that the so-called organ competition can be evaluated thereby extending published pelvic floor computer models. Organ competition is the interaction between the mechanical anterior and posterior compartments which determines whether the one or the other, or both, prolapse. It is important clinically because if a surgeon

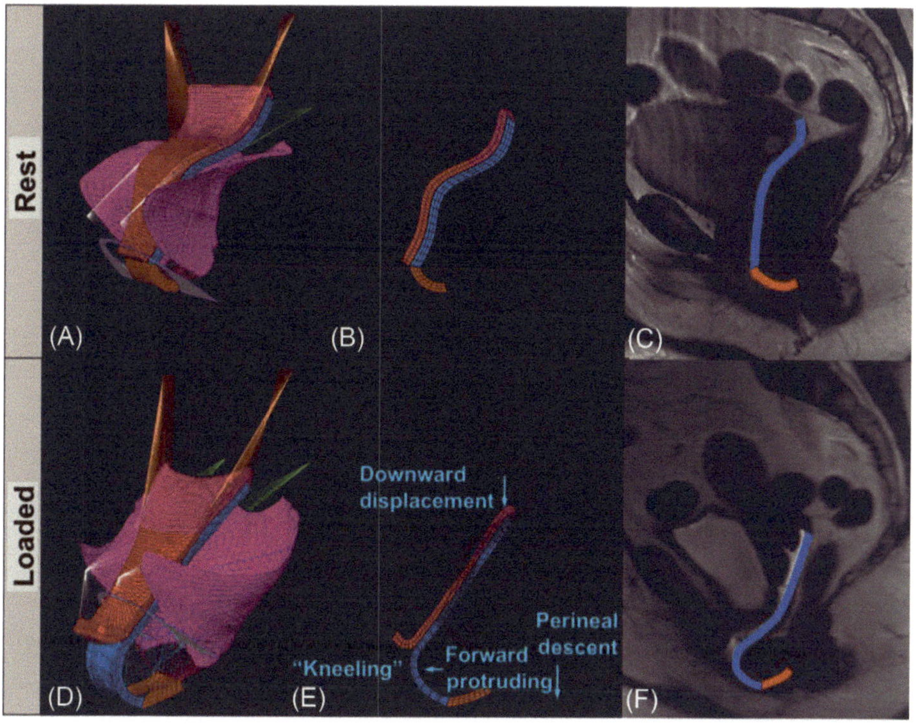

Figure 9 Rectocele characteristics shown in left three-quarter view for the model at rest (A) and under load (D). Panel (B) shows mid-sagittal plane section of model at rest. In (E), "kneeling", downward displacement, forward protruding, and perineal descent behaviors are seen for the rectocele. Static MR (C) and "Stress MR" (F) have posterior vaginal wall outlined in blue and perineal body in brown. *Reproduced from J. Luo, L. Chen, D.E. Fenner, J.A. Ashton-Miller, J.O.A. DeLancey, Multi-compartment 3-D finite element model of rectocele and its interaction with cystocele, J. Biomech. 48 (9) (2015) 1580–1586.*

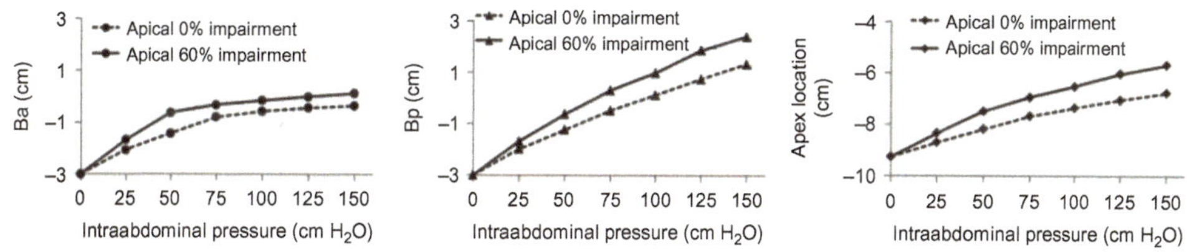

Figure 10 Model-predicted prolapse size is predicted to depend on intraabdominal pressure and apical impairment with levator ani 20% impairment, posterior 85% impairment, and no anterior impairment. *Reproduced from J. Luo, L. Chen, D.E. Fenner, J.A. Ashton-Miller, J.O.A. DeLancey, Multi-compartment 3-D finite element model of rectocele and its interaction with cystocele, J. Biomech. 48 (9) (2015) 1580–1586.*

repairs a cystocele in some women, a rectocele can develop because the cystocele is no longer blocking the hiatus to prevent it being prolapsed. The same can happen when a surgeon repairs a rectocele only to have a cystocele develop. Model simulations (Figs. 11 and 12) demonstrated that this empirical clinical observation can indeed happen. For example

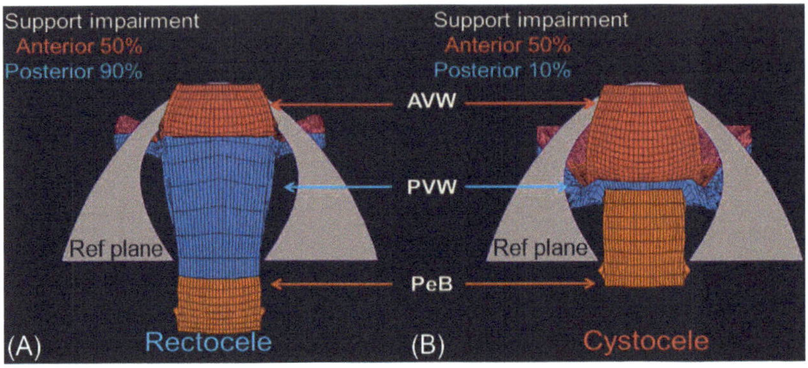

Figure 11 Inferior view of simulation results demonstrating what happens to the anterior compartment when impairments to the posterior compartment are decreased, as they would be after surgical repair under conditions with 140 cm H_2O and impairments of 50% for the levator ani, 80% for apical supports, and 50% for anterior supports. (A) Model-generated rectocele (Ba = −1, Bp = 1 cm) with posterior support impairment of 90% impairment. (B) Reduction of the posterior support impairment to 10% a cystocele develops (points Ba = 0, Bp = −2 cm) despite no change in anterior support stiffness. *Reproduced from J. Luo, L. Chen, D.E. Fenner, J.A. Ashton-Miller, J.O.A. DeLancey, Multi-compartment 3-D finite element model of rectocele and its interaction with cystocele, J. Biomech. 48 (9) (2015) 1580–1586.*

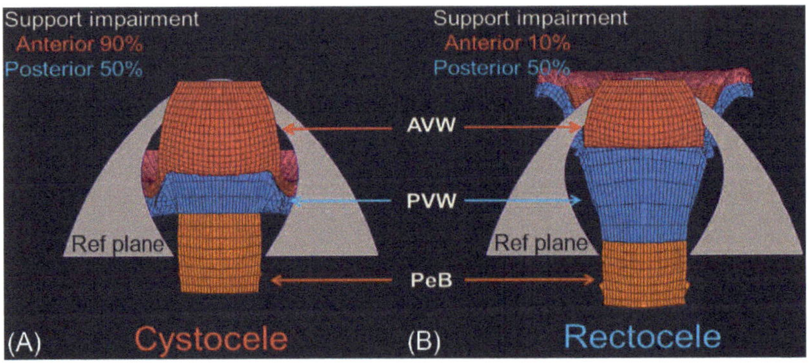

Figure 12 Inferior view of the simulation demonstrating what happens to the posterior compartment when impairments to the anterior compartment are decreased, as they would be after surgical repair under conditions with 140 cm H_2O and impairments of 50% for the levator ani and 50% for posterior supports. Panel (A) shows a model-generated cystocele (Ba = 1, Bp = −1 cm) with anterior support 90%. Panel (B) shows that with reduction of anterior support impairments to 10% a rectocele develops (Ba = 0, Bp = 1 cm) despite no change in posterior support stiffness. *Reproduced from J. Luo, L. Chen, D.E. Fenner, J.A. Ashton-Miller, J.O.A. DeLancey, Multi-compartment 3-D finite element model of rectocele and its interaction with cystocele, J. Biomech. 48 (9) (2015) 1580–1586.*

in Fig. 11, keeping the same simulation conditions for levator ani muscle and apical support, increasing posterior support (from A to B), which is likely to happen after posterior compartment repair, there is a significant reduction in rectocele size, but also an increase in cystocele size. This model helps one understand the conditions under which one compartment can block prolapse of the other compartment.

DISCUSSION

In this chapter, simple and more complex biomechanical models have been used to help to improve our understanding of pelvic floor organ prolapse from a biomechanical point of view. While the focus can be on which particular pelvic organ prolapses (for example, Ref. [14]), we have also seen that the models can address the factors that are responsible for the prolapse. Our approach has been to consider the pelvic organs and their muscular and connective tissue supports as interrelated components of a single system. We rely on detailed anatomical dissections and cross-sectional image review, carefully documented clinical evidence and basic biomechanical theory to guide us in choosing candidate components in the model. The models are then used to test clinically relevant hypotheses. The assumptions made in these models were based on our best understanding of the functional anatomy at the time. We have also taken care to validate our models against clinical data and to refine our assumptions if necessary so that the model behavior is similar to that seen in patients. It has been our experience that making a model has often shown us where our previous assumptions were incorrect, it has sometimes caused us to go back and revisit the real anatomy in a cadaver dissection or on histology, and this has been helpful in leading to a better understanding of structure and function.

We started from the simplest 2D model in order to understand and capture the most important components and how their interactions cause prolapse. Then we analyzed the effects that were missing from 2D model predictions to identify the improvements that would have to be made for a first-generation 3D model to capture those behaviors. Finally, a more complicated multi-compartment model was developed to study the mechanics of rectocele and its interaction with cystocele. Simpler models have fewer components and fewer interactions. Such models are easier to develop and solve numerically and also easier to validate. Yet they can still provide significant insights that help us better understand the main effects of a disease mechanism. It is only when the hypothesis and research questions lay beyond the capability of a simple model to capture the behavior in question that a more complicated model was developed.

It is noteworthy that we also applied simplified loading conditions in our models. The expulsive pressure driving the development of a prolapse is comprised of three components. The first is the hydrostatic pressure represented by the height of the fluid column from the pelvic floor to the diaphragm. While this is approximately twice as high during standing as when lying [15], its absolute value is relatively modest compared to the second component which is intraabdominal pressure generated by the contraction of the diaphragm and abdominal wall muscles, as for example during a heavy-lifting task [16]. This pressure can reach as much as 400 mmHg in an athlete doing a push up with his outstretched fingers [17], but typically is about 100 cm H_2O during a Valsalva and 140 cm H_2O during a cough or sneeze. The third component is the pressure generated by the inertial forces on the pelvic floor when it is called upon to decelerate caudal movement of the abdominal contents. Such forces peak at foot strike when running or when landing a jump. Indeed, landing a parachute jump incorrectly has been associated with pelvic floor muscle avulsion, presumably due to an excessive inertial load caused by failing to flex the knees enough before landing. Since our models were only meant to simulate Valsalva maneuvers which can be reproduced in the MR scanner, we have hitherto only considered rises in intraabdominal pressure in our models, not inertial forces.

Characterizing and correctly modeling the constitutive properties of tissue is another challenge and essential step in these types of biomechanical models. In addition to utilizing material properties that have been characterized by traditional uniaxial tension tests on small excised specimens in vitro (for example, Refs. [18–20]), we have begun to incorporate data taken from in situ in vivo tests [13] in more recent models. Nonetheless, we have had to make simplifying assumptions for the material properties used in our models, and thus far we have only considered isotropic hyperelastic behavior. Effects of fiber direction on pelvic floor soft tissue behavior would be valuable to consider in some instances, but more often may have secondary effects. While some simulations have only showed a small change of prolapse size at a single point in time, one must bear in mind that this small change could be amplified over years of repetitive loading, given tissue hysteresis and time-dependent viscoelastic behavior, including creep. Further work will be needed to include the effect of time-dependent material properties of the levator ani muscle and particularly, the passive connective tissues, in order to better be able to evaluate these effects.

While biomechanical modeling has helped us make progress in understanding the mechanics of prolapse, important knowledge gaps still remain. First, there is a lack of data on the in vivo behavior of many pelvic floor tissues under physiological loading conditions, and this poses a major challenge. An alternative approach is to use inverse finite-element modeling approach to estimate tissue properties based on loading and deformation relationships observed in living women. However the solution of the inverse process is not always unique, especially when the underlying finite-element models become too complicated. Second, the natural history of the disease process can be as long as several decades, so future work is needed to elucidate how tissue adapts to structural impairments over a longer period of time, how obesity affects matters, and how the aging process interacts with disease processes.

CONCLUSIONS

It has been stated that "all models are wrong, but some are useful" [21]. Certainly, our biomechanical models of the pelvic floor support system are simplifications of reality, and they are only as good as the underlying assumptions. Yet, as we have seen, simple models can provide useful insights to help begin to explain, predict, and understand the complex interactions between the many structural components of the pelvic floor. These models have shown that there does not have to be anything innately wrong with the vaginal wall tissue for a prolapse to develop, which was a popular hypothesis when we started this work. Rather, the models show that at least some forms of prolapse can develop when impairments affect the ability of the levator muscles to maintain closure of the levator hiatus. Such insights can be useful in guiding the future treatment of prolapse.

ACKNOWLEDGMENTS

We gratefully acknowledge support from NIH Office for Research on Women's Health SCOR P50 HD 44406 and NICHD R01 HD 38665, R21 HD 79908 for supporting this work.

REFERENCES

[1] J.A. Ashton-Miller, J.O.L. DeLancey, On the biomechanics of vaginal birth and common sequelae, Ann. Rev. Biomed. Eng. 11 (2009) 163–176.
[2] J.O.L. DeLancey, D.M. Morgan, D.E. Fenner, R. Kearney, K. Guire, J.M. Miller, H. Hussain, W. Umek, Y. Hsu, J.A. Ashton-Miller, Comparison of levator ani muscle defects and function in women with and without pelvic organ prolapse, Obstet. Gynecol. 109 (2007) 295–302.
[3] G.E.P. Box, Science and statistics, J. Am. Stat. Assoc. 71 (1976) 791–799.
[4] L. Chen, J.A. Ashton-Miller, Y. Hsu, J.O.L. DeLancey, Interaction among apical support, levatorani impairment, and anterior vaginal wall prolapse, Obstet. Gynecol. 108 (2006) 324–332.
[5] K.D. Bartscht, J.O.L. Delancey, A technique to study the passive supports of the uterus, Obstet. Gynecol. 72 (6) (1988) 940–943.
[6] H. Yamada, Strength of Biological Materials, Waverly Press, Inc., Baltimore, Maryland, USA, 1970. pp. 205–270.
[7] L. Chen, J.A. Ashton-Miller, J.O. DeLancey, A 3D finite element model of anterior vaginal wall support to evaluate mechanisms underlying cystocele formation, J. Biomech. 42 (10) (2009) 1371–1377.
[8] G.V. Rao, C. Rubod, M. Brieu, N. Bhatnagar, M. Cosson, Experiments and finite element modeling for the study of prolapse in the pelvic floor system, Comput. Methods Biomech. Biomed. Engin. 13 (3) (2010) 349–357.
[9] C. Rubod, P. Lecomte-Grosbras, M. Brieu, G. Giraudet, N. Betrouni, M. Cosson, 3D simulation of pelvic system numerical simulation for a better understanding of the contribution of the uterine ligaments, Int. Urogynecol. J. 24 (2013) 2093–2098.
[10] M. Cosson, C. Rubod, A. Valley, V.J.F. Witz, P. Dubois, M. Brieu, Simulation of normal pelvic mobilites in building an MRI-validated biomechanical model, Int. Urogynecol. J. 24 (1) (2013) 105–112.
[11] J. Luo, L. Chen, D.E. Fenner, J.A. Ashton-Miller, J.O.A. DeLancey, Multi-compartment 3-D finite element model of rectocele and its interaction with cystocele, J. Biomech. 48 (9) (2015) 1580–1586.
[12] J. Luo, J.A. Ashton-Miller, J.O.L. DeLancey, A model patient. Female pelvic anatomy can be viewed in diverse 3-dimensional images with a new interactive tool, Am. J. Obstet. Gynecol. 205 (2011) 391.e1–391.e2.

[13] J. Luo, T.M. Smith, J.A. Ashton-Miller, J.O. DeLancey, In vivo properties of uterine suspensory tissue in pelvic organ prolapse, J. Biomech. Eng. 136 (2) (2014) 021016.
[14] Z.W. Chen, P. Joli, Z. Feng, M. Rahim, N. Pirro, M. Bellemare, Female patient-specific finite element modeling of pelvic organ prolapse (POP), J. Biomech. 48 (2015) 238–245.
[15] D.M. Morgan, G. Kaur, Y. Hsu, D.E. Fenner, K. Guire, J. Miller, J.A. Ashton-Miller, J.O. DeLancey, Does vaginal closure force differ in the supine and standing positions? Am. J. Obstet. Gynecol. 192 (5) (2005) 1722–1728.
[16] I.A.F. Stokes, M. Gardner-Morse, S. Henry, Intra-abdominal pressure and abdominal wall muscular function: Spinal unloading mechanism, Clin. Biomech. 25 (2010) 859–866.
[17] N. Eie, Load capacity of the low back, J. Oslo City Hosp. 16 (1966) 73–98.
[18] B. Gabriel, C. Rubod, M. Brieu, B. Dedet, L. de Landsheere, V. Delmas, M. Cosson, Vagina, abdominal skin, and aponeurosis: do they have similar biomechanical properties? Int. Urogynecol. J. 22 (2011) 23–27.
[19] G. Rivaux, C. Rubod, B. Dedet, M. Brieu, B. Gabriel, M. Cosson, Comparative analysis of pelvic ligaments: a biomechanics study, Int. Urogynecol. J. 24 (2013) 135–139.
[20] C. Rubod, M. Brieu, M. Cosson, G. Rivaux, J.C. Clay, L. de Landsheere, B. Gabriel, Biomechanical properties of human pelvic organs, Urology 79 (4) (2012) 968.e17–968.e22.
[21] G.E.P. Box, N.R. Draper, Empirical Model Building and Response Surfaces, John Wiley & Sons, New York, NY, 1987.

CHAPTER TWENTY-ONE

Biomechanical Childbirth Simulations

R. Natal Jorge, S. Brandão, T. Da Roza, M. Parente, T. Mascarenhas
University of Porto, Porto, Portugal

INTRODUCTION

Pregnancy and childbirth are complex processes. Whereas pregnancy itself seems to have its own effects on pelvic floor function due to hormonal changes and other stressors, childbirth also increases the risk for pelvic floor dysfunction. Maternal and fetal anatomy and the dynamic progression of labor are crucial for proper management of delivery. Major trauma to the pelvic floor muscles is a consequence of vaginal childbirth in a considerable number of women [1,2]. Increased incidence of pelvic floor disorders later in life is associated with damage to connective tissue or intolerable degree of muscle stretch or avulsion, and pudendal nerve and urinary or anal sphincter injury during delivery [1,3]. Several studies have demonstrated the ability of ultrasound [4] and magnetic resonance imaging (MRI) to evaluate in detail the pelvic floor muscles defects [5–7] and injury to the anal sphincter, for which prolonged second stage of labor, fetal weight, and head circumference seem to be risk factors [8–10].

In the case of maternal pathology that discourages use of the valsalva maneuver and pushing during delivery — such as a hypertensive crisis, severe cardiac condition, or brain vascular malformations — or in the case of fetal distress due to inadequate fetal progression, assisted/instrumental vaginal delivery may be an option. Forceps or a ventouse (*vacuum* extractor) may be applied to obtain the most adequate presentation of the fetal head and to pull the baby out. Forceps are more successful than a ventouse, but more likely to cause vaginal and muscle tearing [11–14].

Two- and three-dimensional ultrasound may be used to describe objective measures of fetal descent during labor, and thus may be able to predict the success of a vaginal delivery [15,73]. Head direction and angle can be assessed during labor progression [16]. Moreover, midsagittal translabial images can be used to determine additional maternal and fetal landmarks during maternal pushing, such as the widest fetal head diameter and its movement with regard to the infrapubic line [16]. Lack of descent or passage below the infrapubic line and horizontal or downward head direction are poor prognostic signs. Finally, transabdominal ultrasonography can be applied to assess the *vacuum* cup placement before extraction [17].

MRI has been used to evaluate the dynamics of the second stage of labor. *Cine* midsagittal images can be acquired in open-magnets to trace the cardinal movements of the fetus [18], the pattern of uterine contractions, and stretching and compression of the muscles [19]. However, to date, the true mechanical phenomena of childbirth and its effects on the pelvic floor has been demonstrated by using computer simulations [20–23]. Interesting findings have been reported, such as that the pubococcygeal muscle is subjected to a stretch *ratio* of 3.26 [24], which exceeds the largest noninjurious stretch by 217%; pelvic floor muscle activation during the second stage of labor may represent an obstacle to fetal descent, and increased the risk for injury [25,26]; strain in the perineal nerve branch innervating the anal sphincter may reach 33% and cause permanent damage [27]; fetal head shape and large diameter increases the risk of pelvic floor muscle overstretch and injury [20]; and fetal head flexion during vaginal delivery may facilitate birth and protect the pelvic floor [26].

THE MECHANICS OF LABOR

Successful delivery is dependent on the complex interaction between uterine activity, fetal anatomy, and the maternal pelvis. Normal labor is divided into three stages. The first stage is the time between the onset of labor and full cervical dilatation. The second goes through the passage of the fetus along the bony pelvic cavity and the cervix, while the third stage is the period of placenta outflow. The strength of the expulsive forces generated by the uterine contractions together with the abdominal pressure, the pelvic dimensions and configuration, and the size and position of the fetus, dictate the evolution of vaginal labor.

Uterine activity is characterized by the frequency, intensity, and duration of the contractions. During these contractions, intrauterine pressure can reach 8 kPa [2], but it can be as high as 19 kPa during maternal pushing [28]. An adequate pattern of uterine contractions is associated with an increased likelihood of vaginal delivery. If the cervix effaces and dilates, the fetal head descends correctly, but if not, there can be *caput succedaneum* (scalp edema) and molding of the fetal head with overlapping of the skull bones. The latter may also relate to cephalopelvic disproportion, which can be either absolute or relative (when the fetus is malpositioned or in an abnormal fetal head attitude). In essence, delivery proceeds along the line of least resistance, that is, by adaptation of the smallest achievable diameters of the presenting part to the most favorable dimensions and contours of the birth canal. The shape of the true (lesser) pelvis, which is the one involved in birth, influences the diameters of the birth canal [29], and this may in fact predispose to a different fetal approach and therefore change the process of childbirth.

Although obstetric MRI pelvimetry can be applied to pregnant women [30,31], clinical assessment of pelvic diameters (Fig. 1) is the usual method. The pelvic inlet (or true conjugate) is an important anatomical landmark of the pelvis, bounded by a superior circumference. It is the distance from the sacral promontory, the *arcuate* and pectineal

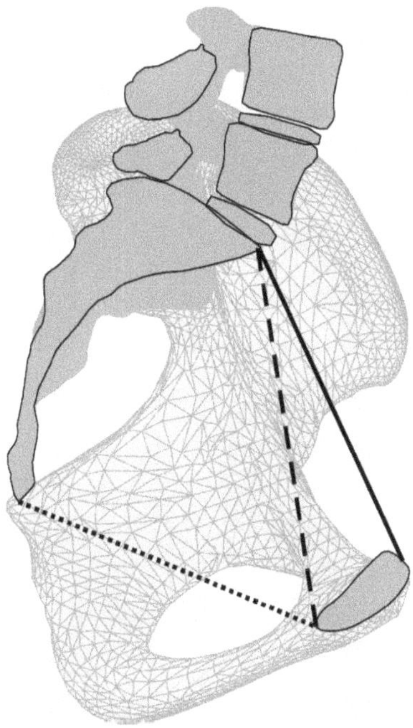

Figure 1 The pelvic inlet and outlet diameters. The pelvic inlet and outlet measurements define the distance through which the head and body must pass while descending through the pelvis. The pelvic inlet (or true conjugate) is the distance from the sacral promontory to the upper margin of the *symphysis pubis (solid line)*, while the diagonal conjugate goes to its lower border *(dashed line)*. The pelvic outlet is defined by the distance between the inferior aspect of the *symphysis pubis* to the coccyx *(dotted line)*.

lines until the upper margin of the *symphysis pubis*, and normally measures 11 cm or more (solid line). The diagonal conjugate is the most easily and commonly assessed. It normally measures 11.5 cm or more, until the lower border of the *symphysis pubis* (dashed line). When any of these diameters is smaller than normal, the inlet is considered as contracted. According to its shape, there are four broad categories that result in different anteroposterior, diagonal, and lateral diameters of the true pelvis [32] (see Fig. 2), which can restrain the position and the cardinal movements of the fetus. For example, the anthropoid pelvis (Fig. 2B) — in which the lower circumference of the pelvis has an irregular shape — seems to be often associated with vaginal delivery in the *occiput* posterior position. The lower circumference of the pelvis has an irregular shape.

During the second stage of labor, the progress of the fetal head descent is measured by determining its position according to the level of the maternal ischial spines, which is classified as the midpoint (0 station). Other variables associated with the fetus influence the course of labor, such as the fetal size; position and presentation; lie; the attitude, that is, head flexion or extension; head position — most commonly cephalic *occiput* anterior — and cranial dimensions (head circumference) [33].

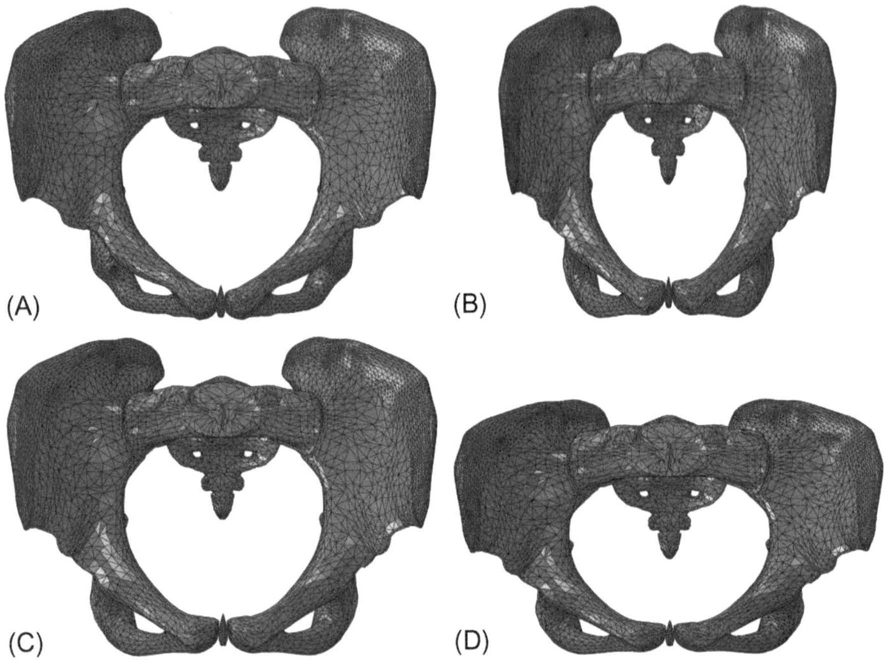

Figure 2 Shape of the osseous pelvis. The shape of the pelvic bones results in different inlet diameters and pelvic categories: android (A), anthropoid (B), gynecoid (C), and platypelloid (D), respectively.

There are seven cardinal movements of the fetal head during its passage through the birth canal (Fig. 3): engagement, descent, flexion, internal rotation, extension, external rotation, and expulsion.

Engagement is the passage of the widest diameter of the presenting part to a level below the plane of the pelvic inlet, and it is considered an important clinical prognostic sign because it demonstrates that the bony pelvis is sufficiently large. The descent refers to the movement of the presenting part through the pelvis. The progress is determined by abdominal palpation and vaginal examination until the presenting part can be seen at the *introitus*. As soon as the descending head meets the resistance offered by the soft tissues of the pelvic floor, the flexion of the fetal head occurs passively as the head descends, due to the shape of the bony pelvis. Although the head flexion onto the chest starts in the beginning of the labor, complete flexion usually occurs only at this stage. The adequate and complete head flexion should present the smallest diameter — the suboccipito-bregmatic diameter (SOBD), 9.5 cm — for optimal passage through the pelvis. As the head descends, the *occiput* of the fetus rotates toward the *symphysis pubis*, and thereby the widest portion of the fetus passes through the widest dimension of the pelvis. With each uterine contraction, the bony pelvis and the muscles of the pelvic floor guide the fetal head. After the head is out, it rotates about 45 degrees to realign with the back and shoulders

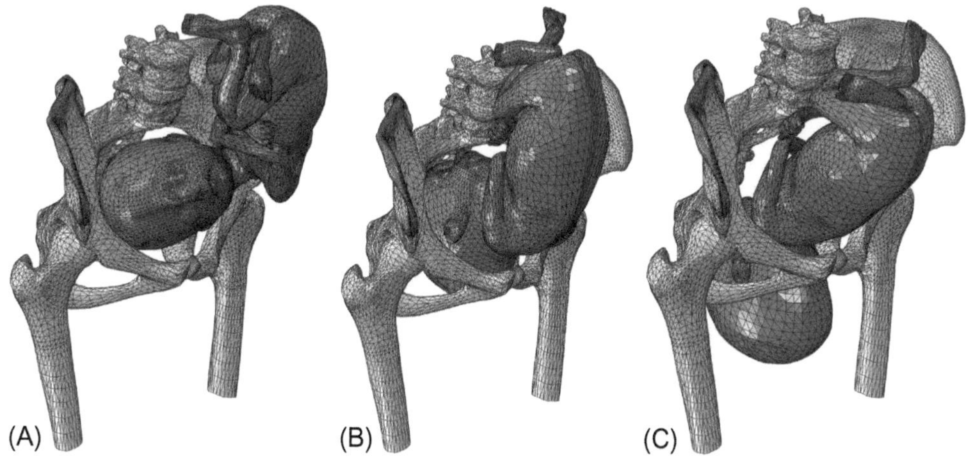

Figure 3 The cardinal movements of the fetus during the second stage of delivery can be reproduced using computational simulation. Engagement (A), rotation (B), and extension (C).

(external rotation). Once the shoulders emerge, the remaining body passes easily (expulsion) and the second stage of labor ends.

The second stage of labor generates high pressure between the fetal head and the vaginal wall and, consequently, great stretching of the pelvic floor muscles and nerves occurs. These obstetrical forces often result in permanent physical and functional changes, which may lead to urinary incontinence or organ prolapse, due to definitive weakening of the pelvic floor. Viktrup et al. identified the length of labor pushing; fetal head circumference; episiotomy; and birth weight as risk factors for postpartum stress urinary incontinence [34,35].

Dystocia may be caused by ineffective uterine contractions or pulling; incompatible fetus-pelvis anatomy; abnormal presentation or position; high weight and size; unadjusted maternal position and preparation [36,37]. Previous studies reported abnormal (dystocic) labor rates of approximately 12–34% of the cases when the fetus is in vertex presentation [38,39]. It often requires emergency cesareans, and *vacuum* or forceps deliveries, which can lead to maternal-fetal trauma [36,73]. Both functional and mechanical dystocia seem to be more prevalent in the first birth, although higher rates of dystocia in the second labor were reported for women with previous dystocia.

COMPUTATIONAL SIMULATION OF CHILDBIRTH

Computational simulations using geometrical and material properties of the pelvic bone and soft tissues, as well as accurate constitutive models to illustrate their behavior are under intensive research [40–42]. The ultimate goal of these computational simulations is to illustrate the whole labor progression, and to act as an adjuvant tool in the clinical setting for subject-specific cases that could result in complicated labor, since they enable prediction of the forces and stresses to which the pelvic soft tissues and the fetal head will be subjected. Accordingly, clinical staff could better prepare and adjust the delivery course by using the live-subject input information from the subject, and the knowledge of the biomechanics of the whole event.

Modeling takes into account not only the geometry of the structures, but also their mechanical properties — viscoelasticity, (an)isotropic mechanical response, or (quasi) incompressibility [43]. The finite element method is one of the most common approaches to model the female pelvis. It allows complex element behavior into a matrix structural analysis to be modeled, by discretizing the *continuum* of a structure into a discrete number of smaller "elements."

Numerical Models for Childbirth Simulation

As previously mentioned in this chapter, the process of birth involves many complicated movements of the fetus through the birth canal [44]. The cardinal movements are simply rough guidelines for understanding the birth mechanism. Until now, the childbirth-related damage to the soft tissues — muscles or nerves — are believed to be a major contributor to female pelvic floors dysfunction [5], but the precise mechanism of injury is still uncertain. In this context, numerical models have evaluated the effects of vaginal delivery on the pelvic floor muscles [20,45,46].

However, to better understand the effects of the childbirth, first it is important know the anatomy of the bony pelvis (size and shape) and the fetal head, which can be used as a guidance in the deliveries [47]. In this sense, Liu et al. used the fetal-pelvic index to determine if the pelvis is larger than the baby, and vice versa. The authors used computed tomographic pelvimetry and ultrasound images of the head of the fetus to develop the model, however, with no information regarding the details of the pelvic bones [48]. Buttin and colleagues presented a biomechanical model based on MR images where the fetus was considered as a rigid body, and the authors included the womb, which is an essential organ during the delivery [49]. The authors simulated the uterine contractions and the forces generated by the abdominal muscles in two different stages. At the first condition, they observed a very small descent (1 mm) with rotation of the fetal body, while in the second stage, the fetus showed a displacement of 20–22 mm in the direction of the cervix. The authors concluded that even using a rigid body for the fetus it is possible to obtain a correct deformation of the geometry of the womb.

In the same way, Lien et al. used MR images from a healthy nulliparous 34-year-old woman to build a numerical model while considering the fetal head as a sphere [24]. In this study, the purpose was to predict levator ani muscle stretch during vaginal birth. The authors found that the most medial portion of the pubococcygeous muscle had the greatest risk for stretch-related injury during vaginal birth, with a stretch *ratio* of 3.26. In 2013, Buttin and colleagues developed another very complete model (ie, the uterus, abdomen, and pelvis of the parturient woman interacting with the fetus) to simulate the second stage of the delivery. They reported that the fetal head displacement occurs essentially in the coronal/sagittal axis and that the fetal head velocity is not linear [50].

In 2012, Yan et al. focused on analyzing if the different shapes of the levator ani muscle influence the progress of the second stage of labor [51]. The results suggested that increased muscle thickness, reduction in pelvic floor circumference and elongated axial span of the muscles increases the maximum stretch of the pelvic floor. In this context, a hands-poised technique performed concurrently with a stereophotogrammetry provided evidence that during perineal deformation, the highest tissue strain was in a transverse direction, more than four times higher than the maximum anteroposterior strain. These results highlighted the relevance of using perineal hands-based protection to reduce perineal laceration [52]. More recently, other authors used the geometry of the pregnant primiparous woman's pelvic system from MRI scans, and simulations were performed by means of varying the size of the fetal head with the inputs from anatomical idealized reconstruction and not real-subject pelvic floor muscles from the MR images [53].

The first stage of labor is much longer than the second and third combined. The full dilation of the cervix may occur in 20 h in primiparous or less than 1 h in multiparous women. This great variability may occur due to the personal characteristics such as age and race [54] in each woman. Only few studies demonstrate the possible effects of the first stage on the pelvic floor [55–57]. The majority of the numerical studies were concentrated on the second stage of labor and the results focused mainly on evaluating pelvic floor deformations. The pubovisceral portion of the levator ani seems to be the one suffering from the largest stretch ratio at the end of the second stage of labor [20,24], and the stretching seems proportional to the size of the fetal head. Fig. 4 illustrates the displacement of the levator ani, which shows that the posterior portion of the pubococcygeous muscle is the one suffering higher dislocation at a fetal vertical displacement of 100 mm.

Considering that after engagement, the fetus will descend and the head flexes as a consequence of the resistance from the cervix walls and pelvic floor muscles, Parente et al. investigated the influence of fetal head flexion on the pelvic floor stress and forces during vaginal delivery. This study provided evidence that the highest head flexion induced the lowest forces in pelvic floor muscles. The maximum values of forces for poor head flexion more than doubled the ones registered for complete fetal head flexion, suggesting that fetal head flexion during vaginal delivery may facilitate birth and protect the pelvic floor [26]. In another paper, the same authors analyzed the stretch and deformation of the pelvic floor induced by the passage of the fetus presenting in an *occiput* posterior malposition [58,59]. As expected, the results demonstrated that head malposition produces higher stretch values when compared to normal position (1.73 for the *occiput* posterior presentation vs. 1.63

Figure 4 Numerical simulation of the fetal descent, showing the displacement field for the pelvic floor muscles (between 0 and 17 mm). At this vertical displacement of 100 mm, the posterior portion of the puboccocygeous muscle evidenced the highest values of displacement (15.6–17 mm).

for the *occiput* anterior presentation). Importantly, in those studies the authors modeled the fetal head as an undeformable structure. The head modeling as a compressive material is discussed in the next section.

Influence of the Material Parameters and Constitutive Models

It is well established that during pregnancy and labor, the soft tissues suffer hormonal changes that contribute to the onset of labor, since they will affect their material response. Until now, the most accurate method for obtaining the mechanical properties of the tissues is through experimental data (uniaxial and biaxial tests). As live-subject analysis is impossible to achieve due to ethical reasons, the majority of the studies have used material properties of cardiac and facial muscles [45,60]. Human pelvic tissue was characterized using in vitro mechanical tests on fresh cadavers [61], and the constitutive parameters of this study were used by Li et al. to compare two forms of the constitutive equations to fit the data: exponential and Neo-Hookean. The results showed a considerable difference in the biomechanical response of the pelvic floor muscles — 56% higher force was required for delivery with the exponential model — with increased nonlinearity in the elastic response [23]. One year later, the same group applied Jing's measurements to study the effect of levator ani muscle anisotropy on the second stage of labor by varying the relative stiffness between the fiber and the matrix components. Results showed a substantial decrease in the magnitude of the force required for delivery with the increase of the fiber anisotropy [62]. In addition, they found the overall maximum principal stretch *ratio* located in the dorso-caudal aspect of the levator ani, corroborating the findings of Parente and colleagues [59].

To evidence how the same numerical simulation can be affected by different parameters, Parente et al. varied the material parameters by introducing two new material sets: one with a stiffer behavior and another with a softer behavior. The constitutive equation adopted for the pelvic floor muscles was a modified form of the incompressible transversely isotropic hyperelastic model proposed earlier by Humphrey and Yin [63]. The findings demonstrated a large difference in the results due to the exponential behavior of the material [59]. In 2012, Jing and colleagues developed a model where the levator ani and the perineal body were assumed to be anisotropic, viscous, and hyperelastic. The authors employed material constitutive relations derived from biaxial tests of human pelvic floor muscle, and the results showed an average of levator hiatus stretch *ratio* of 3.55 [64]. This value was consistent with the range of 1.62–3.76 suggested by a retrospective observational study with 227 women [65], but higher than the maximum stretch *ratios* predicted by Parente et al. and Hoyte et al. [46,66] (1.63 and 2.85, respectively). Importantly, these parameters differ significantly in magnitude, and hence in their stress-strain relationships.

As detailed above, the female pelvic anatomy and physiology responsible for labor and birth are complex. A better characterization of the maternal and fetal anatomic adaptations is still necessary, as well as the material parameters that occur during birth, to understand and better simulate which structures are highly vulnerable to injury during childbirth.

Modeling the Fetus Head

The fetal skull is composed of sutures and fontanels that make it flexible to accommodate the passage through the birth canal. The skull fontanelles allow the parietal and occipital bones to be pressed and displaced thus reducing the skull volume, which is called molding. Head molding may be the result of prolonged compression or mismatch between maternal pelvic shape and fetal head shape, and helps the fetal passage toward the pelvic floor.

Modeling such phenomena enables evaluation of deformation of the fetal skull, and the stress and stretch on the pelvic floor muscles. Some authors have considered the fetus head as having spherical shape geometry taking into account the reduction in skull diameters from the molding [24,66]. However, others used anatomical models built using finite element meshes with the average diameters of the fetal head, since the dimensions of the SOBD, maxillo-vertical diameter (MaVD), or bi-parietal diameter (BPD) are different (Fig. 5), since they will affect not only the decrease in skull volume, but also the mechanical response of muscles and passage outcome [24,45,67].

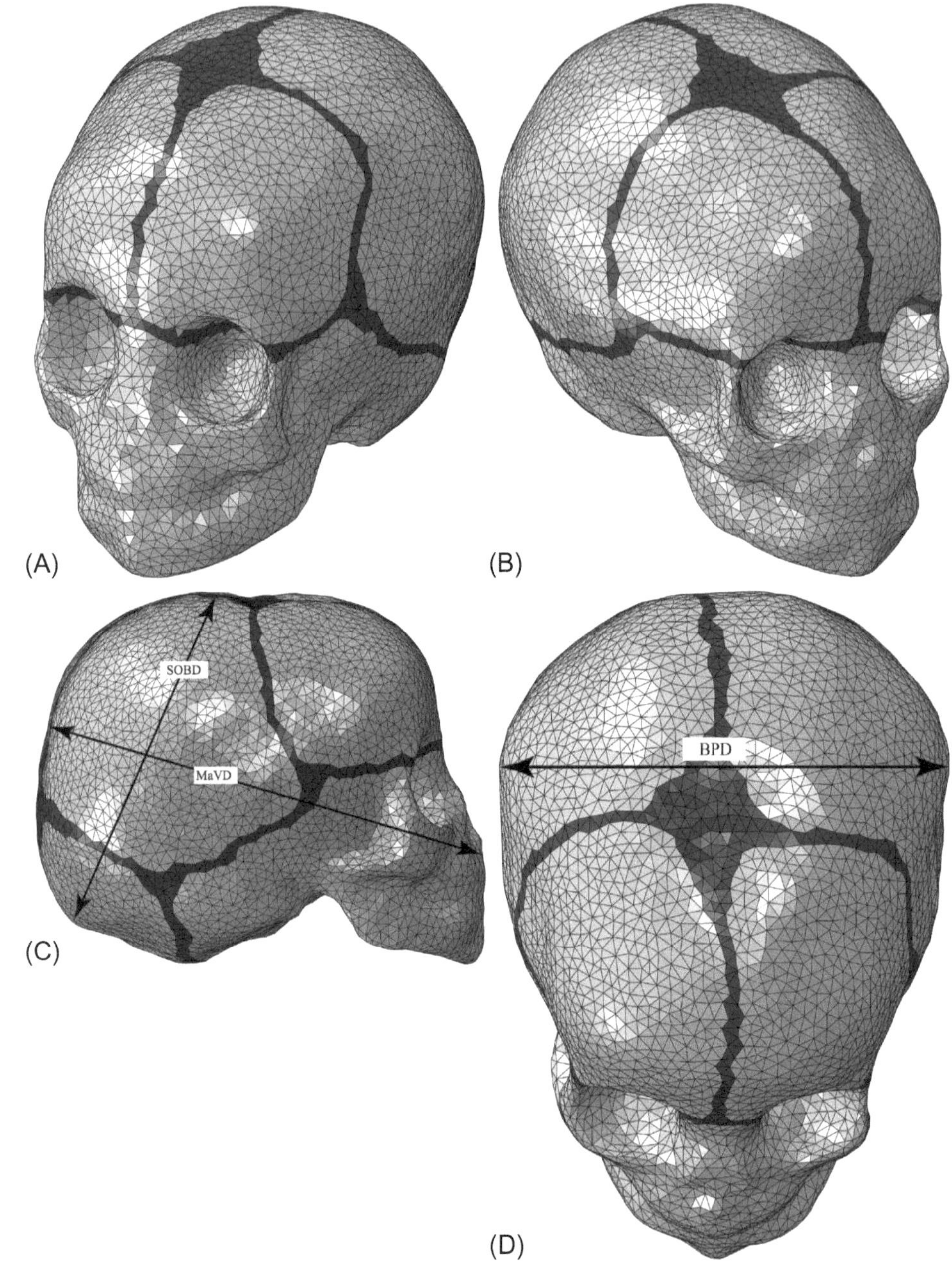

Figure 5 Oblique views of the fetal skull sutures and fontanelles *(upper row, A and B)*. Lateral and superior views of the cranial diameters *(lower row, C and D)*. These structures can be modeled to simulate the molding process through numerical simulation. *SOBD*, suboccipito-bregmatic diameter; *MaVD*, maxillo-vertical diameter; *BPD*, bi-parietal diameter.

Different approaches were chosen to model the fetal head. Li et al. considered the fetal head as a homogeneous and isotropic incompressible material with high stiffness, described by the Neo-Hookean behavior with a material constant of C1 = 0.1 MPa [67], whereas Martins et al. modeled the fetal head as a highly rigid almost undeformable material [45]. Despite the increased computational effort, a few authors have modeled the fetal head as a deformable body with adjusted material properties for the cranial and skull base bones, fontanels, and sutures [57,68]. Lapeer et al. and Pu et al. used nonlinear finite element models to evaluate the effect of the intrauterine (IUP) and head-to-cervix pressures (HCP) and to measure the modified molding index (MMI). The fontanelles and sutures were considered as hyperelastic materials using the Mooney-Rivlin model (with C1 = 1.18 MPa and C2 = 0.295 MPa). Elongation of the MaVD and the orbito-vertical diameters (OrVD), and the compression of the SOBD were described in their models, which is in agreement with clinical evidence. Moreover, the changes of SOBD appear to be more sensitive to increases in IUP and HCP than those of the other diameters. Also, the maximum stress values were higher in the frontal and parietal bones near the anterior fontanelle, which suffered higher strain and displacement, illustrating the definition of molding itself.

Vacuum extraction procedures imply higher forces induced to the skull, around 100 N, and there is evidence of potential hemorrhage when the ventouse is not placed between the anterior and posterior fontanelles, but across the anterior fontanelle [69]. Fig. 6 shows

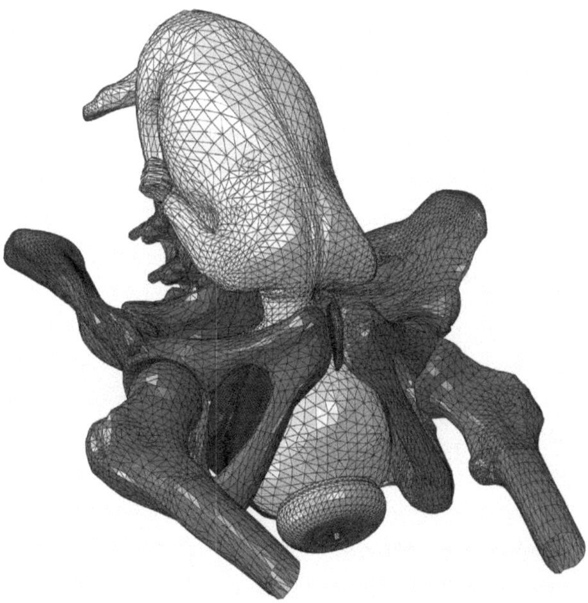

Figure 6 Computational model of the position of a *vacuum* cup used to promote fetal exit.

a numerical model of a *vacuum* cup, correctly positioned along the sagittal suture and about 3 cm ahead of the posterior fontanelle.

Lapeer et al. evaluated the effect of incorrect positioning, and the results showed that it increases overall maximum deformation magnitude and tissue rotation (7.01 mm vs. 3.15 mm; 0.69 rads vs. 0.35 rads) which seems to be aggravated when previous molding of the cranium is considered in the numerical simulation [70].

To highlight the relevance of the material properties of the fetal head on the maternal pelvic floor, in a recent work Silva et al. simulated molding while comparing rigid and deformable head models [71], with the fontanelles and sutures modeled as elastic materials (with $E=3.8$ MPa and $\nu=0.45$) [72]. As expected, the forces and stretch exerted on the pelvic floor muscles were 17.3 and 1.86% higher for the rigid than for the deformable head. The highest values of the forces and stretch appear to be at a vertical displacement between 60 and 70 mm — which corresponds to the moment when the fetus begins head extension — regardless of the material properties attributed to the fetal skull and muscle morphology [24,51,71].

In another work, Buttin et al. used a hyperelastic fetus model (to which the Neo-Hooke's model of constitutive properties was applied), and found a nonlinear pattern of descent of the fetal head, with changes in acceleration and higher displacement in the coronal and sagittal axes [50]. In this model, the frictionless boundary conditions and the fact that the body of the fetus was assumed to be slightly deformable to enable the spine to move freely, facilitated uterine contractions and expulsion forces to guide the fetus. Li et al. found similar displacement directions and, furthermore, a 25% reduction in the reaction forces for a minimally constrained head model (by fixing one degree of freedom), allowing both rotation and translation. The simulation results illustrated that the path of the fetal head influences the force required to expulse the fetus [22], although the stretch ratios for the pelvic floor muscles were similar, as in the study of Silva et al. Further studies should focus on studying the impact of reduced kinematic constraints and frictionless boundary conditions on the stability of these nonlinear systems.

Other obstetric factors such as fetal presentation and position require accurate adaptations to pelvic dimensions, otherwise there may be difficulties in the descending trajectory, as previously presented in the section of this chapter on *Numerical models for childbirth simulation*. Accordingly, it is expected that besides the longer second stage of delivery and the higher deformation and stretch of the pelvic floor muscles [58,59], fetal head molding should be simulated under such conditions, to support a full description of the resultant skull and brain lesions.

Subject-Specific Childbirth Models

A better understanding of the 3D biomechanics of the pelvic soft tissues at the patient-specific level will certainly improve delivery approaches. Although finite element models of the pelvis have been developed, validation by direct comparison with subject-specific data is still not possible which means that each method uses simplifications and assumptions regarding geometry and material properties of the tissues.

Subject-specific MRI datasets are quite useful for the modeling process, due to their inherent tissue contrast, detailed anatomy, and minimal risk. It would be helpful in women with imaging evidence of fetal malposition or in cases where the maternal pelvic anatomy is especially challenging to perform regular vaginal delivery, for which fetal trauma and soft tissue damage are expected. Despite the advantage of individualized geometries, results would still be very dependent on the boundary constraints, estimates of material properties and ground truth comparison or validation, which is a high barrier. The relative importance of model input parameters depend heavily on the constitutive model and on the predictions that are of major interest for investigators.

Moreover, during pregnancy, the loading conditions change as the fetus grows and the pelvis and spine adapt to the increased weight. Furthermore, as stated before in this chapter, the hormonal variations from estrogen and relaxin decrease tissue stiffness and lessen the maximal stretches, but the translation of those dynamic changes in numerical constants is not viable at this time.

CONCLUSIONS

Modeling childbirth is a very complex issue, and any methodology used has its own limitations. The main questions remain: can we predict a pattern of fiber stretch that exceeds the maximum permissible value? Should we model the realistic geometry of the fetus head? Is the head molding imperative to accurately mimic the strain on the skull and fontanelles? Furthermore, a multitude of maternal and fetal variables may affect the maximum muscle stretch ratios, thus affecting the final results. The creation of mathematical models of pelvic floor function and simulation of childbirth are growing and must be enriched. In spite of the progress in biomechanical modeling, the different methodologies applied to the pelvic morphology and fetal head make the studies difficult to compare. Accordingly, further research investigating the process of labor is needed. The ultimate goal would be to build subject-specific models — based on both ultrasound and MRI — that would enable simulations of the best approach to proceed to vaginal, instrumental delivery, or caesarian section by predicting the severity and location of injury to the mother's pelvic floor and to the fetus.

REFERENCES

[1] V.L. Handa, J.L. Blomquist, L.R. Knoepp, K.A. Hoskey, K.C. McDermott, A. Muñoz, Pelvic floor disorders 5–10 years after vaginal or cesarean childbirth, Obstet. Gynecol. 118 (2011) 777–784.

[2] H.U. Memon, V.L. Handa, Vaginal childbirth and pelvic floor disorders, Womens Health (Lond. Engl.) 9 (3) (2013), http://dx.doi.org/10.2217/whe.13.17.

[3] M.A. Bartolii, H.P. Drutz, D. Lovatsis, M. Alarab, Vaginal delivery and pelvic floor dysfunction: current evidence and implications for future research, Int. Uroginecol. J. 21 (8) (2010) 1025–1030, http://dx.doi.org/10.1007/s00192-010-1146-9.
[4] F. Liu, X. Lian, T. Ying, J. Tao, B. Hu, Three-dimensional ultrasound appearance of pelvic floor in nulliparous women and postpartum women one week after their first delivery, Int. J. Med. Sci. 11 (3) (2014) 234–239, http://dx.doi.org/10.7150/ijms.7384.
[5] J.O. DeLancey, R. Kearney, Q. Chou, S. Speights, S. Binno, The appearance of levator ani muscle abnormalities in magnetic resonance images after vaginal delivery, Obstet. Gynecol. 101 (1) (2003) 46–53.
[6] J.M. Miller, C. Brandon, J.A. Jacobson, L.K. Low, R. Zielinski, J. Ashton-Miller, J.O. Delancey, MRI findings in patients considered high risk for pelvic floor injury studied serially after vaginal childbirth, AJR Am. J. Roentgenol. 195 (3) (2010) 786–791, http://dx.doi.org/10.2214/AJR.09.3508.
[7] K. van Delft, R. Thakar, A.H. Sultan, N. Schwertner-Tiepelmann, K. Kluivers, Levator ani muscle avulsion during childbirth: a risk prediction model, BJOG 21 (9) (2014) 1155–1163, http://dx.doi.org/10.1111/1471-0528.12676.
[8] S.M. Baracho, L. Barbosa da Silva, E. Baracho, A. Lopes da Silva Filho, R.F. Sampaio, E. Mello de Figueiredo, Pelvic floor muscle strength predicts stress urinary incontinence in primiparous women after vaginal delivery, Int. Urogynecol. J. 23 (7) (2012) 899–906.
[9] T. Eftekhar, B. Hajibaratali, F. Ramezanzadeh, M. Shariat, Postpartum evaluation of stress urinary incontinence among primiparas, Int. J. Gynaecol. Obstet. 94 (2) (2006) 114–118.
[10] D.V. Valsky, M. Lipschuetz, A. Bord, I. Eldar, B. Messing, D. Hochner-Celnikier, Y. Lavy, S.M. Cohen, S. Yagel, Fetal head circumference and length of second stage of labor are risk factors for levator ani muscle injury, diagnosed by 3-dimensional transperineal ultrasound in primiparous women, Am. J. Obstet. Gynecol. 201 (1) (2009) 91.e1–91.e7, http://dx.doi.org/10.1016/j.ajog.2009.03.028.
[11] C.E. Aiken, A.R. Aiken, J.C. Brockelsby, J.G. Scott, Factors influencing the likelihood of instrumental delivery success. Obstet. Gynecol. 123 (4) (2014) 796–803, http://dx.doi.org/10.1097/AOG.0000000000000188.
[12] A. Ben-Haroush, N. Melamed, B. Kaplan, Y. Yogev, Predictors of failed operative vaginal delivery: a single-center experience, Am. J. Obstet. Gynecol. 197 (3) (2007) 308.e1–308.e5.
[13] H.P. Dietz, Forceps: towards obsolescence or revival? Acta Obstet. Gynecol. Scand. 94 (4) (2015) 347–351, http://dx.doi.org/10.1111/aogs.12592.
[14] F. O'Mahony, G.J. Hofmeyr, V. Menon, Choice of instruments for assisted vaginal delivery, Cochrane Database Syst. Rev. 11 (2010) CD005455, http://dx.doi.org/10.1002/14651858.CD005455.pub2.
[15] L. Yeo, R. Romero, Sonographic evaluation in the second stage of labor to improve the assessment of labor progress and its outcome, Ultrasound Obstet. Gynecol. 33 (2009) 253–258.
[16] W. Henrich, J. Dudenhausen, I. Fuchs, A. Kamena, B. Tutschek, Intrapartum translabial ultrasound (ITU): sonographic landmarks and correlation with successful vacuum extraction, Ultrasound Obstet. Gynecol. 28 (2006) 753–760.
[17] G.Y. Wong, Y.M. Mok, S.F. Wong, Transabdominal ultrasound assessment of the fetal head and the accuracy of vacuum cup application, Int. J. Gynaecol. Obstet. 98 (2007) 120–123.
[18] C. Bamberg, G. Rademacher, F. Güttler, U. Teichgräber, M. Cremer, C. Bührer, C. Spies, L. Hinkson, W. Henrich, K.D. Kalache, J.W. Dudenhausen, Human birth observed in real-time open magnetic resonance imaging, Am. J. Obstet. Gynecol. 206 (6) (2012) 505.e1–505.e6, http://dx.doi.org/10.1016/j.ajog.2012.01.011.
[19] F.V. Güttler, A. Heinrich, J. Rump, M. de Bucourt, B. Schnackenburg, C. Bamberg, B. Hamm, U.K. Teichgräber, Magnetic resonance imaging of the active second stage of labour: proof of principle, Eur. Radiol. 22 (2012) 2020–2026, http://dx.doi.org/10.1007/s00330-012-2455-9.
[20] J.A. Ashton-Miller, J.O. Delancey, On the biomechanics of vaginal birth and common sequelae, Annu. Rev. Biomed. Eng. 11 (2009) 163–176, http://dx.doi.org/10.1146/annurev-bioeng-061008-124823.
[21] X. Li, J.A. Kruger, M.P. Nash, P.M. Nielsen, Modeling childbirth: elucidating the mechanisms of labor, Wiley Interdiscip. Rev. Syst. Biol. Med. 2 (4) (2010) 460–470, http://dx.doi.org/10.1002/wsbm.65.

[22] X. Li, J.A. Kruger, M.P. Nash, P.M.F. Nielsen, Effects of fetal head motion on pelvic floor mechanics, in: K. Miller, P.M.F. Nielsen (Eds.), Computational Biomechanics for Medicine, Springer Science +Business Media, LLC, New York, 2010, p. 129, http://dx.doi.org/10.1007/978-1-4419-5874-7_14.
[23] X. Li, J.A. Kruger, M.P. Nash, P.M. Nielsen, Effects of nonlinear muscle elasticity on pelvic floor mechanics during vaginal childbirth, J. Biomech. Eng. 132 (11) (2010) 111010, http://dx.doi.org/10.1115/1.4002558.
[24] K.C. Lien, B. Mooney, J.O. DeLancey, J.A. Ashton-Miller, Levator ani muscle stretch induced by simulated vaginal birth, Obstet. Gynecol. 103 (2004) 31–40.
[25] M.P. Parente, R.M.N. Jorge, T. Mascarenhas, A.L. Silva-Filho, The influence of pelvic muscle activation during vaginal delivery, Obstet. Gynecol. 115 (4) (2010) 804–808.
[26] M.P. Parente, R.M.N. Jorge, T. Mascarenhas, A.A. Fernandes, A.L. Silva-Filho, Computational modeling approach to study the effects of fetal head flexion during vaginal delivery, Am. J. Obstet. Gynecol. 203 (3) (2010) 217.e1–217.e6.
[27] K.C. Lien, D.M. Morgan, J.O. DeLancey, J.A. Ashton-Miller, Pudendal nerve stretch during vaginal birth: a 3D computer simulation, Am. J. Obstet. Gynecol. 192 (5) (2005) 1669–1676.
[28] A. Rempen, M. Kraus, Pressures on the fetal head during normal labor, J. Perinat. Med. 19 (1991) 199–206.
[29] K. Hanretty, Obstetrics Illustrated, sixth ed., Churchill Livingstone, Edinburgh, 2003.
[30] T.M. Keller, A. Rake, S.C. Michel, B. Seifert, G. Efe, K. Treiber, R. Huch, B. Marincek, R.A. Kubik-Huch, Obstetric MR pelvimetry: reference values and evaluation of inter- and intraobserver error and intraindividual variability, Radiology 227 (1) (2003) 37–43.
[31] M.V. Zaretsky, J.M. Alexander, D.D. McIntire, M.R. Hatab, D.M. Twickler, K.J. Leveno, Magnetic resonance imaging pelvimetry and the prediction of labor dystocia, Obstet. Gynecol. 106 (5 Pt 1) (2005) 919–926.
[32] A. DeCherney, L. Nathan, Current Obstetric & Gynecologic Diagnosis & Treatment, ninth ed., McGraw-Hill, New York, 2003.
[33] S. Gabbe, J. Niebyl, J. Simpson, Obstetrics: Normal and Problem Pregnancies, fifth ed., Elsevier, Philadelphia, 2007, ISBN 0-443-06930-7.
[34] L. Viktrup, G. Lose, The risk of stress incontinence 5 years after first delivery, Am. J. Obstet. Gynecol. 185 (1) (2001) 82–87.
[35] L. Viktrup, G. Rortveit, G. Lose, Risk of stress urinary incontinence twelve years after the first pregnancy and delivery, Obstet. Gynecol. 108 (2) (2006) 248–254.
[36] R. Alijahan, M. Kordi, Risk factors of dystocia in nulliparous women, Iran. J. Med. Sci. 39 (3) (2014) 254–260.
[37] R. Alijahan, M. Kordi, M. Poorjavad, S. Ebrahimzadeh, Diagnostic accuracy of maternal anthropometric measurements as predictors for dystocia in nulliparous women, Iran. J. Nurs. Midwifery Res. 19 (1) (2014) 11–18.
[38] A. Sandström, S. Cnattingius, A.K. Wikström, O. Stephansson, Labour dystocia-risk of recurrence and instrumental delivery in following labour — a population-based cohort study, BJOG 119 (13) (2012) 1648–1656, http://dx.doi.org/10.1111/j.1471-0528.2012.03502.x.
[39] B.P. Zhu, V. Grigorescu, T. Le, M. Lin, G. Copeland, M. Barone, G. Turabelidze, Labor dystocia and its association with interpregnancy interval, Am. J. Obstet. Gynecol. 195 (1) (2006) 121–128.
[40] S. Brandão, M. Parente, T. Mascarenhas, Gomes da Silva AR, I. Ramos, R. Natal Jorge, Biomechanical study on the bladder neck and urethral positions: simulation of impairment of the pelvic ligaments, J. Biomech. 48 (2015) 217–223, http://dx.doi.org/10.1016/j.jbiomech.2014.11.045.
[41] S. Brandão, M. Parente, P. Rocha, T. Mascarenhas, A.R. Gomes da Silva, I. Ramos, R. Natal Jorge, Modeling the contraction of the pelvic floor muscles, Comput. Methods Biomech. Biomed. Engin. (2015), http://dx.doi.org/10.1080/10255842.2015.1028031.
[42] Z.W. Chen, P. Joli, Z.Q. Feng, M. Rahim, N. Pirró, M.E. Bellemare, Female patient-specific finite element modeling of pelvic organ prolapse (POP), J. Biomech. 48 (2) (2015) 238–245, http://dx.doi.org/10.1016/j.jbiomech.2014.11.039.

[43] S. Brandão, T. Da Roza, M. Parente, I. Ramos, T. Mascarenhas, R.M. Natal Jorge, Magnetic resonance imaging of the pelvic floor: from clinical to biomechanical imaging, Proc. Inst. Mech. Eng. H 227 (12) (2013) 1324–1332, http://dx.doi.org/10.1177/0954411913502952.
[44] K.R. Rosenberg, The evolution of modern human childbirth, Am. J. Phys. Anthropol. 35 (1994) 89–124.
[45] J.A. Martins, M.P. Pato, E.B. Pires, R.M. Jorge, M. Parente, T. Mascarenhas, Finite element studies of the deformation of the pelvic floor, Ann. N. Y. Acad. Sci. 1101 (2007) 316–334.
[46] M.P. Parente, R.M. Jorge, T. Mascarenhas, A.A. Fernandes, J.A. Martins, Deformation of the pelvic floor muscles during a vaginal delivery, Int. Urogynecol. J. Pelvic Floor Dysfunct. 19 (2008) 65–71.
[47] G.R. Thurnau, K.A. Hales, M.A. Morgan, Evaluation of the fetal-pelvic relationship [review], Clin. Obstet. Gynaecol. 35 (3) (1992) 570–581.
[48] Y. Liu, M. Scudder, M.L. Gimovsky, CAD modeling of the birth process: Part II, Stud. Health Technol. Inform. 29 (1996) 652–666.
[49] R. Buttin, F. Zara, B. Shariat, T. Redarce, A biomechanical model of the female reproductive system and the fetus for the realization of a childbirth virtual simulator, Conf. Proc. IEEE Eng. Med. Biol. Soc. 2009 (2009) 5263–5266, http://dx.doi.org/10.1109/IEMBS.2009.5334085.
[50] R. Buttin, F. Zara, B. Shariat, T. Redarce, G. Grangé, Biomechanical simulation of the fetal descent without imposed theoretical trajectory, Comput. Methods Prog. Biomed. 111 (2) (2013) 389–401, http://dx.doi.org/10.1016/j.cmpb.2013.04.005.
[51] X. Yan, J.A. Kruger, M.P. Nash, P.M.F. Nielsen, Effects of levator ani muscle morphology on the mechanics of vaginal childbirth, in: P.M.F. Nielsen, et al., (Eds.), Computational Biomechanics for Medicine, Springer Science+Business Media, New York, 2012, , pp. 63–75.
[52] R. Zemčík, J. Karbanova, V. Kalis, L. Lobovský, M. Jansová, Z. Rusavy, Stereophotogrammetry of the perineum during vaginal delivery, Int. J. Gynaecol. Obstet. 119 (2012) 76–80.
[53] J. Lepage, C. Jayyosi, P. Lecomte-Grosbras, M. Brieu, C. Duriez, M. Cosson, C. Rubod, Biomechanical pregnant pelvic system model and numerical simulation of childbirth: impact of delivery on the uterosacral ligaments, preliminary results, Int. Urogynecol. J. 26 (4) (2015) 497–504, http://dx.doi.org/10.1007/s00192-014-2498-3.
[54] L. Albers, The duration of labor in healthy women, J. Perinatol. 19 (2) (1999) 114–119.
[55] M. Bailet, F. Zara, E. Promayon, Shell Finite element model for interactive fetal head deformation during childbirth, Comput. Methods Biomech. Biomed. Engin. 16 (2013) 312–314.
[56] M. Bailet, F. Zara, E. Promayon, Biomechanical model of the fetal head for interactive childbirth simulation, in: Proceedings of Surgetica, 2014, 3–5 December, 2014, Chambéry, France, 2014.
[57] R.J. Lapeer, R.W. Prager, Fetal head moulding: finite element analysis of a fetal skull subjected to uterine pressures during the first stage of labour, J. Biomech. 34 (9) (2001) 1125–1133.
[58] M.P. Parente, R.M. Jorge, T. Mascarenhas, A.A. Fernandes, J.A. Martins, The influence of an occipito-posterior malposition on the biomechanical behavior of the pelvic floor, Eur. J. Obstet. Gynecol. Reprod. Biol. 144 (Suppl. 1) (2009) S166–S169, http://dx.doi.org/10.1016/j.ejogrb.2009.02.033.
[59] M. Parente, R.M. Natal Jorge, T. Mascarenhas, A.A. Fernandes, J.A.C. Martins, The influence of the material properties on the biomechanical behavior of the pelvic floor muscles during vaginal delivery, J. Biomech. 42 (2009) 1301–1306.
[60] S. Lee, A. Darzi, G. Yang, Subject specific finite element modelling of the levator ani, Med. Image Comput. Comput. Assist. Interv. 8 (2005) 360–367.
[61] D. Jing, K. Lien, J.A. Ashton-Miller, J.O.L. DeLancey, Visco-hyperelastic properties of the pelvic floor muscles in healthy women, in: Podium Session 26, North American Congress on Biomechanics, Ann Arbor, Michigan, USA, 5–9 August, 2008.
[62] X. Li, J.A. Kruger, M.P. Nash, P.M. Nielsen, Anisotropic effects of the levator ani muscle during childbirth, Biomech. Model. Mechanobiol. 10 (4) (2011) 485–494, http://dx.doi.org/10.1007/s10237-010-0249-z.
[63] J.D. Humphrey, F.C.P. Yin, On constitutive relations and finite deformations of passive cardiac tissue: a pseudostrain-energy function, J. Biomech. Eng. 109 (1987) 298–304.

[64] D. Jing, J.A. Ashton-Miller, J.O. DeLancey, A subject-specific anisotropic visco-hyperelastic finite element model of female pelvic floor stress and strain during the second stage of labor, J. Biomech. 45 (3) (2012) 455–460, http://dx.doi.org/10.1016/j.jbiomech.2011.12.002.
[65] K. Svabik, K.L. Shek, H.P. Dietz, How much does the levator hiatus have to stretch during childbirth? BJOG 116 (12) (2009) 1657–1662.
[66] L. Hoyte, M.S. Damaser, S.K. Warfield, G. Chukkapalli, A. Majumdar, D.J. Choi, A. Trivedi, P. Krysl, Quantity and distribution of levator ani stretch during simulated vaginal childbirth, Am. J. Obstet. Gynecol. 199 (2) (2008) 198.e1–198.e5, http://dx.doi.org/10.1016/j.ajog.2008.04.027.
[67] X. Li, J.A. Kruger, J.H. Chung, M.P. Nash, P.M. Nielsen, Modelling childbirth: comparing athlete and non-athlete pelvic floor mechanics, Med. Image Comput. Comput. Assist. Interv. 11 (2) (2008) 750–757.
[68] F. Pu, L. Xu, D. Li, S. Li, L. Sun, L. Wang, Y. Fan, Effect of different labor forces on fetal skull molding, Med. Eng. Phys. 33 (5) (2011) 620–625, http://dx.doi.org/10.1016/j.medengphy.2010.12.018.
[69] U. Ali, E. Norwitz, Vacuum-assisted vaginal delivery, Rev. Obstet. Gynecol. 1 (1) (2009) 5–17.
[70] R.J. Lapeer, V. Audiis, Z. Gerkhanov, Simulation of vacuum extraction during childbirth using finite element analysis, in: Conference: SIMULIA Regional User Meeting (RUM 2014), Warrington, UK, 2014.
[71] M.E. Silva, D.A. Oliveira, T.H. Roza, S. Brandão, M.P. Parente, T. Mascarenhas, R.M. Natal Jorge, Study on the influence of the fetus head molding on the biomechanical behavior of the pelvic floor muscles, during vaginal delivery, J. Biomech. 48 (9) (2015) 1600–1605, http://dx.doi.org/10.1016/j.jbiomech.2015.02.032.
[72] B. Coats, S. Margulies, Material properties of human infant skull and suture at high rates, J. Neurotrauma 23 (2006) 1222–1232.
[73] F.S. Molina, K.H. Nicolaides, Ultrasound in labor and delivery, Fetal Diagn. Ther. 27 (2) (2010) 61–67, http://dx.doi.org/10.1159/000287588.

SECTION 5

Conclusions

L. Hoyte*, M. Damaser†‡§

*Director, Urogynecology and Female Pelvic Reconstructive Surgery, USF College of Medicine, Tampa, FL, United States of America;
Medical Director, Urogynecology and Robotic Surgery, Tampa General Hospital, Tampa, FL, United States of America
†Professor of Molecular Medicine, Cleveland Clinic Lerner College of Medicine, Case Western Reserve University, Cleveland, OH, United States of America
‡Lerner Research Institute, Cleveland Clinic, Cleveland, OH, United States of America
§Advanced Platform Technology Center, Cleveland VA Medical Center, Cleveland, OH, United States of America

CHAPTER TWENTY-TWO

What the Future Holds

L. Hoyte*, M. Damaser†,‡,§

*Director, Urogynecology and Female Pelvic Reconstructive Surgery, USF College of Medicine, Tampa, FL, United States of America; Medical Director, Urogynecology and Robotic Surgery, Tampa General Hospital, Tampa, FL, United States of America
†Professor of Molecular Medicine, Cleveland Clinic Lerner College of Medicine, Case Western Reserve University, Cleveland, OH, United States of America
‡Lerner Research Institute, Cleveland Clinic, Cleveland, OH, United States of America
§Advanced Platform Technology Center, Cleveland VA Medical Center, Cleveland, OH, United States of America

Investigation of pelvic floor biomechanics is a field in its infancy, just beginning to stretch its legs, and curious about what might be possible. Although the chapters in this book are very thorough, they leave many questions unanswered since many aspects of pelvic floor biomechanics have yet to be investigated. As we look ahead to a future of greater precision in imaging, advanced multiscale computational analysis methods, automatic segmentation techniques, increased computing power, improved tissue modeling, and the development of noninvasive biomechanical measurement methods, we can imagine what might be possible in the growing field of pelvic floor biomechanics.

Physical therapy techniques could be improved if they are informed by biomechanical models and simulations demonstrating the effect of specific methods on the biomechanics of the pelvic floor. Personalized medicine would be achievable with the development of patient-specific rehabilitation protocols based on models created from each patient's specific anatomy, physiology, and pathophysiology, which could predict the rehabilitation protocol most likely to be effective for that patient. Inclusion of genetic and biomechanical influencers of biomechanics would make the models only stronger and potentially more accurate for patient-specific applications.

Given the unfortunate recent history of vaginal mesh for pelvic organ prolapse, development of improved prosthesis design methods, materials, and indications for use are essential to improve treatment outcomes. The powerful combination of improved and realistic animal models along with development of accurate tissue models and simulations will undoubtedly enable design of optimal prostheses for patient-specific treatment of pelvic organ prolapse. These may be biologic, synthetic, or some combination of these, potentially with cells or other biologic therapy seeded onto a scaffold. Sophisticated models built with accurate biomechanical inputs could be used to predict which designs are most likely to be successful, potentially in a personalized medicine or patient-specific manner.

Advanced computational analysis methods offer the promise of multiscale modeling methodologies to model and simulate organs and tissues at the ultra-, micro-, and macroscales concurrently. These methods, along with increased computing power, would enable the quantitative testing of competing theories of biomechanical function of the pelvic floor, such as the hammock and integral theories. The outcomes of such models would allow the scientific and medical communities to objectively compare competing theories of pelvic floor function and potentially develop improved treatments. Treatments based on these theories could likewise be tested quantitatively and, via the use of sensitivity analyses, improved and/or evolved into patient-specific therapies.

The promise of automatic segmentation technologies could enable application of computational analysis methods to test biomechanical hypotheses across large populations of subjects in reasonable timeframes. These could include hypotheses regarding the effect of pelvis shape on pelvic floor biomechanics and function; as well as changes with aging, parity, and other risk factors, or pathology in controlled comparisons between populations. Non-invasive imaging combined with automated segmentation and increased computational capabilities would enable longitudinal studies to investigate the natural progression of pelvic floor disorders in women as they age. Such investigations would not only increase our knowledge in the field but would undoubtedly lead to development of improved treatments and prophylactic therapies to slow the advance of pelvic floor disorders with aging and other risk factors.

We imagine a future scenario in which a patient with pelvic organ prolapse would undergo medical imaging and biomechanical diagnostic tests, the result of which would be fed into a patient-specific biomechanical model that could rapidly simulate multiple treatment options, including prosthetic designs and native tissue repairs, and suggesting the optimal one for that patient. If indicated, the optimal prostethic design, in terms of shape, size, material, and biologic components could then be produced on a 3D printer, providing an optimal clinical solution tailor-made for each patient only a short period after the diagnostic tests. Similar medical imaging and biomedical testing might be offered to a prepregnant woman, and computational analysis applied to determine optimal labor and delivery scenarios to minimize the risk of childbirth related pelvic floor injuries.

This future would only be possible with development of sophisticated models, precise imaging techniques, and novel methods of biomechanical assessment, along with increased computing power. Also essential to being able to realize this bright future for women worldwide is an increase in the expert workforce focusing on pelvic floor biomechanics, including both bioengineers and physicians with interest and expertise in the topic.

Thus, this textbook represents the first step in a journey of empowering experts who will build a bright future for women via their knowledge of pelvic floor biomechanics. We have no doubt that they will then create a full spectrum of improved durable patient-specific therapies for women with pelvic floor disorders. As the potential of this young field is realized, optimal interventions ranging from physical therapy and native tissue repairs to prosthetic implants could be confidently offered to women, based on their specific pathology and biomechanical characteristics.

INDEX

Note: Page numbers followed by "*f*" indicate figures, and "*t*" indicate tables.

D

E

P

R

W

Y

FSC
www.fsc.org
MIX
Papier | Fördert gute Waldnutzung
FSC® C083411

Kolibri 360 GmbH
Ferdinand-Jühlke-Straße 7
99095 Erfurt, Deutschland
produktsicherheit@kolibri360.de